An Elementary Approach To Mathematics

Burton Rodin, Series Editor
Goodyear Series in Mathematics

Finite Mathematics: An Elementary Approach
Lawrence G. Gilligan and Robert B. Nenno

Introduction to Mathematics
Ramakant Khazanie and Daniel Saltz

An Elementary Approach to Mathematics
Thomas Koshy

An Elementary Approach To Mathematics

Thomas Koshy

Framingham State College

Goodyear Publishing Company, Inc.

Pacific Palisades, California

Library of Congress Cataloging in Publication Data

Koshy, Thomas
   An elementary approach to mathematics
   (Goodyear series in mathematics)
   1. Mathematics 1961–
   I. Title

QA39.2.K67 510

Library of Congress Catalog Card Number: 75–11271

ISBN 0-87620-274-1

In Loving Memory Of My Father

Contents

Preface

This textbook has been designed primarily with two groups of students in mind: elementary education majors who need an understanding of the fundamental concepts and principles of the so-called modern mathematics and the liberal arts majors who wish a basic knowledge of mathematics so that they may relate those ideas to their everyday life. However, it can successfully be used for in-service courses, too. Pages 569 through 573 present suggested course outlines for one- and two-semester courses and for one-, two-, and three-quarter courses.

What is the motivation to write a book like this for a market that is already flooded with textbooks? It is the desire for a text that can be easily read by a student with an average background in mathematics and for a text that ties mathematics to the real world. This book grew out of a set of notes successfully class-tested over a period of three full years. A survey of the Table of Contents reveals that the choice of topics in the text has been greatly influenced by the latest CUPM (Committee on the Undergraduate Program of the Mathematical Association of America) recommendations for a teacher-training program.

An Elementary Approach to Mathematics is not in the "theorem-proof-example," "definition-theorem-proof" or "definition-theorem-proof-example" style. Instead, concepts, definitions, and theorems are nearly always presented through thoughtfully selected examples and then illustrated by further examples for reinforcement. There are over 400 worked-out examples in the text. Proofs of most of the theorems have been omitted in favor of placing greater emphasis on understanding through applications.

The text is divided into 16 chapters, each containing several small sections. Every chapter opens with a list of carefully chosen objectives for the chapter, a brief preview of the chapter—and some historical notes whenever possible. Many step-by-step algorithms are provided to clarify computational procedures. The section on summary and comments at the end of every chapter has been found extremely useful and helpful by students. Every chapter is followed by a carefully selected list of suggested readings that may be used by the student for further enrichment of the subject matter.

The sets of exercises with varying degrees of difficulty at the end of almost every section are a salient feature of this text. There are exercises for conceptual reinforcement and for developing computational skills and problem-solving techniques. There are about 4150 exercises in the text. The majority of them can be done by most of the students in class. Most of the exercise sets contain a set of thought-provoking true-false questions. Starred (★) sections and starred (★) exercises are aimed to challenge those students who have a better background in mathematics and may be omitted without any loss of continuity.

The organization of the chapters offers great flexibility to the instructor; he may select topics to suit the needs of his students. Fun topics like the well-known rabbit problem, exotic numbers, magic squares, etc., are included to hold the student's interest.

Even though it may seem mathematically correct and there is great beauty in beginning with a chapter on logic, I feel that it might discourage some students. Consequently, I have placed the chapter on sets first. However, these two chapters—sets and logic—are written in such a way that they can be taught in any order, according to the discretion of the instructor.

Chapter 1, which discusses the algebra of sets, introduces the basic vocabulary and symbols needed to study nearly every chapter in the text, except perhaps Chapter 2. The fundamental concepts, terms, and symbols are explained carefully and illustrated via well-chosen examples. An important application of sets in survey analysis is also included.

Chapter 2 treats the algebra of logic. It is presented to help the student think logically and to express himself systematically in precise terms. An interesting application of logic in switching circuits is discussed.

Chapters 3–7 discuss the familiar number systems—the system of whole numbers, the system of integers, the system of rational numbers, and the system of real numbers. Whole numbers are introduced as an application of sets. Each system, except the first, is developed from the preceding system after exhibiting a deficiency of that system. Several subsets of the set of natural numbers and some of their fascinating properties are presented in Chapter 5.

Chapter 8 is devoted to discussion of some classic numeration systems. It points out the merits of the Hindu-Arabic system over the others and illustrates the fundamental arithmetic algorithms in different bases.

Chapter 9 takes the student back to number systems. It gives a brief description of the richest number system: the system of complex numbers.

Chapter 10 introduces one of the fundamental concepts in mathematics: the concept of a function. Several special relations and special functions are presented and illustrated in this chapter. It also contains a small section on linear programming.

For the first time in the text, the student gets an opportunity to study and enjoy some finite mathematical systems and their application in Chapter 11. A study of algebraic systems culminates in Chapter 12, where the student is introduced to new systems in which elements and operations no longer have to be the familiar numbers and the familiar operations.

In Chapter 13, the student is intuitively introduced to some of the fundamental concepts, terms, and symbolism in geometry. The language of sets discussed in Chapter 1 plays a significant role in the presentation.

Chapters 14 and 15 provide a concise coverage of some fairly standard topics in probability and statistics, respectively, giving practical applications from everyday life.

A brief historical survey of the origin of high-speed computers, names of a few artificial languages, and a brief introduction to a few number systems used by these incredible machines are the topics of discussion in Chapter 16.

Answers to nearly one-half of the exercises are given at the end of the book. The other answers can be found in the Instructor's Manual; which also contains a test for every chapter. Most of these tests have been class-tested with gratifying results.

In preparing this book, I have been immensely influenced by the thoughts and criticisms of my students and colleagues. My gratitude goes to my colleagues Dr. Alfred Bown, Dr. Walter Czarnec, Prof. Anita Goldner, Mr. Thomas Leonard, Prof. Edna Robinson, and Dr. Bernard Rosman; to my student June Jones, who read the manuscript and made several valuable suggestions for the improvement of the manuscript; to the reviewers Burton Rodin (UCSD), Gerald Bradley (Claremont College), and Robert Herrera (UCLA); and to my students Karen Rosado and Elizabeth White, who checked the answers of most of the exercises in the text.

The following people, institutions, or companies have been very helpful in securing various pieces of art, and I sincerely appreciate their cooperation: Dr. Paul Bateman, Dr. Arthur Doyle, Dr. Bryant Tuckerman, Dr. Charles Zapsalis, The Boston Globe, Dover Publications, Inc., W. H. Freeman and Company, International Business Machines Corporation, Keuffel and Esser Company, Oxford University Press, Sun Oil Company, The Torsion Balance Company, University of Illinois at Urbana, and Worcester Art Museum.

I am grateful to the people at Goodyear Publishing Company for their sincere work in this project—especially to Jack Pritchard, for his interest and cooperation in the project; and to Jeri Walsh and Susan Gerstein for their diligence, patience, and understanding in editing the manuscript.

Finally, and most importantly, I wish to thank my wife and fulltime secretary, Gracy, for her patience and understanding during the last three long years and for her excellent typing of the manuscript. A word of thanks to my children, Suresh and Sheeba, who missed me a lot during these years.

Framingham   Thomas Koshy
Massachusetts

A Word To
The Student

Do you find mathematics a difficult subject to learn? Do you find mathematics texts, in general, hard to read? If you belong to one of those two categories, here is a book that is written especially for you.

In this text, concepts, definitions, and theorems are almost always motivated through thoughtfully selected examples and then illustrated by further examples for reinforcement. Almost every section contains a set of exercises that vary in difficulty: from very simple to fairly simple to difficult. Problem solving is a cornerstone of studying mathematics, so do as many exercises as you can. However, do not try to do them before studying that section. Also, do the exercises in order from the beginning, as they are graded in difficulty. Some of the exercises will develop your computational skills and problem-solving techniques, while some will reinforce concepts discussed in that section. Do not get discouraged if you cannot do a problem in your first attempt. Read the problem very carefully and ask yourself if you have understood what is given and what is to be found. Try again! If you still have difficulty, do not hesitate to refer back to the relevant paragraph in the section. Try again.

Starred (★) sections and starred (★) exercises are included as a challenge for those of you to whom mathematics comes relatively easily.

The first two chapters—sets and logic—especially Chapter 1, provide the basic tools for discussions in subsequent chapters. Consequently, please pay careful attention to the new terms and symbols you come across in these chapters. Also, review these chapters whenever necessary.

In this text, you will find many applications of mathematics to everyday life: survey analysis, switching circuits, the post office function, linear programming, casting out nines, and binary codes, to name a few.

Fun topics like the rabbit problem (Section 5.7) and magic squares (Section 5.8) are included especially for you and to make the subject more enjoyable.

Mathematics is no more difficult than any other subject. Do you have the willingness, patience, and time to sit down and do the work? If the answer is yes, you will find this text worth reading and mathematics worth studying; you will say later that mathematics can be fun and fun can be mathematics. Give it a try!

A word of caution: please do not read this book as if you are reading a newspaper, a novel, or a magazine. If you are tired or in a bad mood, do not get frustrated by trying to do the problems. That is not the best time to do mathematics. Whenever you read this book or do the exercises, please make sure that you have a pencil and enough sheets of paper at your side to write the definitions and theorems in your own words and to do the exercises.

Finally, if you have any questions, comments, or suggestions, I shall be very pleased and grateful to hear from you.

An Elementary Approach To Mathematics

"The essence of mathematics lies in its freedom."

G. CANTOR

1/Algebra of Sets

After studying this chapter, you should be able to:
- *describe a set in two ways, whenever possible*
- *check if a set is a subset of another set*
- *check if two sets are equal*
- *show relationships between sets via Venn diagrams*
- *find the union, intersection, relative complement, and cartesian product of two sets*
- *find the complement of a set*
- *identify the properties of these operations*
- *apply Venn diagrams to simple problems in survey analysis*
- *show if there exists a 1-1 correspondence between two sets*
- *identify finite and infinite sets*

1.0 INTRODUCTION

In this chapter, we shall give a brief introduction to that area of mathematics that is usually referred to as the theory of sets. The concept of a set is so essential and fundamental that it plays a very significant role in all the branches of mathematics and its applications to other branches of learning. The advent of this theory revolutionized the mathematics programs in schools, giving rise to the so-called modern mathematics programs. One of the major contributions of this area of mathematics is to make the mathematical language more precise and concise, enabling both the student and the teacher to express themselves with precision.

The concept of a set was first introduced by the German mathematician Georg Cantor (1845–1918) during the latter part of the nineteenth century. In

1856 the Cantors moved to Frankfurt, Germany, from St. Petersburg, Russia, where he was born. From early childhood, Cantor showed keen interest in mathematics and wanted to be a mathematician. But this was against the will of his father, who wanted him to become an engineer. However, he pursued his interest in mathematics, receiving his Ph.D. degree at the age of 22 from the University of Berlin. At the age of 30, he published his outstanding work on abstract properties in set theory. It was too abstract for his contemporary mathematicians, who did not recognize the significance of his results. The continuous attacks by them on Cantor and his work caused him to have a nervous breakdown. He finally died in a mental hospital in Halle at the age of 73.

Our aim in this chapter is to define the fundamental concepts and operations of set theory. Some applications are also discussed.

1.1   THE CONCEPT OF A SET

Knowingly or unknowingly, we use the concept of a set in our everyday life; for example, set of dishes, collection of books, group of girls, list of names, herd of cattle, flock of sheep. What is a set? We consider a *set* to be a collection of well-defined objects. By "well-defined" we mean that there is some criterion that enables us to determine whether or not any given object is in the set. The objects of the set are called *members* or *elements* of the set. Thus, given any object, there are only two possibilities: either it is a member of the set or it is not.

Example 1.1   Consider the following sets:
1. The set of letters of the English alphabet
2. The set of Presidents of the United States since 1960
3. The set of days of the week
4. The set of integers 1, 3, 5, and 7
5. The set of all monkeys in Africa
6. The set of all Nobel laureates in physics

The reader is warned that a set cannot be described simply by saying that it consists of all the old cars in Boston or good students on the campus. The reason is that the objects are not well defined; two different people will not necessarily come up with the same list of old cars or good students. A car labeled "old" by one person need not be labeled "old" by another.

Observe that a set need not consist of objects of the same kind. For instance, we can have a set containing a girl, a car, and a book.

Sets are usually denoted by capital letters $A, B, C, \ldots$, and their elements by lower case letters $a, b, c, \ldots$. If an object x is an element of a set A, we write $x \in A$, which may be read, "x is an element of A" or "x belongs to A." If x is not an element of A, we write $x \notin A$.

Example 1.2   If A is the set of numbers 1, 3, and 5, then $3 \in A$, but $2 \notin A$.
Example 1.3   Let B be the set of all months of the year. Then June $\in B$, whereas $5 \notin B$, Sunday $\notin B$.

A set can be described in several ways. Suppose we wish to consider the set A consisting of all days of the week. The members of the set are Sunday, Monday, Tuesday, Wednesday, Thursday, Friday, and Saturday. A convenient way of writing this set is

$A = \{$Sunday,Monday,Tuesday,Wednesday,Thursday,Friday,Saturday$\}$

We read this, "A is the set consisting of Sunday, Monday, Tuesday, Wednesday, Thursday, Friday and Saturday." Here we have actually listed or exhibited all members of the set. Accordingly, this is called *listing method* or *roster method*. The symbols "$\{$" and "$\}$" are called *braces*.
Example 1.4   If A is the set consisting of the elements x, y, and z, then

$$A = \{x,y,z\}$$

Observe that the roster method has a serious limitation. If the set contains a fairly large number of elements, it is tedious or even impossible to tabulate all the elements of the set. For example, it is difficult to write all the elements of the set of all cars in Chicago and impossible to write all the integers.

A second way of describing a set is by stating the precise property or properties possessed by the members of the set. If we are given some object, in order to determine if it is a member of the set, we need only check whether the object has the required properties.
Example 1.5   Let A be the set of New England states. Let x be an arbitrary object. Then $x \in A$ if and only if x is a New England state. In this case we write,

$$A = \{x \mid x \text{ is a New England state}\}$$

We read this, "A is the set of all those objects x such that x is a New England state." The vertical line "$\mid$" is read "such that." This way of describing a set is called *set-builder notation*. Thus in the set-builder notation, a set is defined as

$$A = \{x \mid x \text{ has a given property}\}$$

Example 1.6   If B is the set of all states in the United States, then in the set-builder notation, B is described as

$$B = \{x \mid x \text{ is a state in the United States}\}$$

Example 1.7   Let C be the set of all months of the year with exactly 30 days. In the roster method,

$$C = \{\text{September,April,June,November}\}$$

In the set-builder notation,

$$C = \{x \mid x \text{ is a month of the year with exactly 30 days}\}$$

Exercise 1.1

1. Which of the following are (well defined) sets?

a) all cars in Maine

b) all books in the college library

c) all great Presidents of the United States

d) all tall people in Europe

e) all students on the campus

f) all readable books

g) all long rivers in Canada

2. Describe the following sets by the roster method.

a) set of all days of the week

b) set of months that begin with the letter A {August, April}

c) set of days that begin with the letter S

d) set of all letters of the word "googol" - {g,o,l}

e) set of all letters of the word "hillbill"

3. Describe the following sets in the set-builder notation.

a) set of all nations in the world {x/x is a nation in the world}

b) set consisting of Tuesday and Thursday {x/x is a day of week beginning with the letter T}

c) set consisting of January, February, May and July {x/x is a month of the year that is in d*

d) set consisting of all members of the United Nations {x/x is a member of the UNations}

★e) set consisting of Britain, China, France, Russia, and the United States {x/x is a permanent member of the UN Security Council}

4. Describe the following sets in two different ways.

a) set A of all digits {1,2,3,4,5,6,7,8,9} {x/x is a digit}

b) set B of all months of the year B: {Jan, Feb, Mar, April, May, Jun, July, Aug, Sept, Oct, Nov} B = {x/x is a month of a year}

c) set C of all countries bordering the U.S. Mexico, Canada} C = {x/x is a country bordering the US}

d) set D of all integers between 0 and 4 D = {1,2,3} {x/x is a integer between 0 and 4}

★e) set E of all planets E = {some thing} E = {x/x is an planet}

5. Give an example of a set that can be written in the roster method but not in the set-builder notation. {5, Jim}

6. Give an example of a set that can be written in the set-builder notation but not in the roster method can't be labeled (members) {x/x is an imaginary no.}

7. Mark *true* or *false*.

a) The great leaders of Europe form a set.

b) $10 \in \{x \mid x$ is a digit$\}$.

c) Cam $\notin \{$Bam,Lam,Cam$\}$.

d) There cannot exist a set with no elements.

e) Elements of a set can also be sets.

f) The sets $A = \{1,2,3\}$ and $B = \{0,2,3\}$ have common elements.

g) Each element of $A = \{a,b\}$ is also an element of $B = \{x,a,b,c\}$.

1.2 RELATIONS BETWEEN SETS

Consider the sets $A = \{$Sam,Mary$\}$ and $B = \{$Sam,Mary,Fred,John$\}$. We observe that the sets A and B are related in a certain fashion. Sam and Mary, which are already members of the set A, are also members of the set B. That is, every element of A is also an element of B. In that case, we say that A is a *subset* of B.

Definition 1.1   A set A is said to be a *subset* of a set B if every element of A is also an element of B. We denote this by writing $A \subseteq B$. If A is not a subset of B, we write $A \nsubseteq B$. Clearly $A \nsubseteq B$ if A contains an element not found in B.

Example 1.8   If $A = \{a,c\}$, $B = \{a,d\}$, and $C = \{a,b,c,d\}$, then $A \subseteq C$, $B \subseteq C$, but $A \nsubseteq B$ and $B \nsubseteq A$. (Why?)

Example 1.9   Let $A = \{0,2,4,6,8\}$ and $B = \{x \mid x$ is a digit$\}$. Then $A \subseteq B$ and $B \nsubseteq A$. (Why?)

Consider the sets $B = \{w,x,y,z\}$ and $A = \{x,y\}$. Clearly $A \subseteq B$. But B contains the elements w and z, which are not in A. Accordingly, we make the following definition:

Definition 1.2   Set A is called a *proper subset* of set B, denoted by $A \subset B$, if $A \subseteq B$ and B contains at least one element that is not in A.

Observe that in example 1.8, both A and B are proper subsets of C.

Example 1.10   Consider the sets A, B, and C given by

$$A = \{x \mid x \text{ is a capital city}\}$$

$$B = \{x \mid x \text{ is a city}\}$$

$$C = \{\text{Chicago,London,Paris}\}$$

Then $A \subset B$, $C \subset B$, but $A \nsubseteq C$. (Why?)

Example 1.11   If A is the set of all women students taking mathematics 113 and B the set of all students taking mathematics 113, then A need not be a proper subset of B. (Why?)

Definition 1.3   Observe that, by definition, every set is a subset of itself. Such a subset is called an *improper subset.*

Example 1.12   If $A = \{x,y,z\}$, then $\{x\}, \{y\}, \{z\}, \{x,y\}, \{y,z\}, \{x,z\}, \{x,y,z\}$ are all subsets of A.

Let A be the set of all months of the year that end in the letter r and $B = \{\text{September,October,November,December}\}$. We note that the sets A and B contain exactly the same elements. So it seems natural to say that they must be the same set.

Definition 1.4   Two sets A and B are said to be *equal*, denoted by $A = B$, if A and B contain exactly the same elements. That is, $A = B$ if every element of A is an element of B and every element of B is an element of A.

Consequently, in order to show that two sets are the same, we must show that each is a subset of the other.

Example 1.13   If $D = \{1,2,3,4\}$ and $E = \{x \mid x$ is a counting number less than 5$\}$, then $D = E$.

Example 1.14   Let $A = \{1,2,3\}$, $B = \{3,1,2\}$, and $C = \{2,1,3\}$. Then $A = B = C$. (Why?)

This example shows that the order in which the elements of a set are written is immaterial.

Example 1.15   If $A = \{1,2,3\}$, $B = \{1,1,2,3\}$, and $C = \{2,1,2,3,2\}$, then by the equality of sets, $A = B = C$.

This shows that a set does not change if its elements are repeated. Hence, we make the convention that the elements of a set should not be repeated.

Does every set have to contain some element? Can there exist a set with no elements at all? Suppose John went hunting in a nearby jungle and returned with great tales, but not animals. Therefore, the set of animals John caught contains no elements. Accordingly, we make the following definition:

Definition 1.5   The set containing no elements is called the *empty set* or the *null set*, usually denoted by the symbol $\varnothing$ or the symbol $\{\Box\}$.

Example 1.16   The set of all two-legged dogs is empty. So is the set of all women Presidents of the United States. The set $\{x \mid x + 1 = x\}$ is empty. (Why?) The set of all green elephants is empty.

We remark that the empty set is a subset of every set, i.e., $\varnothing \subseteq A$ for every set A.

For every meeting we have some kind of agenda, so every major topic to be discussed at the meeting comes from the agenda. In a similar fashion, for every discussion on sets, it is always possible to choose a set such that every set under consideration is a subset of that set. Such a set is called a universal set for that discussion.

Definition 1.6   A set with the property that every set in a given discussion is a subset of that set is called a *universal set*, denoted by U.

Example 1.17   If one wishes to discuss the sets $A = \{a,b\}$, $B = \{c\}$, and $C = \{a,d,e\}$, then $U = \{a,b,c,d,e,f\}$ may be taken as a universal set.

Example 1.18   In a discussion about different sets of dogs, the set of all dogs or the set of all animals will serve as a universal set.

Consider the set of all Presidents of France and the set of all Presidents of the United States. We note that those two sets have no elements in common. Thus, it is possible to have two sets with no common elements.

Definition 1.7   Two sets are said to be *disjoint* if they have no elements in common.

Example 1.19   The sets $A = \{1,2\}$ and $B = \{3,4\}$ are disjoint. So are the sets $C = $ set of all cats and $D = $ set of all dogs.

Exercise 1.2

1. Mark *true* or *false*.
   a) $3 \in \{1,2,3\}$.
   b) $2 \subseteq \{1,2,3\}$.
   c) $\{1\} \subset \{1,2\}$.
   d) $\{1,3\} \subseteq \{1,3,5\}$.
   e) $\{x\} \in \{\{x\},y\}$.
   f) $\{x\} \subseteq \{\{x\},y\}$.
   g) $\{0\} = \emptyset$.
   h) $0 \in \emptyset$.
   i) $\emptyset \in \{1,2,3\}$. *false*
   j) $\emptyset \subset \{1,2,3\}$. *true*
   k) $\emptyset \subseteq \emptyset$. *true*
   l) $\emptyset \in \{\emptyset\}$. *true*
   m) $\emptyset \subset \{\emptyset\}$. *true*
   n) $\{x,\{y\}\} = \{x,y\}$. *false*
   o) $\{\emptyset\} = \emptyset$. *false*
   p) $\{x \mid x \neq x\} = \emptyset$. *true*
   q) $\emptyset$ is a subset of every set. *true*
   r) Every set is a subset of itself. *true*
   s) $\emptyset$ is a proper subset of every set. *false*
   t) Every nonempty set has at least two subsets. *true*
   u) The set of all vowels is a proper subset of the set of all letters in the English language. *true*
   v) The set of all squares is a proper subset of the set of all rectangles. *true*
   w) {Canada,India,Poland} is a subset of {Canada,Hungary,Indonesia, Poland}. *false*
   x) The set of all members of the U.N. is a proper subset of the set of all nations in the world. *true*

2. a) What is the set of all letters of the word "ALABAMA"?
   b) Same as (a) with the word "APOLLO." $\{A,P,O,L\}$
   c) Same as (a) with the word "TALLAHASSEE." $\{T,A,L,H,S,E\}$
   d) What is the set of all digits of the number 12,321?
   e) What is the set of all countries that border with the United States? $\{Mexico, Canada\}$

3. If $A = \{1,2\}$, $B = \{2,3,5\}$, and $C = \{3,6,7\}$, which of the following are true?
   a) $A \subseteq B$.
   b) $A \subset B$.
   c) $\emptyset \subset B$.
   d) B and C have no common elements.
   e) A and C are disjoint. *true*
   f) B and C are not disjoint. *true*
   g) The set of elements belonging to both B and C is {3}. *true*
   h) The set of elements belonging to B or C is {2,3,5,6,7}. *true*

4. Which of the following are subsets of the set $\{a,b,2,0,x,z\}$?
   a) $A = \{2,a,b\}$. *yes*
   b) $B = \{a\}$. *yes*
   c) $C = \{a,b,c\}$. *no*
   d) $D = \{x,z\}$. *yes*
   e) $E = \emptyset$. *yes*
   f) $F = \{\emptyset\}$. *no*
   g) $G = \{0\}$. *yes*
   h) $H = \{a,x,z\}$. *yes*
   i) $I = \{a,0,\{x\}\}$. *no*

5. Which of the following sets are equal?
   a) $A = \{0,1\}$.
   b) $B = \{a,b,c,e,g\}$. *garbage*
   c) $C = \{a,b,e,g,r\}$. *garbage*
   d) $D = \{x \mid x^2 = x\}$.
   $0 \mid 0^2 = 0$
   $1 \mid 1^2 = 1$

   e) E = set of all weeks with eight days. ✓
   f) F = set of all letters of the word "garbage".
   g) G = set of all months of the year that begin with the letter R. ∅
   h) H = set of all letters of the word "cabbage".

6. If $A = \{x, \{y\}, z\}$, which of the following are true?
   a) $x \in A$. True
   b) $\{y\} \subset A$. True
   c) $\{y\} \in A$. True
   d) $\{x, y\} \subseteq A$. False
   e) $\{\{y\}\} \subseteq A$. True
   f) $\{\{y\}, z\} \subseteq A$. True
   g) $\{x, z\} \subset A$. True ? h) $y \in A$. False
   i) $\varnothing \subset A$. True

7. a) Is $\{0\} = \varnothing$? NO
   b) Is $\{\varnothing\} = \varnothing$? NO
   c) Is $\{x \mid x^2 = x\} = \{0, 1\}$? Yes
   d) Is $\{x \mid x^2 = 1\} = \{0, 1\}$? NO

8. a) Is $\{1, 2, 4\} \subset \{4, 1, 2, 4\}$? NO
   b) Is $\{x, y\} = \{y, x\}$? Yes
   c) Is $\{x, \{y\}\} = \{x, y\}$? NO

9. If $A \subseteq B$ and $B \subseteq C$, is $A \subseteq C$? Yes subset   proper subset

10. What is the difference between $A \subseteq B$ and $A \subset B$?

11. If $\{a, b\} = \{b, c\}$, is $a = c$? yes   DISJOINT SET

12. If B and C are disjoint sets and $A \subseteq B$, what can you say about A and C?

13. If $A = \{1, \{2\}, \{2, 3\}\}$, which of the following are true?
   a) $\{2\} \in A$. True
   b) $2 \in A$. False
   c) $3 \in A$. False
   d) $\{2\} \subseteq A$. False
   e) $\{2, 3\} \in A$. True
   f) $\{\{2\}\} \subseteq A$. True
   g) $\varnothing \subseteq A$. True
   h) $\varnothing \in A$. False
   i) $A = \{1, 2, 3\}$. False

14. If A = set of all triangles, B = set of all equilateral triangles, and C = set of all isosceles triangles, which of the following are true?
   a) $B \subset A$.
   b) $B \subseteq C$.
   c) $C \subset A$.
   d) $C \subset B$.
   e) C is the set of elements belonging to both B and C.
   f) B is the set of elements belonging to B or C.

15. Which of the following are null sets?
   a) set of all people 6 feet tall set
   b) set of all months with 32 days ∅
   c) set of all years with 365 days set
   d) set of all letters of the word "banana" set
   ★e) set of all birds that cannot fly set

16. For each of the following sets, find a suitable universal set.
   a) set of all people in Canada {people in world}
   b) set of all capital cities {all cities}
   c) set consisting of Algebra, Calculus, and Geometry {maths}
   d) set consisting of Britain, China, and France {all countries}
   e) set consisting of Kennedy, Johnson, and Nixon {all presidents}

17. What is the difference between the sets $\{a, b\}$ and $\{\{a\}, \{b\}\}$?
   $\{a, b\}$ $\{\{a\}, \{b\}\}$
   a is element $\{a\}$ set
   b is element $\{b\}$ set

★**18.** Find the set of all subsets of A (called the *power set* of A, denoted by $p(A)$) if
  a) $A = \{1,2\}$.   **b)** $A = \{1,\{2\}\}$.   **c)** $A = \{x,\{x,y\}\}$.
  d) $A = \varnothing$.   **e)** $A = \{\varnothing\}$.   **f)** $A = \{1\}$.
  g) $A = \{\varnothing,\{\varnothing\}\}$.   **h)** $A = \{a,b,c\}$.   **i)** $A = \{a,b,c,d\}$.
★**19.** List all the nonempty proper subsets of
  a) $A = \{x,y\}$.   **b)** $A = \{a,b,c\}$.   **c)** $A = \{\varnothing\}$. none
  $\{x\}$ $\{y\}$

1.3 VENN DIAGRAMS

A simple and very useful device for illustrating relationships between sets is by means of diagrams called *Venn diagrams*, named after the English logician John Venn (1834–1923), who gave up priesthood in 1859 to pursue his study in symbolic logic. In a Venn diagram, the universal set U is usually represented by the set of points inside a rectangle and any subset A of U by the set of points enclosed by a circle inside the rectangle. If $x \in A$, then x is represented by a point inside the circle, as shown in Fig. 1.1.

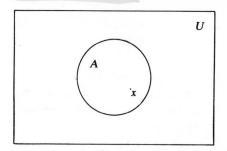

Fig. 1.1

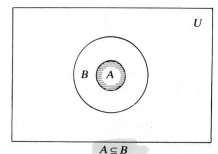

$A \subseteq B$
Fig. 1.2

Figure 1.2 shows A is a subset of B. Figure 1.3 shows the sets A and B with some common elements; the shaded region represents the set of such elements. In Fig. 1.4, the two circles are nonintersecting, that is, the sets A and B have no elements in common and hence are disjoint.

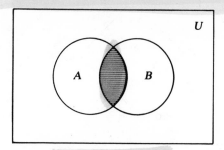

A and B not disjoint
Fig. 1.3

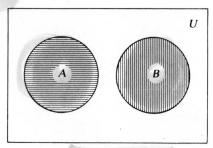

A and B disjoint
Fig. 1.4

We remark that Venn diagrams are just a pictorial representation of relationships between sets and not a substitute for proofs of theorems.

★1.4 LINE DIAGRAMS

In the previous section, we discussed a pictorial way of representing relationships between sets. A more sophisticated way of illustrating relationships between sets is by the use of *line diagrams*. If $A \subseteq B$, then in such a diagram we write A just below B and connect them by a straight line, as shown in Fig. 1.5. The larger set is always placed at the top of the diagram.

$A \subseteq B$

Fig. 1.5

$A \subseteq B \subseteq C$

Fig. 1.6

If $A \subseteq B$ and $B \subseteq C$, then the corresponding line diagram looks as in Fig. 1.6.

Line diagrams are sometimes more useful than Venn diagrams, especially when there are several sets under consideration. By looking at the diagram one can easily determine if one set is a subset of another.

Example 1.20   If $A = \{1,2,4\}$, $B = \{1,2,3\}$, $C = \{1,2\}$, and $D = \{1\}$, the corresponding line diagram looks like Fig. 1.7 below. The reader should note that

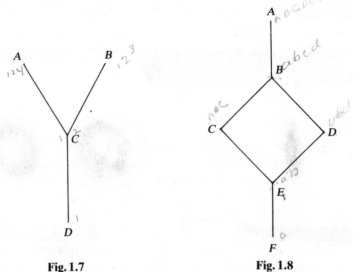

Fig. 1.7

Fig. 1.8

$D \subseteq C$: hence the line running downward from C to D. The set C is a subset of both A and B: hence the descending lines from A and B to C. The descending lines from A and B meet at C, which means that the elements of C are common to A and B. This may be observed by looking at the sets A, B, and C.

Example 1.21   Let $A = \{a,b,c,d,e,f\}$, $B = \{a,b,c,d\}$, $C = \{a,b,c\}$, $D = \{a,b,d\}$, $E = \{a,b\}$, and $F = \{a\}$. The line diagram of the sets A, B, C, D, E, and F is given in Fig. 1.8. The descending line from E to F indicates that $F \subseteq E$. Sets C and D contain the set E: hence the descending lines from C and D to E. Sets C and D are both subsets of B, while B is a subset of A: hence the lines from B down to C and D, and the line from A to B.

Example 1.22   If $A = \{1,2,3\}$, then the set of all subsets of A, called the *power set* $p(A)$ of A, is given by $p(A) = \{\varnothing,\{1\},\{2\},\{3\},\{1,2\},\{2,3\},\{3,1\},A\}$. The line diagram of the family of all subsets of A is shown in Fig. 1.9.

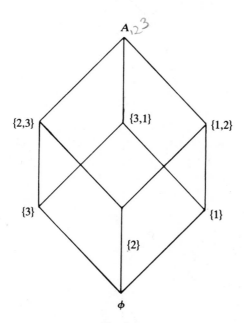

Fig. 1.9

Exercise 1.4

1. Draw a line diagram for each of the following families of sets.

   a) $A = \{\text{Eisenhower,Kennedy,Johnson,Nixon}\}$, $B = \{\text{Eisenhower,Kennedy, Johnson}\}$, $C = \{\text{Kennedy,Johnson}\}$, and $D = \{\text{Nixon}\}$.

   b) $A = \{\text{Britain,Canada,China,France,Russia}\}$, $B = \{\text{Britain,Canada, France,Russia}\}$, $C = \{\text{Britain,Canada,China,France}\}$, $D = \{\text{Britain, Canada,France}\}$, and $E = \{\text{Canada,France}\}$

   c) $A = \{a,b,c,d,e,g\}$, $B = \{a,b,c,d,e,f\}$, $C = \{a,b,c,d,e\}$, $D = \{a,b,c,d\}$, $E = \{c,d\}$, and $F = \{a,b\}$.

d) $A = \{a,b,c,d,e,f,g\}$, $B = \{a,b,c,d,e\}$, $C = \{a,b,c,e\}$, $D = \{a,b,c,d\}$, $E = \{a,b,c\}$, $F = \{c\}$, $G = \{b\}$, and $H = \{a\}$.

e) $A = \{0,1,2,3,4,5,6,7,8\}$, $B = \{0,1,2,3,4,6\}$, $C = \{0,1,2,3,5\}$, $D = \{0,1,2,3\}$, $E = \{0,1,2\}$, $F = \{0,2\}$, $G = \{0,1\}$, and $H = \{0\}$.

2. Find suitable sets to illustrate the following line diagrams.

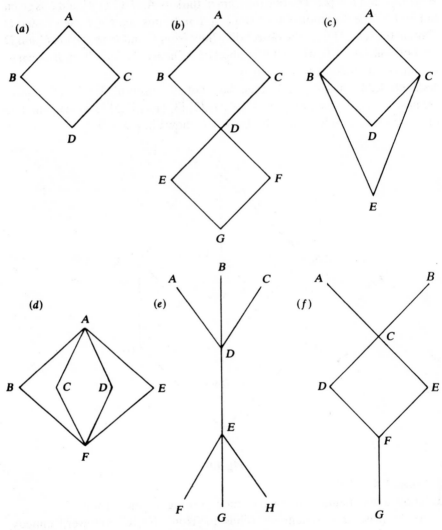

3. Find a line diagram for the power set $p(A)$ if
   a) $A = \{1\}$ **b)** $A = \{1,2\}$
   c) $A = \{\varnothing,\{\varnothing\}\}$ **d)** $A = \{1,2,\{3\}\}$.

1.5   UNION

In the previous section, we discussed relations between sets. We now concentrate on how to construct new sets from given sets.

Consider the sets A and B given by $A = \{1,2,3,4\}$ and $B = \{2,4,6\}$. By putting all the elements of both A and B together, we get a new set, $C = \{1,2,3,4,2,4,6\} = \{1,2,3,4,6\}$. Recall that no element in a set should be repeated. What set do we get if we put all the elements of the sets $D = \{\text{London, Paris}\}$ and $E = \{\text{England, France, Canada}\}$ together? It is the set $\{\text{London, Paris, England, France, Canada}\}$. This is one way of forming a new set from two given sets. The set C is called the *union* of sets A and B.

Definition 1.8   The *union* of two sets A and B, usually denoted by $A \cup B$, is the set of all elements belonging to either A or B. We read $A \cup B$ as "A union B."

The shaded region in Fig. 1.10 represents the set $A \cup B$.

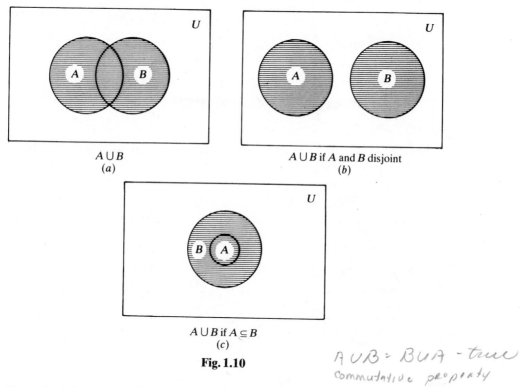

$A \cup B$
(a)

$A \cup B$ if A and B disjoint
(b)

$A \cup B$ if $A \subseteq B$
(c)

Fig. 1.10

$A \cup B = B \cup A$ - true
commutative property

Example 1.23   If $A = \{a,b,c\}$ and $B = \{b,c,d,e\}$, then

$$A \cup B = \{a,b,c,d,e\} = B \cup A$$

Example 1.24   If $A = \{x \mid x \text{ is a Senator of the United States}\}$ and $B = \{x \mid x \text{ is a Representative of the United States}\}$ then $A \cup B$ is the set of all members of the Congress of the United States. Notice that $B \cup A$ is the same set as $A \cup B$. (Why?)

From the above examples and the Venn diagrams, it would seem that $A \cup B = B \cup A$ for any two sets A and B. Indeed, it is true!

Theorem 1.1   (***Commutative Property***).   *For any two sets A and B,*

$$A \cup B = B \cup A$$

This theorem says that we can take the union of two sets in any order we like.

We observe that $A \cup \emptyset = A$, $A \cup A = A$, and $A \cup U = U$ for any set A. Also, if $A \subseteq B$, then $A \cup B = B$, as illustrated in Fig. 1.10c. Clearly both A and B are subsets of $A \cup B$, but they need not be proper!

Example 1.25   Consider the sets A, B, and C given by $A = \{1,2,3,4\}$, $B = \{2,4,6\}$, and $C = \{1,3,5\}$. What are the sets $(A \cup B) \cup C$ and $A \cup (B \cup C)$?

Solution: $A \cup B = \{1,2,3,4\} \cup \{2,4,6\} = \{1,2,3,4,6\}$

$$(A \cup B) \cup C = \{1,2,3,4,6\} \cup \{1,3,5\} = \{1,2,3,4,5,6\}$$

$$B \cup C = \{2,4,6\} \cup \{1,3,5\} = \{1,2,3,4,5,6\}$$

$$A \cup (B \cup C) = \{1,2,3,4\} \cup \{1,2,3,4,5,6\} = \{1,2,3,4,5,6\}$$

Thus for these sets A, B, and C, $A \cup (B \cup C) = (A \cup B) \cup C$. Is this true in general? This is answered by the following theorem:

Theorem 1.2   (***Associative Property***)   *For any three sets A, B, and C,*

$$(A \cup B) \cup C = A \cup (B \cup C)$$

We remark that the associative property for the union operation tells us that the expression $A \cup B \cup C$ is meaningful. We can take the union of A and B first and then take its union with C to get $(A \cup B) \cup C$, or take the union of B and C first and then its union with A to get $A \cup (B \cup C)$. The above theorem says both sets are the same. The shaded region in Fig. 1.11 represents the set $A \cup B \cup C$.

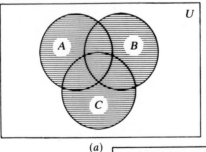

(a)

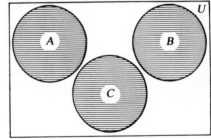

(b)

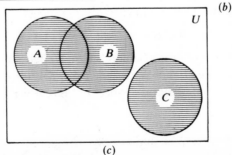

(c)

Fig. 1.11

Exercise 1.5

1. If $A = \{1,2,3,4\}$, $B = \{2,4,6,8\}$, and $C = \{1,3,4,8,9\}$, find:

   a) $A \cup B$ **b)** $B \cup C$ **c)** $A \cup \varnothing$

   d) $A \cup C$ **e)** $A \cup (B \cup C)$ **f)** $\varnothing \cup \varnothing$

2. If $A = $ set of all letters of the word "Barbara," $B = $ set of all letters of the word "Banana," and $C = $ set of all letters of the word "Canada," find the following sets. Make no distinction between capital and lower case letters.

   a) $A \cup B$ *{bar ns}* *{a b c d n e}*   **b)** $B \cup C$ *{a b c n d}*   **c)** $A \cup (B \cup C)$ *{a b c d n r}*

   d) $B \cup (C \cup A)$ **e)** $B \cup B$ *{b a n}* ?   **f)** $B \cup \varnothing$ *{b a n}* ?

3. Mark *true* or *false* (A and B are any two sets, and U is the universal set).

   a) $A \cup A = A$.

   b) $A \cup \varnothing = \varnothing$.

   c) $A \cup U = U$.

   d) $A \cup B = B \cup A$.

   e) $(A \cup B) \cup C = A \cup (B \cup C)$.

   f) If $A \cup B = \varnothing$, then $A = B = \varnothing$.

   g) If $A \subseteq B$, then $A \cup B = A$. *false (B)*

   h) If $A \cup B = B$, then $A \subseteq B$. *true*

4. Find the set of all common elements of the sets A and B if

   a) $A = \{x,y,z\}$ and $B = \{b,x,z\}$. *{x z}*

   b) $A = \{0,1\}$ and $B = \{1,3,5\}$. *{1}*

   c) $A = \{2,4,6\}$ and $B = \{1,3\}$. *$\varnothing$*

1.6   INTERSECTION

Consider the sets $A = \{w,x,y,z\}$ and $B = \{x,y,z,a,b\}$. We observe that A and B have the elements x, y, z in common and they form the set $\{x,y,z\}$. What is the set of all elements common to the sets $A = \{$Fred, Ted, Ned$\}$ and $B = \{$Paul, Ted, Fred$\}$? The answer is the set $\{$Fred, Ted$\}$. This illustrates a second way of constructing a new set from two given sets.

Definition 1.9   The *intersection* of two sets A and B, denoted by $A \cap B$, is the set of all elements belonging to both A and B. We read $A \cap B$ as "A intersection B."

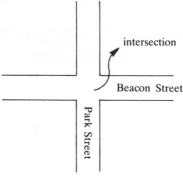

Fig. 1.12

Example 1.26 If $A = \{1,2,3,4\}$ and $B = \{2,4,6,8\}$, then
$$A \cap B = \{2,4\} = B \cap A$$

Example 1.27   If A is the set of all governors in the United States and B the set of all United States Senators, then $A \cap B = \varnothing = B \cap A$.

The shaded region in Fig. 1.13 represents the set $A \cap B$.

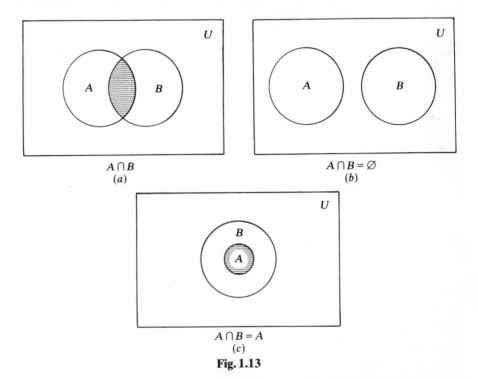

$A \cap B$
(a)

$A \cap B = \varnothing$
(b)

$A \cap B = A$
(c)

Fig. 1.13

Recall that two sets are disjoint if and only if they have no common elements. Thus, A and B are disjoint if and only if $A \cap B = \varnothing$.

In examples 1.26 and 1.27, we notice that $A \cap B = B \cap A$. So it is only natural to ask if this is true in general.

Theorem 1.3 (**Commutative Property**)   *For any two sets A and B,*
$$A \cap B = B \cap A$$

Notice that as in the case of the union operation, this theorem tells us that the order in which we take the intersection of two sets is immaterial.

Example 1.28   If $A = \{1,2,3,5\}$, $B = \{2,4,5,6\}$, and $C = \{3,5,6\}$, then
$$B \cap C = \{2,4,5,6\} \cap \{3,5,6\} = \{5,6\}$$
$$A \cap B = \{1,2,3,5\} \cap \{2,4,5,6\} = \{2,5\}$$

$$A \cap (B \cap C) = \{1,2,3,5\} \cap \{5,6\} = \{5\} = \{2,5\} \cap \{3,5,6\} = (A \cap B) \cap C$$

This example now suggests the following theorem.

Theorem 1.4   (*Associative Property*)   For any three sets A, B, and C,

$$A \cap (B \cap C) = (A \cap B) \cap C$$

It now follows from this theorem that $A \cap B \cap C$ makes sense and $A \cap B \cap C = A \cap (B \cap C) = (A \cap B) \cap C$.

Example 1.29   Draw a Venn diagram to represent the set $A \cap B \cap C$.

Solution: We know that $A \cap B \cap C = (A \cap B) \cap C$. In Fig. 1.14, the set $A \cap B$ is represented by the region shaded by horizontal lines and the set C by vertical lines. Therefore the set $(A \cap B) \cap C$ is represented by the cross-shaded region, i.e., the region shaded both horizontally and vertically.

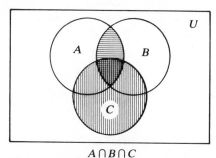

$A \cap B \cap C$

Fig. 1.14

Observe that $A \cap \emptyset = \emptyset$, $A \cap A = A$ and $A \cap U = A$. Also, if $A \subseteq B$, then $A \cap B = A$, as illustrated in Fig. 1.13c. The set $A \cap B$ is a subset of both A and B, but it need not be proper!

Exercise 1.6

1. Use (set-theoretic) symbols to rewrite the following in terms of symbols.
   a) The number 7 is an element of the set consisting of 1,3,5,7.
   b) Every element of A is also an element of B.
   c) The set A is a proper subset of the set B.
   d) Sets A and B have no common elements.
   e) Sets A and B have some common elements.

2. If $A = \{1,2,3,4\}$, $B = \{2,4,6,8\}$, and $C = \{1,3,4,8,9\}$, find the following sets.
   a) $A \cap B$ **b)** $B \cap C$ **c)** $A \cap \emptyset$
   d) $A \cap C$ **e)** $A \cap (B \cap C)$ **f)** $\emptyset \cap \emptyset$

3. If A, B, and C denote the sets of letters of the words "Alabama," "lamb," and "Canada," respectively, find the following sets.
   a) $A \cap B$ **b)** $B \cap C$ **c)** $A \cap (B \cap C)$
   d) $(A \cap B) \cap C$ **e)** $A \cap A$ **f)** $A \cap \emptyset$

4. Mark *true* or *false* (*A* and *B* are any two sets and *U* the universal set).
  a) $A \cap A = A$.
  b) $A \cap B = B \cap A$.
  c) $A \cap \varnothing = U$.
  d) $A \cap U = \varnothing$.
  e) $A \cup B = A \cap B$.
  f) If $A \subseteq B$, then $A \cap B = B$. *false*
  g) If $A \cap B = A$, then $A \subseteq B$. *true*
  h) If $A \cap B = \varnothing$, then $A = B = \varnothing$. *false*

1.7 DISTRIBUTIVE PROPERTIES

Assume that we are given three sets, *A*, *B*, and *C*. We can take the union of *B* and *C* to get $B \cup C$ and then take the intersection of *A* with $B \cup C$ to get $A \cap (B \cup C)$. It will be interesting to see how this set is related to the sets $A \cap B$ and $A \cap C$ and, similarly, how the set $A \cup (B \cap C)$ is related to the sets $A \cup B$ and $A \cup C$.

Example 1.30 Consider the sets $A = \{1,2,4,6\}$, $B = \{2,3,4,5\}$, and $C = \{2,4,6\}$. We have

$$B \cup C = \{2,3,4,5\} \cup \{2,4,6\} = \{2,3,4,5,6\}$$
$$A \cap B = \{1,2,4,6\} \cap \{2,3,4,5\} = \{2,4\}$$
$$A \cap C = \{1,2,4,6\} \cap \{2,4,6\} = \{2,4,6\}$$
$$A \cap (B \cup C) = \{1,2,4,6\} \cap \{2,3,4,5,6\} = \{2,4,6\}$$
$$(A \cap B) \cup (A \cap C) = \{2,4\} \cup \{2,4,6\} = \{2,4,6\} = A \cap (B \cup C)$$

Similarly, we can verify that $A \cup (B \cap C) = \{1,2,4,6\} = (A \cup B) \cap (A \cup C)$.

Example 1.31 Draw a Venn diagram to represent the set $A \cap (B \cup C)$.

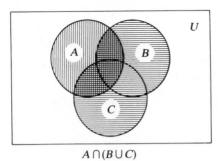

$A \cap (B \cup C)$

Fig. 1.15

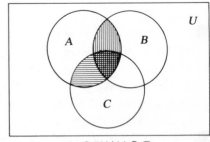

$(A \cap B) \cup (A \cap C)$

Fig. 1.16

Solution: The horizontally shaded region in Fig. 1.15 represents the set $B \cup C$ and the vertically shaded region represents the set *A*. The cross-shaded region represents the set $A \cap (B \cup C)$.

Example 1.32 Draw a Venn diagram to represent the set $(A \cap B) \cup (A \cap C)$.

Solution: In Fig. 1.16 the sets $A \cap B$ and $A \cap C$ are represented by the regions shaded vertically and horizontally, respectively. Thus the union of these sets, $(A \cap B) \cup (A \cap C)$, is represented by the shaded region in the figure.

Observe that the cross-hatched region in Fig. 1.15 and the shaded region in Fig. 1.16 represent the same set. This together with example 1.30 now suggests that $A \cap (B \cup C) = (A \cap B) \cup (A \cap C)$! Is it true for any three sets, A, B, and C? This is answered in the following theorem.

Theorem 1.5   (**Distributive Property**)   *For any three sets A, B, and C,*

$$A \cap (B \cup C) = (A \cap B) \cup (A \cap C)$$

and $\qquad\qquad\qquad A \cup (B \cap C) = (A \cup B) \cap (A \cup C)$

In other words, the intersection operation distributes over the union operation and vice versa.

Exercise 1.7

1. If $A = \{1,2,3,4\}$, $B = \{2,4,6,8\}$, and $C = \{1,3,4,8,9\}$, find the following sets.
   a) $A \cup (B \cap C)$
   b) $A \cap (B \cup C)$
   c) $A \cup (B \cap \varnothing)$
   d) $A \cap (B \cup \varnothing)$
   e) $(A \cup B) \cap (A \cup C)$
   f) $(A \cap C) \cup B$
2. Draw Venn diagrams to represent the following sets.
   a) $A \cup (B \cap C)$
   b) $(A \cap B) \cup C$
   c) $(A \cup B) \cap C$
   d) $(A \cap B) \cup (C \cap B)$
3. Find the set of all elements of A that are not in B, if
   a) $A = \{1,2,3,4\}$ and $B = \{0,2,4\}$.
   b) $A = \{a,b,c\}$ and $B = \{b,c\}$.
   c) $A = \{x,y\}$ and $B = \{x,y,z\}$.

1.8   COMPLEMENTS

Consider the sets $A = \{a,b,c\}$ and $B = \{a,x,y\}$. What are the elements of A that are not in B? Clearly, they are the elements b and c, which form the set $\{b,c\}$. Similarly, the elements of B that are not in A form the set $\{x,y\}$. This suggests a third method of constructing a new set from two given sets.

Definition 1.10   Let A and B be any two sets. The *relative complement* of B, with respect to A, denoted by $A - B$, is the set of all those elements of A that are not in B. This is also called the *difference* $A - B$. We read $A - B$ as "A minus B."

Example 1.33   If $A = \{w,x,y,z\}$ and $B = \{w,z,a,b\}$, then

$$A - B = \{x,y\} \quad \text{and} \quad B - A = \{a,b\}$$

Example 1.34   Let A be the set of all students in a certain course and B the set of all female students in that course. Then, $A - B$ is the set of all male students, if any, in that course. Note that in this case $B - A = \varnothing$. (Why?)

The shaded region in Fig. 1.17 represents the set $A - B$.

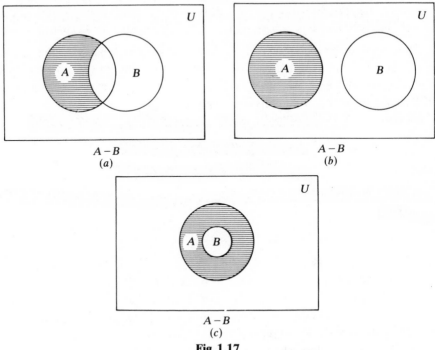

$A - B$
(a)

$A - B$
(b)

$A - B$
(c)

Fig. 1.17

Examples 1.33 and 1.34 tell us that the difference operation is not commutative. That is, in general, $A - B \neq B - A$. In Fig. 1.18, the horizontally shaded region represents $A - B$ and the vertically shaded region represents $B - A$.

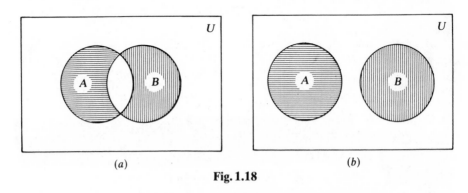

(a)

(b)

Fig. 1.18

It follows from the definition that $A - \emptyset = A$ and $A - A = \emptyset$, for any set A. Also, if $A \subseteq B$, then $A - B = \emptyset$. Notice that $A - B \subseteq A$ always!

Definition 1.11   Let A be any set and U the universal set. Then the *absolute complement of A* or simply the *complement* of A, denoted by A', is the set of all those elements of U that do not belong to A. We read A' as "A prime." Thus A' consists of all those elements in U that lie outside the set A.

Example 1.35 Consider the sets $A = \{1,3,5,7,9\}$, $B = \{1,4,8,9\}$, $C = \{0,2,3,6,9\}$, $D = \{0,2,4,6,8\}$, and $U = \{0,1,2,3,4,5,6,7,8,9\}$. Then

$$A' = \{0,2,4,6,8\} \qquad B' = \{0,2,3,5,6,7\}$$
$$C' = \{1,4,5,7,8\} \qquad D' = \{1,3,5,7,9\}$$

The shaded region in Fig. 1.19 represents the set A'.

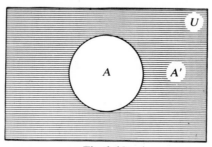

Fig. 1.19

Clearly, $A \cap A' = \varnothing$, and $\varnothing' = U$, and $U' = \varnothing$. From Definition 1.11, we conclude that $A - B = A \cap B'$, which is useful in simplifying expressions of sets involving the difference operation.

Given a set A, we know how to find A'. So it is only natural to ask: how is the complement $(A')'$ of A' related to A? Let's take an example.

Example 1.36 Let $A = \{1,3,5\}$ and $U = \{1,2,3,4,5\}$. Then

$$A' = \{2,4\} \qquad \text{and} \qquad (A')' = \{1,3,5\} = A$$

Figure 1.19 also suggests that $(A')' = A$ for any set A. Indeed, it is true for any set A.

Given two sets A and B, we can form their union $A \cup B$, and their intersection $A \cap B$. How are the complements $(A \cup B)'$ and $(A \cap B)'$ of $A \cup B$ and $A \cap B$ related to A' and B'?

Example 1.37 Consider the sets $A = \{a,b,c\}$, $B = \{b,c,d,e\}$, and $U = \{a,b,c,d,e,f\}$. We have,

$$A \cup B = \{a,b,c,d,e\} \qquad (A \cup B)' = \{f\}$$
$$A \cap B = \{b,c\} \qquad (A \cap B)' = \{a,d,e,f\}$$
$$A' = \{d,e,f\} \qquad B' = \{a,f\}$$

Thus
$$A' \cap B' = \{f\} = (A \cup B)'$$
and
$$A' \cup B' = \{a,d,e,f\} = (A \cap B)'$$

Example 1.38 Draw Venn diagrams to represent the sets $(A \cup B)'$ and $A' \cap B'$. The shaded region in Fig. 1.20a represents the set $(A \cup B)'$. In Fig.

1.20b, the horizontally shaded region represents the set A' and the vertically shaded region the set B'. Hence the cross-shaded region represents the set $A' \cap B'$. In these two figures, we notice that the sets $(A \cup B)'$ and $A' \cap B'$ represent the same region. This would suggest that $(A \cup B)' = A' \cap B'$.

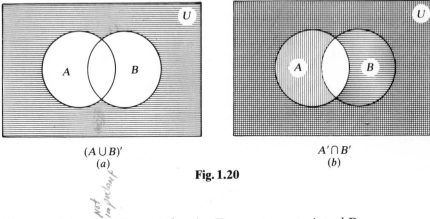

(A∪B)'
(a)

A'∩B'
(b)

Fig. 1.20

Theorem 1.6   (*De Morgan's laws*)   *For any two sets A and B,*

$$(A \cup B)' = A' \cap B'$$

and

$$(A \cap B)' = A' \cup B'$$

Exercise 1.8

1. If $A = \{1,2,3,4\}$, $B = \{2,4,6,8\}$, and $C = \{1,3,4,8,9\}$, find the following sets.
   a) $A - B$
   b) $B - C$
   c) $A - \varnothing$
   d) $\varnothing - B$
   e) $(A - C) \cap B$
   f) $(A - B) - C$
   g) $A - (B - C)$
   h) $(A \cup B) - C$
   i) $A - (B \cap C)$

2. If A = set of all letters of the word "Tallahassee" and B = set of all letters of the word "Tennessee," find the following.
   a) $A \cup B$
   b) $A \cap B$
   c) $A - B$
   d) $B - A$

3. If $U = \{a,b,c,x,y\}$, find the complement of the following sets.
   a) $A = \{a,b,x\}$.
   b) $B = \{x,y\}$.
   c) $C = \{a,y\}$.
   d) $D = \varnothing$.
   e) $E = \{x\}$.
   f) $F = U$.

4. If $\quad A = \{0,1,3,4,5\}$, $\quad B = \{1,3,6,7\}$, $\quad C = \{2,4,5,6\}$, $\quad$ and $U = \{0,1,2,3,4,5,6,7\}$, find:
   a) A'.
   b) $B \cap C'$.
   c) $A - B'$.
   d) $(A \cap B)'$.
   e) $A' \cap B'$.
   f) $(A \cup B)'$.
   g) $A' \cup B'$.
   h) $(A \cap \varnothing)'$.
   i) $(A \cap C)'$.
   j) $(A' \cap C')'$.
   k) $A \cap A'$.
   l) $(A \cap U)'$.
   m) $(A - B) \cup (B - A)$
   n) $(A - B) \cup (B - C)$
   o) $(A - A') \cup (B - B')$

5. Let A and B be any two sets and U the universal set. If $A \cap B = \varnothing$ and $A \cup B = U$, what can be said about A' and B'?

6. Fill in the following blanks (A and B are any sets).

a) $A \cup A' =$ _____   **b)** $A \cap A' =$ _____   **c)** $A - A' =$ _____

d) $A' - A =$ _____   **e)** $A \cap \varnothing =$ _____   **f)** $A - \varnothing =$ _____

g) $\varnothing - A =$ _____   **h)** $A - \varnothing' =$ _____   **i)** $A \cup U =$ _____

j) $A \cap \varnothing' =$ _____   **k)** $A - A =$ _____   **l)** $A \cap U =$ _____

m) $\varnothing - \varnothing =$ _____   **n)** $A \cap U' =$ _____   **o)** $(A \cup B)' =$ _____

7. Find the following sets, if $A \subseteq B$ and $B \subseteq A$.

a) $A \cup B$ **b)** $A \cap B$ **c)** $A \cap B'$

d) $A' \cap B$ **e)** $(A \cup B)'$ **f)** $(A \cap B)'$

g) $A' \cup B'$ **h)** $A' \cap B'$ **i)** $A \cup \varnothing$

8. Draw Venn diagrams to represent the following sets.

a) $(A \cup B) - C$ **b)** $(A \cap B) - C$ **c)** $(A - B) - C$

d) $A \cap (B' \cup C)$ **e)** $A \cap (B' \cup C')$ **f)** $A - (B - C)$

9. In the following Venn diagrams, use set language to describe the shaded region:

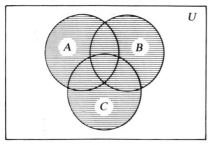

(a)

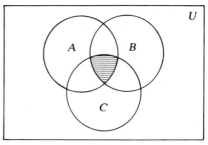

(b)

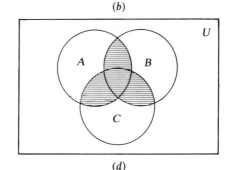

(c)

(d)

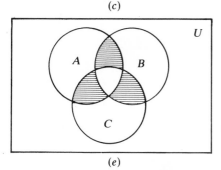

(e)

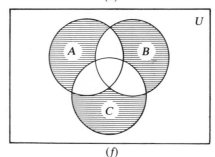

(f)

10. Mark *true* or *false* (A and B are any sets, and U the universal set).
   a) $A - \emptyset = A$.
   b) $\emptyset - A = A$.
   c) $\emptyset - \emptyset = U$.
   d) $A - A' = \emptyset$.
   e) $A - (A - \emptyset) = A$.
   f) $A - B = B - A$.
   g) $A - B = A \cap B'$.
   h) If $A \subseteq B$, then $A' \subseteq B'$.
   i) If $A = B$, then $A' = B'$.
   j) If $A' = B'$, then $A = B$.

11. Verify that $(A \cap B)' = A' \cup B'$, using Venn diagrams.

12. Let $U =$ set of all parallelograms, $A =$ set of all rectangles, and $B =$ set of all squares. Find:
   a) $A \cup B$ **b)** $A \cap B$ **c)** $A - B$ **d)** A'

★**13.** Let A and B be any two sets. The *symmetric difference* between A and B is defined by $A \Delta B = (A - B) \cup (B - A)$. Prove that
   a) $A \Delta \emptyset = A$ **b)** $A \Delta A = \emptyset$
   c) $A \Delta U = A'$ **d)** $A \Delta A' = U$.

★**14.** Use Venn diagrams to illustrate the following.
   a) $C - (A \cap B) = (C - A) \cup (C - B)$.
   b) $C - (A \cup B) = (C - A) \cap (C - B)$.

1.9 APPLICATIONS OF VENN DIAGRAMS

Here we give applications of Venn diagrams in survey analysis. Let's start with an example to illustrate this.

Example 1.39 In a survey of 100 college students, it was found that 50 students read the Times, 43 read the Post, 28 read the Globe, 13 read the Times and the Post, 11 read the Post and the Globe, 12 read the Globe and the Times, and 5 read all of these. How many of these students read

a) the Times only?
b) the Post only?
c) the Globe only?
d) the Times or the Post?
e) the Post or the Globe?
f) the Times, but not the Globe?
g) the Post, but not the Globe?
h) the Times and the Post, but not the Globe?
i) the Times or the Post, but not the Globe?
j) none of these?

Solution: Let's consider the set of 100 students interviewed for the survey as

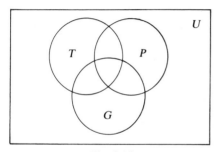

Fig. 1.21

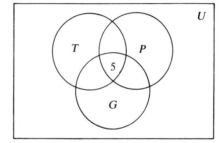

Fig. 1.22

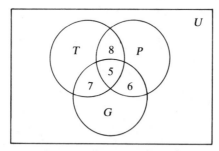

Fig. 1.23

our universal set U. Let T, P, and G represent the set of students reading the Times, the Post, and the Globe, respectively.

Step 1   We now represent the sets T, P, and G in a Venn diagram, as shown in Fig. 1.21. The universal set is divided into eight smaller regions. Our aim is, using the given information, to distribute the 100 students surveyed into these regions so that we can read the necessary answers from the diagram. A careful study of the data tells us that we should use it in the reverse order.

Step 2   The fact that five students read the Times, the Post, and the Globe implies that the set $T \cap P \cap G$ contains 5 elements, as shown in Fig. 1.22.

Step 3   That 12 students read the Globe and the Times means the set $G \cap T$ contains 12 elements. But from step 1, five of these 12 students read the Post also. Thus the region $(T \cap G) - P$ contains 7 elements. Since 11 students read the Post and the Globe, the set $(P \cap G) - T$ contains 6 elements. Similarly, the set $(T \cap P) - G$ contains 8 elements. This is exhibited in Fig. 1.23.

Step 4   Since 28 students read the Globe and we have, by now, accounted for only $5 + 6 + 7 = 18$ of those students, the region $G - (T \cup P)$ contains 10 elements. Similarly, the regions $T - (P \cup G)$ and $P - (G \cup T)$ contain 30 and 24 elements, respectively. The resultant Venn diagram is shown in Fig. 1.24.

Step 5   Since $T \cup P \cup G$ contains only $30 + 8 + 5 + 7 + 6 + 24 + 10 = 90$ elements, we have so far accounted for only 90 students. Therefore, the remaining 10 elements must lie outside the set $T \cup P \cup G$, as shown in Fig. 1.25.

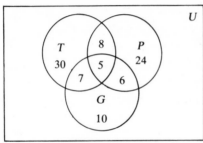

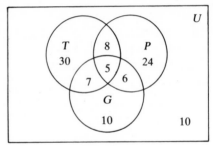

Fig. 1.24   **Fig. 1.25**

Thus, from the diagram:

a) The set $T-(P\cup G)$ contains 30 elements. That is, 30 students read only the Times.

b) The set $P-(G\cup T)$ contains 24 elements, showing that there are 24 students who read only the Post.

c) Since the set $G-(T\cup P)$ contains 10 elements, 10 students read only the Globe.

d) The set $T\cup P$ is all students reading the Times or the Post and contains $30+8+5+7+6+24=80$ elements.

e) Similarly, there are $24+8+5+7+6+10=60$ students who read either the Post or the Globe, represented by the region $P\cup G$.

f) There are $30+8=38$ students who read the Times but not the Globe, represented by $T-G$.

g) There are $24+8=32$ students who read the Post but not the Globe, represented by $P-G$.

h) The set of students who read the Times and the Post, but not the Globe, is $(T\cap P)-G$ and there are 8 such students.

i) There are $30+8+24=62$ students who read the Times or the Post but not the Globe, represented by the region $(T\cup P)-G$.

j) The region $(T\cup P\cup G)'$ represents the set of students who read none of the three papers and, accordingly, there are 10 such students.

Exercise 1.9

1. a) Find the sets A and B if $A\cap B=\{c,d\}$, $A\cap B'=\{a,b\}$, and $A'\cap B=\{e\}$.
   b) Find the universal set U for part (a) if $B'=\{a,b,f,g\}$.

2. In a survey of a certain number of housewives, of the two kinds of laundry detergents "Lex" and "Rex," it was found that 15 women like only "Lex," 10 like "Lex" and "Rex," 20 like only "Rex," and 5 women like neither. How many people were interviewed for the survey?

3. Find the universal set U if $A\cap B=\{3,4\}$, $A'=\{5,6,7,8\}$, and $B'=\{1,2,7,8\}$.

4. Find the sets A, B, C if $A\cap B=\{b,c,d\}$, $B\cap C=\{b,c\}$, $C\cap A=\{a,b,c\}$, $A\cap(B\cup C)'=\{h\}$, $B\cap(C\cup A)'=\{f\}$, $C\cap(A\cup B)'=\{g\}$, and the universal set $U=\{a,b,c,d,e,f,g,h\}$.

5. a) Find the sets A, B, and C if $A \cap B = \{2,5\}$, $B \cap C = \{4,5\}$, $C \cap A = \{3,5\}$, $A \cup (B \cap C) = \{1,2,3,4,5\}$, $B \cup (C \cap A) = \{2,3,4,5,6\}$, and $C \cup (A \cap B) = \{2,3,4,5,7\}$.

   b) Find the universal set U for part (a) if $A' = \{0,4,6,7\}$.

6. In a survey of 160 college students, it was found that 95 students take a course in English, 72 take a course in French, 67 take a course in German, 35 take a course in English and French, 37 take a course in French and German, 40 take a course in German and English, and 25 take a course in all the three.

   a) How many students take a course in English only?

   b) How many take a course in German only?

   c) How many take a course in English or French?

   d) How many take a course in English or French or German?

   e) How many take a course in English, but not in French?

   f) How many take a course in French, but not in German?

   g) How many take a course in English or French, but not in German?

   h) How many take a course in English and French, but not in German?

   i) How many take a course in English, but in neither French nor German?

   j) How many take a course in neither English, French, nor German?

7. Of the 200 people interviewed as to whether they like cheeseburgers, hot dogs, or sandwiches for lunch, it was found that 95 people like cheeseburgers, 85 like hot dogs, 95 like sandwiches, 55 like both cheeseburgers and hot dogs, 45 like hot dogs and sandwiches, 50 like both sandwiches and cheeseburgers, and 30 like all the three. How many people like

   a) cheeseburgers only?

   b) cheeseburgers but not sandwiches?

   c) cheeseburgers or hot dogs, but not sandwiches?

   d) cheeseburgers and sandwiches, but not hot dogs?

   e) none of these?

1.10   ONE-TO-ONE CORRESPONDENCE

Consider the sets $A = \{$Jack,Jim,Joe$\}$ and $B = \{$Carol,Kathy,Mary$\}$, and the sets $C = \{$Bill,Frank,John$\}$ and $D = \{$Jane,Jean$\}$. Our aim in this section is to discuss an important property shared by the sets A and B but not by the sets C and D. We want to introduce the notions "has the same number of elements as," "has more elements than," and "has fewer elements than" without using the technique of counting. The idea of one-to-one correspondence, usually written as 1-1 correspondence, is nowadays introduced in schools as early as kindergarten, as a child tries to pair each finger on his right hand with exactly one protrusion on the glove for the right hand.

Assume that each member of the set A is allowed to date exactly one member of the set B, and each member of the set B exactly one member of the set A. What are the possible ways of *pairing* or *matching*, as they are called in mathematics, the boys of set A with the girls of set B? For instance, we could match Jack with Carol, Jim with Kathy, and Joe with Mary, and vice versa. In this case, we have paired each element of A with a unique element of B and vice versa. In other words, corresponding to every element of A there is a unique element of B, and corresponding to every element of B there exists a unique element of A. We can exhibit this pairing in a diagram, as shown in Fig. 1.26a.

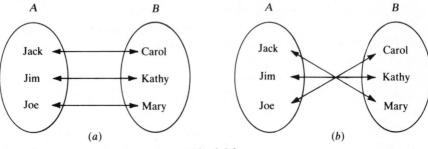

Fig. 1.26

But notice that this is not the only way we can pair the elements of A with the elements of B uniquely. A second possible pairing is shown in Fig. 1.26b. Such a pairing or matching of the elements of set A with those of set B is called a one-to-one correspondence. More precisely, we make the following definition:

Definition 1.12   A *one-to-one correspondence* between sets A and B is a pairing of the elements of A with those of B such that
1. every element of A is paired with a unique element of B and
2. every element of B is paired with a unique element of A.
If a in A is paired with b in B, we write $a \leftrightarrow b$.

Now, do the sets $C = \{\text{Bill,Frank,John}\}$ and $D = \{\text{Jane,Jean}\}$ have the property we have been discussing? Let's see if we can pair each element of C with a unique element of D and conversely. If Bill dates one of the two girls, then Frank or John must date the other girl. This leaves one of the two boys with no girls to date. In this case, every element of D can be paired with a

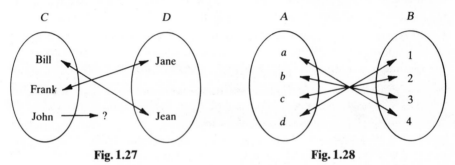

Fig. 1.27   **Fig. 1.28**

unique element of *C*, but not conversely. Figure 1.27 illustrates this situation. Hence there does not exist a 1-1 correspondence between *C* and *D*. In such a situation, we say that there are "more elements in *C* than in *D*" or "fewer elements in *D* than *C*."

Example 1.40 Consider the sets $A = \{a,b,c,d\}$ and $B = \{1,2,3,4\}$. Each element of *A* can be paired with exactly one element of *B* and each element of *B* with exactly one element of *A*, as shown in Fig. 1.28. Thus, there does exist a 1-1 correspondence between *A* and *B*.

Definition 1.13 Two sets *A* and *B* are said to be *equivalent* if there exists a 1-1 correspondence between them; we write $A \sim B$. If *A* is not equivalent to *B*, we write $A \not\sim B$.

Example 1.41 The sets $A = \{\text{Kennedy,Johnson,Nixon}\}$ and $B = \{x,y,z\}$ are equivalent, since there is a 1-1 correspondence between them.

Example 1.42 If *A* = set of all months of the year that begin with the letter *M* and *B* = set of all days of the week that begin with the letter *S*, then $A \sim B$. (Why?)

Example 1.43 If $A = \{\text{Frank,Fred,Ed}\}$ and $B = \{\text{London,Paris,Chicago,} \text{Boston}\}$, then $A \not\sim B$.

Observe that if two sets are equal, they are clearly equivalent; however, if two sets are equivalent, they need not be equal.

The reader should notice that equivalent sets share an interesting property. What is it? We will discuss more about equivalent sets in Chapter 3.

Exercise 1.10
1. Mark *true* or *false*.
   a) Equal sets are equivalent.
   b) Equivalent sets are equal.
   c) If $A \sim B$, then there is a 1-1 correspondence between *A* and *B*.
   d) $\{\varnothing\}$ and $\varnothing$ are equivalent.
   e) $\{\varnothing\}$ and $\{0\}$ are equivalent.
   f) $\{0,1\}$ and $\{\varnothing,\{\varnothing\}\}$ are not equivalent.
   g) There exists a 1-1 correspondence between the set of letters of the word "India" and that of the word "Indiana."
   h) There exists a 1-1 correspondence between the set of letters of the word "Alabama" and that of the word "Canada."
2. What is the difference between $A = B$ and $A \sim B$?
3. List all the possible 1-1 correspondences between the sets $A = \{\text{Johnson,Nixon}\}$ and $B = \{\text{Boston,Chicago}\}$. There are two of them!
4. Exhibit all the possible 1-1 correspondences of the set $A = \{x,y\}$ with itself.
5. List all the possible 1-1 correspondences between the sets $A = \{a,b,c\}$ and $B = \{x,y,z\}$. There are six of them!

6. Same as Problem 4 but with $A = \{x,y,z\}$.

7. List three possible 1-1 correspondences between the sets $A = \{$Albany,Boston,Chicago,Detroit$\}$ and $B = \{w,x,y,z\}$. There are 24 of them!

8. Are the sets $A = \{$John,Jean,Jerry,Joe$\}$ and $B = \{$Maine,Massachusetts,Michigan$\}$ equivalent?

9. Find a set that is equivalent to each of the following sets.

a) $A = \{0,1,2\}$.

b) $B =$ set of all months of the year that end in r.

c) $C =$ set of all letters of the word "statistics".

d) $D =$ set of all digits of the number 1,230,321.

10. Which of the following sets are equal and which are equivalent?

a) $A =$ set of all letters of the word "Alabama".

b) $B =$ set of all letters of the word "balm".

c) $C =$ set of all letters of the word "lamb".

d) $D =$ set of all letters of the word "Mississippi".

e) $E = \{$Boston,Chicago,London,New York,Paris$\}$.

f) $F =$ set of all letters of the word "Calcutta".

g) $G =$ set of all letters of the word "calculus".

11. Does there exist a 1-1 correspondence between the set of sides of a triangle and the set of its angles?

1.11 INFINITE AND FINITE SETS

In the previous section, we defined two sets to be equivalent if and only if there exists a 1-1 correspondence between them. It is clear that every set is equivalent to itself. So it is challenging to ask: "Can there be a set that is equivalent to a proper subset of itself?" This is answered affirmatively, as shown below:

Consider the sets $A = \{1,2,3,...,n,...\}$ and $B = \{2,4,6,...,2n,...\}$. Notice that $B \subset A$. We can, however, exhibit a 1-1 correspondence between these two sets as follows:

$$\begin{array}{ccccccc} 1 & 2 & 3 & 4 & 5 & \ldots & n & \ldots \\ \updownarrow & \updownarrow & \updownarrow & \updownarrow & \updownarrow & & \updownarrow & \\ 2 & 4 & 6 & 8 & 10 & \ldots & 2n & \ldots \end{array}$$

Here we have paired each element n in A with the unique element $2n$ in B and conversely; hence, by definition, $A \sim B$. Thus we have shown the existence of at least one set that is equivalent to a proper subset of itself. In fact, many such sets exist! A set that is equivalent to a proper subset of itself naturally deserves special attention. Such a set is called an *infinite* set.

Definition 1.14 A set is said to be *infinite* if it is equivalent to a proper subset of itself. Intuitively, a set is infinite if the set, when written in the roster method, has no last element.

What do we have to do in order to show that a certain set A is infinite? Find a suitable proper subset B of A and then establish a suitable 1-1 correspondence between them.

Consider now the set $A = \{a,b,c,d\}$. No matter what proper subset B of A we choose, we cannot exhibit a 1-1 correspondence between them. Accordingly, the set A is not infinite. We are now in a position to make the following definition:

Definition 1.15   A set is said to be *finite* if it is not infinite.

This is clearly not a convenient definition to determine the finiteness of a set. Roughly speaking, a set is finite if it has an "end" element when written in the roster method. Clearly the sets $\{w,x,y,z\}$ and $\{a,b,c\}$ are finite but not equivalent. Any two randomly chosen finite sets are not necessarily equivalent. Are all the infinite sets equivalent? Although it might seem somewhat paradoxical, the fact of the matter is that they are not!

We will encounter some very important infinite and finite sets in the next few chapters.

Exercise 1.11

1. Mark *true* or *false*.
   a) The set of letters of the English alphabet is finite.
   b) The set of vowels of the English alphabet is finite.
   c) The set of all people in the world is infinite.
   d) The set of all words in this book is infinite.
   e) The empty set is neither finite nor infinite. *false*
   f) The empty set is finite. *true*
   g) If $A \sim B$ and A is finite, then B is finite. *true*
   h) If A is infinite, then every proper subset of A is infinite. *false*
   i) If $A \subseteq B$ and A is infinite, then B is infinite. *true*
   j) There are finite sets that are infinite. *false*
   k) Some infinite sets are finite. *false*
2. Which of the following sets are finite and which are infinite?
   a) set of all countries of the world
   b) set of all people in the United States
   c) set of all United States Senators
   d) set of all grains of sand on the beaches of Canada
   e) set of all cars in Europe
   f) set of all Presidents of the United States
   g) set of all points on a straight line
   h) set of all words in the Bible
   i) set of all turnpikes in the United States
   j) set of all governors in your state
3. Explain why the sets $A = \{a,b,c\}$ and $B = \{a,b,c,d\}$ are finite.
★4. Show that the sets $E = \{2,4,6,8,...\}$ and $D = \{5,10,15,20,...\}$ are infinite.

1.12 CARTESIAN PRODUCT

Recall that, thus far, we defined three different operations on sets: union, intersection, and difference. In this section we define a new operation, called *cartesian product*, after the French mathematician René Descartes (1596–1650).

Consider the set $\{a,b\}$. Here we do not assume that the element a should be written in the first place and the element b in the second place. The set could very well look like $\{b,a\}$. Now let's assign a position to each element of the set. To be precise, assume that the element a should be treated as the first element and the element b as the second in the set. Such a set is called an *ordered pair*.

Definition 1.16   An *ordered pair* is a set of two elements x and y, in which one element, say x, is designated as the first element and the other element, y, as the second element. An ordered pair is usually denoted by (x,y).

Ordered pairs and the ordered pair notation were systematically used for the first time in mathematics by Descartes. By definition, (a,b) is an ordered pair, whereas $\{a,b\}$ is an unordered pair. The reader should observe that although $\{a,b\} = \{b,a\}$, in general $(a,b) \neq (b,a)$ unless $a = b$.

Definition 1.17   Two ordered pairs (a,b) and (c,d) are said to be *equal* if their corresponding elements are the same; that is, if $a = c$ and $b = d$.

We shall make use of ordered pairs to form new sets. Ordered pairs will be used in greater detail in Chapter 10. The reader probably has used ordered pairs in school when drawing graphs of curves and certainly when using road maps or a world atlas! Let's give a less trivial example here.

Example 1.44   Ron wants to make a pleasure trip from Boston to London via New York. He can travel from Boston to New York by car, plane, or ship and from New York to London by plane or ship. The way he can travel from Boston to New York and then to London can be represented by an ordered pair (x,y), where x represents the means of transportation from Boston to New York and y that from New York to London. For instance, (car, plane) means he will go from Boston to New York by car and then to London by plane. Thus the different possible ordered pairs will give all the possible ways he could make his trip. Now, what are the possible ordered pairs? They are all of the form (x,y) with x in $A = \{$car, plane, ship$\}$ and y in $B = \{$plane, ship$\}$, namely:

$\{$(car,plane), (car,ship), (plane,plane), (plane,ship), (ship,plane), (ship,ship)$\}$

Observe that this is a set whose elements are all ordered pairs.

Tree diagrams can conveniently be used to represent the modes of transportation Ron can use for his trip, as shown in Fig. 1.29.

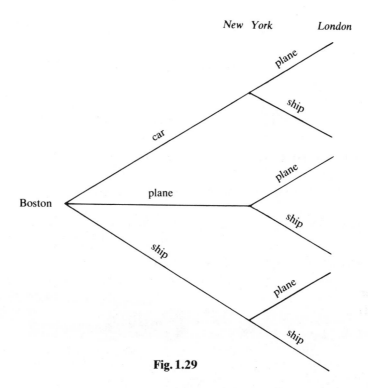

New York   London

plane

ship

car

plane

plane

Boston

ship

plane

ship

Fig. 1.29

Consider the sets $A = \{a,b\}$ and $B = \{1,2,3\}$. What are the possible ordered pairs (x,y) with x in A and y in B? The first-place x has two choices and the second-place y has three choices. Notice that $(a,1)$, $(a,2)$, $(a,3)$ are the ordered pairs we can form with a as the first element; $(b,1)$, $(b,2)$, $(b,3)$ are the ordered pairs we can form with b as the first element. Thus the set of all possible ordered pairs is $\{(a,1),(a,2),(a,3),(b,1),(b,2),(b,3)\}$. Such a set is called the cartesian product of the two sets A and B. Let's now make the following definition:

Definition 1.18   The *cartesian product* (or *cross product*) of two sets A and B is the set of all ordered pairs (a,b) with a in A and b in B, denoted by $A \times B$. We read $A \times B$ as "the cartesian product of A and B" or "A cross B."

Observe that $A \times B$ shows all possible ways of pairing the elements of A with those of B.

Example 1.45   If $A = \{a,b\}$ and $B = \{1,2,3\}$, then

$$A \times B = \{(a,1),(a,2),(a,3),(b,1),(b,2),(b,3)\}$$

and $$B \times A = \{(1,a),(1,b),(2,a),(2,b),(3,a),(3,b)\}$$

There is a convenient way of constructing the cartesian product of two sets. We write the elements of A in a column (vertically) and the elements of B in a row (horizontally), as shown in Fig. 1.30.

$$B$$

×	1	2	3
A a			
b			

Fig. 1.30

$$B$$

×	1	2	3
A a	$(a,1)$	$(a,2)$	$(a,3)$
b	$(b,1)$	$(b,2)$	$(b,3)$

Fig. 1.31

Now, with a as the first element, write down all the possible ordered pairs (a,n) under each n in B, all in the same row containing the element a. Similarly with the second element b in A. Then the diagram looks like the one in Fig. 1.31. This way of arranging the elements of a product set is called *lattice arrangement*.

Observe that example 1.45 tells that the Cartesian product of two sets lacks a property that two of the three previous operations, union and intersection, possess, namely, it is not *commutative*. That is, $A \times B \neq B \times A$ in general. Notice that $A \times B = B \times A$ if and only if $A = B$. However, $A \times B \sim B \times A$ always! Also, by definition, $A \times \emptyset = \emptyset = \emptyset \times A$ for any set A.

When we defined the product of two sets, we did not assume that A and B are different. Even if A and B are equal, we can still take the cartesian product of A with itself.

Example 1.46   If $A = \{1,2,3\}$, then

$$A \times A = \{(1,1),(1,2),(1,3),(2,1),(2,2),(2,3),(3,1),(3,2),(3,3)\}$$

Example 1.47   Consider the sets $A = \{1,2\}$, $B = \{3,4\}$, and $C = \{2,3\}$. We have

$$B \cup C = \{2,3,4\} \qquad B \cap C = \{3\}$$
$$A \times B = \{1,2\} \times \{3,4\} = \{(1,3),(1,4),(2,3),(2,4)\}$$
$$A \times C = \{1,2\} \times \{2,3\} = \{(1,2),(1,3),(2,2),(2,3)\}$$
$$A \times (B \cup C) = \{1,2\} \times \{2,3,4\}$$
$$= \{(1,2),(1,3),(1,4),(2,2),(2,3),(2,4)\} = (A \times B) \cup (B \times C)$$
$$A \times (B \cap C) = \{1,2\} \times \{3\} = \{(1,3),(2,3)\} = (A \times B) \cap (A \times C)$$

This example suggests the following theorem:

Theorem 1.7   (*Distributive Properties*)   *For any three sets A, B, and C,*

$$A \times (B \cup C) = (A \times B) \cup (A \times C)$$

and $$A \times (B \cap C) = (A \times B) \cap (A \times C)$$

In other words, the cartesian product distributes over both union and intersection.

Exercise 1.12

1. a) Find $A \times B$ if

   (1) $A = \{x\}$ and $B = \{x,y\}$.

   (2) $A = \{a,b,c\}$ and $B = \{1\}$.

   (3) $A = \{$Kennedy,Johnson,Nixon$\}$ and $B = \{$Massachusetts, Texas,California$\}$.

   b) Represent each $A \times B$ in part (a) by a lattice arrangement.

2. Find $A \times A$ if

   (1) $A = \{a,b\}$.   (2) $A = \{x,y,z\}$.   (3) $A = \{1,2,3,4\}$.

3. Find the sets A and B if

   a) $A \times B = \{(0,1),(1,1),(0,2),(1,2)\}$.

   b) $A \times B = \{(a,x),(a,y),(b,x),(b,y),(c,x),(c,y),(a,z),(b,z),(c,z)\}$.

   c) $A \times B = \varnothing$.

4. Mark *true* or *false* (A,B,C,D are any sets).

   a) $\{a,b\} = (a,b)$.

   b) $(a,b) = (b,a)$.

   c) (a,b) is a set. *true*

   d) Elements of $A \times B$ are sets. *true*

   e) The operation of cross product is commutative. *false*

   f) The operation of cross product is associative. *false*

   g) If $A \times B = \varnothing$, then $A = \varnothing$.

   h) $A \times B \sim B \times A$.

   i) If $A \times B = B \times A$, then $A = B$. *true*

   j) If $A = B$, then $A \times B = B \times A$. *true*

   ★k) If $A \sim B$, then $A \times C \sim B \times C$ for any set C. *true*

   ★l) If $A \sim B$ and $C \sim D$, then $A \times C \sim B \times D$. *true*

   ★m) $A \sim A \times \{1\}$ *true*

   ★n) $A \times \{1\}$ and $A \times \{2\}$ are disjoint sets. *true*

5. If $\{a,b\} = \{a,c\}$, show that $(a,b) = (a,c)$.   *b-c equal*

6. If $(a,b) = (a,c)$, show that $\{a,b\} = \{a,c\}$.   *b-c equal*

7. If $\{a,b\} = \{b,c\} = \{c,d\}$, show that $(a,b) = (c,d)$.   *a equal, b-b equal b-a, a equal c*

8. If $(a,b) = (b,c)$, what can you say about the elements a,b, and c?   *equal c equal b*

9. a) If $A = \{a,b\}$, $B = \{b,c\}$, and $C = \{c,d\}$, verify the following:

   (1) $A \times (B \cup C) = (A \times B) \cup (A \times C)$

   (2) $A \times (B \cap C) = (A \times B) \cap (A \times C)$

   ★b) Is $A \cap (B \times C) = (A \cap B) \times (A \cap C)$?

   ★c) Is $A \cup (B \times C) = (A \cup B) \times (A \cup C)$?

10. Assume a person has 4 shirts and 3 ties. In how many ways can he choose a shirt and a tie to go to work on Monday? Draw a tree diagram to illustrate your answer.

11. Suppose there are two boys and three chairs in a room. In how many possible ways can these boys be seated on these chairs? Illustrate your answer by drawing a tree diagram.

12. If the boys Ted, Fred, and Ed can date one of the girls Sue and Mary, what are the possible ways in which a boy can date a girl? Draw a tree diagram to illustrate your answer.

★13. Find $A \times (B \times C)$ and $(A \times B) \times C$ if
   (1) $A = \{a,b\}$, $B = \{c,d\}$, and $C = \{e,f\}$.
   (2) $A = \{1,2,3\}$, $B = \{3,4\}$, and $C = \{x\}$.
   (3) $A = \{1,3\}$, $B = \{a,b,c\}$, and $C = \varnothing$.

1.13 SUMMARY AND COMMENTS

Basic definitions and symbols in set theory are introduced. We discussed two types of operations on sets: *unary and binary operations.* Unary operations are performed on one set to obtain a second set, while binary operations allow us to combine two sets to form a third set. The operations of union, intersection, relative complement, and cartesian product are binary, whereas the operation of complementation is unary. We shall discuss some fundamental unary and binary operations on special sets and their properties in greater detail in the coming chapters. Following is a list of new terms and symbols introduced in the chapter.

Roster method (listing method)
Set-builder notation $\{x \mid ...\}$
Subset $\subseteq$
Proper subset $\subset$
Improper subset $=$
Universal set U
Empty set (null set) $\varnothing$
Disjoint sets
Power set of a set A $p(A)$
Unary and binary operations
Union $\cup$
Intersection $\cap$
Difference $-$
Complement of a set A A'
Venn diagrams
Line diagrams
One-to-one correspondence
Equivalent sets $\sim$
Finite and infinite sets
Ordered pair (a,b)
Cartesian product of two sets A and B $A \times B$

Also, the following important properties are discussed:
Commutative properties: $A \cup B = B \cup A$, $A \cap B = B \cap A$

Associative properties:	$A \cup (B \cup C) = (A \cup B) \cup C$
	$A \cap (B \cap C) = (A \cap B) \cap C$
Distributive properties:	$A \cap (B \cup C) = (A \cap B) \cup (A \cap C)$
	$A \cup (B \cap C) = (A \cup B) \cap (A \cup C)$
De Morgan's laws:	$(A \cup B)' = A' \cap B', \quad (A \cap B)' = A' \cup B'$

We ask the reader to become familiar with the new symbols, terms, and properties presented and discussed in this chapter, since they are so basic that we will be using several of them in the coming chapters.

We have discussed two types of simple applications of sets in this chapter:
1. applications of Venn diagrams in survey analysis
2. applications of cartesian product in counting problems via tree diagrams.
In Section 1.12, the reader should have noticed how tree diagrams are useful and helpful in counting techniques, as well as in discussing the possible ways of arranging elements of two sets.

Whole numbers, used in our every day life and discussed in Chapter 3, are also an application of sets. Even though whole numbers and their properties have not yet been introduced formally, it is assumed that the reader is familiar with them in order to do the applications.

The reader has probably noticed by now that the set language is useful and powerful in expressing mathematical ideas in precise and concise terms.

SUGGESTED READING

Christian, Robert R., *Introduction to Logic and Sets*, Blaisdell Publishing Co., Waltham, Mass. (1965), pp. 45–65.

Harmon, F. L., and D. E. Dupree, *Fundamental Concepts of Mathematics*, Prentice-Hall, Inc., Englewood Cliffs, N.J. (1964), pp. 29–42.

Johnson, P. E., "The Early Beginnings of Set Theory," *The Mathematics Teacher*, vol. 63 (Dec. 1970), pp. 690–692.

Lipschutz, S., *Set Theory and Related Topics*, Schaum's Outline Series, McGraw-Hill Book Co., New York (1964), pp. 1–29.

Miller, C. D., and V. E. Heeren, *Mathematical Ideas, an Introduction* (2nd ed.), Scott, Foresman and Co., Glenview, Ill. (1973), pp. 43–70.

Sanders, W. J., "Equivalence and Equality," *The Arithmetic Teacher*, vol. 16 (April 1969), pp. 317–322.

Triola, Mario F., *Mathematics and the Modern World*, Cummings Publishing Co., Menlo Park, Calif. (1973), pp. 18–39.

Wheeler, R. E., *Modern Mathematics: An Elementary Approach* (3rd ed.), Brooks/Cole Publishing Co., Monterey, Calif. (1973), pp. 39–72.

"Symbolic logic has been disowned by many logicians on the plea that its interest is mathematical and by many mathematicians on the plea that its interest is logical."

A. N. WHITEHEAD

2/Algebra of Logic

After studying this chapter, you should be able to:
- *indicate if a sentence is a statement*
- *construct truth tables of statements*
- *identify the basic laws of logic*
- *show if two statements are logically equivalent via truth tables*
- *check if a statement is a tautology*
- *check the validity of an argument*
- *recognize direct and indirect proofs*
- *to disprove a statement via counterexample*

2.0 INTRODUCTION

The study of logic was originated among the ancient Greeks, led by Aristotle (384–322 B.C.). It was not until the seventeenth century that symbols were used in the development of logic. The German philosopher and mathematician Gottfried Wilhelm Leibniz (1646–1716) has been given the credit for introducing symbolism in logic. No further significant contributions in symbolic logic were made until those of George Boole (1815–1864), an English mathematician. Boole was the son of a shopkeeper. Later he became professor of mathematics at Queen's College, Cork, Ireland, in 1849, where he published his outstanding work, "An Investigation of the Laws of Thought," in symbolic logic.

Bertrand Russell (1872–1970) and Alfred North Whitehead (1861–1947) contributed considerably in the development of mathematical logic through their significant work "Principia Mathematica."

Logic directly or indirectly plays a central role in the development of every branch of learning. Consequently, it is the purpose of this chapter to introduce the fundamental concepts of logic via its symbols and language so that they may help the reader to think logically and systematically, to make decisions and inferences sensibly, and finally to express himself in more precise and concise terms.

2.1 STATEMENTS

Observe that the English language consists of four types of sentences: declarative, interrogative, exclamatory, and imperative. Not every sentence is acceptable in the study of logic. We are concerned only with a special type of declarative sentence. Every branch of learning makes use of sentences that can be classified as true or false, but not both. For example, "Kennedy was the first Catholic President of the United States" is a true statement; "Lincoln was the first President of the United States" is a false statement; "Johnson was a great President of the United States" may or may not be true (why?). We now make the following definition.

Definition 2.1 A *statement* is a declarative sentence that is either true or false but not both; it is usually denoted by the letters p, q, or r.

The following are examples of statements:
a) Paris is the capital of France.
b) It is snowing.
c) There were 29 days in February, 1972.
d) Australia is part of Europe.
e) $1+1=0$
f) If $x = 2$, then $x^2 = 4$.
g) A triangle is equilateral if and only if all its sides are equal.

We now give examples of sentences that are not statements according to our definition:
1. Let me out!
2. x is a dog.
3. Kennedy was a great President of the United States.
4. "What's My Line?" is a good show.
5. What a fine boy!
6. What is silent in "calm"?

Recall that the fundamental property of a statement is that it is either true or false and excludes any "middle situation." The truthfulness or falsity of a statement is called its *truth value*, denoted by T and F, respectively. For example, the truth value of the statement "Chicago is part of Illinois" is T (true), whereas that of the statement "Paris is in Russia" is F (false).

The preceding statements (*a*) through (*e*) are examples of *simple statements.* A *compound statement* is formed by combining two or more simple statements, called *components.* The truth value of a compound statement depends on the truth value of its components.

Exercise 2.1

1. Which of the following are statements?
   a) Texas is the largest state of the United States.
   b) Who is the silent member of "parliament"?
   c) The earth is flat.
   d) George Washington was the first President of the United States.
   e) John flunked the course.
   f) Moscow is the capital of Russia.
   g) What a great man!
2. Classify the following statements as simple or compound statements.
   a) Ted is smart and rich.
   b) Boston is in Canada.
   c) $1 + 1 = 0$
   d) $2 \cdot 3 = 7$ implies that $4 \cdot 3 = 14$.
   e) If Linda is tall, then she is happy.
   f) A square has four sides.
   g) The house is white if and only if the car is green.
   h) Not all birds can fly.
3. What are the simple statements in each of the following compound statements?
   a) Mathematics can be fun and fun can be mathematics.
   b) Fred likes hunting or swimming.
   c) If $2 = 3$, then $5 = 5$.
   d) The house is blue implies the car is red.
   e) Chicago is in Canada if and only if Paris is in England.

2.2   LOGICAL CONNECTIVES

Recall that, in the preceding chapter, we discussed several methods of combining sets to construct new sets. This section is motivated by similar considerations: how to combine statements to make new statements. You will find several similarities between the way we combine sets to form new sets and the way we combine statements to obtain new statements.

Example 2.1   The statement, "Boston is the capital of Canada and $1 + 1 = 0$" is a compound statement. Its components are "Boston is the capital of Canada" and "$1 + 1 = 0$."

Example 2.2   "It is raining or cold" is a compound statement with components, "It is raining" and "It is cold."

The rest of this section is devoted to the construction of compound statements and the discussion of their truth values. The words "and," "or," "not," "if . . . then," "if and only if" are used to connect statements to form compound statements. Accordingly, they are called *connectives*.

Example 2.3 Consider the statements

p: John likes cheeseburgers for lunch
q: Mary likes hot dogs for lunch

These two statements can be connected using the word "and" to form the compound statement, "John likes cheeseburgers and Mary likes hot dogs for lunch."

Definition 2.2 Let p and q be any two statements. Then the compound statement obtained by writing the word "and" in between them is called the *conjunction* of the statements p and q, denoted by $p \wedge q$.

Example 2.4 Let p: India is part of Indiana and q: $1+1=0$. Then the conjunction of p and q is given by

$p \wedge q$: India is part of Indiana and $1+1=0$

Let's now discuss the truth values of the compound statement $p \wedge q$. The statement p may be true or false; the same is the case with the statement q. Consequently, there exist four possible pairs of truth values for the components p and q:

Table 2.1

1. p is true and q is true
2. p is true and q is false
3. p is false and q is true
4. p is false and q is false

p	q	$p \wedge q$
T	T	T
T	F	F
F	T	F
F	F	F

When is the statement $p \wedge q$ true? When is it false? If both p and q are true, then $p \wedge q$ is true; if p is true and q is false, then $p \wedge q$ is false; if p is false and q is true, still $p \wedge q$ is false; if both p and q are false, then $p \wedge q$ is false. A convenient way of exhibiting the truth values of $p \wedge q$ in terms of those of its components is by means of a *truth table*, as shown in Table 2.1. There are as many rows in the truth table as there are possible pairs of truth values for p and q, namely, four. The first two columns of the table contain the possibilities of truth values of p and q. The third column contains the corresponding truth values of $p \wedge q$.

Example 2.5 Consider the following statements:
a) Paris is the capital of France and $1+1=2$.
b) Paris is the capital of France and $1+1=0$.
c) Boston is the capital of Canada and $1+1=2$.

d) Boston is the capital of Canada and $1 + 1 = 0$.

It follows clearly that statement (a) is true; the other three are false since at least one component in each case is false.

Consider the statements:

p: Paris is in France

q: Boston is in Canada

These two statements can be combined using the connective "or" to form the compound statement:

Paris is in France or Boston is in Canada

Definition 2.3   Let p and q be any two statements. The compound statement obtained by writing the word "or" in between them is called the *disjunction* of the statements p and q, denoted by $p \vee q$.

Example 2.6   Consider the statements

p: Bill wants cheeseburgers for lunch

q: Bill wants hot dogs for lunch

Then $p \vee q$:   Bill wants cheeseburgers or hot dogs for lunch

Example 2.7   The statement "Mary will go to Boston or to Chicago at 2 p.m. today" is the disjunction of the statements "Mary will go to Boston at 2 p.m. today" and "Mary will go to Chicago at 2 p.m. today."

Notice that the usage of the word "or" in the ordinary language is somewhat ambiguous. For instance, in example 2.6, the word "or" is used in the sense of the legal "and/or" to mean at least one: Bill wants cheeseburgers or hot dogs or both for lunch, since he will not refuse if he is given both for lunch. In this case, we say that the "or" is used in the *inclusive* sense. Accordingly, this type of disjunction is called an *inclusive disjunction*. In example 2.7, the word "or" is used in the *exclusive* sense, to mean at least one, but not both: Mary will go to Boston at 2 p.m. today or to Chicago at 2 p.m. today, but not both. Accordingly, this type of disjunction is called an *exclusive disjunction*. In our remaining discussions, we will be concerned only with inclusive disjunctions, hence from now on disjunction means inclusive disjunction.

Consider the compound statement $p \vee q$. The statement p may be true or false; the statement q also may be true or false. Accordingly, there are four possible cases to be investigated:

1. p is true and q is true

2. p is true and q is false

3. p is false and q is true

4. p is false and q is false

Table 2.2

p	q	$p \vee q$
T	T	T
T	F	T
F	T	T
F	F	F

These four cases are entered in the first two columns of Table 2.2. The disjunction $p \lor q$ of the two statements p and q is true if at least one of them is true; otherwise it is false. Consequently, the disjunction of two statements is false if and only if both components are false. This is summarized in Table 2.2.

Example 2.8   Consider the following statements:

a) Paris is the capital of France or $1 + 1 = 2$.

b) Paris is the capital of France or Boston is in Canada.

c) Paris is in England or Boston is in Massachusetts.

d) Paris is in England or Boston is in France.

The truth value of the statements (a), (b), and (c) is T, whereas that of statement (d) is F (why?).

Let's now discuss a third way of forming statements from given ones.

Definition 2.4   The *negation* of a statement p is obtained by writing a phrase such as "it is false that" or "it is not the case that" in front of the statement p, or by writing the word "not" in a suitable location in the given statement p. The negation of p is denoted by $\sim p$.

Example 2.9   Consider the statements

p: Charlie Brown was a President of the United States

q: The sun rises in the east

r: Chicago is the capital of Canada

Then $\sim p$:   Charlie Brown was not a President of the United States

$\sim q$:   The sun does not rise in the east

$\sim r$:   Chicago is not the capital of Canada

It is clear that if p is true then $\sim p$ is false, and if p is false then $\sim p$ is true. This is summarized in Table 2.3.

Table 2.3

p	$\sim p$
T	F
F	T

Consider the following statements:

a) If you wax my car, then I will pay you five dollars.

b) If a triangle is equilateral, then it is isosceles.

c) If John studies hard, then he shall pass the course.

d) If $x = 10$, then $x^2 = 100$.

Observe that each of these statements, which involves a condition, is of the form, "If p, then q." Such a statement is called a *conditional statement* or *statement of implication*, denoted by $p \to q$. The statements p and q are called the *hypothesis* and *conclusion*, respectively, of the implication $p \to q$. Statements of the form $p \to q$ are very common in mathematics.

Example 2.10   Let p: Roses are red and q: Violets are blue. Then

$p \to q$:   If roses are red, then violets are blue

and $q \to p$:   If violets are blue, then roses are red

Example 2.11   Consider the statements p: All sides of a triangle are equal and q: The triangle is equilateral. Then

$p \to q$:   If all the sides of a triangle are equal,
          then it is equilateral

and $q \to p$:   If a triangle is equilateral, then all its
          sides are equal

The implication $p \to q$ is also read as:
1. p implies q
2. p is sufficient for q
3. p only if q
4. q is necessary for p

Let's now discuss the truth values of the conditional statement $p \to q$. Since statements p and q can be true or false, there are four cases to be discussed, as before:
1. p is true and q is true
2. p is true and q is false
3. p is false and q is true
4. p is false and q is false

Table 2.4

p	q	$p \to q$
T	T	T
T	F	F
F	T	T
F	F	T

In order that we may better understand the discussion of the truth values of $p \to q$, it seems desirable that we consider an example. Notice that we can very well regard an implication as a conditional promise and that the implication is false if and only if the promise is violated. Consider the conditional statement:

$p \to q$:   If you wax my car, then I will pay you five dollars

If you wax my car (p true) and if I give you five dollars (q true), then clearly the implication $p \to q$ is true. If you wax my car (p true) and I do not pay you five dollars (q false), then my promise is violated and hence the implication is false. What happens if you do not wax my car (p false)? Then, I may or may not pay you five dollars (q may be true or false). In any case, my promise has not been tested and hence was not violated. Consequently, the implication $p \to q$ is not false and is true. In other words, if p is false, then $p \rightarrow q$ is true regardless of the truth value of q. Thus, $p \rightarrow q$ is

a) true if p is true and q is true,
b) false if p is true and q is false,
c) true if p is false and q is true,
d) true if p is false and q is false.

This discussion is summarized in Table 2.4.

Example 2.12   Consider the following statements:

a) If $1+1=2$, then Lyndon Johnson was a President of the United States.
b) If $1+1=2$, then Sir Winston Churchill was a President of the United States.
c) If $1+1=0$, then Lyndon Johnson was a President of the United States.
d) If $1+1=0$, then Sir Winston Churchill was a President of the United States.

From our discussion above, it follows that statements (a), (c), and (d) are true, whereas statement (b) is false (why?).

From the conditional statement $p \rightarrow q$, we can now form three new conditional statements. The statement $q \rightarrow p$, obtained by interchanging the hypothesis and the conclusion, is called the *converse* of the implication $p \rightarrow q$.

Example 2.13   Consider the statements:

$$p \rightarrow q: \quad \text{If a triangle is equilateral, then it is isosceles}$$
$$r \rightarrow s: \quad \text{If Harry marries Mary, then she is happy}$$

The converse of each of these statements is given by

$$q \rightarrow p: \quad \text{If a triangle is isosceles, then it is equilateral}$$
$$s \rightarrow r: \quad \text{If Mary is happy, then Harry marries her}$$

The implication $\sim p \rightarrow \sim q$ is called the *inverse* of the conditional statement $p \rightarrow q$.

Example 2.14   The inverse of each of the statements $p \rightarrow q$ and $r \rightarrow s$ in example 2.13 is given by

$$\sim p \rightarrow \sim q: \quad \text{If a triangle is not equilateral, then it is not}$$
$$\text{isosceles}$$
$$\sim r \rightarrow \sim s: \quad \text{If Harry does not marry Mary, then she is not happy}$$

The conditional statement $\sim q \rightarrow \sim p$ is called the *contrapositive* of the implication $p \rightarrow q$.

Example 2.15   The contrapositives of the statements $p \rightarrow q$ and $r \rightarrow s$ in example 2.13 are given by

$$\sim q \rightarrow \sim p: \quad \text{If a triangle is not isosceles, then it is not}$$
$$\text{equilateral}$$
$$\sim s \rightarrow \sim r: \quad \text{If Mary is not happy, then Harry does not marry her}$$

Observe that the converse of the inverse of the statement $p \rightarrow q$ is the same as the inverse of the converse of the statement $p \rightarrow q$, namely, the contrapositive, $\sim q \rightarrow \sim p$.

Statements of the form "p if and only if q" are common in mathematics. For example, consider the following statements:

a) A triangle is equilateral if and only if all its sides are equal.

b) Tomorrow is Thursday if and only if today is Wednesday.

c) An integer is even if and only if its last digit is even.

Each of these statements is the conjunction of two conditional statements of the form $p \rightarrow q$ and $q \rightarrow p$. Statements of the form $(p \rightarrow q) \wedge (q \rightarrow p)$, which involve two implications, are called *biconditional statements*, denoted by $p \leftrightarrow q$. We may read $p \leftrightarrow q$ also as "p implies q and q implies p" or "p is necessary and sufficient for q." It follows from Table 2.5 that a biconditional statement $p \leftrightarrow q$ is true if p and q are either both true or both false; consequently, $p \leftrightarrow q$ is false if exactly one of the statements p and q is false.

Table 2.5

p	q	$p \rightarrow q$	$q \rightarrow p$	$(p \rightarrow q) \wedge (q \rightarrow p)$	$p \leftrightarrow q$
T	T	T	T	T	T
T	F	F	T	F	F
F	T	T	F	F	F
F	F	T	T	T	T

Example 2.16   Consider the following statements:

a) London is in England if and only if Paris is in France.

b) London is in England if and only if Boston is in France.

c) Boston is in England if and only if Paris is in France.

d) Paris is in England if and only if London is in France.

The statements (a) and (d) are true, whereas the statements (b) and (c) are false (why?).

So far in our discussion of statements, we have been using the connectives $\wedge$, $\vee$, $\sim$, $\rightarrow$, and $\leftrightarrow$ one at a time. Now, two or more connectives can be used to form more complicated compound statements, as shown by the following examples.

Example 2.17   Construct a truth table for the statement $(p \wedge q) \vee (\sim p)$.

Solution: Notice that the statement $(p \wedge q) \vee (\sim p)$ is the disjunction of the statements $p \wedge q$ and $\sim p$, where $p \wedge q$ is the conjunction of p and q, and $\sim p$ is the negation of p.

Step 1   Since $(p \wedge q) \vee (\sim p)$ can be constructed from the simple statements p and q, the first two columns of the truth table are reserved for the possible pairs of truth values of p and q. Since there are two choices for the truth value of p and two choices for that of q, the table consists of $2 \cdot 2 = 4$ rows; if there are three components p, q, and r, then $2 \cdot 2 \cdot 2 = 2^3 = 8$ rows are needed. More generally, if there are n components, then $2 \cdot 2 \ldots 2$ to n times $= 2^n$ rows are required. From the statements p and q we can form the statements $p \wedge q$ and $\sim p$, requiring two more columns, a total of four columns so far. Now we can form the disjunction $(p \wedge q) \vee (\sim p)$ of the statements $p \wedge q$ and $\sim p$; hence we need one more column, requiring a total of five columns altogether, as shown in Table 2.6.

Table 2.6. Step 1

p	q	$p \wedge q$	$\sim p$	$(p \wedge q) \vee (\sim p)$
T	T			
T	F			
F	T			
F	F			

Step 2   Let's now fill in the third column, headed by $p \wedge q$, by using the first two columns. Recall that the conjunction $p \wedge q$ of two statements p and q is true if and only if both of its components are true (see Table 2.1). Now the table looks as in Table 2.7.

Table 2.7. Step 2

p	q	$p \wedge q$	$\sim p$	$(p \wedge q) \vee (\sim p)$
T	T	T		
T	F	F		
F	T	F		
F	F	F		

Step 3 ·   Since $\sim p$ is the negation of p, we can easily fill in the fourth column, headed by $\sim p$. Recall that $\sim p$ is false if p is true and true if p is false (see Table 2.3). After completing the fourth column, the table looks as in Table 2.8.

Table 2.8. Step 3

p	q	$p \wedge q$	$\sim p$	$(p \wedge q) \vee (\sim p)$
T	T	T	F	
T	F	F	F	
F	T	F	T	
F	F	F	T	

Step 4   It now remains to fill in the last column, headed by $(p \land q) \lor (\sim p)$, which is the disjunction of $p \land q$ and $\sim p$. Recall that the disjunction of two statements is true if and only if at least one of them is true (see Table 2.2). Using the entries in the third and fourth columns, the last column can easily be filled in. The truth table of $(p \land q) \lor (\sim p)$ now looks as in Table 2.9. It follows from Table 2.9 that $(p \land q) \lor (\sim p)$ is always true except when p is true and q is false (see last column). Columns 3 and 4 are the intermediate steps required to complete the last column.

Table 2.9. Step 4

p	q	$p \land q$	$\sim p$	$(p \land q) \lor (\sim p)$
T	T	T	F	T
T	F	F	F	F
F	T	F	T	T
F	F	F	T	T

Example 2.18   Construct a truth table for the statement $(p \land q) \rightarrow (p \lor q)$.

Solution: Notice that to construct a truth table for the statement $(p \land q) \rightarrow (p \lor q)$, we need a column for each of the statements p, q, $p \land q$, $p \lor q$, and $(p \land q) \rightarrow (p \lor q)$, preferably in that order. Using the truth tables for the conjunction, disjunction, and implication of two statements, we can construct the truth table for $(p \land q) \rightarrow (p \lor q)$, as illustrated in Table 2.10.

Table 2.10

p	q	$p \land q$	$p \lor q$	$(p \land q) \rightarrow (p \lor q)$
T	T	T	T	T
T	F	F	T	T
F	T	F	T	T
F	F	F	F	T

Example 2.19   Construct a truth table for the statement $(p \rightarrow q) \leftrightarrow (\sim p \lor q)$.
Solution: The truth table consists of 4 rows and 6 columns, one for each of the statements p, q, $p \rightarrow q$, $\sim p$, $\sim p \lor q$, and $(p \rightarrow q) \leftrightarrow (\sim p \lor q)$. The truth table of $(p \rightarrow q) \leftrightarrow (\sim p \lor q)$ is exhibited in Table 2.11. Observe from the table (see last

Table 2.11

p	q	$p \rightarrow q$	$\sim p$	$\sim p \lor q$	$(p \rightarrow q) \leftrightarrow (\sim p \lor q)$
T	T	T	F	T	T
T	F	F	F	F	T
F	T	T	T	T	T
F	F	T	T	T	T

column) that the given statement is always true, regardless of the truth values of p and q.

Exercise 2.2

1. Rewrite the following sentences in symbols, where

$$p: \quad \text{John is wealthy}$$

$$q: \quad \text{John is tall}$$

a) John is not wealthy.
b) John is wealthy and tall.
c) John is wealthy or tall.
d) If John is wealthy, then he is tall.
e) John is wealthy if and only if he is short.
f) John is wealthy only if he is short.
g) John is neither wealthy nor tall.
h) John is not wealthy, but tall.
i) If John is not wealthy, then he is not tall.
j) It is not true that John is either wealthy or tall.

2. Translate the following into words, where

$$p: \quad \text{It is cold}$$

$$q: \quad \text{It is snowing}$$

a) $\sim p$ **b)** $p \wedge q$ **c)** $p \vee q$
d) $\sim p \wedge q$ **e)** $p \vee \sim q$ **f)** $\sim(p \wedge q)$
g) $\sim(p \vee q)$ **h)** $\sim p \vee \sim q$ **i)** $\sim p \wedge \sim q$

3. Write each of the following in "If-then" form.
a) $x = 2$ implies $x^2 = 4$.
b) The house is not white if it is mine.
c) Every equiangular triangle is isosceles.
d) Lines perpendicular to the same line are parallel.
e) $x^2 = 16$ is necessary for $x = 4$.
f) The house is on Washington street only if it is mine.
g) $x = 1$ is sufficient for $x^2 = 1$.

4. Negate the following statements.
a) Canada is the largest country in the world.
b) Linda is not smart.
c) John is tall and smart.
d) Dick is happy or rich.
e) It is not true that Ron is handsome.
f) Either Paris is in England or London is in France.
g) Neither $1 + 1 = 0$ nor $2 + 3 = 6$.

5. What is the truth value of the following statements?

    a) Paris is not the capital of Russia and $1+1=2$.

    b) Paris is the capital of Canada or France.

    c) If $1+1=0$, then London is in Tokyo.

    d) If Paris is in Russia, then France is the largest country in the world.

    e) Today is Saturday if and only if tomorrow is Sunday.

    f) Boston is not the largest city in the world if and only if $1+1=0$.

6. Let t be a true statement. What is the truth value of

    a) $p \vee t$ **b)** $p \vee \sim p$ **c)** $p \wedge \sim p$ **d)** $p \wedge \sim t$

    e) $\sim p \vee t$ **f)** $\sim p \wedge \sim t$ **g)** $\sim t \to p$ **h)** $p \to t$

7. Find the truth value of

    a) $p \wedge q$ if q is not true

    b) $p \wedge q$ if $\sim q$ is not false

    c) $p \vee q$ if $\sim p$ is false

    d) $p \to q$ if q is true

    e) $(p \wedge q) \to (p \vee q)$ if $\sim q$ is true

    f) $(p \vee q) \to (p \wedge q)$ if p and $\sim q$ are true

    g) $p \to (p \vee q)$ if q is false

    h) $(p \vee q) \to p$ if $\sim q$ is not false

8. Construct a truth table for each of the following statements.

    a) $p \wedge \sim q$ **b)** $\sim p \wedge \sim q$ **c)** $\sim p \vee \sim q$

    d) $\sim(p \vee q)$ **e)** $\sim(p \wedge q)$ **f)** $p \to p \vee q$

    g) $\sim p \to \sim q$ **h)** $\sim(\sim p \vee q)$ **i)** $p \wedge (q \wedge r)$

9. Write down the converse, the inverse and the contrapositive of the following statements.

    a) If London is in England, then Paris is in France.

    b) If $1+1=3$, then $2+2=6$.

    c) If the house is not white, then it is mine.

★10. Which of the following statements contain words that indicate quantity, for example: "all," "every," "each," "none," "some," "one," etc.?

    a) Some people have blue eyes.

    b) Every board is black.

    c) Lyndon Johnson was a President of the United States.

    d) All dogs have two legs.

    e) There are some girls with brown eyes.

    f) Boston is the capital of France.

    g) No babies are logical.

    h) If Paris is in England, then London is in France.

Words that indicate quantity are called *quantifiers*. The quantifier "all" is called the *universal quantifier*, denoted by $\forall$ (an inverted A). The quantifier "some," which indicates the existence of something, is called the *existential quantifier*, denoted by $\exists$ (a reversed E).

★**11.** Negate the following statements.
   a) All men are mortal.
   b) Everybody likes cheeseburgers.
   c) Some people like mathematics.

2.3 LAWS OF LOGIC

This section is devoted to the discussion of some important laws of logic that play a central role in later chapters. First, let's consider the following example.

Example 2.20   Construct a truth table for $p \rightarrow q$ and $\sim p \vee q$.

Solution: We can combine the truth tables of $p \rightarrow q$ and $\sim p \vee q$ into one truth table, as shown in Table 2.12. Notice that the columns headed by $p \rightarrow q$ and $\sim p \vee q$ are identical. In other words, for all possible combinations of truth values of p and q, the statements $p \rightarrow q$ and $\sim p \vee q$ are both true simultaneously or both false simultaneously.

Table 2.12

p	q	$p \rightarrow q$	$\sim p$	$\sim p \vee q$
T	T	T	F	T
T	F	F	F	F
F	T	T	T	T
F	F	T	T	T

Now we are in a position to make the following definition.

Definition 2.5   Two statements p and q are said to be *logically equivalent* if they are both true simultaneously or false simultaneously, denoted by $p \equiv q$. Thus $p \equiv q$ if the columns headed by p and q in their truth tables are identical.

It now follows from Table 2.12 that $p \rightarrow q \equiv \sim p \vee q$.

Example 2.21   Show that $p \rightarrow q \equiv \sim q \rightarrow \sim p$.

Solution: Since the columns headed by $p \rightarrow q$ and $\sim q \rightarrow \sim p$ in Table 2.13 are identical, $p \rightarrow q \equiv \sim q \rightarrow \sim p$. Thus, the contrapositive of the statement $p \rightarrow q$ is

Table 2.13

p	q	$p \rightarrow q$	$\sim q$	$\sim p$	$\sim q \rightarrow \sim p$
T	T	T	F	F	T
T	F	F	T	F	F
F	T	T	F	T	T
F	F	T	T	T	T

logically equivalent to the statement itself. It follows that $q \to p \equiv \sim p \to \sim q$; that is, the converse and the inverse of the conditional statement $p \to q$ are also logically equivalent. These facts play an important part in mathematics.

Theorem 2.1   (***Commutative Laws***)   *For any two statements p and q,*

$$p \land q \equiv q \land p \qquad \text{and} \qquad p \lor q \equiv q \lor p$$

Proof: That $p \land q \equiv q \land p$ and $p \lor q \equiv q \lor p$ follows clearly from Table 2.14.

Table 2.14

p	q	$p \land q$	$q \land p$	$p \lor q$	$q \lor p$
T	T	T	T	T	T
T	F	F	F	T	T
F	T	F	F	T	T
F	F	F	F	F	F

This theorem tells us that the order in which we take the conjunction or disjunction of two statements does not affect the truth value of the compound statement.

Assume that we wish to find the truth value of the statement $p \land q \land r$. Since we can take the conjunction of only two statements at a time, we can take their conjunction either as $(p \land q) \land r$ or as $p \land (q \land r)$. Do these statements have the same truth value? Are they logically equivalent? These questions are answered in the affirmative by the following theorem:

Theorem 2.2   (***Associative Laws***)   *For any three statements p, q, and r,*

$$p \land (q \land r) \equiv (p \land q) \land r \qquad \text{and} \qquad p \lor (q \lor r) \equiv (p \lor q) \lor r$$

Since $p \land (q \land r) \equiv (p \land q) \land r$ and $p \lor (q \lor r) \equiv (p \lor q) \lor r$, it follows that the expressions $p \land q \land r$ and $p \lor q \lor r$ are not ambiguous, but do make sense.

Theorem 2.3   (***Distributive Laws***)   *For any three statements p, q, and r,*

$$p \land (q \lor r) \equiv (p \land q) \lor (p \land r) \qquad \text{and} \qquad p \lor (q \land r) \equiv (p \lor q) \land (p \lor r)$$

This theorem tells us that conjunction is distributive over disjunction and vice versa.

Theorem 2.4   (***De Morgan's Laws***)   *For any two statements p and q,*

$$\sim(p \land q) \equiv \sim p \lor \sim q \qquad \text{and} \qquad \sim(p \lor q) \equiv \sim p \land \sim q$$

Proof: That $\sim(p \wedge q) \equiv \sim p \vee \sim q$ follows clearly from Table 2.15.

Table 2.15

p	q	$p \wedge q$	$\sim(p \wedge q)$	$\sim p$	$\sim q$	$\sim p \vee \sim q$
T	T	T	F	F	F	F
T	F	F	T	F	T	T
F	T	F	T	T	F	T
F	F	F	T	T	T	T

We ask the reader to verify the other half of the theorem.

De Morgan's laws tell us that the negation of the conjunction of two statements is logically equivalent to the disjunction of their negations; also, the negation of the disjunction of two statements is logically equivalent to the conjunction of their negations. Consequently, we must be careful while negating a conjunction or disjunction.

Example 2.22   Consider the following statements:

p: John graduated from school
q: John got admitted to Harvard

Then $p \wedge q$:   John graduated from school and got admitted to Harvard

$p \vee q$:   John graduated from school or got admitted to Harvard

The negations of these compound statements are given by

$\sim(p \wedge q) \equiv$ Either John did not graduate from school or did not get admitted to Harvard

$\sim(p \vee q) \equiv$ Neither did John graduate from school nor get admitted to Harvard

Exercise 2.3

1. If $p \equiv q$ and q is not true, what is the truth value of p?
2. If $p \equiv q$ and $q \equiv r$, where r is true, what is the truth-value of p?
3. If the statements p and $p \rightarrow q$ are both true, what can you say about the truth value of q? Why? [*Hint*: use Table 2.4.] This gives the *law of detachment.*
4. Find the truth value of
   a) $p \wedge q$ if $q \equiv t$ and t is false.
   b) $p \vee q$ if $p \equiv t$ and t is true.

    c) $p \wedge (q \wedge r)$ if $r \equiv t$ and t is not true.

    d) $p \vee (q \vee r)$ if $q \equiv t$ and t is not false.

5. Find the truth value of

    a) $q \wedge p$ if $p \wedge q$ is true.

    b) $p \vee q$ if $q \vee p$ is false.

    c) $(p \wedge q) \wedge r$ if $p \wedge (q \wedge r)$ is not false.

    d) $p \vee (q \vee r)$ if $(p \vee q) \vee r$ is not true.

    e) $p \wedge (q \vee r)$ if $(p \wedge q) \vee (p \wedge r)$ is false.

6. Mark *true* or *false* (p, q, r are any statements).

    a) $p \vee \sim p$ is a true statement.

    b) $p \wedge \sim p$ is a false statement.

    c) $p \wedge q \equiv q \wedge p$.

    d) $p \wedge q \equiv p \vee q$.

    e) $\sim(p \wedge q) \equiv \sim p \wedge \sim q$.

    f) $\sim(p \vee q) \equiv \sim p \vee \sim q$.

    g) $p \to q \equiv q \to p$.

    h) If $p \wedge q \equiv p \wedge r$, then $q \equiv r$.

    i) If $p \vee q \equiv p \vee r$, then $q \equiv r$.

    j) If $p \to q$ is true, then p is true.

7. Prove each of the following.

    a) $\sim(\sim p) \equiv p$ **b)** $p \wedge p \equiv p$

    c) $p \vee p \equiv p$ **d)** $\sim(p \vee q) \equiv \sim p \wedge \sim q$

    e) $\sim(p \to q) \equiv p \wedge \sim q$ **f)** $p \wedge (q \wedge r) \equiv (p \wedge q) \wedge r$

8. Let t be a true statement. Prove that

    a) $p \vee t \equiv t$ **b)** $p \vee \sim p \equiv t$ **c)** $p \wedge \sim p \equiv \sim t$

    d) $\sim p \vee t \equiv t$ **e)** $\sim p \wedge \sim t \equiv \sim t$ **f)** $p \to t \equiv t$

★9. Use De Morgan's laws to prove each of the following.

    a) $\sim(\sim p \wedge \sim q) \equiv p \vee q$ **b)** $\sim(\sim p \vee q) \equiv p \wedge \sim q$

    c) $\sim(\sim p \vee \sim q) \equiv p \wedge q$ **d)** $\sim(p \wedge \sim q) \equiv \sim p \vee q$

    e) $\sim(p \to q) \equiv p \wedge \sim q$ **f)** $p \to \sim q \equiv \sim(p \wedge q)$

★10. Without using truth tables show that

    a) $p \wedge (q \wedge \sim p)$ is always a false statement.

    b) $p \vee (q \vee \sim p)$ is always a true statement.

    c) $p \to (p \vee q)$ is always a true statement.

       [*Hint*: $r \to s \equiv \sim r \vee s.$]

2.4 TAUTOLOGIES AND ARGUMENTS

In the previous section, we observed that there are statements that have exactly the same truth values at all possible cases. In this section, we concentrate on a special type of statement that is closely related to the concept of logical equivalence.

Example 2.23 Construct a truth table for $p \vee \sim p$.

Table 2.16

p	$\sim p$	$p \vee \sim p$
T	F	T
F	T	T

Table 2.17

p	q	$p \vee q$	$p \rightarrow p \vee q$
T	T	T	T
T	F	T	T
F	T	T	T
F	F	F	T

Notice from Table 2.16 that the statement $p \vee \sim p$ is always true. Thus, there exist statements that are always true, irrespective of the truth or falsity of their components. Accordingly, we have the following definition:

Definition 2.6   A statement that is always true is called a *tautology*.

It follows from Table 2.10 that the implication $(p \wedge q) \rightarrow (p \vee q)$ is a tautology, and from Table 2.16 that the disjunction $p \vee \sim p$ is also a tautology. The tautology $p \vee \sim p$ is called the *law of excluded middle*, since a statement is either true or false so that there exists no "middle possibility."

Example 2.24   Show that $p \rightarrow p \vee q$ is a tautology.

Solution: Since the column headed by $p \rightarrow p \vee q$ contains only T's in Table 2.17, it follows that $p \rightarrow p \vee q$ is a tautology.

Theorem 2.5   (*Law of Syllogism*)   *Let p, q, r be any three statements. If $p \rightarrow q$ and $q \rightarrow r$, then $p \rightarrow r$. In other words, $[(p \rightarrow q) \wedge (p \rightarrow r)] \rightarrow (p \rightarrow r)$ is a tautology.*

The law of syllogism plays a key role in proving mathematical results. It tells us that if one statement p implies a second statement q and the second statement q implies a third statement r, then we can always conclude that the first statement p implies the third statement r (see Fig. 2.1). More generally, if $p \rightarrow q, q \rightarrow r, r \rightarrow s, \ldots, \rightarrow u, u \rightarrow v$, then $p \rightarrow v$. The law of syllogism is also called the *chain rule*.

Let's now use some of the ideas and concepts discussed so far to study and check the validity of our reasoning in a discussion. We start with some statements called hypotheses; we would like to logically conclude some other statements called conclusions, through reasoning or arguments. What does it mean to say that the reasoning is logical or the argument is valid?

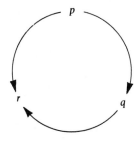

Fig. 2.1

Definition 2.7 An argument is said to be *valid* if the conjunction of the hypotheses implies the conclusion, otherwise it is *invalid* (a *fallacy*).

An argument is valid if and only if the conclusion is a logical consequence of the hypothesis. In other words, if the hypothesis is assumed true, then the conclusion must also be true. We warn the reader that a true conclusion does not assure the validity of the argument! It is possible to draw a true conclusion from a false hypothesis by invalid reasoning! In other words, a true conclusion does not require a true hypothesis. However, a true hypothesis leads us to a true conclusion by a valid argument and a false conclusion necessarily requires a false hypothesis.

Example 2.25 Test the validity of the following argument:

$$\left.\begin{matrix} p \vee q \\ \sim p \end{matrix}\right\} \quad \text{hypotheses}$$

$$\therefore q \ \} \quad \text{conclusion}$$

Solution: It is given that the statements $p \vee q$ and $\sim p$ are true. That $\sim p$ is true simply means that p is false. Since p is false and $p \vee q$ is true, it follows that q is true (why?). That q is true is in fact the given conclusion. Consequently, the argument is valid. In fact, that $[(p \vee q) \wedge \sim p] \rightarrow q$ is a tautology can be seen from Table 2.18.

Table 2.18

p	q	$p \vee q$	$\sim p$	$(p \vee q) \wedge \sim p$	$[(p \vee q) \wedge \sim p] \rightarrow q$
T	T	T	F	F	T
T	F	T	F	F	T
F	T	T	T	T	T
F	F	F	T	F	T

Example 2.26 Test the validity of the following argument:

If Peter lives on Main Street, he lives in the heart of the city.
Peter does not live in the heart of the city.

∴ Peter does not live on Main Street.

Solution: Let's first translate this argument into symbols, for convenience. Let

p: Peter lives on Main Street
q: Peter lives in the heart of the city

Then the argument can be rewritten as

$$p \to q$$
$$\frac{\sim q}{\therefore \sim p}$$

Since $\sim q$ is true, q is false. Since the implication $p \to q$ is true and q is false, p cannot be true (why?). Consequently, p is false, which agrees with the given conclusion. Therefore, the argument is valid. We ask the reader to verify that $[(p \to q) \wedge \sim q] \to \sim p$ is a tautology.

Example 2.27 Test the validity of the following argument:

If it did not rain Saturday afternoon, Mary went to a movie.
Mary went to a movie or took a nap Saturday afternoon.
Mary did not take a nap Saturday afternoon.

∴ It did not rain Saturday afternoon.

Solution: Let

p: It did not rain Saturday afternoon
q: Mary went to a movie Saturday afternoon
r: Mary took a nap Saturday afternoon

Now the given argument can symbolically be written as follows:

$$p \to q$$
$$q \vee r$$
$$\frac{\sim r}{\therefore p}$$

Let's use the given data in the reverse order. Since $\sim r$ is true, r is false. Since $q \vee r$ is true and r is false, it follows that q is true. Since the implication $p \to q$ is true and q is true, it does not necessarily follow that p is true: p may be true or false (see Table 2.4). Therefore, the argument is invalid.

Exercise 2.4

1. Prove that the following statements are tautologies.

a) $\sim(\sim p) \leftrightarrow p$ **b)** $\sim(p \wedge \sim p)$ **c)** $p \wedge q \rightarrow p$

d) $p \wedge q \rightarrow q$ **e)** $[p \wedge (p \rightarrow q)] \rightarrow q$

f) $\sim(p \wedge q) \leftrightarrow (\sim p \vee \sim q)$ **g)** $[(p \vee q) \wedge \sim p] \rightarrow q$

h) $(p \rightarrow q) \leftrightarrow (\sim p \rightarrow \sim p)$ **i)** $[(p \rightarrow q) \wedge (q \rightarrow r)] \rightarrow (p \rightarrow r)$

2. Test the validity of the following arguments.

a) $p \rightarrow q$

$\underline{\qquad q \qquad}$

$\therefore \ p$

b) $p \vee q$

$q \vee r$

$\underline{\quad \sim r \quad}$

$\therefore \ p$

c) $p \leftrightarrow q$

$\sim p \vee r$

$\underline{\quad \sim r \quad}$

$\therefore \sim q$

d) If Bill likes cats, then he dislikes dogs.

Bill likes dogs.

$\therefore$ Bill dislikes cats.

e) If Ron passes this course, then he will graduate this year.

Ron does not pass this course.

$\therefore$ Ron will not graduate this year.

f) Frank bought a new car or a new house.

If he bought a new house, then he raked the lawn.

But he did not rake the lawn.

$\therefore$ Frank bought a new car.

g) Democrats are not wealthy or liberal.

Democrats are wealthy or illogical.

Democrats are logical.

$\therefore$ Democrats are liberal.

h) If Peter is married, then he is happy.

If he is happy, then he does not read the Times.

He does read the Times.

$\therefore$ Peter is unmarried.

★2.5   MATHEMATICAL PROOF

The word "proof" in the English language means different things to different people. To people in photography, "proof" means a trial print of a negative; to people in printing, it means a trial impression; to lawyers, it means all the facts, admissions, etc., that together contribute to a verdict. But a mathematical proof means none of these; it simply means the evidence of establishing the validity of a statement.

Every result in mathematics can be stated in the form H → C. There are statements that are assumed true (hypotheses) and statements that are to be established (conclusions). A *theorem* in mathematics consists of three parts: hypothesis (H), implication, and conclusion (C). Thus, *proving a theorem* H → C simply means establishing the statement C as a logical consequence of the statement H and/or previously known results.

There are two types of proofs in mathematics: *direct proof* and *indirect proof.* A direct proof makes use of the repeated application of the law of syllogism: $(p \to q) \wedge (q \to r) \to (p \to r)$. In the case of a direct proof of a theorem H → C, we start with the given hypothesis H and end with the desired conclusion C as the final step of a chain of implications: $H \to C_1, C_1 \to C_2, \ldots, C_n \to C$, where each step must be justified. Using the law of syllogism, it then follows that H → C. We now give a few examples where theorems are proved directly.

Example 2.28 Prove directly the following theorem:

> If Mr. Smith is a Republican, then he is wealthy.
> He is a Republican or illogical.
> He is logical.
> _____
> ∴ Mr. Smith is wealthy.

Proof: Let

p: Mr. Smith is a Republican
q: Mr. Smith is wealthy
r: Mr. Smith is logical

Then, the given theorem can be written in symbols as

$$\left. \begin{array}{c} p \to q \\ p \vee \sim r \\ r \end{array} \right\} \text{ hypotheses}$$

$$\therefore q \quad \} \quad \text{conclusion}$$

r is true	*given*
Therefore, $\sim r$ is false	*definition of negation of a statement*
$p \vee \sim r$ is true	*given*
Therefore, p is true	*definition of disjunction of two statements*
$p \to q$ is true	*given*
Therefore, q is true	*definition of implication*

Thus we conclude that Mr. Smith is wealthy. Hence the theorem.

Example 2.29 Prove directly the following theorem:

> It is cold or raining.
> It is not raining or the sun is shining.
> The sun is not shining.
> _____
> ∴ It is cold.

Proof: Let p: It is cold
                                           q: It is raining
                                           r: The sun is shining

Now, the theorem can be translated in symbols as follows:

$$p \vee q$$
$$\sim q \vee r$$
$$\underline{ \sim r }$$
$$\therefore p$$

Assuming that $p \vee q$, $\sim q \vee r$, and $\sim r$ are true, we have to prove that the statement p is true.

$\sim r$ is true	*given*
Therefore, r is false	*definition of negation of a statement*
$\sim q \vee r$ is true	*given*
Therefore, $\sim q$ is true	*definition of disjunction of two statements*
q is false	*definition of negation of a statement*
$p \vee q$ is true	*given*
Therefore, p is true	*definition of disjunction of two statements*

Consequently, we conclude that it is cold. Hence the theorem.

So far, we have been proving theorems using the direct method. Let's now turn our attention to the second kind of proof: indirect proof or *proof by contradiction*. The essence of indirect proof lies in the fundamental principle that a false conclusion can be reached through valid arguments only if the hypothesis itself is false (see Table 2.4). Suppose we wish to prove the theorem $H \rightarrow C$. Recall that $H \rightarrow C \equiv \sim C \rightarrow \sim H$ (see Table 2.13). Consequently, in order to prove the theorem $H \rightarrow C$, we need only establish the validity of its contrapositive, $\sim C \rightarrow \sim H$. Therefore, assuming that the desired conclusion C is false, we try to contradict the hypothesis, H. In general, the contradiction need not be of H, but can be of some well-known facts or previously proven results. Thus, the cornerstone in the indirect method lies in negating the desired conclusion and arriving at some contradiction. Let's now discuss a few sample theorems where the indirect method is used to prove them.

Example 2.30   Prove indirectly the following theorem:

> If Mr. Smith is a Republican, then he is wealthy.
> He is a Republican or illogical.
> He is logical.
>
> ---
> $\therefore$ Mr. Smith is wealthy.

Proof: Let p: Mr. Smith is a Republican
                                           q: He is wealthy
                                           r: He is logical

The theorem can now be symbolized:

$$\left. \begin{array}{c} p \to q \\ p \vee \sim r \\ r \end{array} \right\} \quad \text{hypotheses}$$

$$\overline{\therefore \; q \quad \} \quad \text{conclusion}}$$

Our aim is to establish the truthfulness of the statement q. Assume, if possible, that q is false.

$p \to q$ is true	*given*
p is false	*definition of implication*
$p \vee \sim r$ is true	*given*
$\sim r$ is true	*definition of disjunction of two statements*
r is false	*definition of negation*
But r is true	*given*

Therefore, r is both true and false, which is a contradiction. Consequently, our assumption that q is false is wrong. Thus q is true. Hence the theorem.

Example 2.31 Prove indirectly the following theorem:

Linda has a job.
If she is married, then she has no job.
If she is not married, then she is lonely.
$\overline{\therefore \text{ Linda is lonely.}}$

Proof: Let

p: Linda has a job
q: She is married
r: She is lonely

The theorem can now be stated in symbols in the form

$$\left. \begin{array}{c} p \\ q \to \sim p \\ \sim q \to r \end{array} \right\} \quad \text{hypotheses}$$

$$\overline{\therefore \; r \quad \} \quad \text{conclusion}}$$

Assume that the desired conclusion r is false.

$\sim q \to r$ is true	*given*
$\sim q$ is false	*definition of implication*
q is true	*definition of negation*
$q \to \sim p$ is true	*given*
$\sim p$ is true	*definition of implication*
p is false	*definition of negation*

But this contradicts the given assumption that p is true. Therefore, what we assumed at the beginning of the proof must be wrong. Thus, p is true and hence the theorem holds.

Before we close this section, we warn the reader that a (randomly chosen) collection of examples for which a theorem is true does not constitute a proof of the theorem, unless the theorem can be verified for all possible cases, which will not be practical in general. However, in order to show that a theorem is false, i.e., to disprove a theorem, we need only produce an example, called a *counterexample*, for which the theorem is false.

Example 2.32   Consider the statement

$$p: \quad \text{Every girl is a brunette}$$

In order to disprove this statement, we need only find at least one girl who is not a brunette. This is essentially the same as showing the truth of the negation

$$\sim p: \quad \text{Some girls are not brunettes}$$

Example 2.33   "The square of every integer is even" is a false statement, since we have a counter example, 3. For,

$$3^2 = 3 \cdot 3 = 9$$

which is not even.

Exercise 2.5

1. Prove the following theorems directly.

a) $p \lor q$
$\dfrac{\sim q}{\therefore \ p}$

b) $p \leftrightarrow q$
$\sim q \rightarrow r$
$\dfrac{\sim r}{\therefore \ p}$

c) $p \lor q$
$p \rightarrow \sim r$
$\dfrac{r}{\therefore \ q}$

d) $p \rightarrow q$
$r \rightarrow \sim q$
$\dfrac{p}{\therefore \sim r}$

e) $p \rightarrow \sim q$
$q \lor \sim r$
$\dfrac{r}{\therefore \sim p}$

f) $p \leftrightarrow \sim q$
$\sim p \lor r$
$\dfrac{\sim r}{\therefore \ q}$

g) If Mr. Rogers is a Republican, he is wealthy.
   He is not wealthy.
   $\overline{\therefore \text{ Mr. Rogers is not a Republican.}}$

h) Alice is rich if and only if she is beautiful.
   Either she is married or not beautiful.
   She is not married.
   $\overline{\therefore \text{ Alice is not rich.}}$

i) John likes music or hates driving.
   Either he does not like music or does not enjoy reading.
   He enjoys reading.
   $\overline{\therefore \text{ He hates driving.}}$

2. Prove the following theorems indirectly.

a) $p \lor q$
$\underline{\quad \sim p \quad}$
$\therefore q$

b) $p \leftrightarrow q$
$\sim q \rightarrow r$
$\underline{\quad \sim r \quad}$
$\therefore p$

c) $p \rightarrow q$
$p \lor \sim r$
$\underline{\quad\quad r \quad}$
$\therefore q$

d) p
$p \rightarrow \sim q$
$\underline{\sim r \rightarrow q}$
$\therefore r$

e) $p \leftrightarrow q$
$\underline{\quad\quad p \quad}$
$\therefore q$

f) $p \rightarrow q$
$q \rightarrow r$
$\underline{\quad \sim r \quad}$
$\therefore \sim p$

g) Joe is handsome if and only if he is rich.
But he is not rich.
$\overline{\therefore \text{ He is not handsome.}}$

h) Carol is a baby if and only if she is illogical.
Either she is illogical or unhappy.
But she is happy.
$\overline{\therefore \text{ Carol is a baby.}}$

i) If it does not rain today, then I will go to a film.
If I am not happy, then I will not go to a film.
I am not happy.
$\overline{\therefore \text{ It rains today.}}$

3. Give a counterexample to disprove the following statements.
   a) Every city is a capital city.
   b) If x is a month of the year, then x has 30 days.
   c) If x is a country, then x is part of Europe.
   d) All integers are even.

★2.6   MATHEMATICAL SYSTEMS

Throughout this book, we will be coming across different "mathematical systems." What is a mathematical system? What does it consist of? What do we mean by the structure of a mathematical system?

A *mathematical system* consists of a set of elements, one or more relations among the elements, one or more operations on the elements, undefined terms, defined terms, axioms (postulates), and theorems. The study of the properties of the elements, operations, and relations that are a consequence of undefined and defined terms constitutes the *structure* of the system.

What are undefined terms? Not every term, in general, can be defined, except perhaps by going in circles. For example, suppose we look up for the meaning of the adjective "main" in a dictionary. Each time we look up its meaning, we may be subjected to circular definitions (see Fig. 2.2), where one

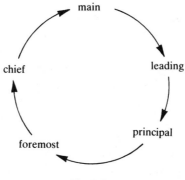

Fig. 2.2

word is defined in terms of another and finally we come back where we started. Consequently, in mathematics, as everywhere else, we accept certain terms as undefined. For example, the terms "point" and "line" are left undefined in geometry; the term "set" is usually treated as an undefined term in set theory. New terms are then formally defined in terms of undefined or previously defined terms in a system.

Once we have enough tools consisting of defined and undefined terms, we are in a position to construct sentences and statements. Recall that a statement is a declarative sentence that can be classified as true or false, but not both. As in the case of undefined terms, certain statements called *axioms* (*postulates*) are accepted as true at the beginning. An axiom need not be a self-evident truth but is simply regarded as a true statement without any formal proof. Also, one man's axiom need not be the same as another man's axiom. Axioms are usually kept at a bare minimum. They are chosen in such a way that one does not contradict another (*consistency*) and does not follow as a logical consequence of another (*independence*). Consistency is an essential and fundamental property of a set of axioms, while independence is not necessary but desirable.

Finally, theorems are concluded logically using undefined terms, defined terms, axioms, and laws of logic.

The student is already familiar with a number of mathematical systems. He will become more familiar with some old and new ones in the chapters to follow. A few important systems are the following:

1. The system consisting of sets and subsets together with the relations "is a subset of," "is equal to," "is equivalent to," etc.; the operations of union, intersection, and complement; the axioms and laws for relations and operations, as discussed in the preceding chapter.

2. A collection of statements with the operations of conjunction, disjunction, and negation, as we have just considered; truth values, laws of logic, tautologies, theorems, proofs, etc.

3. The system of integers with the operations of addition, subtraction, and multiplication; the equality and order relations; several properties.
4. The system of rational numbers with addition, subtraction, multiplication, and division; the equality and order relations; many properties.
5. The system of real numbers and the system of complex numbers, which contains as a subsystem the system of real numbers.
6. A finite modular system of remainders with addition and multiplication illustrated by convenient tables.
7. A geometry of points, lines, curves, and regions in a plane; with incidence, similarity, and congruence relations; metric properties.

★2.7 SWITCHING NETWORKS

This section is devoted to the discussion of an important application of symbolic logic. The algebra of logic plays a significant role in the design of switching networks. No prior knowledge of electricity is assumed here. In switching networks, we are concerned about the flow of current through various switches and wires.

A *switching network* is simply an arrangement of wires and switches connecting two terminals, as shown in Fig. 2.3. A switch that permits the flow of current is said to be *closed* (*on*), as shown in Fig. 2.3b; otherwise it is said to be *open* (*off*), as shown in Fig. 2.3a. A switching network is *closed* if it will let current flow from one end of the network to the other end; otherwise it is *open*.

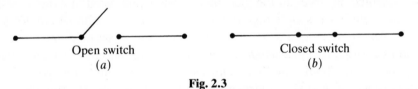

Open switch   Closed switch
(a)   (b)

Fig. 2.3

Two switches *A* and *B* can be connected either *in series*, as shown in Fig. 2.4, or *in parallel*, as shown in Fig. 2.5. Observe that the circuit in Fig. 2.4 is closed if both switch *A* and switch *B* are closed, and open if at least one of them is open. Notice the similarity between this and the logical operation of conjunction of two statements. Recall that the conjunction $p \wedge q$ of two statements p and q is true if both p and q are true, and false if at least one of them is false. Accordingly, the switching circuit in Fig. 2.4 is denoted by $A \wedge B$.

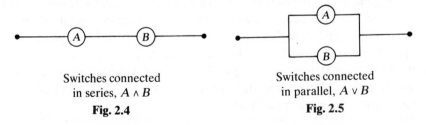

Switches connected   Switches connected
in series, $A \wedge B$ in parallel, $A \vee B$
Fig. 2.4   **Fig. 2.5**

It is clear from Fig. 2.5 that the network will permit the flow of current if either A is closed or B is closed or if both are closed. Notice the analogy between this and the logical operation of disjunction. Accordingly, the switching circuit in Fig. 2.5 is denoted by $A \vee B$.

It is possible for an electrical network to have two switches A and A' (A prime) with the property that if one is closed, then the other is open and vice versa.

A switching network, in general, consists of several series and parallel connections and can be described symbolically using the connectives $\wedge$, $\vee$, and $'$, as shown by the following examples.

Example 2.34   Describe the circuit in Fig. 2.6 symbolically.

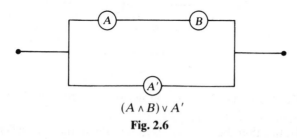

$$(A \wedge B) \vee A'$$

Fig. 2.6

Solution: Since the switches A and B are connected in series, we write $A \wedge B$. Since the switch A' is in parallel with the circuit $A \wedge B$, the given circuit is represented symbolically by $(A \wedge B) \vee A'$. It now follows that the network is closed if

1. both A and B are closed,
2. A is open and B is closed, and
3. A is open and B is open.

Example 2.35   Describe the circuit in Fig. 2.7 symbolically.

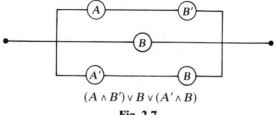

$$(A \wedge B') \vee B \vee (A' \wedge B)$$

Fig. 2.7

Solution: Switches A and B' are connected in series and hence we write $A \wedge B'$. The switch B is in parallel with $A \wedge B'$ and we have $(A \wedge B') \vee B$. That A' and B are connected in series is described by $A' \wedge B$. Since $(A \wedge B') \vee B$ and $A' \wedge B$ are connected in parallel, the given circuit is described symbolically by $[(A \wedge B') \vee B] \vee (A' \wedge B)$.

Example 2.36 Construct a switching circuit represented symbolically by $A \wedge [(B' \wedge C) \vee B]$.

Solution: Notice that the series circuit $B' \wedge C$ is connected in parallel with B to obtain the circuit $(B' \wedge C) \vee B$. Now switch A is connected with $(B' \wedge C) \vee B$ in series to obtain the given circuit $A \wedge [(B' \wedge C) \vee B]$. The corresponding switching network is exhibited in Fig. 2.8.

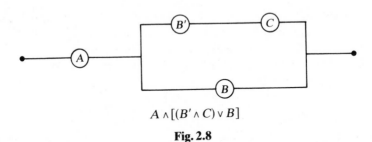

$$A \wedge [(B' \wedge C) \vee B]$$

Fig. 2.8

Example 2.37 Construct a switching network represented symbolically by $[A \wedge (B \vee C')] \vee [(A' \vee B') \wedge C]$.

Solution: Notice that the switch A is in series with the parallel circuit $B \vee C'$. The parallel circuit $A' \vee B'$ is connected in series with C. Finally, the circuit $A \wedge (B \vee C')$ is in parallel with $(A' \vee B') \wedge C$. Figure 2.9 shows the corresponding electrical network.

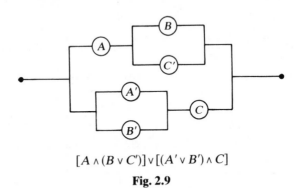

$$[A \wedge (B \vee C')] \vee [(A' \vee B') \wedge C]$$

Fig. 2.9

Two switching networks are said to be *equivalent* if they have exactly the same electrical behavior: either closed simultaneously or open simultaneously. One of the simplest, most useful applications of the algebra of logic is to replace an electrical network by a simpler equivalent network so that it is possible to minimize the cost of labor, wastage of time, etc., resulting in considerable savings! Observe that the circuit $A \vee A' \equiv T$ is always closed whereas the circuit $A \wedge A' \equiv F$ is always open for any switch A. (Why?) Also, $A \vee T \equiv T$ and $A \wedge T \equiv A$ for any switch A.

Example 2.38 Replace the circuit in Fig. 2.10 by an equivalent, simpler circuit.

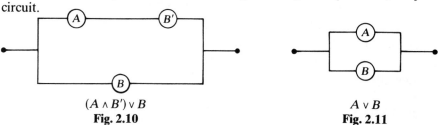

$(A \wedge B') \vee B$
Fig. 2.10

$A \vee B$
Fig. 2.11

Solution: The network in Fig. 2.10 is represented symbolically by $(A \wedge B') \vee B$. Let's now simplify the expression $(A \wedge B') \vee B$, using the laws of logic.

$(A \wedge B') \vee B \equiv (A \vee B) \wedge (B' \vee B)$ *distributive law*

           $\equiv (A \vee B) \wedge T$ $A' \vee A \equiv T$ *for any switch A*

           $\equiv A \vee B$ $A \wedge T \equiv A$ *for any switch A*

Consequently, the given network can be replaced by the circuit shown in Fig. 2.11.

Exercise 2.7

1. Represent the following networks symbolically.

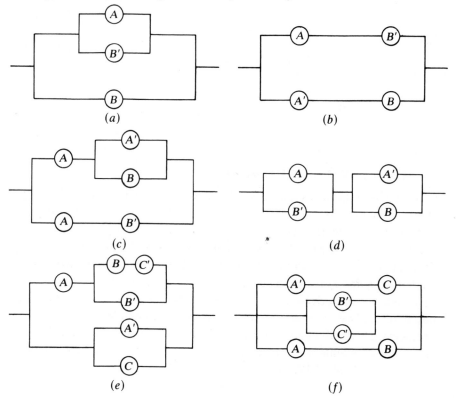

(a)

(b)

(c)

(d)

(e)

(f)

2. Construct a switching circuit represented symbolically by

    a) $(A \vee B) \wedge A'$ **b)** $A \vee (B \wedge C)$

    c) $(A \vee B) \wedge (A \vee C)$ **d)** $(A \wedge B') \vee (A \vee B)$

    e) $(A \wedge B') \vee (A' \wedge B)$ **f)** $(A \wedge B) \vee (A' \wedge B) \vee (B' \wedge C)$

3. If A is any switch and T is a switch that is always closed, what can you say about the electrical behavior of

    a) $A \vee T$ **b)** $A \vee A'$ **c)** $A \wedge A'$

    d) $A \wedge T'$ **e)** $A' \vee T$ **f)** $A' \wedge T'$

4. Construct a simple, equivalent circuit for each of the following circuits.

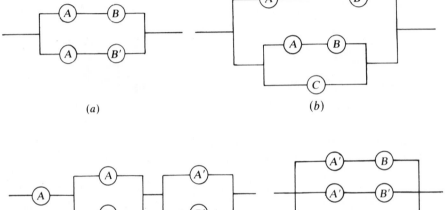

(a)   (b)

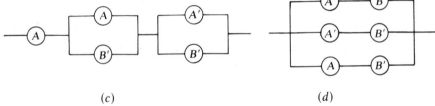

(c)   (d)

2.8 SUMMARY AND COMMENTS

This chapter introduces some of the basic concepts and laws of symbolic logic with the hope that they will help the reader to think, argue, and express himself logically and systematically.

A statement in logic is a declarative sentence that can be identified as true or false, but not both. The truth or falsity of a statement is called its truth value, denoted by T and F, respectively.

The logical connectives "and" ($\wedge$), "or" ($\vee$), "not" ($\sim$), "if ... then" ($\rightarrow$), and "if and only if" ($\leftrightarrow$) are used to combine statements to form compound statements. Truth tables are a convenient way of discussing the truth values of compound statements. Truth tables 2.1–2.5 can be combined, for convenience, into one, as in Table 2.19.

Table 2.19

p	q	$p \wedge q$	$p \vee q$	$\sim p$	$p \rightarrow q$	$p \leftrightarrow q$
T	T	T	T	F	T	T
T	F	F	T	F	F	F
F	T	F	T	T	T	F
F	F	F	F	T	T	T

From the conditional statement $p \rightarrow q$, we can form three new statements:
1. $q \rightarrow p$ (converse of $p \rightarrow q$)
2. $\sim p \rightarrow \sim q$ (inverse of $p \rightarrow q$)
3. $\sim q \rightarrow \sim p$ (contrapositive of $p \rightarrow q$)

Two statements p and q are logically equivalent if they have the same truth values at all possible cases, denoted by $p \equiv q$.

We discussed the following important laws of logic:
1. *commutative laws*: $p \wedge q \equiv q \wedge p$ and $p \vee q \equiv q \vee p$
2. *associative laws*: $p \wedge (q \wedge r) \equiv (p \wedge q) \wedge r$ and $p \vee (q \vee r) \equiv (p \vee q) \vee r$
3. *distributive laws*: $p \wedge (q \vee r) \equiv (p \wedge q) \vee (p \wedge r)$ and
   $p \vee (q \wedge r) \equiv (p \vee q) \wedge (p \vee r)$
4. *De Morgan's laws*: $\sim(p \wedge q) \equiv \sim p \vee \sim q$ and
   $\sim(p \vee q) \equiv \sim p \wedge \sim q$
5. *law of detachment*:   if $p \wedge (p \rightarrow q)$ is true then q is true
6. *law of excluded middle*: $p \vee \sim p$ is always a true statement
7. *law of syllogism*: $(p \rightarrow q) \wedge (q \rightarrow r) \rightarrow (p \rightarrow r)$ is always a true statement

A statement that is always true irrespective of the truth values of its components is called a tautology.

The statements p and q are the hypothesis and the conclusion of the implication $p \rightarrow q$. An argument in a certain discussion is said to be valid if the given conclusion is a logical consequence of the hypothesis; otherwise it is invalid. The symbols and laws of logic play an important part in discussing the validity of arguments.

Every theorem in mathematics can be stated in the form $H \rightarrow C$. Proving a theorem simply means establishing the desired conclusion C as a logical consequence of the given hypothesis H or previously known results or both.

There exist two kinds of proofs in mathematics: direct proof and indirect proof (proof by contradiction). A direct proof of a theorem $H \rightarrow C$ involves in arriving at the desired conclusion C from the hypothesis H as the final stage of a sequence of implications. An indirect proof consists of negating the desired conclusion and arriving at some contradiction.

In order to disprove a theorem H→C, we need only exhibit an example called counterexample, for which the hypothesis H holds and the conclusion C fails.

A mathematical system consists of a set of elements, one or more relations among the elements, one or more operations on the set, undefined terms, defined terms, axioms, and theorems.

Symbolic logic plays a central role in the design of electrical networks.

Finally, we advise the reader to become familiar with the symbols and laws of logic, since they play a vital role in the remaining chapters.

SUGGESTED READING

Bartley, W. W., III, "Lewis Carrol's Lost Book on Logic," *Scientific American*, vol. 227 (July 1972), pp. 39–46.

Boyer, L. E., "On Validity and the Use of Truth Tables," *School Science and Mathematics*, vol. 69 (March 1969), pp. 553–560.

Carrol, L., *Symbolic Logic, Part 1. Elementary*, Dover Publications, Inc., New York (1958).

Christian, R. R., *Introduction to Logic and Sets* (2nd ed.), Blaisdell Publishing Co., Waltham, Mass. (1965), pp. 1–44.

Desmonde, W. H., *Computers and Their Uses*, Prentice-Hall, Inc., Englewood Cliffs, N.J. (1964), pp. 62–104.

Lipschutz, S., *Set Theory and Related Topics*, Schaum's Outline Series, McGraw-Hill Book Co., New York (1964), pp. 187–231.

Maxfield, J. E., and M. W. Maxfield, *Key to Mathematics*, W. B. Saunders Co., Philadelphia (1973), pp. 101–126.

Miller, C. D., and V. E. Heeren, *Mathematical Ideas, an Introduction* (2nd ed.), Scott, Foresman and Co., Glenview, Ill. (1973), pp. 95–125.

Roberge, J. J., "Negation in the Major Premise as a Factor in Children's Deducting Reasoning," *School Science and Mathematics*, vol. 69 (March 1969), pp. 715–722.

"God created the natural numbers; all else is the work of man."
L. Kronecker

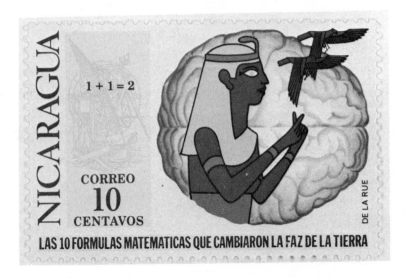

3/The System of Whole Numbers

After studying this chapter, you should be able to:
- *find the cardinal number of a finite set*
- *evaluate the sum, the product, and the difference (whenever possible) of two whole numbers*
- *identify the properties of addition and multiplication of whole numbers*
- *check if a whole number is less than another*
- *identify the properties of the order relation*
- *solve equations of the form ax + b = cx + d*

3.0 INTRODUCTION

We discussed sets and their properties under some basic operations in Chapter 1. Our aim in the present chapter is to concentrate on a particular set, the set of whole numbers, which is infinite.

One of the most widely used mathematical concepts we learn early in life is that of counting. We use the "numbers" 1, 2, 3, . . . , in our day-to-day life for counting. What actually are numbers? How are they related to sets? Where do they come from? What are their properties? Is the set of whole numbers "large enough" for all purposes? We shall try to answer these questions in this chapter.

In Section 1.10, we discussed the concepts of 1-1 correspondence, "more than," and "less than." Primitive man seems to have had these concepts. The concept of number, which is very closely related, was known to him. For instance, assume he wanted to keep track of the number of sheep he owned. As each sheep left the pen in the morning, he put a pebble in his bag for each sheep.

In the evening, as the sheep returned to the pen, he removed one pebble from the bag for each sheep entering the pen. If he was left with no pebbles in the bag and no sheep outside the pen, he knew that he had all his sheep in the pen. In this case he established, in mathematical terms, a 1-1 correspondence between his flock of sheep and the bag of pebbles. On the other hand, if he was left with some pebbles, but no sheep, he knew that he had lost some of his sheep. In this case, he could not establish a 1-1 correspondence between the two sets. Accordingly, as we discussed in Section 1.10, the bag of pebbles contained "more elements than" the flock of sheep. If he ran out of pebbles and more sheep were still coming to the pen, then he concluded that he had gained more sheep than he owned. That is, the set of pebbles contained "fewer elements than" the flock of sheep.

Thus, we observe that the technique of counting is very closely related to the concept of 1-1 correspondence.

3.1 THE CARDINALITY OF A SET

Consider the sets $A = \{x,y,z\}$ and $B = \{Kennedy, Johnson, Nixon\}$. Clearly $A \neq B$; however $A \sim B$. More generally, if A and B are any two sets such that $A \sim B$, we observe that they have something in common. If $A \sim B$, we can establish a 1-1 correspondence between them. Intuitively, this means that the two sets contain the same number of elements.

Example 3.1 If $A = \{Boston, Chicago, Detroit\}$ and $B = \{a,b,c\}$, then $A \sim B$. Each set contains "three" elements. These two sets have the common property of *threeness*.

Example 3.2 If $A = \{a,b,c,d\}$ and $B = \{Jack, Jerry, Jim, Joe\}$, then $A \sim B$. Each set contains "four" elements. The sets A and B have the common property of *fourness*.

Now we are in a position to make the following definition:

Definition 3.1 Two sets A and B are said to have the same *cardinality* or *cardinal number* if $A \sim B$. We then write $c(A) = c(B)$. Observe that "cardinal number" is an abstract property common to all equivalent sets.

Recall that the set $\varnothing$ contains no elements. Accordingly, we define the cardinal number of $\varnothing$ as *zero*, denoted by the familiar symbol "0." Now consider the set $\{0\}$. We define the cardinality of any set equivalent to the set $\{0\}$ as *one*, denoted by the symbol "1." Similarly, "2" denotes the cardinality, *two*, of every set equivalent to the set $\{0,1\}$. In a similar manner, we define the cardinal numbers, denoted by the symbols, $3, 4, 5, \ldots, n, \ldots$, where "$n$" denotes the cardinality of any set equivalent to the set $\{0,1,2,\ldots,n-1\}$. Thus we have the set of cardinal numbers represented by the symbols $0,1,2,3,\ldots,n,\ldots$, called the set of *whole numbers*, denoted by W. Its proper subset $\{1,2,3,...,n,...\}$ is called the set of *natural numbers* or *counting numbers*, denoted by *N*.

Observe that "whole number" is an abstract property common to all finite equivalent sets, and the definition associates each finite set with a unique whole number.

What do we usually do in order to count the number of elements (find the cardinality) of a finite set *A*? We establish a 1-1 correspondence between the set *A* and a suitable set $\{1,2,3,...,n\}$, usually called a *reference set.* If *A* is equivalent to $\{1,2,3,...,n\}$, then *A* contains *n* elements; that is, $c(A)=n$. Observe that the last counting number in the set $\{1,2,3,...,n\}$ gives the cardinal number of the set *A*, provided the numbers $1,2,3,\ldots,n$ are written in increasing order.

Example 3.3 Find the cardinality of the set $\{a,b,c\}$.

Solution: Let's establish a 1-1 correspondence between $\{a,b,c\}$ and the set $\{1,2,3\}$ as follows:

$$
\begin{array}{ccc}
a & b & c \\
\updownarrow & \updownarrow & \updownarrow \\
1 & 2 & 3
\end{array}
\qquad c(A)=3
$$

Thus $\{a,b,c\} \sim \{1,2,3\}$. Therefore, the cardinality of the set $\{a,b,c\}$ is 3. Recall, from Section 1.10, that this is not the only 1-1 correspondence we could display between these sets.

Example 3.4 Find the cardinal number of the set *A* of all letters of the word "Massachusetts."

Solution: Notice that $A=\{m,a,s,c,h,u,e,t\} \sim \{1,2,3,4,5,6,7,8\}$, and hence $c(A)=8$.

Example 3.5 Consider the sets $A=\{1,3,5\}$, $B=\{0,2\}$, and $C=\{1,2,3\}$. Notice that $A \sim C$, since $c(A)=3=c(C)$, and that $c(B)=2$.

$$A \cup B = \{0,1,2,3,5\} \qquad c(A \cup B)=5$$

$$A \cap B = \varnothing \qquad c(A \cap B)=0$$

$$A - C = \{5\} \qquad c(A-C)=1$$

$$A \cap (B \cup C)=\{1,3\} \qquad c(A \cap (B \cup C))=2$$

$$A \times B = \{(1,0),(1,2),(3,0),(3,2),(5,0),(5,2)\} \qquad c(A \times B)=6$$

$$B \times B = \{(0,0),(0,2),(2,0),(2,2)\} \qquad c(B \times B)=4$$

Recall that whole numbers are cardinal numbers of finite sets. But there are sets that are not finite! What about the cardinality of such a set? In particular, what is the cardinality of the set *N* of natural numbers? In Section 1.11, we showed that the set of natural numbers is infinite. Consequently, the cardinality of the set of natural numbers is not a whole number.

Exercise 3.1

1. What is the cardinal number of each of the following sets?
   a) $A = \{0,1,x\}$.
   b) $B = \{$Kennedy,Johnson,Nixon$\}$.
   c) $C =$ set of all days of the week.
   d) $D =$ set of all letters of the English alphabet.
   e) $E = (a,b)$.
   f) $F =$ set of all letters of the word "tweedledee".
   g) $G =$ set of all letters of the word "Cincinnati".
   h) $H =$ set of all letters of the word "Alabama"
   i) $I =$ set of all digits in the number 12,321.
   j) $J =$ set of all letters in the word "Daddy".

2. If $A = \{1,2,4,5\}$, $B = \{0,1,3\}$, $C = \{0,2,4\}$, and $U = \{0,1,2,3,4,5,6,7\}$, find the cardinality of each of the following sets.
   a) A $c(A):4$
   b) $A \cup B$ $c(A \cup B)6$
   c) $A \cap B$ $c(A \cap B):1$
   d) $A - B$ $c(A-B):2$
   e) $B - C$ $c(B-C)2$
   f) $A - (B \cap C)$ $c(A-(B \cap C)):4$
   g) $(A \cup B) - C$ $c((A \cup B)-c:3$
   h) $A \cup B \cup C$ $c(A \cup B \cup C):6$
   i) $A \cap B \cap C$ $c(A \cap B \cap C):0$
   j) $(A \cup B) \cap C$ $c((A \cup B) \cap C:$
   k) $A \cap (B \cup C)$ $c(A \cap (B \cup C))3$
   l) $A \cap B'$ $c(A \cap B'):3$
   m) $A \cup C'$ $c(A \cup C'):7$
   n) $A - B'$ $c(A-B'):1$
   o) $A \cup (B \cap C)'$ $c(A \cup (B \cap C)):6$

3. What is the cardinal number of the set A if
   a) $A = \{\varnothing\}$ $:1$
   b) $A = \{0,1\}$ $:2$
   c) $A = \{\varnothing,\{\varnothing\}\}$ 2
   d) $A = \{\{\varnothing\}\}$ $:1$
   e) $A = \{0,\{\varnothing\}\}$ $:2$
   f) $A = \{1,\{2,3\},3\}$ 3

4. What is the reference set used to find the cardinal number of each of the following sets? What is the cardinal number of each of them?
   a) $A =$ set of all letters of the word "Tennessee".
   b) $B = \{0,2,4,6\}$
   c) $C = \{x,y,z\}$
   d) $D =$ set of all New England states.
   e) $E =$ set of all letters of the word "Honolulu".
   f) $F =$ set of all governors in your state.
   g) $G = \{x \mid x$ is a state of the United States$\}$.
   h) $H =$ set of all letters of the word "tintinnabulation".

5. If $A = \{1,2\}$, $B = \{a,b,c\}$, and $C = \{1\}$, find the cardinality of each of the following sets.
   a) $A \times B$ $c(A \times B)6$
   b) $A \times A$ $c(A \times A):4$
   c) $A \times C$ $c(A \times C):2$
   d) $A \times (B \cup C)$ $c(A \times (B \cup C)):6$
   e) $A \times (B \cap C)$ $c(A \times (B \cap C)):0$
   f) $A \times (B - C)$ $c(A \times (B-C)):6$
   g) $A \times \varnothing$ $c(A \times \varnothing):0$
   h) $\varnothing \times \varnothing$ $c(\varnothing \times \varnothing):0$
   i) $(A \times B) \times C$ $c((A \times B) \times C):6$

6. Mark *true* or *false* (A and B are any sets).
   a) Equivalent sets have the same cardinal number.
   b) If $A \sim B$ and $c(A) = 6$, then $c(B) = 6$.
   c) If $c(A) = 0$, then $A = \varnothing$.
   d) If $c(A) = c(B)$, then $A = B$. ~

e) If $c(A \cap B) = 0$, then $A = B = \emptyset$.

f) The cardinality of (a, b) is 2.

g) "Zero" is not nothing.

h) Every whole number is a cardinal number.

i) Every natural number is a cardinal number.

j) Every whole number is the cardinality of a nonempty set.

7. Give a counterexample to each of the false statements in problem 6, whenever possible.

★8. Find the next three letters of this sequence: ottffsse⋯.

3.2 THE EQUALITY RELATION

Recall that a whole number is defined to be the cardinality of a finite set. If A and B are two sets and $A \sim B$, then $c(A) = c(B)$. This permits us to make the following definition:

Definition 3.2   Let a and b be any two whole numbers. Then, there exist finite sets A and B such that $c(A) = a$ and $c(B) = b$. The whole numbers a and b are said to be *equal*, denoted by $a = b$, if and only if A is equivalent to B. That is, $a = b$ if and only if $A \sim B$. We read $a = b$ as "a is equal to b" or "a is the same as b." If a is not equal to b, we write $a \neq b$ (a not equal to b). That $a = b$ simply means that a and b are different names of the very same number.

The set of whole numbers clearly satisfies the following properties with respect to the equality relation, where a,b,c are any whole numbers:

1. $a = a$ for every whole number a (reflexive property)
2. If $a = b$, then $b = a$ (symmetric property)
3. If $a = b$ and $b = c$, then $a = c$ (transitive property)

These three properties of the equality relation on the set of whole numbers play an important role in the discussion of the fundamental properties of the set of whole numbers under the operations of addition, multiplication, and subtraction, as we shall see in the following sections.

3.3 ADDITION OF WHOLE NUMBERS

With the concept of cardinality of a set, we are now in a position to discuss the development of the whole number system. We do not assume that the reader is not familiar with the four fundamental operations: addition, subtraction, multiplication, and division. But our aim is to present them more systematically so that the reader will have a better understanding of the development of the number system.

How is the operation of addition defined on the set of whole numbers? Let's start with an example. Consider the sets $A = \{a,b\}$ and $B = \{x,y,z\}$. Notice that $A \sim \{1,2\}$, $B \sim \{1,2,3\}$ and hence $c(A) = 2$, $c(B) = 3$. Also notice that A and B are disjoint sets. Let's take the union of these sets. We have $A \cup B = \{a,b,x,y,z\}$

and $c(A \cup B) = 5$. We were all taught in school how to add 2 and 3. The result, from our experience, is 5 and we write $2 + 3 = 5$. That is,

$$c(A) + c(B) = c(A \cup B)$$
$$\downarrow \qquad \downarrow \qquad \downarrow$$
$$2 \ + \ 3 \ = \ \quad 5$$

Is $c(A \cup B) = c(A) + c(B)$ true for any two sets A and B? The answer is no, as shown by the following example.

Example 3.6 Consider the sets $A = \{a,b\}$ and $B = \{b,c,d\}$.

$$A \cup B = \{a,b,c,d\} \qquad \text{and} \qquad c(A \cup B) = 4$$

But

$$c(A) + c(B) \neq c(A \cup B)$$
$$\downarrow \qquad \downarrow \qquad \downarrow$$
$$2 \ + \ 3 \ \neq \ \quad 4$$

What makes it work in the above example, but not here? Observe that in the former case, $A \cap B = \varnothing$, whereas in this example, $A \cap B = \{b\} \neq \varnothing$.

These two examples suggest to us how the operation of addition should be defined on the set of whole numbers.

Definition 3.3 Let a and b be any two whole numbers. Let A and B be two finite disjoint sets such that $a = c(A)$ and $b = c(B)$. Then the *sum* of the whole numbers a and b, denoted by $a + b$, is defined as $c(A \cup B)$. That is,

$$a + b = c(A \cup B)$$

We read $a + b$ as "*a plus b*," as usual. The operation we have defined is called *addition*, and $a + b$ means adding b to a.

The reader is warned that in our definition we assumed that it is always possible to find two such disjoint sets. Is it actually true? The answer is yes! This should leave no ambiguity with our definition.

The above definition of addition of whole numbers might seem a little sophisticated. The advantage is that most of the basic properties of addition of whole numbers can be derived from this definition. Notice that it is the union operation on sets together with their disjointness that helps us to define the operation of addition on the set of whole numbers. So, it looks as if these two operations, union and addition, should have similar properties.

If a and b are two whole numbers, what can we say about $a + b$? Is it also a whole number? That the sum of two whole numbers is a whole number follows from the fact that the union of two finite sets is again a finite set. Thus, the addition operation helps us to associate a unique whole number $a + b$ with the two whole numbers a and b. That is, addition is a *binary operation* on the set of whole numbers. We also say that the set of whole numbers is *closed* under the operation of addition. We can now state the first property of the set of whole numbers under the operation of addition.

Theorem 3.1 (***Closure Property***) *The set of whole numbers is closed under addition.*

Notice that although the set of whole numbers is closed under addition, this does not imply that its subsets are also closed under addition. For instance, the set of odd natural numbers $\{1,3,5,...\}$ is not closed under addition (why?), while the set of even natural numbers $\{2,4,6,...\}$ is. The set $\{10,20,30,...\}$ is closed under addition, while the set $\{5,15,25,...\}$ is not (why?).

Given two whole numbers, for instance, 2 and 3, we can add them in two different ways, either as $2+3$ or $3+2$. From our experiences, we know that $2+3=5=3+2$.

Example 3.7 Verify that $2+4=4+2$.

Solution: Consider two disjoint sets A and B such that $c(A)=2$ and $c(B)=4$. For example, let $A=\{a,b\}$ and $B=\{w,x,y,z\}$.

$$A \cup B = \{a,b,w,x,y,z\}$$

$$B \cup A = \{w,x,y,z,a,b\}$$

Thus $A \cup B = B \cup A$ and is equivalent to $\{1,2,3,4,5,6\}$. Therefore,

$$c(A \cup B) = c(B \cup A) = 6$$

That is,

$$c(A)+c(B) = c(A \cup B) = c(B \cup A) = c(B)+c(A)$$
$$\downarrow \quad\quad \downarrow \quad\quad\quad \downarrow \quad\quad\quad\quad \downarrow \quad\quad\quad \downarrow \quad\quad \downarrow$$
$$2 \ + \ 4 \ = \quad\quad 6 \quad\quad = \quad\quad 6 \quad\quad = \ 4 \ + \ 2$$

Is this true in general? That is, is $a+b=b+a$ for any two whole numbers a and b? The answer is yes!

Theorem 3.2 (***Commutative Property***) *For any two whole numbers a and b,*

$$a+b=b+a$$

Proof: Since a and b are whole numbers, there exist finite sets A and B such that $c(A)=a$ and $c(B)=b$. By the remark that followed Definition 3.3, we can very well assume that $A \cap B = \emptyset$. Then

$a+b = c(A \cup B)$	*definition of addition of whole numbers*
$\quad = c(B \cup A)$	*commutative property of the union operation*
$\quad = b+a$	*definition of addition of whole numbers*

Thus, $a+b=b+a$

Observe that in order to establish the commutative property of addition, we used the commutative property of the union operation on sets. This theorem tells us that the order in which we add two whole numbers is no longer important.

Assume we want to find the sum $2+3+5$. Recall that addition is a binary operation. However, we can find the sum either as $(2+3)+5$ or as $2+(3+5)$.

Do they have the same sum now? Well, $(2+3)+5=5+5=10=2+8=2+(3+5)$. More generally, we now have the following result:

Theorem 3.3 (*Associative Property*) *For any whole numbers a, b, and c,*

$$(a+b)+c = a+(b+c)$$

The associative property of addition tells us that if we add three or more whole numbers, the way we group them is immaterial. With the result, the expression $a+b+c$, without any parentheses, is not ambiguous.

Example 3.8 Show that $(2+3)+(5+8)=(2+5)+(3+8)$ without evaluating the sums.

Proof:

$(2+3)+(5+8)$	$=[(2+3)+5]+8$	*associative property of addition*
	$=[2+(3+5)]+8$	*associative property of addition*
	$=[2+(5+3)]+8$	*commutative property of addition*
	$=[(2+5)+3]+8$	*associative property of addition*
	$=(2+5)+(3+8)$	*associative property of addition*

Therefore, $(2+3)+(5+8)=(2+5)+(3+8)$

Example 3.9 If a,b,c,d are any whole numbers, prove that $(a+b)+(c+d)=(a+d)+(b+c)$, justifying the steps.

Proof: $(a+b)+(c+d)$

	$=(a+b)+(d+c)$	*commutative property of addition*
	$=a+[b+(d+c)]$	*associative property of addition*
	$=a+[(b+d)+c]$	*associative property of addition*
	$=a+[(d+b)+c]$	*commutative property of addition*
	$=a+[d+(b+c)]$	*associative property of addition*
	$=(a+d)+(b+c)$	*associative property of addition*

Therefore, $(a+b)+(c+d)=(a+d)+(b+c)$

Notice that the commutative and the associative properties allow us to add any finite number of whole numbers in any order we like.

Example 3.10 Let $A=\{a,b,c\}$. Then $c(A)=3$, $c(\varnothing)=0$, and

$$A \cup \varnothing = \{a,b,c\} = A = \varnothing \cup A$$

$$c(A \cup \varnothing) = c(A)+c(\varnothing) = 3+0 = 3 = 0+3 = c(\varnothing)+c(A) = c(\varnothing \cup A)$$

Let a be any whole number. Then, by definition, there exists a finite set A such that $a=c(A)$. For any set A, we observed in Section 1.5 that $A \cup \varnothing = A = \varnothing \cup A$. This tells us that $a+0=a=0+a$ for any whole number a (why?). That is, the addition of zero to any whole number a has no effect on the whole number a and it remains the same. Thus, we observe that the whole number *zero* possesses a very special property and hence needs special consideration.

Theorem 3.4 (***Existence of a Unique Additive Identity***) *For any whole number a,*

$$a + 0 = a = 0 + a$$

The whole number zero is called the *identity element* with respect to addition or the *additive identity* for the set of whole numbers.

Observe that the whole number zero plays a role under addition similar to the role played by $\varnothing$ under the union operation.

Theorem 3.5 (***Addition Property***) *Let a,b,c,d be whole numbers. If a = b and c = d, then a + c = b + d. In other words, equal numbers can be added to equal numbers to get equal numbers.*

The addition property is very useful in our development of the number system.

Example 3.11 Let $A = \{a,b,c\}$, $B = \{b,c,d\}$, $C = \{x,y\}$, and $D = \{u,v\}$. Then $A \sim B$ and $C \sim D$, since

$$c(A) = 3 = c(B) \qquad \text{and} \qquad c(C) = 2 = c(D)$$

Further, $A \cup C = \{a,b,c,x,y\}$ and $B \cup D = \{b,c,d,u,v\}$. Clearly, $A \cup C \sim B \cup D$:

$$c(A) + c(C) = c(A \cup C) = c(B \cup D) = c(B) + c(D)$$

$$\downarrow \quad \downarrow \qquad \downarrow \qquad\qquad \downarrow \qquad\quad \downarrow \quad \downarrow$$

$$3 + 2 = \quad 5 \quad = \quad 5 \quad = 3 + 2$$

Theorem 3.6 (***Cancellation Property***) *If a,b,c are whole numbers such that a + b = a + c, then b = c.*

Observe that the cancellation property is the converse of the addition property. Also, the cancellation property is a very useful one as shown by the following example.

Example 3.12 If $2 + x = 2 + y$, then by the cancellation property we have $x = y$. If $4 + x = 10$, then $x = 6$ since $10 = 4 + 6$.

Before we close this section, we would like to make a point clearer. Recall that when we defined the operation of addition of two whole numbers a and b, we assumed that the sets A and B associated with them are disjoint. If A and B are not disjoint, how are their cardinal numbers related to the cardinal number of the set $A \cup B$? Let's consider the following examples.

Example 3.13 Consider the sets $A = \{a,b,1,2\}$ and $B = \{a,b,2,3,4\}$. Then $A \cup B = \{a,b,1,2,3,4\}$ and $A \cap B = \{a,b,2\}$. Further, $c(A) = 4$, $c(B) = 5$, $c(A \cup B) = 6$ and $c(A \cap B) = 3$.

Thus,

$$c(A)+c(B) \quad = \quad c(A \cup B)+c(A \cap B)$$

$$\downarrow \quad \downarrow \qquad\qquad \downarrow \qquad\quad \downarrow$$

$$4 \ + \ 5 = 9 = \qquad 6 \quad + \quad 3$$

Example 3.14   Consider the sets $A = \{x,y,z\}$ and $B = \{a,b\}$. Then $A \cup B = \{x,y,z,a,b\}$ and $A \cap B = \varnothing$. Also, $c(A) = 3$, $c(B) = 2$, $c(A \cup B) = 5$, and $c(A \cap B) = 0$.

Thus,

$$c(A)+c(B) \quad = \quad c(A \cup B)+c(A \cap B)$$

$$\downarrow \quad \downarrow \qquad\qquad \downarrow \qquad\quad \downarrow$$

$$3 \ + \ 2 \ = 5 = \qquad 5 \quad + \quad 0$$

Accordingly, we have the following result:

Theorem 3.7   *Let A and B be two finite sets. Then*

$$c(A)+c(B)=c(A \cup B)+c(A \cap B)$$

Exercise 3.3

1. Use the definition of addition of whole numbers to illustrate each of the following:

a) $1+3=4$ b) $3+2=5$ c) $2+4=6$
d) $3+5=8$ e) $5+2=7$ f) $2+2=4$

2. Verify the following.
a) $2+(3+5)=(2+3)+5$
b) $1+(2+3)=(3+1)+2$
c) $6+(8+0)=8+6$
d) $(3+5)+(8+10)=3+[(5+10)+8]$
e) $2+[3+(4+0)]=(2+4)+3$
f) $3+(4+5)=(5+3)+4$

3. For which of the following pairs of sets A and B is it true that $c(A)+c(B)=c(A \cup B)$?
a) $A = \{0,1\}$ and $B = \{2,3,4\}$.
b) $A = \{w,x,y\}$ and $B = \{x,y,z\}$.
c) $A = \{$Britain,France,China$\}$ and $B = \{$Canada,Poland,Russia$\}$.
d) $A = \{$Kennedy,Johnson,Nixon$\}$ and $B = \{$Massachusetts,Texas, California$\}$.

4. For the following pairs of sets A and B, verify that $c(A)+c(B)= c(A \cup B)+c(A \cap B)$.
a) $A = \{0,1\}$ and $B = \{x,y,z\}$.

b) $A = \{a,b,c\}$ and $B = \{b,c,d\}$.

c) $A =$ set of all days of the week and $B = \{\text{Saturday,Sunday}\}$.

d) $A =$ set of all letters of the word "Alabama" and $B =$ set of all letters of the word "Tennessee".

e) $A =$ set of all letters of the word "Daddy" and $B =$ set of all letters of the word "Zoo".

5. State the property of addition used in each of the following statements, where each symbol represents a whole number.

a) $3+5$ is a whole number.

b) $2+5 = 5+2$

c) $2+(5+7) = 2+(7+5)$

d) $(a+b)+(c+d) = (c+d)+(a+b)$

e) $4+0 = 4$

f) If $2+x = 2+y$, then $x = y$

g) If $a = b$, then $4+a = 4+b$.

h) $3+0 = 0+3$

i) $(a+b)+(c+d) = [(a+b)+c]+d$

j) $a+[b+(c+d)] = (a+b)+(c+d)$

6. a) Is the set $\{0,1\}$ closed under addition? Why?

b) Is the set $\{0\}$ closed under addition? Why?

7. For what value of x in W is each of the following true?

a) $2+x = 4$

b) $3+x = 3$

c) $7+x = 10$

d) $x+0 = x$

e) $8+x = 16$

f) $8+x = 5$

8. Mark *true* or *false* (A,B are any sets; a,b,c are any whole numbers).

a) The equality relation is symmetric.

b) The equality relation is transitive.

c) Addition is a binary operation on the set of natural numbers.

d) The set $\{1,3,5,7,9,...\}$ is closed under addition.

e) The set $\{10,20,30,40,...\}$ is closed under addition.

f) The set of natural numbers is closed under the operation of "adding 1."

g) The set of natural numbers contains a unique additive identity.

h) If $b = c$, then $a+b = a+c$.

i) $a+b+c = a+(b+c)$.

j) $0+0$ is defined.

k) $x+0 = x$ for every whole number x.

9. Give a counterexample to each of the false statements in problem 8, whenever possible.

10. a) Find $c(A \cup B)$ if $c(A) = 10$, $c(B) = 15$, and $c(A \cap B) = 5$.

b) Find $c(A \cup B)$ if $c(A) = 8$, $c(B) = 7$, and $A \cap B = \varnothing$.

c) Find $c(A)$ if $c(A) = c(B)$, $c(A \cup B) = 15$, and $c(A \cap B) = 5$.

d) Find $c(A \cap B)$ if $c(A) = c(B) = 10$ and $c(A \cup B) = 15$.

e) Find $c(A \cap B)$ if $c(A) = a$, $c(B) = b$, and $c(A \cup B) = a+b$.

f) Find $c(B)$ if $c(A) = 15$, $c(A \cup B) = 25$, and $c(A \cap B) = 4$.

11. Let x be a natural number. Can $x+x = 0$?

12. Let x be a whole number such that $x+x = x$. Prove that $x = 0$.

13. Let x and y be whole numbers such that $x+y = x$. Prove that $y = 0$.

14. Prove each of the following, justifying your steps, where each symbol represents a whole number.

a) $a + (b + c) = c + (a + b)$

b) $a + b + c = c + b + a$

c) $a + (b + 0) = b + a$

d) $[(a + b) + c] + d = [b + (a + c)] + d$

3.4 MULTIPLICATION OF WHOLE NUMBERS

In the preceding section, we defined and discussed the properties of the binary operation of addition of whole numbers. In this section, we will focus our attention on a second binary operation, called multiplication. Let's consider the following example.

Example 3.15 Assume Ron has 3 bags of apples, each bag containing 2 apples. He wants to find how many apples he has altogether. Also, assume he does not know how to multiply two numbers but knows how to add them. He takes the two apples from one bag, the two from the second bag and then the two from the third bag. Now he finds that he has $2 + 2 + 2 = 6 = 3 \cdot 2$ apples. This example shows that the operation of multiplication could be defined in terms of addition, namely as repeated addition.

Let's now solve Ron's problem by another method, which will in turn suggest how multiplication could be defined using sets. Let's label the three bags a, b, and c; the two apples in each bag as 1 and 2. Then we could list all the apples as

$$(a,1), (a,2), (b,1), (b,2), (c,1), (c,2)$$

where $(a,1)$ denotes the first apple in the bag labeled a, $(b,2)$ denotes the second apple from the bag marked b, etc. Observe that this is the cartesian product $A \times B = \{a,b,c\} \times \{1,2\}$ of the stwo sets $A = \{a,b,c\}$ and $B = \{1,2\}$, as discussed in Section 1.12. Accordingly, the cardinality of $A \times B$ gives the answer to Ron's problem, namely 6 apples. Observe that $c(A) = 3$, $c(B) = 2$, and $c(A \times B) = 6$.

Example 3.16 Consider the sets $A = \{1,2\}$ and $B = \{a,b,c,d\}$:

$$A \times B = \{(1,a),(1,b),(1,c),(1,d),(2,a),(2,b),(2,c),(2,d)\}$$

Thus $c(A) = 2$, $c(B) = 4$, and $c(A \times B) = 8$.

These two examples now suggest to us how multiplication should be defined on the set of whole numbers.

Definition 3.4 Let a and b be any two whole numbers. Let A and B be sets such that $a = c(A)$ and $b = c(B)$. Then the *product* of a and b, denoted by $a \cdot b$ or $a \times b$, is defined as the cardinal number of $A \times B$. That is,

$$a \cdot b = c(A \times B) \qquad \text{or} \qquad c(A) \cdot c(B) = c(A \times B)$$

The operation we defined is called *multiplication* of whole numbers. The product $a \cdot b$ of a and b is also denoted as ab, provided there is no confusion.

This definition of multiplication of whole numbers might seem somewhat artificial, as in the case of addition. Again, the advantage is that most of the fundamental properties of multiplication can be established from this definition. Notice that in this definition, we do not assume that $A \cap B = \varnothing$. (Why?)

This definition depends on the fact that the cartesian product of two finite sets is finite, which in turn tells us that $a \cdot b$ is also a whole number. Thus, we have the first property of multiplication:

Theorem 3.8 (***Closure Property***)   *The set of whole numbers is closed under multiplication.*

Since multiplication is repeated addition, it is only natural that these two operations share similar properties. In the preceding section, we found that addition on W is commutative. Is it true in the case of multiplication? Let's consider the following example:

Example 3.17   Consider the sets $A = \{1,2\}$ and $B = \{a,b,c\}$:

$$c(A) = 2 \quad \text{and} \quad c(B) = 3$$

$$A \times B = \{(1,a),(1,b),(1,c),(2,a),(2,b),(2,c)\}$$

$$B \times A = \{(a,1),(a,2),(b,1),(b,2),(c,1),(c,2)\}$$

Thus $A \times B \neq B \times A$, but $A \times B \sim B \times A$. Therefore, $c(A \times B) = c(B \times A)$ and hence by definition,

$$2 \cdot 3 = 3 \cdot 2 = 6$$

This example suggests the following property:

Theorem 3.9 (***Commutative Property***)   *For any two whole numbers a and b,*

$$a \cdot b = b \cdot a$$

Proof: Since a and b are whole numbers, there exist finite sets A and B such that $a = c(A)$ and $b = c(B)$.

$a \cdot b = c(A \times B)$ *definition of multiplication of whole numbers*

$\quad = c(B \times A)$ $A \times B \sim B \times A$

$\quad = b \cdot a$ *definition of multiplication of whole numbers*

Thus, $a \cdot b = b \cdot a$

The commutative property of multiplication of whole numbers, as in the case of addition, tells us that the order in which two whole numbers are multiplied is immaterial.

We found, in the preceding section, that addition is associative. That is, we can, for instance, evaluate $2 + 3 + 5$ without any ambiguity. Since multiplication is also a binary operation, does $2 \cdot 3 \cdot 5$ make any sense? In other words, does

$2 \cdot (3 \cdot 5) = (2 \cdot 3) \cdot 5$? We have,

$$2 \cdot (3 \cdot 5) = 2 \cdot 15 = 30$$

and

$$(2 \cdot 3) \cdot 5 = 6 \cdot 5 = 30$$

Therefore, $2 \cdot (3 \cdot 5) = (2 \cdot 3) \cdot 5$ and this is true in general.

Theorem 3.10 (**Associative Property**) *For any whole numbers a, b, and c,*

$$a \cdot (b \cdot c) = (a \cdot b) \cdot c$$

The associative property of multiplication of whole numbers simply tells us that in a product of three or more whole numbers, they can be grouped in any fashion we like.

The commutative and the associative properties of multiplication allow us to evaluate a product of three or more whole numbers in any order.

Example 3.18 Show that $(2 \cdot 3)(5 \cdot 8) = (2 \cdot 5)(3 \cdot 8)$ without evaluating the products.

Proof:
$$\begin{aligned}
(2 \cdot 3)(5 \cdot 8) &= 2 \cdot [3 \cdot (5 \cdot 8)] && \textit{associative property of multiplication} \\
&= 2 \cdot [(3 \cdot 5) \cdot 8] && \textit{associative property of multiplication} \\
&= 2 \cdot [(5 \cdot 3) \cdot 8] && \textit{commutative property of multiplication} \\
&= 2 \cdot [5 \cdot (3 \cdot 8)] && \textit{associative property of multiplication} \\
&= (2 \cdot 5)(3 \cdot 8) && \textit{associative property of multiplication}
\end{aligned}$$

Thus, $(2 \cdot 3)(5 \cdot 8) = (2 \cdot 5)(3 \cdot 8)$

Example 3.19 Prove that $(ab)(cd) = (ad)(bc)$.

Proof:
$$\begin{aligned}
(ab)(cd) &= (ab)(dc) && \textit{commutative property of multiplication} \\
&= a[b(dc)] && \textit{associative property of multiplication} \\
&= a[(bd)c] && \textit{associative property of multiplication} \\
&= a[(db)c] && \textit{commutative property of multiplication} \\
&= a[d(bc)] && \textit{associative property of multiplication} \\
&= (ad)(bc) && \textit{associative property of multiplication}
\end{aligned}$$

Thus, $(ab)(cd) = (ad)(bc)$

Recall that we observed the existence of a special element, zero, with respect to addition in the set of whole numbers. It has the unique property that the effect of adding zero to any whole number is null. Does the set W possess such an element with respect to multiplication?

Example 3.20 Consider the sets $A = \{a\}$ and $B = \{b,c,d\}$:

$$c(A) = 1 \quad \text{and} \quad c(B) = 3$$

$$A \times B = \{(a,b),(a,c),(a,d)\} \sim B$$
$$B \times A = \{(b,a),(c,a),(d,a)\} \sim B$$
$$c(A \times B) = c(B) = c(B \times A)$$

That is, $\qquad\qquad 1 \cdot 3 = 3 = 3 \cdot 1$

More generally, we have the following property:

Theorem 3.11 (***Existence of a Unique Multiplicative Identity***) *For any whole number a,*

$$a \cdot 1 = a = 1 \cdot a$$

The whole number 1 is called the *identity for multiplication* or the *multiplicative identity.*

Observe that the multiplicative identity plays a role similar to that of the universal set under intersection.

Now there is a unique property possessed by the whole number 0 under multiplication.

Theorem 3.12 (***Zero Property of Multiplication***) *For any whole number a,*

$$a \cdot 0 = 0 = 0 \cdot a$$

Proof: Let A be a set such that $a = c(A)$:

$\quad a \cdot 0 = c(A \times \varnothing)$ *definition of multiplication of whole numbers*

$\qquad\quad = c(\varnothing)$ $A \times \varnothing = \varnothing$ *for any set A*

$\qquad\quad = 0$ *definition of the whole number 0*

Since $a \cdot 0 = 0 \cdot a$ by commutativity, it follows that $a \cdot 0 = 0 = 0 \cdot a$.

The above theorem tells that the product of any whole number with zero is zero. But when can the product of two whole numbers be zero? This is answered by the following theorem.

Theorem 3.13 (***Product Law***) *If a and b are whole numbers such that ab = 0, then either a = 0 or b = 0.*

The product law says that if the product of two whole numbers is zero, then at least one of them must be the whole number zero. Equivalently, the theorem states that the product of two nonzero whole numbers is nonzero. (Why?)

Example 3.21
$$5 \cdot 0 = 0 = 0 \cdot 5$$
$$100 \cdot 0 = 0 = 0 \cdot 100$$
$$0 \cdot 0 = 0 = 0 \cdot 0$$
$$2 \cdot 5 = 10 = 5 \cdot 2 \neq 0$$

Corresponding to the addition property of whole numbers, we now have the following result with respect to multiplication:

Theorem 3.14 (***Multiplication Property***) *Let a,b,c,d be whole numbers. If $a = b$ and $c = d$, then $ac = bd$.*

The multiplication property tells us that if equal numbers are multiplied by equal numbers, their products remain equal.

Example 3.22   Let $A = \{0,1\}$, $B = \{a,b\}$, $C = \{2,3,4\}$, and $D = \{x,y,z\}$. Clearly, $A \sim B$ and $C \sim D$, since

$$c(A) = 2 = c(B)$$
$$c(C) = 3 = c(D)$$

Also,

$$A \times C = \{(0,2),(0,3),(0,4),(1,2),(1,3),(1,4)\}$$
$$B \times D = \{(a,x),(a,y),(a,z),(b,x),(b,y),(b,z)\}$$

hence

$$A \times C \sim B \times D$$
$$c(A \times C) = c(B \times D)$$
$$c(A \times C) = 2 \cdot 3 = 6 = 2 \cdot 3 = c(B \times D)$$

Thus

$$c(A) \cdot c(C) = c(B) \cdot c(D)$$

Theorem 3.15 (***Cancellation Property***) *If b and c are whole numbers and a is any nonzero whole number such that $ab = ac$, then $b = c$.*

The cancellation property simply states that the same number a can be dropped from the statement $ab = ac$, provided it is nonzero.

Example 3.23 If x and y are whole numbers such that $3x = 3y$, then the cancellation property for multiplication tells us that $x = y$. Also, if $2x = 2 \cdot 5$, then $x = 5$.

The reader is warned that the cancellation property does not say that if a,b,c are whole numbers such that $ab = ac$, then $b = c$. For instance, $0 \cdot 2 = 0 \cdot 4$, but $2 \neq 4$. The condition that $a \neq 0$ is crucial!

Exercise 3.4

1. Use the definition of multiplication of whole numbers to illustrate the following.

a) $3 \cdot 1 = 3$ b) $2 \cdot 4 = 8$ c) $3 \cdot 4 = 12$
d) $3 \cdot 5 = 15$ e) $5 \cdot 3 = 15$ f) $3 \cdot 3 = 9$

2. Verify each of the following.

   a) $2 \cdot (3 \cdot 4) = (2 \cdot 4) \cdot 3$

   b) $2 \cdot (3 \cdot 6) = (6 \cdot 2) \cdot 3$

   c) $2 \cdot [(4 \cdot 5) \cdot 6] = [(2 \cdot 5) \cdot 4] \cdot 6$

   d) $2 \cdot (4 \cdot 5) = 4 \cdot (5 \cdot 2)$

   e) $(3 \cdot 5) \cdot (4 \cdot 8) = (4 \cdot 3) \cdot (8 \cdot 5)$

   f) $2 \cdot [4 \cdot (5 \cdot 6)] = (2 \cdot 6) \cdot (4 \cdot 5)$

3. For the following pairs of sets A and B, verify that $c(A \times B) = c(A) \cdot c(B)$.

   a) $A = \{0\}$ and $B = \{0,1\}$.

   b) $A = \{a,b\}$ and $B = \{x,y,z\}$.

   c) $A = \{1,\{2\}\}$ and $B = \{x,y\}$.

   d) $A = \{\varnothing\}$ and $B = \{1,\{x\}\}$.

   e) $A = \{1,2,3\}$ and $B = \{a,b,c,d\}$.

   f) $A =$ set of all letters of the word "Alabama" and $B =$ set of all letters of the word "zoo".

   g) $A =$ set of all letters of the word "Mississippi" and $B =$ set of all letters of the word "Tennessee".

4. For each of the following sets A, verify that $c(A \times A) = c(A) \cdot c(A)$.

   a) $A = \{0,1\}$

   b) $A = \{\varnothing\}$

   c) $A =$ set of all letters of the word "Cincinnati".

   d) $A =$ set of all letters of the word "calculus".

   e) $A =$ set of all months of the year that begin with the letter J.

5. State the property of multiplication used in each of the following statements, where each symbol represents a whole number.

   a) $x \cdot y$ is a whole number.

   b) $2 \cdot (3 \cdot 5) = 2 \cdot (5 \cdot 3)$

   c) $(ab)(cd) = (cd)(ab)$

   d) $x \cdot 1 = x$

   e) $(ab)(cd) = a[b(cd)]$

   f) If $3x = 3y$, then $x = y$.

   g) If $2x = 0$, then $x = 0$.

   h) If $x = y$, then $2x = 2y$.

   i) If $3x = 3 \cdot 2$, then $x = 2$.

   j) If $x = 3$, then $2x = 2 \cdot 3$.

6. **a)** Is the set $\{1\}$ closed under multiplication? Why?

   b) Is the set $\{0,1\}$ closed under multiplication? Justify your answer.

7. For what value of x in W is each of the following true?

   a) $2x = 4$ **b)** $3x = 3$ **c)** $2x = 0$

   d) $8x = 16$ **e)** $1 \cdot x = x$ **f)** $2x = 3$

8. Mark *true* or *false*.
   a) Multiplication is a binary operation on the set of natural numbers.
   b) Addition is repeated multiplication.
   c) Multiplication of whole numbers is associative.
   d) Zero is a multiplicative identity for the set of whole numbers.
   e) If $ab = ac$ then $b = c$, where $a,b,$ and c are any whole numbers.
   f) If $ab = ac$ then $b = c$, where $a,b,$ and c are any natural numbers.
   g) $0 \cdot 0 = 0$
   h) N is closed under the operation "multiplying by 1."
   i) W has an identity with respect to multiplication.
   j) If $ab = 0$ then $a = b = 0$, where a and b are whole numbers.
9. Give a counterexample to each of the false statements in problem 8.
10. If x is a whole number such that $x \cdot 1 = 1$, prove that $x = 1$.
11. If x is a natural number such that $x \cdot x = x$, prove that $x = 1$.
12. Let x be a whole number such that $x \cdot x = 0$. What is your conclusion?
13. If x is a natural number and y a whole number such that $x \cdot y = 0$, prove that $y = 0$.
14. If x is a natural number, can $x \cdot x = 0$?
15. Let a and b be whole numbers such that $a = b$. Prove that $ac = bc$ for every whole number c.
★16. Prove each of the following, where a,b,c,d are any whole numbers.
   a) $a(bc) = (ca)b$
   b) $(ab)c = a(cb)$
   c) $a(b \cdot 1) = ba$
   d) $(ab)(cd) = (ac)(bd)$

3.5 THE DISTRIBUTIVE PROPERTY

So far in this chapter, we have been discussing either the operation of addition or the operation of multiplication, but not both at the same time. In the case of sets, we found that we could have expressions of sets containing both union and intersection operations. They give rise to the distributive properties:

$$A \cap (B \cup C) = (A \cap B) \cup (A \cap C)$$

and

$$A \cup (B \cap C) = (A \cup B) \cap (A \cup C)$$

for any sets $A,B,$ and C. That is, intersection distributes over union and vice versa. Are there similar results in the set of whole numbers? For instance, is multiplication distributive over addition? Is addition distributive over multiplication? The student is reminded that $a(b + c)$ means, add b and c first and then multiply that sum by a.

Example 3.24 Evaluate $2(3 + 5)$ and $2 \cdot 3 + 2 \cdot 5$

Solution:
$$2(3 + 5) = 2 \cdot 8 = 16$$
$$2 \cdot 3 + 2 \cdot 5 = 6 + 10 = 16$$

Thus, we observe that

$$2(3+5) = 2 \cdot 3 + 2 \cdot 5$$

Example 3.25 Without evaluating $2(3+4)$ and $2 \cdot 3 + 2 \cdot 4$, show that $2(3+4) = 2 \cdot 3 + 2 \cdot 4$.

Solution: Consider the sets $A = \{0,1\}$, $B = \{a,b,c\}$, and $C = \{w,x,y,z\}$. Notice that $c(A) = 2$, $c(B) = 3$, and $c(C) = 4$. Then, since $B \cup C = \{a,b,c,w,x,y,z\}$,

$$A \times (B \cup C) = \{(0,a),(0,b),(0,c),(0,w),(0,x),(0,y),(0,z),(1,a),(1,b),$$

$$(1,c),(1,w),(1,x),(1,y),(1,z)\}$$

$$A \times B = \{(0,a),(0,b),(0,c),(1,a),(1,b),(1,c)\}$$

$$A \times C = \{(0,w),(0,x),(0,y),(0,z),(1,w),(1,x),(1,y),(1,z)\}$$

Clearly, $A \times (B \cup C) = (A \times B) \cup (A \times C)$

where $(A \times B) \cap (A \times C) = \varnothing$. Therefore,

$$c[A \times (B \cup C)] = c(A \times B) + c(A \times C)$$

That is, $2(3+4) = 2 \cdot 3 + 2 \cdot 4$

Let a,b, and c be whole numbers. Then so are $a \cdot b$, $a \cdot c$, $b+c$, and $a(b+c)$. (Why?) How are these four numbers related? The above examples suggest that $a(b+c) = a \cdot b + a \cdot c$. That this is true in general is stated in the following result:

Theorem 3.16 (***Distributive Property***)   *For any whole numbers a,b, and c,* $a \cdot (b+c) = a \cdot b + a \cdot c$.

Example 3.26   Prove that $(a+b)c = ac + bc$.

Proof: $(a+b)c = c(a+b)$ *commutative property of multiplication*

$$= ca + cb \qquad \text{\textit{distributive property}}$$

$$= ac + bc \qquad \text{\textit{commutative property of multiplication}}$$

$$(a+b)c = ac + bc$$

Notice the difference between this result and the one in Theorem 3.16. In Theorem 3.16, the multiplication sign is on the left of the plus sign, whereas in example 3.26 addition precedes multiplication. This is also a distributive property.

Theorem 3.16 and Example 3.26 show that multiplication distributes over addition. Now, is addition distributive over multiplication? Is $a + (b \cdot c) = (a+b) \cdot (a+c)$ for any whole numbers a, b, and c? Well, here is a counterexample to show that it is not!

Example 3.27   Let $a = 1$, $b = 2$, and $c = 3$:

$$a + (b \cdot c) = 1 + (2 \cdot 3)$$
$$= 1 + 6$$
$$= 7$$

whereas

$$(a + b) \cdot (a + c) = (1 + 2)(1 + 3)$$
$$= 3 \cdot 4$$
$$= 12$$

Thus, $a + (b \cdot c) \neq (a + b) \cdot (a + c)$, in general.

However, this example does not say that there do not exist values for a, b, and c such that $a + (b \cdot c) = (a + b) \cdot (a + c)$. Let's consider the following example.

Example 3.28   If $a = 0$, $b = 1$, and $c = 2$, then

$$a + (b \cdot c) = 0 + 1 \cdot 2 = 2 = 1 \cdot 2 = (0 + 1) \cdot (0 + 2) = (a + b) \cdot (a + c)$$

In fact, there exist infinitely many such sets of values for a, b, and c!

The distributive property is extremely important in our discussion and development of the number system. For example, it can conveniently be used to find the product of large numbers, as shown in the following example:

Example 3.29   Use the distributive property to evaluate the product $25 \cdot 32$.

Solution:
$$25 \cdot 32 = 25(30 + 2)$$
$$= 25 \cdot 30 + 25 \cdot 2$$
$$= 750 + 50$$
$$= 800$$

Example 3.30   Prove that $(a + b)(c + d) = ac + ad + bc + bd$.

Proof: $(a + b)(c + d)$

$= a(c + d) + b(c + d)$	*distributive property*
$= (ac + ad) + (bc + bd)$	*distributive property*
$= ac + ad + bc + bd$	*associative property of addition*

Exercise 3.5

1. Evaluate the following:

   a) $2 + 3 \cdot 5$

   b) $2(3 + 5)$

   c) $2 \cdot 3 + 5$

   d) $2(3 + 5) + 4 \cdot 2$

   e) $2 \cdot 3 + 5 \cdot 7$

   f) $2 \cdot 3 + 2 \cdot 5$

   g) $3 + 2 \cdot 5$

   h) $2 + 3(5 + 7)$

2. Verify that $a(b+c) = ab + ac$ if $a = c(A)$, $b = c(B)$, and $c = c(C)$, where
   a) $A = \{1,2\}$, $B = \{0\}$, and $C = \{x\}$.
   b) $A = \{\emptyset\}$, $B = \{1,2\}$, and $C = \{x,y,z\}$.
   c) $A = \emptyset$, $B = \{x,y,z\}$, and $C = \{1\}$.
   d) $A = \{1,x\}$, $B =$ set of all letters of the word "Atlanta," and $C =$ set of all letters of the word "Canada".

3. Use the distributive property to evaluate the following:
   a) $23 \cdot 42$ **b)** $15 \cdot 22$ **c)** $20 \cdot 63$
   d) $53 \cdot 30$ **e)** $121 \cdot 25$ **f)** $30 \cdot 43$

4. State the property used in each of the following statements, where each symbol represents a whole number.
   a) $2(3+5) = 2(5+3)$ **b)** $(2+3) \cdot 5 = (2+5) \cdot 3$
   c) $4(6+8) = 4 \cdot 6 + 4 \cdot 8$ **d)** $(2+3)5 = 2 \cdot 5 + 3 \cdot 5.$
   e) $(2+5)7 = 7(2+5)$ **f)** $2+3 \cdot 4 = 3 \cdot 4 + 2$

5. Mark *true* or *false*.
   a) Multiplication is distributive over addition.
   b) Addition is distributive over multiplication.
   c) $a(1+0) = a$
   d) $a(b+0) = ab$
   e) $a(b+1) = ab + b$
   f) $(2+a)b = 2a + ab$
   g) $(a+1)(b+1) = a+b+ab$
   h) There exist whole numbers a, b, and c such that $a(b+c) = ab + ac$.
   i) $a + (bc) = (a+b)(a+c)$ for any whole numbers a, b, and c.

6. Give a counterexample to each of the false statements in problem 5.

7. Show the following:
   a) $3x + 5x = 8x$ **b)** $4a + 5a = 9a$
   c) If $2x = 3x$, then $x = 0$.   **d)** $a(b+0) = ab$

8. For what value of x in W is each of the following true?
   a) $2+3x = 5$ **b)** $2x + 3x = 10$
   c) $3+5x = 3$ **d)** $2 + (3 \cdot 5)x = 32$
   e) $(2+3)x = 3+2$ **f)** $(2 \cdot 3)x + (2 \cdot 5)x = (10+2)4$

9. Find three sets of whole numbers a,b,c such that $a + bc = (a+b)(a+c)$.

10. Use the distributive property to prove that $a \cdot 0 = 0$ for every whole number a. (*Hint*: $0+0 = 0$.)

11. Prove each of the following (x^2 means $x \cdot x$).
   a) $a(a+b) = a^2 + ba$ **b)** $(a+b)(a+c) = a^2 + ab + ac + bc$
   c) $(a+b)^2 = a^2 + 2ab + b^2$ **d)** $a(b+c+d) = ab + ac + ad$

3.6 THE ORDER RELATION

The reader probably has some idea about the concepts "less than" and "greater than." The primitive man did! We made a passing remark on these concepts in Section 1.10. Our aim in this section is to present them a little more carefully. Let's start with an example.

Example 3.31   Consider the sets $A = \{a,b,c\}$ and $B = \{1,2,3,4,5\}$. Clearly, $c(A) = 3$ and $c(B) = 5$. Also, neither $A = B$ nor $A \sim B$. However, we can find a proper subset C of B—in fact several proper subsets exist!—with the property that $A \sim C$. For instance, C could very well be the set $\{1,2,3\}$. In other words, there exists a 1-1 correspondence between the set A and a proper subset of B, as shown below:

$$A = \{a,b,c\}$$

$$\updownarrow \updownarrow \updownarrow$$

$$B = \{1,2,3,4,5\}$$

What does this really mean? The set B contains *more* elements than the set A and the set A contains *fewer* elements than the set B. In that case, we say that $c(A)$ is less than $c(B)$, that is, 3 is less than 5.

Let's now translate these ideas in terms of whole numbers. We have the whole numbers 3 and 5. By adding the suitable natural number 2 to 3, we get the number $3 + 2 = 5$, and say that 3 is less than 5. Now, we are in a position to make the following definition:

Definition 3.5   Let a and b be two whole numbers. The whole number a is said to be *less than* the whole number b if there exists a natural number x such that $a + x = b$. If a is less than b, we write $a < b$. If a is not less than b, we write $a \not< b$. If $a < b$, we also say that b is *greater than* a and write $b > a$.

Example 3.32 $2 < 8$ since we can find the natural number $x = 6$ such that $2 + 6 = 8$. Similarly, $10 < 15$ because $10 + 5 = 15$, and $6 < 20$ because $6 + 14 = 20$. But $13 \not< 5$ since there exists no natural number x such that $13 + x = 5$.

A very remarkable property of the set of whole numbers is that every pair of whole numbers falls in exactly one of three mutually disjoint categories. That is, if a and b are whole numbers, then a is related to b in precisely one of the following ways: $a < b$ or $a = b$ or $b < a$.

Theorem 3.17   (Law of Trichotomy)   *If a and b are any two whole numbers, then exactly one of the following holds:*

$$(1) \ \ a < b \qquad (2) \ \ a = b \qquad (3) \ \ b < a$$

The law of trichotomy says that any two whole numbers are either equal or one is less than the other.

In Section 3.2, we found that the equality relation " $=$ " is both reflexive and symmetric. The order relation " $<$ " is neither reflexive nor symmetric. That is, $a \not< a$, and $a < b$ does not imply that $b < a$, where a and b are any whole numbers. But is it transitive, as in the case of the equality relation?

Theorem 3.18 (**_Transitive Property_**)   *If a,b,c are whole numbers such that* $a < b$ *and* $b < c$, *then* $a < c$.

Proof: Since $a < b$ and $b < c$, there exist natural numbers x and y such that $a + x = b$ and $b + y = c$. Then

$$(a + x) + y = c \qquad substitution$$

$$a + (x + y) = c \qquad associative\ property\ of\ addition$$

$$x + y\ is\ a\ natural\ number \qquad N\ is\ closed\ under\ addition$$

$$a < c \qquad definition\ of\ the\ order\ relation$$

Example 3.33 $2 < 5$ since $2 + 3 = 5$ and $5 < 11$ since $5 + 6 = 11$. Observe that $2 < 11$, since $2 + 9 = 2 + (3 + 6) = 11$.

In the case of addition of whole numbers, we found that if equals are added to equals, the sums remain equal. Is there a similar result in the case of the order relation?

Example 3.34 $3 < 8$ since $3 + 5 = 8$. Let's add the same number 7 to both sides of the inequality $3 < 8$. Then the two whole numbers we get are 10 and 15, and they are related as $10 < 15$. Thus $3 < 8$ and $3 + 7 < 8 + 7$.

Theorem 3.19   *Let a and b be any whole numbers. Then* $a < b$ *if and only if* $a + c < b + c$ *for every whole number c.*

Observe that the theorem states that if we add the same number to both sides of an inequality, the inequality remains of the same form. That is, " $<$ " remains the same as " $<$ " and " $>$ " remains the same as " $>$." Also, we can cancel the same whole number c from both sides of the inequality $a + c < b + c$ to get $a < b$.

A similar result exists under multiplication, with the extra condition that c be nonzero.

Theorem 3.20   *Let a and b be any whole numbers. Then* $a < b$ *if and only if* $ac < bc$ *for every natural number c.*

This theorem states that if both sides of an inequality are multiplied by a natural number, it remains of the same type. Also, we can cancel the same natural number c from both sides of the inequality $ac < bc$ and it remains of the same form.

Example 3.35 $3 < 5$ since $3 + 2 = 5$. Multiplying both sides of the inequality $3 < 5$ by 4, we get the numbers 12 and 20, and $12 < 20$. Therefore, $3 < 5$ implies that $3 \cdot 4 < 5 \cdot 4$. Also, $12 < 20$ implies that $3 < 5$.

Let a and b be whole numbers. If a is less than or equal to b, we write $a \leq b$. Similarly $a \geq b$ implies $a > b$ or $a = b$. For example, $\{x \in W \,|\, x \leq 3\} = \{0,1,2,3\} =$

$\{x \in W \,|\, x < 4\}$. Now, $a < x < b$ means $x > a$ and $x < b$ at the same time. For example, $\{x \in W \,|\, 2 < x < 5\} = \{3,4\} = \{x \in W \,|\, 2 < x \le 4\} = \{x \in W \,|\, 3 \le x < 5\}$.

We remark that Theorems 3.18, 3.19, and 3.20 all remain true if "$<$" is replaced by "$\le$," and then in Theorem 3.20, c could very well be a whole number. It need not be a natural number. (Why?)

Theorem 3.21   *If a and b are any whole numbers, then $a \le b$ if and only if $a^2 \le b^2$.*

Example 3.36   Find the set of all whole numbers x such that $x^2 \le 25$.

Solution: Since $x^2 \le 25$, by Theorem 3.21, $x \le 5$. Thus, the set of whole numbers x such that $x^2 \le 25$ is $\{x \in W \,|\, x \le 5\} = \{0,1,2,3,4,5\}$.

Exercise 3.6

1. Use the definition of the order relation to verify the following:
   a) $3 < 8$ **b)** $11 < 14$ **c)** $3 < 13$ **d)** $18 < 31$
2. Find $A \cup B$, $A \cap B$, $A - B$, and $B - A$ if $A = \{x \in W \,|\, x \le 4\}$ and $B = \{x \in N \,|\, x < 6\}$.
3. Let $A = \{x \in W \,|\, 0 \le x < 5\}$, $B = \{x \in W \,|\, 4 \le x \le 6\}$, $C = \{x \in W \,|\, 3 < x < 8\}$, and $U = \{x \in W \,|\, 0 \le x \le 9\}$. Find the following sets and their cardinal numbers.
   a) $A \cup B$ **b)** $A \cap B$ **c)** $A - B$
   d) $B - A$ **e)** A' **f)** $A \cap B'$
   g) $A' - C$ **h)** $(A \cup B) \cap C'$ **i)** $(A \cup B)' \cup C'$
   j) $(A \cup B) - C$ **k)** $(A' \cup B') \cap C'$ **l)** $A' \cap B' \cap C'$
4. For what value of x in N is each of the following true?
   a) $2 + x < 4$ **b)** $2x \le 4$
   c) $(2+3)x \le 15$ **d)** $2 + (3 \cdot 4)x \le 26$
   e) $2 \cdot 3 + 4x \le 18$ **f)** $(2 + 3 \cdot 4)x \le 3 + 2 \cdot 7$
5. Find the smallest element in each of the following sets.
   a) $A = \{3,5,2\}$ **b)** $B = \{4,5,6\}$
   c) $C = \{8,6,7,5\}$ **d)** $D = \{x \in W \,|\, 3 < x < 6\}$
   e) $E = \{x \in W \,|\, 3 \le x < 6\}$ **f)** $F = \{x \in W \,|\, 3 \le x \le 6\}$
6. Mark *true* or *false* (a,b,c,d,x are any whole numbers).
   a) Every natural number is less than itself.
   b) If $a < b$ and $b < c$, then $a < c$. transitive
   c) If $a < c$ and $b < c$, then $a < b$.
   d) If $a < b$, then $b < a$. symatric
   e) If $a < b$, then $ac < bc$ for every whole number c.
   f) If $a \le b$, then $b \le a$.
   g) Every whole number is greater than zero.
   h) Every natural number is greater than zero.
   i) Every whole number is less than or equal to itself.
   j) If $a \le b$ and $b \le c$, then $a \le c$. transitive

k) If $a \le b$ and $b \le a$, then $a = b$.

l) If $a < b$ and $c < d$, then $a + c < b + d$.

7. Give a counterexample to each of the false statements in problem 6.

8. Let x and y be natural numbers such that $x < y$. Does it follow that

    a) $4 + x < 4 + y$? Why?

    b) $4x < 4y$? Why?

9. Why do we have to assume that c is a natural number in Theorem 3.20?

10. Let x be a whole number such that $2x \le x$. What is your conclusion?

11. Let x be a natural number such that $x^2 \le x$. What is your conclusion?

12. Is there a whole number x such that $2x < x$?

13. Is there a natural number x such that $x^2 < x$?

14. Find the set of all whole numbers x such that

    a) $x^2 < 4$ **b)** $x^2 \le 0$ **c)** $2x^2 \le 32$

    d) $3x^2 + 5 < 32$ **e)** $x(x+1) < x + 16$ **f)** $x(x+3) < 3(27 + x)$

15. If the minimum and the maximum temperatures of a certain day were 27 degrees and 79 degrees, respectively, what were the possible temperatures in degrees during the day?

16. If the area of a square does not exceed 25 square feet, what are the possible values for its side in feet?

17. If three times a whole number is 13 more than 23, find the whole number.

3.7 THE WHOLE NUMBER LINE

The less-than relation "$<$" and its properties help us to compare the "sizes" or "magnitudes" of whole numbers. In Section 1.3, we discussed how Venn diagrams are useful to represent relationships between sets pictorially. In a similar manner, whole numbers can conveniently be represented by points on a straight line called the *whole number line*, as follows:

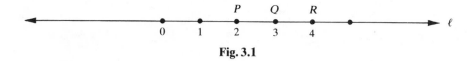

Fig. 3.1

Choose an arbitrary point on a straight line ℓ and label that point as 0, called the *origin*. Now select another point on ℓ, on the right side of 0, and mark that point as 1. The distance between these two points, called the *unit length* or the *scale*, helps us to determine the points P, Q, R, etc., on ℓ to the right of 1. Label those points by 2, 3, 4, etc. The numbers 2 and 3 are called the *coordinates* of the points P and Q, respectively. Using the less-than relation, we have assigned a position to each whole number. If $x < y$, then the point representing the whole number x lies on the left of the point representing the whole number y. Thus, we have established a 1-1 correspondence between the set W of whole numbers and a set A of points on the line. Observe that A is only a proper subset of the set of points on the line. For example, there is no whole number to represent any point between the points P and Q on ℓ or any point on the left of the point 0. What numbers represent those points? This will be answered in the next few chapters. We used to say that the whole number line is full of "holes" or "gaps."

Let's now represent the operations of addition and multiplication on the whole number line. Assume that we want to evaluate $2+3$ from the number line. First we draw an arrow 2 units long starting at the origin and pointing to its right. The arrow stops at the point labeled 2. Now we draw an arrow 3 units long starting at the point with coordinate 2, to the right. It terminates at the point with coordinate 5, as shown in Fig. 3.2. Thus $2+3=5$.

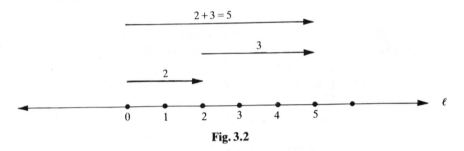

Fig. 3.2

The commutative property of addition is illustrated on the whole number line in Fig. 3.3.

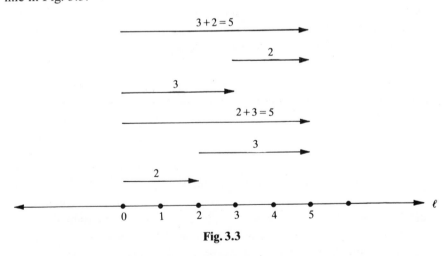

Fig. 3.3

Suppose we wish to evaluate $2 \cdot 3$ using the whole number line. Recall that multiplication is repeated addition. Draw an arrow 3 units long originating at the origin, pointing to the right and terminating at the point with coordinate 3.

Now draw another arrow of the same length starting at this point. It terminates at the point with coordinate 6, as shown in Fig. 3.4. Thus $2 \cdot 3 = 6$.

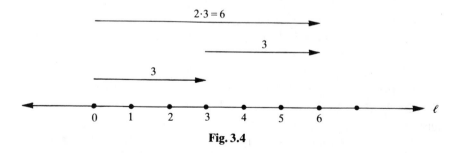

Fig. 3.4

The commutative property of multiplication of whole numbers is illustrated geometrically in Fig. 3.5.

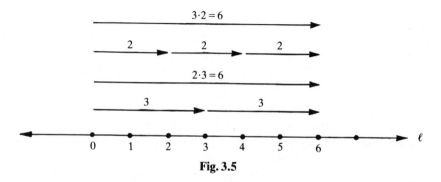

Fig. 3.5

If we take any two points on the whole number line, either they are the same or one point lies on the left of the other. This illustrates the law of trichotomy.

Since $2 < 3$ and $3 < 4$, the transitive property of the order relation tells us that $2 < 4$. What does this mean geometrically? On the number line, if the point labeled 2 lies to the left of the point labeled 3, and the point labeled 3 lies to the left of the point labeled 4, the point marked 2 lies to the left of the point marked 4, illustrating the transitive property, as shown in Fig. 3.6.

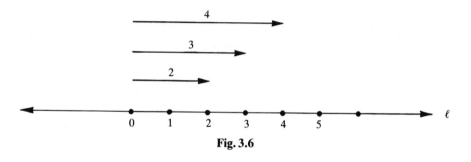

Fig. 3.6

Similarly, Theorems 3.20, and 3.21 can be illustrated geometrically, using the whole number line.

3.8 EQUATIONS AND SOLUTIONS

In Section 2.1, we discussed what is meant by a statement. A statement may be true or false, but not both. A statement that two expressions are equal is called an *equation*. For example, $2+3=5$, $3 \cdot 8 = 2+11 \cdot 2$, $2x+3=11$, $3y+2=13$, $3x = (2+1)x$, $3z = z+2z$, $(x+y)^2 = x^2 + 2xy + y^2$ are all equations.

Any symbol in an equation whose value is not known is called a *variable* or *unknown*. For example, x is a variable in the equation $2x+3=7$. In the equation $ax+b=0$, a is called the *coefficient* of x, and b the *constant term*.

Notice that the equation $2x+3=7$ is true for only one value of x, namely, for $x=2$, whereas the equations $2x=x+x$ and $3x=2x+x$ are true for all permissible values of the variable x. The equation $(x+y)^2 = x^2 + 2xy + y^2$ is true for all values of x and y.

An equation that is true for all permissible values of the unknown(s) is called an *identity*, otherwise it is a *conditional equation*. For instance, $2x=x+x$ is an identity, whereas $2x=4$ is a conditional equation. (Why?)

Any value t of the unknown that makes the equation true is called a *solution* of the equation. If t is a solution, then t is said to *satisfy* the equation. For example, 2 is a solution of the equation $3x=6$. The set of all solutions of an equation is called the *solution set* of the equation. The solution sets of the equations $3x=6$ and $2x=x+x$ are $\{2\}$ and W, respectively. By *solving* an equation, we mean finding the solutions of the equation. The *graph* of a set S of numbers is the set of points on the number line whose coordinates are in S.

Example 3.37 Solve the equation $x+3=10$.

Solution: To solve the equation $x+3=10$, we must isolate the x and get rid of 3 from the left–hand side.

$$x+3=10 \qquad \textit{given}$$
$$x+3=7+3 \qquad \textit{definition of addition of whole numbers}$$
$$x=7 \qquad \textit{cancellation property for addition}$$

Now we can actually check if 7 is a solution by substituting 7 for x in the original equation. Then we get $7+3=10$, which is a true statement. Hence 7 is a solution. Notice that the equation has a unique solution. Thus 7 is the solution of the equation $x+3=10$, and the solution set is $\{7\}$.

Example 3.38 Find and graph the solution set of the equation $4x+3=5+3x$.

Solution: To solve the given equation, we must isolate x using the properties of addition of whole numbers.

$$4x+3=5+3x \qquad \textit{given}$$
$$4x+3=3x+5 \qquad \textit{commutative property of addition}$$

$$3x + x + 3 = 3x + 5 \qquad \textit{definition of addition of whole numbers}$$
$$x + 3 = 5 \qquad \textit{cancellation property for addition}$$
$$x + 3 = 2 + 3 \qquad \textit{definition of addition of whole numbers}$$
$$x = 2 \qquad \textit{cancellation property for addition}$$

It may be easily checked that 2 is the solution of the equation. Hence the solution set is {2}. Its graph is the point with coordinate 2 on the whole number line, marked by a heavy dot, as shown in Fig. 3.7.

Fig. 3.7

Example 3.39 Find the solution set of the equation $7x + 5 = 17 + 3x$.
Solution: $7x + 5 = 17 + 3x \qquad \textit{given}$

$$7x + 5 = 3x + 17 \qquad \textit{commutative property of addition}$$
$$7x + 5 = 3x + 12 + 5 \qquad \textit{definition of addition of whole numbers}$$
$$7x = 3x + 12 \qquad \textit{cancellation property of addition}$$
$$3x + 4x = 3x + 12 \qquad \textit{definition of addition of whole numbers}$$
$$4x = 12 \qquad \textit{cancellation property of addition}$$
$$4x = 4 \cdot 3 \qquad \textit{definition of multiplication of whole numbers}$$
$$x = 3 \qquad \textit{cancellation property of multiplication}$$

Thus {3} is the solution set of the given equation.
Example 3.40 The sum of the ages of Ron and his brother Don is 29. Find their age if Don is 3 years older than Ron.
Solution: Let x be Ron's age. Then Don is $x + 3$ years old. Notice that their ages will be known provided the value of x can be found. Therefore, our aim is to get an equation in x using the given data and to solve the equation. Since the sum of their ages is 29, we have

$$x + (x + 3) = 29$$
$$(x + x) + 3 = 29$$
$$2x + 3 = 29$$
$$2x + 3 = 26 + 3$$
$$2x = 26$$
$$2x = 2 \cdot 13$$
$$x = 13$$

Therefore, Ron is 13 years old and Don is $13+3=16$ years old. We ask the reader to justify each of the above steps.

Example 3.41   Find and graph the solution set of the inequality $x^2 \leq 25$, where x is a whole number.

Solution: By Theorem 3.21, $x^2 \leq 25$ if and only if $x \leq 5$. Thus, the solution set is $\{x \in W \mid x \leq 5\} = \{0,1,2,3,4,5\}$. The graph of the solution set is exhibited in Fig. 3.8.

Fig. 3.8

Exercise 3.8

1. Find the solution set of each of the following equations. Graph each solution set on the whole number line.
   a) $2x+7=13$
   b) $5x=15+2x$
   c) $2+(3 \cdot 4)x = 2x + 3 \cdot 4$
   d) $(3+2 \cdot 3)x = (2 \cdot 3)x$
   e) $3x+5x=8x$
   f) $(2 \cdot 5)x + 3 = 8 \cdot 3 + 7x$

2. Which of the following equations are solvable in W? in N?
   a) $3x+5=20$
   b) $8x+13=13$
   c) $3x+14=2(x+7)$

3. Anita bought 13 pounds of candy. What is the total price she had to pay if one pound of candy costs 2 dollars?

4. If the sum of 3 consecutive natural numbers is 366, find them.

5. The sum of two numbers is 69. Find them if one is larger than the other by 7.

6. The perimeter of a rectangle is 46 feet. Find the dimensions of the rectangle if its length is 3 feet more than its width.

7. If Ron and Don are respectively 12 and 46 years of age, in how many years will Don's age be three times that of Ron?

8. Ron bought 3 pounds of one kind of candy and 7 pounds of another. The first kind of candy costs 50 cents more per pound than the second kind. What is the price of each kind of candy he bought, if he had to pay 15 dollars?

9. The area of the floor of a room is 3645 square inches. If the area of a tile is 81 square inches, how many tiles are needed to cover the floor of the room?

10. The length of a rectangle is 3 feet more than its width. Find the dimensions of the rectangle if its area is 18 times its width.

11. Harry is allowed 12 cents a mile as mileage allowance. How many miles did he travel if he was given 5 dollars and 16 cents?

12. Frank has employed a certain number of people in his company. Each is given 20 dollars a day. If he gives out 460 dollars a day as wages, how many employees does he have?

13. Find and graph the solution set of the following equations in the set of whole numbers.

a) $x^2 \leq 9$

b) $5x^2 < 125$

c) $3x^2 + 7 \leq 34$

d) $(x+7)x < 7x + 49$

e) $x(x+4) \leq 4(x+16)$

f) $x^2 \leq 0$

3.9 SUBTRACTION OF WHOLE NUMBERS

So far we have defined and discussed the properties of two binary operations, addition and multiplication, on the set of whole numbers. In this section, we will focus our attention on a third operation, called *subtraction.*

The concept of subtraction is formed very early in life, as illustrated by the following example. Assume that Ron has seven toy trucks. His brother Don asks him if he can have three of them. Ron very reluctantly says yes and wants to know how many he will have left then. He puts the toy trucks together and removes three of them for Don. He now finds that he has only "$7 - 3 = 4$" toy trucks.

Let's put Ron's problem in mathematical terms. Label his toy trucks by $a,b,c,d,e,f,$ and g. Let $A = \{a,b,c,d,e,f,g\}$ be the set of his toy trucks; then $c(A) = 7$. Assume Ron gives the trucks $e,f,$ and g to Don, that is, the set $B = \{e,f,g\}$. Now "remove" the set B from A to get the new set $A - B = \{a,b,c,d\}$. Then $A - B$ is the set of toy trucks Ron possesses and $c(A - B) = 4$.

Example 3.42   Consider the sets $A = \{1,2,3,4,5\}$ *and* $B = \{4,5\}$.

Then $A - B = \{1,2,3\}$, $(A - B) \cap B = \varnothing$, and $(A - B) \cup B = A$.

Therefore, $$c(A - B) + c(B) = c(A)$$

That is, $$c(A - B) + 2 = 5$$

Thus, to evaluate $c(A - B)$, we need only solve the equation $x + 2 = 5$. Using the cancellation property of addition, we find that $c(A - B) = x = 3$. The solution of the equation $x + 2 = 5$ is the number to be added to 2 to get 5.

Let's make use of these examples to define the operation of subtraction on the set of whole numbers, as follows:

Definition 3.6   Let a and b be whole numbers. The *difference* of a and b, denoted by $a - b$, is defined as the solution of the equation $a = b + x$. The operation "$-$" is called *subtraction.* We read $a - b$ as "*a minus b*" or "*b subtracted from a.*" In $a - b$, the number a is called the *minuend* and b the *subtrahend.*

Thus, in order to evaluate $a - b$, we have to solve the equation $a = b + x$ and hence $a - b$ exists in W if and only if the equation $a = b + x$ is solvable in the set of whole numbers.

Example 3.43 Evaluate $5 - 3$.

Solution: $5 - 3 = x$ exists in W if and only if $5 = 3 + x$ is solvable in W. Using the cancellation property of addition, we have $x = 2$. Therefore, $5 - 3 = 2$.

Example 3.44 Find the difference $7 - 4$.

Solution: $7 - 4 = x$ exists in W if and only if $7 = 4 + x$ can be solved in W. But $7 = 4 + x$ is true if and only if $x = 3$. Therefore, $7 - 4 = 3$.

Example 3.45 Evaluate $3 - 3$.

Solution: Solving the equation $3 = 3 + x$, we find that $x = 0$ and hence $3 - 3 = 0$.

Example 3.46 Evaluate $3 - 5$.

Solution: $3 - 5 = x$ exists in W if and only if $3 = 5 + x$ is solvable in W. But $3 = 5 + x$ if and only if $0 = 2 + x$, and there is no whole number x that when added to 2 gives 0. That is, $2 + x = 0$ is not solvable in W and hence $3 - 5$ does not exist in the set of whole numbers. This occurs because $3 < 5$.

What do we conclude from this example? If $a < b$, then $a - b$ is not defined. Thus, the equation $a = b + x$ is solvable in the set of whole numbers if and only if $a \geq b$.

Example 3.46 shows that W is not closed under subtraction. Consequently, not every equation of the form $a = b + x$ with a,b in W is solvable in the set of whole numbers. This is a deficiency of the set of whole numbers.

Example 3.47 Verify that $5(13 - 7) = 5 \cdot 13 - 5 \cdot 7$.

Solution: $5(13 - 7) = 5 \cdot 6 \quad = 30$

$$5 \cdot 13 - 5 \cdot 7 = 65 - 35 = 30$$

Therefore, $5(13 - 7) = 30 \quad = 5 \cdot 13 - 5 \cdot 7$

This example suggests the following result:

Theorem 3.22 (***Distributive Property***) *If a,b,c are whole numbers such that $b \geq c$, then $a(b - c) = ab - ac$.*

This theorem tells us that multiplication distributes over subtraction (when subtraction is possible).

Let's now consider the operation of subtraction on the whole number line. Assume we wish to evaluate $5 - 3$ using the number line. Draw an arrow 5 units long starting at the origin, 0, and stopping at the point labeled 5, on the right of 0. Now draw an arrow 3 units long to the left, starting at the point marked 5 and terminating at the point marked 2. Then $5 - 3 = 2$, as illustrated in Fig. 3.9. Observe that if addition involves moving to the right on the number line, subtraction involves moving to the left on the number line. That is, subtraction "undoes" what addition "does." Accordingly, subtraction is called the *inverse* of addition.

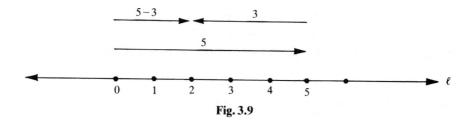

Fig. 3.9

Exercise 3.9

1. Solve each of the following equations in W, if possible.

a) $8 = 5 + x$ **b)** $4 = 2 + y$ **c)** $7 = 11 + z$

d) $13 = 8 + y$ **e)** $3 = 8 + x$ **f)** $5 = 13 + x$

2. Verify each of the following.

a) $2(5 - 2) = 2 \cdot 5 - 2 \cdot 2$ **b)** $(3 + 5) - 2 = 3 + (5 - 2)$

c) $(8 - 3) + 5 = (8 + 5) - 3$ **d)** $(8 - 3) + (11 - 8) = (8 + 11) - (3 + 8)$

3. Mark *true* or *false* (a,b,c are any whole numbers).

a) W is closed under subtraction.

b) Subtraction is commutative on the set of whole numbers.

c) Subtraction is associative on the set of whole numbers.

d) There are whole numbers a,b, and c such that $(a - b) - c = a - (b - c)$.

e) Every equation of the form $a = b + x$ is solvable in the set of whole numbers.

f) Every equation of the form $a = a + x$ is solvable in the set of natural numbers.

g) Every equation of the form $a = a + x$ is solvable in the set of whole numbers.

h) $a - b$ is a whole number if and only if $a \geq b$.

i) $a - a$ is a whole number.

j) $a - 0 = a$

4. Give a counterexample to each of the false statements in problem 3.

5. Are there whole numbers a,b,c such that $a - (b \cdot c) = (a - b)(a - c)$?

6. Are there whole numbers a,b,c such that $(a - b) + c = a - (b + c)$?

7. John got married in 1956. How old was he then if he was born in 1931?

8. If the temperature at 8 a.m. of a certain day was 15 degrees and at 11 a.m. was 36 degrees, what was the rise in temperature during that time?

9. If the temperature rose 25 degrees in 6 hours starting at 7 a.m., and if at 1 p.m. it is 67 degrees, what was the temperature at 7 a.m.?

10. If Linda withdrew 125 dollars from the bank where she had a balance of 345 dollars, what is her present balance at the bank?

11. If 5 less than twice a number is 7, find the number.

12. Peter and Paul are 19 and 23 years old, respectively. How old were they when Paul was twice as old as Peter?

3.10 SUMMARY AND COMMENTS

We introduced the concept of cardinality of a set and observed that there are two types of cardinal numbers: cardinal numbers of finite sets and of infinite sets. We called the set of cardinal numbers of finite sets the set of whole numbers, denoted by W, and its proper subset $N = W - \{0\}$ is the set of natural numbers. Consequently, the cardinal number of the set of natural numbers, which is an infinite set, is not a whole number. It is denoted by $\aleph_0$ (aleph naught); $\aleph$ is the first letter of the Hebrew alphabet.

The equality relation, we found, is reflexive, symmetric, and transitive.

We defined three operations on the set of whole numbers: addition, multiplication, and subtraction. We have the following properties of addition and multiplication:

	addition	*multiplication*
closure property:	$a + b \in W$	$a \cdot b \in W$
commutative property:	$a + b = b + a$	$a \cdot b = b \cdot a$
associative property:	$a + (b + c) = (a + b) + c$	$a(bc) = (ab)c$
unique identity element:	$a + 0 = a = 0 + a$	$a \cdot 1 = a = 1 \cdot a$

where a, b, and c are any whole numbers. Under subtraction, W is neither closed, commutative, associative, nor has an identity element. Multiplication distributes over both addition and subtraction, but not vice versa. We now have the following additional properties:

a) If $a = b$ and $c = d$, then $a + c = b + d$. (*addition property*)

b) If $a + b = a + c$, then $b = c$. (*cancellation property*)

c) If $a = b$ and $c = d$, then $a \cdot c = b \cdot d$. (*multiplication property*)

d) If $a \cdot b = a \cdot c$ and $a \neq 0$, then $b = c$. (*cancellation property*)

e) If $a \cdot b = 0$, then $a = 0$ or $b = 0$. (*product law*)

The whole number a is less than the whole number b, denoted by $a < b$, if we can find a suitable natural number x such that $a + x = b$. The relation $a \leq b$ implies $a < b$ or $a = b$.

Law of trichotomy: for any two whole numbers a and b, exactly one of the following holds: either $a < b$, $a = b$, or $b < a$.

The order relation satisfies the following properties:

1. If $a < b$ and $b < c$, then $a < c$. (*transitive property*)

2. $a < b$ if and only if $a + c < b + c$ for every whole number c.

3. $a < b$ if and only if $ac < bc$ for every natural number c.

4. $a < b$ if and only if $a^2 < b^2$.

Since $N \subset W$, we have the following line diagram:

Fig. 3.10

Is the "tree" in Fig. 3.10 large enough for all our purposes? Well, in Section 3.9, we observed that not every equation of the form $a = b + x$ is solvable in the set of whole numbers. So we conclude that our "tree" does not have enough "branches" and has not "grown" enough.

SUGGESTED READING

Brainerd, C. J., "The Origins of Number Concepts," *Scientific American*, vol. 228 (March 1973), pp. 101–109.

Brumfiel, C., and I. Vance, "On Whole Number Computation," *The Arithmetic Teacher*, vol. 16 (April 1969), pp. 253–257.

Engen, H. V., "Counting Is Not Basic!" *School Science and Mathematics*, vol. 68 (Nov. 1968), pp. 720-722.

Haines, M., "Concepts to Enhance the Study of Multiplication," *The Arithmetic Teacher*, vol. 10 (Feb. 1963), pp. 95–97.

McFarland, D., and E. M. Lewis, *Introduction to Modern Mathematics* (2nd ed.), D. C. Heath and Co., Lexington, Mass. (1973), pp. 135–158.

Ohmer, M. M., C. V. Aucoin, and M. J. Cortez, *Elementary Contemporary Mathematics* (2nd ed.), Xerox College Publishing, Lexington, Mass. (1972), pp. 69–106.

Pribnow, J. R., "Why Johnny Can't 'Read' Word Problems," *School Science and Mathematics*, vol. 69 (March 1969), pp. 591–598.

Wheeler, R. E., *Modern Mathematics: An Elementary Approach* (3rd ed.), Brooks/Cole Publishing Co., Monterey, Calif. (1973), pp. 79–120.

*"Integral numbers are the foun-
tainhead of all mathematics."*
H. MINKOWSKI

4/The System of Integers

After studying this chapter, you should be able to :
- *find the sum of two integers*
- *identify the properties of addition*
- *find the difference of two integers*
- *evaluate the product of two integers*
- *identify the properties of multiplication*
- *identify and use the distributive property*
- *check if an integer is less than another integer*
- *identify the properties of the order relation*
- *divide an integer* a *by an integer* b, *whenever possible*

4.0 INTRODUCTION

In Chapter 3, we observed a serious deficiency of the set W of whole numbers: that it is not closed under subtraction. That is, not every equation of the form $a = b + x$ with a,b in W is solvable in W. If $a < b$, then $a - b$ does not exist in the set of whole numbers. This shows that we should enlarge or extend the set of whole numbers to a "larger" set, where we can solve every equation of the form $a = b + x$, to meet the deficiency of the set of whole numbers. This is done by adjoining extra elements called "negative integers" to the set of whole numbers. The resulting set is the set of integers.

Is there any need for negative integers in our day-to-day life? Several times on the radio or television we hear the weatherman making statements like: "The temperature now is 5 degrees below zero outside our studios," "The temperature rose from 7 degrees below zero to zero degrees by 9 o'clock this morning," "The temperature tonight will go down to 15 degrees below zero," etc. Negative integers are very convenient to express these physical situations. The problem of finding the present temperature that is 5 degrees below zero amounts to solving the equation $x + 5 = 0$.

It is no wonder that it is difficult to introduce and teach the concept of a negative integer in school. Even the famous mathematician, Diophantus of Alexandria, called them "absurd."

4.1 INTEGERS

Let a and b be any whole numbers. We observed that $a - b$ is a whole number only if $a \geq b$; otherwise it doesn't exist in the set of whole numbers. Intuitively, $x = a - b$ is a solution of the equation $a = b + x$. This suggests to us how we should define an integer.

Definition 4.1 The difference $a - b$ of two whole numbers a and b is called an *integer*.

According to our definition, $3 - 5$, $5 - 3$, $1 - 4$, $0 - 8$ are all integers.

Consider the equation $a + x = 0$ with a in W. This equation can be solved in W if and only if $a = 0$. In any case, $x = 0 - a$ is a solution of the equation for every whole number a. Let's denote $0 - a$ by $-a$ for convenience and call it the *negative* of a or the *additive inverse* of a. Thus, $-a$ is the number that when added to a gives the number zero, that is, $(-a) + a = 0 = a + (-a)$. For example, -1 is the additive inverse of 1 and -5 is the additive inverse of 5.

Let a and b be whole numbers. Notice that $a - b$ and $b - a$ are not the same unless $a = b$. Consequently, we can very well treat the integer $a - b$ as an ordered pair (a,b). Thus, $(a,b) = a - b$ and $(b,a) = b - a$. Even though the properties of integers can be discussed conveniently using the ordered pair notation, we will be using only the familiar notation: $0, \pm 1, \pm 2, \pm 3, \ldots$. The set $\{0, \pm 1, \pm 2, \pm 3, \ldots\}$ is the *set of integers*, denoted by I. The set of numbers $\{-1, -2, -3, \ldots\}$ is called the set of *negative integers*. The set of numbers $\{1, 2, 3, \ldots\}$ is also called the set of *positive integers*. Some times the positive integer n is denoted as $+n$ for emphasis. We remark that the integer 0 is neither positive nor negative.

Observe that there exists a 1-1 correspondence between the set of positive integers and the set of negative integers, as shown below:

$$
\begin{array}{ccccc}
1 & 2 & 3 & \ldots & n & \ldots \\
\updownarrow & \updownarrow & \updownarrow & & \updownarrow \\
-1 & -2 & -3 & \ldots & -n & \ldots
\end{array}
$$

Recall that when we constructed the whole number line in Section 3.7, we did not use any point on the left of the origin. The number line we discussed there is shown in Fig. 4.1. Now we assign a special position to each negative

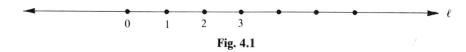

Fig. 4.1

integer on the number line. Let's make use of the unit length that we used to determine the points labeled 1, 2, 3, ... on the right of the origin 0, and mark similar points on its left. Let's label them −1, −2, −3, ..., as illustrated in Fig. 4.2. The coordinates of the points P and Q are −3 and 3, respectively.

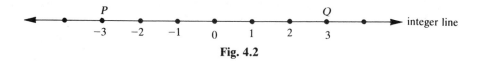

Fig. 4.2

Thus, we have exhibited a 1-1 correspondence between the set of integers and a set of points on the line. Such a line is called the *integer line.*

The integers 4 and −3 are represented by the tips of the arrows drawn to the right and left of the origin, respectively, as shown in Fig. 4.3.

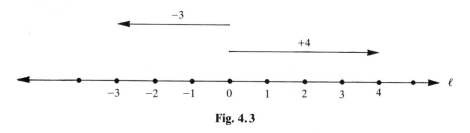

Fig. 4.3

Exercise 4.1

1. What is the additive inverse of each of the following integers?

   a) 5   b) 7   c) 0
   d) $3+2$ e) $2+(3+5)$ f) $5-3$
   g) $7-4$ h) $3+(-3)$ i) $5-(3-2)$
   j) $-(3+5)+(5+3)$ k) $(-5)+5$ l) -5

2. Solve each of the following equations.

   a) $x+3=0$ b) $x+0=0$ c) $(-5)+x=0$
   d) $y+(-8)=0$ e) $z+0=-8$ f) $x=(-4)+4$

3. Mark *true* or *false.*

   a) Every integer has an additive inverse.
   b) Every whole number has an additive inverse in W.
   c) The integer 0 has an additive inverse.
   d) −0 is defined.
   e) Every equation of the form $a+x=0$ with a in I is solvable in I.

4. Give a counterexample to each of the false statements in problem 3.

4.2 ADDITION OF INTEGERS

Recall that we denoted the solution of the equation $a + x = 0$ as $x = -a$. Thus, $a + (-a) = 0 = (-a) + a$. This tells us that the sum of a whole number a and its additive inverse $-a$ is zero. Now we define the operation of addition on the set of integers by extending the definition of addition on the set of whole numbers, given in Section 3.3, so that all the properties of the set of whole numbers under addition are preserved.

Let a and b be two integers. If a and b are whole numbers, we discussed in Section 3.3 how to find their sum, $a + b$, which is a whole number and hence an integer. If $b = -a$, then $a + (-a) = 0$.

Example 4.1 How should $(-2) + 5$ be defined if addition is to be associative?

Solution: $(-2) + 5 = (-2) + (2 + 3)$

$$= [(-2) + 2] + 3 \qquad \textit{if addition is to be associative}$$

$$= 0 + 3$$

$$= 3$$

Thus, we should define $(-2) + 5 = 3 = 5 - 2$, where we observe that $2 < 5$.

Example 4.2 How should we define $5 + (-2)$ if the commutative property is to be preserved?

Solution: $5 + (-2) = (-2) + 5 \qquad \textit{if addition is to be commutative}$

$$= 3 \qquad \textit{using example } 4.1$$

This suggests that we define $5 + (-2) = 3 = 5 - 2$, where we notice that $2 < 5$.

Example 4.3 How should $(-2) + (-5)$ be defined if addition is to be associative?

Solution: $[(-2) + (-5)] + (2 + 5) = [(-2) + (-5)] + (5 + 2)$

$$= (-2) + \{[(-5) + 5] + 2\} \qquad \textit{if addition is to}$$
$$\textit{be associative}$$

$$= (-2) + (0 + 2)$$

$$= (-2) + 2$$

$$= 0$$

Therefore, $(-2) + (-5)$ is the additive inverse of $2 + 5 = 7$, and hence we should define $(-2) + (-5) = -7 = -(2 + 5)$.

Recall that 0 is the additive identity for the set of whole numbers. So we should have $(-3) + 0 = -3 = 0 + (-3)$, if 0 is to be an additive identity for the set of integers.

Example 4.4 How should we define $(-5) + 2$ if addition is to be associative?

Solution: $(-5) + 2 = [(-3) + (-2)] + 2 \qquad \textit{using example } 4.3$

$$= (-3) + [(-2) + 2] \qquad \textit{if addition is to be associative}$$

$$= (-3) + 0$$

$$= -3 \qquad \textit{if 0 is to be an additive identity}$$

Thus, we should define $(-5)+2=-3=-(5-2)$, where we observe that $5>2$.

Examples 4.1–4.4 now suggest how the operation of addition should be defined on the set of integers.

Definition 4.2 Let a and b be any whole numbers. The *sum* of a and b, denoted by $a+b$, is defined as follows:

1. $a+b$ is defined as in Definition 3.3.
2. $a+(-b)=a-b=(-b)+a$ if $a \geq b$.
3. $a+(-b)=-(b-a)=(-b)+a$ if $a<b$.
4. $(-a)+(-b)=-(a+b)=(-b)+(-a)$.

Example 4.5
$$2+3=5=3+2$$
$$(-2)+0=-2=0+(-2)$$
$$3+(-2)=3-2=1=(-2)+3$$
$$2+(-3)=-(3-2)=-1=(-3)+2$$
$$(-2)+(-3)=-(2+3)=-5=-(3+2)=(-3)+(-2)$$

Let's now illustrate examples 4.1–4.4 on the integer line. To find $(-2)+5$, we draw an arrow 2 units long to the left from the origin and then an arrow 5 units long in the opposite direction (to the right) from the point marked -2. It terminates at the point labeled 3. Therefore, we have $(-2)+5=3$. Now, to find $5+(-2)$, draw an arrow 5 units long to the right from the origin to terminate at the point marked 5, then draw an arrow 2 units long from that point in the opposite direction to terminate at the point labeled 3. Therefore, we have $5+(-2)=3$. Thus, $(-2)+5=3=5+(-2)$

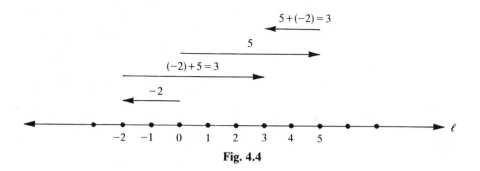

Fig. 4.4

To find $(-5)+(-2)$ from the integer line, draw an arrow 5 units long from the origin to the left. It terminates at the point marked -5. Now draw an arrow of 2 units long in the same direction from that point, to terminate at the point marked -7. Thus, $(-5)+(-2)=-7$. Also observe from Fig. 4.5 that $(-2)+(-5)=-7$.

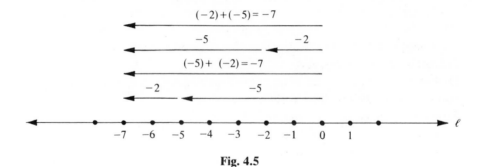

Fig. 4.5

Example 4.4 is illustrated in Fig. 4.6.

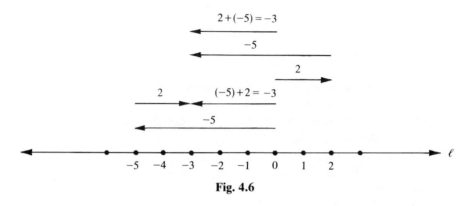

Fig. 4.6

Let's now discuss the fundamental properties of addition on the set of integers. Observe from Definition 4.2 that the sum of any two integers, positive, zero, or negative, is a unique integer. That is, addition is a binary operation on the set of integers.

Theorem 4.1   (***Closure Property***)   *If a and b are integers, then a + b is also an integer.*

It follows from Definition 4.2 that the sum of two positive integers is a positive integer and the sum of two negative integers is a negative integer.

Theorem 4.2   (***Commutative Property***)   *For any two integers a and b,*

$$a + b = b + a$$

This property of addition states that the order in which we add two integers does not affect the sum.

Theorem 4.3   (***Associative Property***)   *For any integers a, b, and c,*

$$a + (b + c) = (a + b) + c$$

The associative property of addition states that when we add three or more integers, we can group them in any order we like. The grouping does not affect the result. Consequently, expressions like $a+b+c$, $a+b+c+d$, etc., have meaning and are not confusing or undefined. The commutative and associative properties together tell us that when finding the sum of three or more integers, we can rearrange and group them in any order.

Example 4.6   Verify that $2+[(-3)+(-5)]=[2+(-3)]+(-5)$.

Solution:

$$
\begin{aligned}
2+[(-3)+(-5)] &= 2+[-(3+5)] && \textit{definition of addition of integers}\\
&= 2+(-8) && \textit{definition of addition of whole numbers}\\
&= -(8-2) && \textit{definition of addition of integers}\\
&= -6 && \textit{definition of subtraction of whole numbers}\\
[2+(-3)]+(-5) &= [-(3-2)]+(-5) && \textit{definition of addition of integers}\\
&= (-1)+(-5) && \textit{definition of subtraction of whole numbers}\\
&= -(1+5) && \textit{definition of addition of integers}\\
&= -6 && \textit{definition of addition of whole numbers}
\end{aligned}
$$

Thus, $2+[(-3)+(-5)]=[2+(-3)]+(-5)$

Example 4.7   Verify that $a+b+(-c)=(-c)+b+a$.

Solution:

$$
\begin{aligned}
a+b+(-c) &= a+[b+(-c)] && \textit{associative property of addition}\\
&= a+[(-c)+b] && \textit{commutative property of addition}\\
&= [a+(-c)]+b && \textit{associative property of addition}\\
&= [(-c)+a]+b && \textit{commutative property of addition}\\
&= (-c)+(a+b) && \textit{associative property of addition}\\
&= (-c)+(b+a) && \textit{commutative property of addition}\\
&= (-c)+b+a && \textit{associative property of addition}
\end{aligned}
$$

The integer 0 has the unique property that the addition of 0 to any integer a has no effect on it and the sum remains the same as a, as in the case of whole numbers. Accordingly, we have the following result:

Theorem 4.4   (***Existence of a Unique Additive Identity***)   *For any integer a,*

$$a+0=a=0+a$$

The integer 0 is an additive identity for the set of integers. It is not hard to show that it is unique, so we can talk about *the* additive identity of the set of integers.

We observed in the previous section that for every integer a, there exists an integer $-a$ such that $a+(-a)=0=(-a)+a$. The quantity $-a$ is an additive inverse of a and it is unique; that is, every integer has a unique additive inverse.

Theorem 4.5   (*Existence of a Unique Additive Inverse*)   *For every integer a, there exists an integer $-a$ such that*

$$a+(-a)=0=(-a)+a$$

Since every integer a has a unique additive inverse, from now on we can talk about *the* additive inverse of a. Notice that a and $-a$ are additive inverses of each other; that is, $-(-a)=a$. (Why?) For instance, -5 is the additive inverse of 5, and the additive inverse $-(-5)$ of -5 is 5. This is demonstrated in Fig. 4.7. Also, $-(a+b)=(-a)+(-b)$ for any two integers a and b. This simply means that the additive inverse of the sum of two integers is the same as the sum of their additive inverses. For example, $-(2+3)=(-2)+(-3)$.

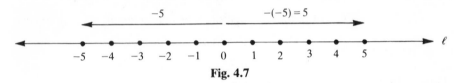

Fig. 4.7

We observed in the previous chapter that if a,b,c,d are whole numbers such that $a=b$ and $c=d$, then $a+c=b+d$. This property can now be extended to the set of integers.

Theorem 4.6   (*Addition Property*)   *If a,b,c,d are integers such that $a=b$ and $c=d$, then $a+c=b+d$.*

Example 4.8   Solve the equation $x+15=5$.

Solution: To solve the equation $x+15=5$, we must remove 15 from the left-hand side of the equation.

$x+15=5$	*given*
$(x+15)+(-15)=5+(-15)$	*addition property*
$x+[15+(-15)]=5+(-15)$	*associative property of addition*
$x+0=-10$	*definition of addition*
$x=-10$	*0 is the additive identity*

Let's now check if -10 is actually the solution of $x+15=5$.

$$x+15=(-10)+15$$
$$=(-10)+(10+5)$$
$$=[(-10)+10]+5$$
$$=0+5$$
$$=5$$

Consequently, -10 is the solution of the given equation.

The addition property states that if equals are added to equals, the sums remain equal. Conversely, we have the following result:

Theorem 4.7   (***Cancellation Property***)   *If a,b,c are integers such that* $a + b = a + c$, *then* $b = c$.

The cancellation property tells us that the same integer can be cancelled or dropped from both sides of an equality. This property is very useful in solving problems, as shown in the following example:

Example 4.9   If $3 + x = 7$, then $x = 4$ since $7 = 3 + 4$. If $(-8) + x = -17$, then since $-17 = (-8) + (-9)$, we have $x = -9$.

Exercise 4.2

1. Evaluate the following sums:
   a) $5 + (-3)$
   b) $(-2) + 8$
   c) $(-7) + 15$
   d) $(-8) + (-7)$
   e) $(-3) + 0$
   f) $(-0) + 0$
   g) $(a + 3) + (-4)$
   h) $[(-2) + 3] + [5 + (-3)]$

2. Verify the following:
   a) $2 + 3 + 5 = 5 + 3 + 2$
   b) $(-2) + (-3) + 4 = -1$
   c) $3 + (-4) + 5 = 5 + 3 + (-4)$
   d) $[2 + (-3)] + (2 + 3) = (2 + 2) + [(-3) + 3]$

3. Solve the following equations:
   a) $x + 7 = 15$
   b) $x + 8 = 3$
   c) $x + (-5) = 3$
   d) $x + (-12) = -15$
   e) $x + (-7) = -3$
   f) $x + (-5) = 10$

4. If a and b are integers such that $a + b = a$, what is your conclusion?

5. Mark *true* or *false* (a,b,c are any integers).
   a) The set of integers is closed under addition. T
   b) Addition on the set of integers is commutative. T
   c) The set of negative integers is closed under addition. T
   d) Every nonnegative integer is a positive integer. F `0`
   e) $(-2) + 5 = 5 - 2$ T
   f) $-(-a) = a$ T
   g) $-a$ is a negative integer. F $-(-2)$ +4/
   h) I contains a unique additive identity. T $a + 0 = a = 0 + a$
   i) Every integer has a unique additive inverse. T $3 : -3$ $-/ : /$
   j) If $a = b$, then $a + c = b + c$.
   k) $-(a + b) = (-a) + (-b)$

6. Give a counterexample to each of the false statements in problem 5.

7. State the property of addition used in each of the following statements, where each symbol represents an integer.
   a) $(-a) + 0 = -a$ additive identity
   b) $x + y$ is an integer.   closure property + I
   c) $x + y = y + x$ commut + I

d) If $x = y$, then $3 + x = 3 + y$.
e) $(-a) + 0 = 0 + (-a)$ commut. + I
f) $(a+b)+(c+d) = [(a+b)+c]+d$ assoc. + J
g) If $(-3)+x = (-3)+y$, then $x = y$.
h) $a+[(-b)+(c+d)] = [a+(-b)]+(c+d)$
i) $[x+(-y)]+z = x+[(-y)+z]$
j) $[a+(-b)]+[c+(-d)] = [c+(-d)]+[a+(-b)]$

8. The temperature at 7 p.m. was 3 degrees below zero. What is the temperature at 8 p.m. if it went down by 4 degrees in that hour? 7° below

9. The temperature at 6 p.m. was 5 degrees below zero. What is the temperature at 9 p.m. if it went down by 3 degrees each hour? 14° below

10. One day Ron noticed his account at the bank overdrawn by 32 dollars. If he deposited 217 dollars the next day in his account, how much money did he then have in his account? $185.00

11. Suppose Frank traveled 85 miles towards east from home and then 35 miles back. How far will he be then from home? 80 miles

12. Suppose Charlie borrowed 725 dollars from his friend. If he returned 175 dollars a month later and 350 dollars three months later, how much money does Charlie still owe the friend?

13. Simplify the following expressions.
   a) $[a+(-b)]+b$ **b)** $[a+(-b)]+(-a)$
   c) $(-b)+(a+b)$ **d)** $(-b)+[(-a)+b]$
   e) $(a+b)+[(-c)+(-b)]$ **f)** $(a+b)+[(-a)+(-b)]$

14. Prove each of the following.
   a) $a+b+c = c+a+b$
   b) $a+[(-b)+(-c)] = -(b+c)+a$
   c) $(a+b)+[(-a)+(-b)] = 0$
   d) $(a+b)+[c+(-d)] = [a+(-d)]+(b+c)$

15. Define an operation $*$ on I by $a * b = a+b+1$.
   a) Is the set of integers closed under $*$?
   b) Is $*$ commutative?
   c) Is $*$ associative?
   d) Does I have an identity with respect to $*$?
   e) Does every integer have an inverse with respect to $*$?

★16. Prove that the set of integers has a unique additive identity. (*Hint:* Assume there are two additive identity elements 0 and $0'$.)

★17. Using the ordered pair notation for integers, define the operation of addition on I as $(a,b)+(c,d) = (a+c, b+d)$. Also, $(a,b) = (c,d)$ if and only if $a+d = b+c$.
   a) Evaluate $(2,2)+(8,5)$.
   b) Evaluate $(5,3)+(3,5)$.

   c) What is the inverse of (a,b)?

   d) Find an additive identity for I.

   e) For what value of x is $(x,3)+(12,5)=(9,3)$?

4.3 SUBTRACTION OF INTEGERS

The operation of subtraction of whole numbers was introduced in Section 3.9. The set of whole numbers, we observed, is not closed under subtraction. This is why we had to extend the set of whole numbers to the set of integers. We now define subtraction of integers by extending the definition of subtraction of whole numbers so that the properties of whole numbers under subtraction are preserved.

Definition 4.3   Let a and b be integers. The solution of the equation $a=b+x$ is called the *difference* of a and b, denoted by $a-b$. The operation we defined is called *subtraction*. We read $a-b$ as "a minus b," or "b subtracted from a." Thus, $a-b=x$ if and only if $a=b+x$.

Example 4.10

$$5-3=\quad 2 \quad \text{since} \quad 5=3+2$$
$$5-(-3)=\quad 8 \quad \text{since} \quad 5=(-3)+8$$
$$(-5)-3=-8 \quad \text{since} \quad -5=3+(-8)$$
$$(-5)-(-3)=-2 \quad \text{since} \quad -5=(-3)+(-2)$$

Recall that $a-b$ is a whole number and hence an integer if $a \geq b$. We want to check if $a-b$ is an integer when $a<b$. The above example suggests that the difference of any two integers is also an integer. This is a consequence of the following theorem.

Theorem 4.8   *If a and b are any integers, then $a-b=a+(-b)$.*

Proof: Observe that by Definition 4.3, we need only show that $a=b+[a+(-b)]$.

$$b+[a+(-b)]=b+[(-b)+a] \qquad commutative\ property\ of\ addition$$
$$=[b+(-b)]+a \qquad associative\ property\ of\ addition$$
$$=0+a \qquad -b\ is\ the\ additive\ inverse\ of\ b$$
$$=a \qquad 0\ is\ the\ additive\ identity$$

Example 4.11

$$5-3=5+(-3)=2$$
$$(-3)-2=(-3)+(-2)=-5$$
$$3-(-2)=3+[-(-2)]=3+2=5$$
$$(-3)-(-2)=(-3)+[-(-2)]=(-3)+2=-1$$

The operation of subtraction of integers can be illustrated on the integer line, using Theorem 4.8. Suppose we wish to evaluate $5-3$, using the integer line.

Notice that $5-3=5+(-3)$. Draw an arrow 5 units long from the origin to the right. It stops at the point labeled 5. Now draw an arrow 3 units long in the opposite direction from that point, to terminate at the point marked 2. Thus, $5-3=5+(-3)=2$.

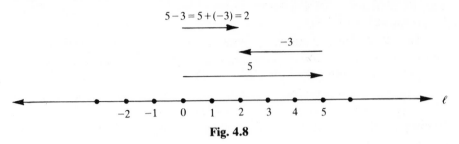

Fig. 4.8

That $3-(-2)=5$, $(-3)-2=-5$, and $(-3)-(-2)=-1$ are illustrated in Figs. 4.9, 4.10, and 4.11, respectively.

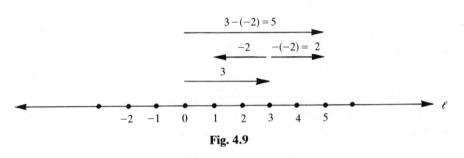

Fig. 4.9

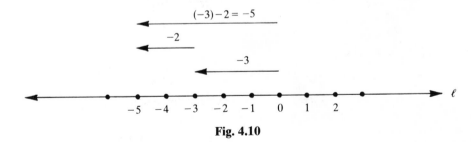

Fig. 4.10

Fig. 4.11

Theorem 4.8 is a very useful property. It expresses subtraction in terms of addition. It states that subtracting b from a is the same as adding the additive inverse $-b$ of b to a. Therefore, $a-(-b)=a+[-(-b)=a+b$. Also observe that $-b+a=(-b)+a=a+(-b)=a-b$. Since the set of integers is closed under addition, it follows from Theorem 4.8 that the set of integers is closed under subtraction. (Why?)

Example 4.12   John spent 28 dollars at the grocery store out of his weekly salary. What is his weekly salary if he had a balance of 135 dollars after shopping?

Solution: Assume John's weekly salary is x dollars. Since he spent 28 dollars for shopping, his balance is $x-28$ dollars. Therefore,

$$x-28=135$$
$$x=28+135 \quad \textit{by the definition of subtraction}$$
$$x=163$$

Thus, John's weekly salary is 163 dollars.

Observe that we have used the same symbol "$-$" with two different meanings. In $a-b$, it refers to the operation of subtraction. In $a+(-b)$, the symbol $-b$ means the additive inverse of b. The use of the same minus sign "$-$" with different meanings is justified by Theorem 4.8 and leaves no confusion.

Example 4.13 $5-3=2$ and $3-5=3+(-5)=-2$
Therefore, $\qquad\qquad\qquad 5-3 \neq 3-5$

$$2-(3-5)=2-(-2)=2+[-(-2)]=2+2=4$$

whereas $\qquad\qquad (2-3)-5=-1-5=-(1+5)=-6$

This example shows that subtraction is neither commutative nor associative.

Exercise 4.3

1. Evaluate the following:
   a) $(5-3)-2$
   b) $-5+(+3)$
   c) $[(-5)+3]-7$
   d) $5-(3-2)$
   e) $-3-(5+8)$
   f) $5-[(-3)+(-7)]$
   g) $-8-[(-3)+5]$
   h) $-7-[(-5)+(-3)]$
   i) $-8-[(-7)+(+11)]$

2. Solve the following equations:
   a) $x-3=8$
   b) $x+(-8)=-5$
   c) $x+8=5$
   d) $x-(-3)=-10$
   e) $x+7=-(3+5)$
   f) $x+11=(-8)+(-3)$
   g) $x+2=(-3)-(-5)$
   h) $4-x=4$
   i) $3-x=6$

3. Find three sets of values for a, b, and c such that
   a) $a-b=b-a$
   b) $(a-b)-c=a-(b-c)$

4. If a is an integer such that $a-0=0-a$, what is your conclusion?

5. If b is its own additive inverse, show that $a+b=a-b$.

6. Mark *true* or *false* (a,b,c are any integers).
   a) The set of integers is closed under subtraction.
   b) Subtraction on the set of integers is commutative.

c) Subtraction on the set of integers is associative. F.

d) The set of negative integers is closed under subtraction. F

e) $0-0$ is defined. T

f) $a-b=(-b)+a$ T

g) If $a = b$, then $a-c=b-c$.

h) If $2=3+x$, then $x=2-3$.

i) $a-0=a$ T

j) 0 is an identity with respect to subtraction.

7. Give a counterexample to each of the false statements in problem 6.

8. If a and b are any integers such that $a+b=a-b$, what is your conclusion?

9. If a, b, and c are any integers such that $a = b$, prove that $a-c=b-c$.

10. If a,b,c,d are integers such that $a = b$ and $c = d$, prove that $a-c=b-d$.

11. Simplify the following expressions.

a) $a-(-b)$ b) $(-a)-(-b)$ c) $a+(b-a)$

d) $(-a)+(b+a)$ e) $(a-b)+b$ f) $(a-b)+(-a)$

g) $(a+b)+(-a)$ h) $(a+b)+(-b)$ i) $(a+b)+[-a-b]$

12. The temperature at 5 a.m. was 5 degrees below zero. If the temperature rose to 10 degrees by 9 a.m., what is the rise in temperature during that period?

13. The temperature at 8 p.m. was 7 degrees. If the temperature went down to 10 degrees below zero, what was the fall in temperature?

14. On Monday Frank found that his account in the bank was overdrawn by 23 dollars. On the next Friday he noticed that his account was overdrawn by 148 dollars. How many dollars did he overdraw from the account between Monday and Friday, if he did not deposit any money in that account?

15. John borrowed 37 dollars from his friend Don and 13 dollars from another friend, Ron. After returning these from his weekly salary, he has a balance of 135 dollars. What is his weekly salary?

16. Judy has 28 dollars and wants to buy a dress that costs 51 dollars. How much additional money should she have, to buy the dress?

17. The temperature at 5 a.m. was 7 degrees below zero. If the temperature rose 19 degrees in 6 hours, what is the temperature at 11 a.m.?

18. Prove the following:

a) $a+(b-c)=(a-c)+b$

b) $(a-b)-c=a-(b+c)$

c) $a-(b-c)=a-b+c$

19. Define an operation $\oplus$ on I as $a \oplus b = a+b-1$.

a) Is the set of integers closed under $\oplus$?

b) Is $\oplus$ commutative?

c) Is $\oplus$ associative?

d) Does I have an identity with respect to $\oplus$?

e) Does every integer have an inverse with respect to $\oplus$?

★20. Considering an integer as an ordered pair of whole numbers, define the operation of subtraction on I as $(a,b)-(c,d)=(x,y)$ if and only if $(a,b)=(c,d)+(x,y)$. Evaluate:

a) $(15,8)-(7,3)$ **b)** $(3,8)-(5,7)$ **c)** $(7,5)-(5,7)$

4.4 MULTIPLICATION OF INTEGERS

We defined the binary operations of addition and subtraction on the set of integers simply by extending these operations defined in Chapter 3 on the set of whole numbers, so that the properties of W under these operations remain valid. Likewise, the definition of multiplication of integers is motivated by our desire that the properties of W under multiplication are preserved.

We observed in Section 3.4 that multiplication is repeated addition. How should we define the product $a \cdot b$ of two integers a and b? If a and b are whole numbers, then we can use Definition 3.4 to evaluate the product $a \cdot b$. But a may be a negative integer or b may be a negative integer or both may be negative integers. That is, how should the products $(-a) \cdot b$, $a \cdot (-b)$, and $(-a) \cdot (-b)$ be defined, where a and b are whole numbers? If we consider multiplication as repeated addition, then we must have

$$a \cdot (-b) = (-b)+(-b)+\cdots+(-b) \qquad a \ times$$

$$= -(a \cdot b) \qquad\qquad by \ Definition \ 4.2$$

and also $(-b) \cdot a = -(b \cdot a) = -(a \cdot b)$, if multiplication is to be commutative. Now, observe that $a \cdot (-b)+(-a) \cdot (-b)=[a+(-a)] \cdot (-b)$, for multiplication to be distributive over addition. Thus,

$$a \cdot (-b)+(-a) \cdot (-b) = 0 \cdot (-b)$$

$$= 0$$

for the zero property of multiplication to hold. Thus, if multiplication is to have these three properties, then we notice that $(-a) \cdot (-b)$ must be the additive inverse of $a \cdot (-b) = -(a \cdot b)$. But the additive inverse of $-(a \cdot b)$ is $a \cdot b$. This suggests that we should define $(-a) \cdot (-b)$ as $a \cdot b$. Now, we are in a position to make the following definition of multiplication on the set of integers.

Definition 4.4 Let a and b be whole numbers. The *product $a \cdot b$* of a and b is defined as follows:

1. $a \cdot b$ is defined as in Definition 3.4.
2. $a \cdot (-b) = -(a \cdot b) = (-a) \cdot b$
3. $(-a) \cdot (-b) = a \cdot b$

Example 4.14

$$2 \cdot 3 \quad = \quad 6 \quad = \quad 3 \cdot 2$$

$$2 \cdot (-3) = -(2 \cdot 3) = -6$$

$$3 \cdot (-2) = -(3 \cdot 2) = -6$$

$$(-2) \cdot (-3) = \quad 2 \cdot 3 \ = \ 6$$
$$(-3) \cdot (-2) = \quad 3 \cdot 2 \ = \ 6$$
$$(-1) \cdot 5 \quad = -(1 \cdot 5) = -5$$
$$5 \cdot (-1) = -(5 \cdot 1) = -5$$

As before, we denote the product $a \cdot b$ by ab if it creates no confusion. It follows from the definition that the product of two positive integers is positive, the product of two negative integers is positive and the product of a positive integer and a negative integer is negative. This is summarized as follows:

(positive)(positive) = positive

(positive)(negative) = negative

(negative)(negative) = positive

The definition of multiplication of integers can be illustrated on the integer line, as shown below:

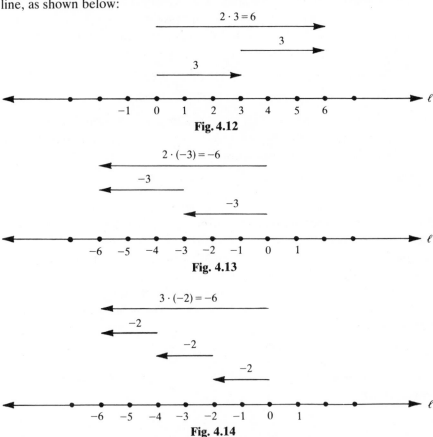

Fig. 4.12

Fig. 4.13

Fig. 4.14

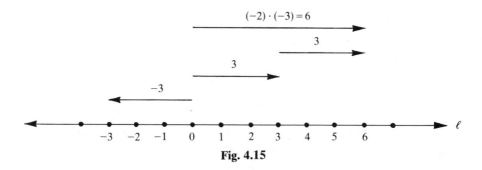

Fig. 4.15

It is clear from Definition 4.4 that the product of any two integers is a unique integer. In other words, multiplication is a binary operation on the set of integers.

Theorem 4.9   (*Closure Property*)   *The set of integers is closed under multiplication.*

Notice that the set of negative integers is not closed under multiplication. (Why?)

Theorem 4.10   (*Commutative Property*)   *For any integers a and b,*

$$a \cdot b = b \cdot a$$

As in the case of addition, this property tells that the product of two integers is not affected by the order in which they are multiplied.

Theorem 4.11   (*Associative Property*)   *For any integers a, b, and c, $a(bc) = (ab)c$.*

This property of multiplication states that when multiplying three or more integers, we can group them in any order we like without changing the value of the product. Accordingly, expressions like abc, $abcd$, etc., do not create any confusion. We can now use the commutative and the associative properties to rearrange or group the integers in a product of three or more integers in any order we like without changing the result.

Example 4.15 $3 \cdot 5 \cdot 8 = 3 \cdot (5 \cdot 8) = 3 \cdot 40 = 120$

$3 \cdot 5 \cdot 8 = (3 \cdot 5) \cdot 8 = 15 \cdot 8 = 120$

Example 4.16   Verify that $3 \cdot [(-5) \cdot (-7)] = [3 \cdot (-7)] \cdot (-5)$.

Solution: $3 \cdot [(-5) \cdot (-7)]$

$= 3 \cdot [(-7) \cdot (-5)]$ *commutative property of multiplication*

$= [3 \cdot (-7)] \cdot (-5)$ *associative property of multiplication*

Observe that $3 \cdot [(-5) \cdot (-7)] = 105 = [3 \cdot (-7)] \cdot (-5)$.

Example 4.17   Verify that $(ab)(-c) = (ac)(-b)$.

Solution: $(ab)(-c)$

$$= -[(ab)c] \qquad \text{\textit{definition of multiplication of integers}}$$
$$= -[a(bc)] \qquad \text{\textit{associative property of multiplication}}$$
$$= -[a(cb)] \qquad \text{\textit{commutative property of multiplication}}$$
$$= -[(ac)b] \qquad \text{\textit{associative property of multiplication}}$$
$$= (ac)(-b) \qquad \text{\textit{definition of multiplication of integers}}$$

In Section 4.2 we observed the existence of a unique identity element with respect to addition. It is clear from Definition 4.4 that $1 \cdot a = a = a \cdot 1$ for any integer a. That is, the multiplication of any integer a by 1 does not change the integer. The integer 1 is a *multiplicative identity* for the set of integers. It is not hard now to show that it is unique, so that we can talk about *the* multiplicative identity of the set of integers.

Theorem 4.12 (***Existence of a Unique Multiplicative Identity***) *For any integer a,*

$$a \cdot 1 = a = 1 \cdot a$$

Example 4.18 Observe that $3 = 3$ and $3 \cdot 8 = 3 \cdot 8$, multiplying both sides by 8. Also, $5 = 5$ and $5 \cdot (-7) = 5 \cdot (-7)$, multiplying both sides by -7.

This example suggests the following result, which is an extension of the multiplication property for whole numbers.

Theorem 4.13 (***Multiplication Property***) *If a,b,c,d are integers such that $a = b$ and $c = d$, then $ac = bd$.*

The multiplication property states that an equality of integers is not affected if both sides of the equality are multiplied by the same integer. We now have the following result, which is almost the converse of the multiplication property.

Theorem 4.14 (***Cancellation Property***) *If b and c are any integers and a is any nonzero integer such that $ab = ac$, then $b = c$.*

The cancellation property states that the same nonzero integer can be cancelled from both sides of an equation involving the operation of multiplication. Observe that the assumption that a be nonzero is crucial! For instance, $0 \cdot 3 = 0 \cdot 5$, but $3 \neq 5$. The cancellation property is very useful in solving problems.

Example 4.19 Solve the equation $5x - 8 = 12$.

Solution: To solve the equation $5x - 8 = 12$, we must isolate x, using the properties of addition and multiplication.

$$5x - 8 = 12 \qquad \text{\textit{given}}$$
$$5x + (-8) = 12 \qquad \text{\textit{Theorem 4.8}}$$
$$[5x + (-8)] + 8 = 12 + 8 \qquad \text{\textit{addition property of integers}}$$

$$5x + [(-8) + 8] = 12 + 8 \qquad \textit{associative property of addition}$$
$$5x + 0 = 12 + 8 \qquad -8 \textit{ is the additive inverse of } 8$$
$$5x = 12 + 8 \qquad 0 \textit{ is the additive identity for I}$$
$$5x = 20 \qquad \textit{definition of addition of integers}$$
$$5x = 5 \cdot 4 \qquad \textit{definition of multiplication of integers}$$
$$x = 4 \qquad \textit{cancellation property of multiplication}$$

It can easily be checked that 4 is actually the solution of the given equation.

Example 4.20   If $3x = 12$, then $x = 4$, since $12 = 3 \cdot 4$. If $-8x = 24$, then $x = -3$, since $24 = (-8) \cdot 3$.

Theorem 4.15   (**Product Law**)   *Let a and b be integers. If $ab = 0$ then either $a = 0$ or $b = 0$.*

The product law states that if a product of two integers is zero, then at least one of them must be zero. Equivalently, the product of two nonzero integers is nonzero.

Example 4.21   Find and graph the solution set of the equation
$$(x + 3)(x - 2) = 0.$$

Solution:
$$(x + 3)(x - 2) = 0 \qquad \textit{given}$$
$$x + 3 = 0 \quad \text{or} \quad x - 2 = 0 \qquad \textit{product law}$$
$$(x + 3) + (-3) = 0 + (-3) \quad \text{or}$$
$$(x - 2) + 2 = 0 + 2 \qquad \textit{addition property}$$
$$(x + 3) + (-3) = -3 \quad \text{or}$$
$$(x - 2) + 2 = 2 \qquad 0 \textit{ is the additive identity}$$
$$(x + 3) + (-3) = -3 \quad \text{or}$$
$$[x + (-2)] + 2 = 2 \qquad \textit{Theorem 4.8}$$
$$x + [3 + (-3)] = -3 \quad \text{or}$$
$$x + [(-2) + 2] = 2 \qquad \textit{associative property of addition}$$
$$x + 0 = -3 \quad \text{or} \quad x + 0 = 2 \qquad \textit{additive inverses}$$
$$x = -3 \quad \text{or} \quad x = 2 \qquad 0 \textit{ is the additive identity}$$

Therefore, the solution set is $\{2, -3\}$. Its graph is illustrated on the integer line, in Fig. 4.16.

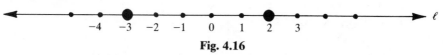

Fig. 4.16

In the previous chapter, we observed that the set of whole numbers satisfies the zero property of multiplication; that is, $a \cdot 0 = 0 = 0 \cdot a$ for any whole number a. This property holds in the set of integers also.

Theorem 4.16 (***Zero Property of Multiplication***)   *For any integer a,*

$$a \cdot 0 = 0 = 0 \cdot a$$

Exercise 4.4

1. Evaluate the following products:
   a) $5 \cdot (-3) = -15$
   b) $(-2) \cdot 7 = -14$
   c) $(-5) \cdot (-7) = +35$
   d) $(-5) \cdot 0 = 0$
   e) $(-1) \cdot (-1) = +1$
   f) $(-3) \cdot (-5) \cdot (-8) = -120$
   g) $[(-2) \cdot 3] \cdot [(-3) \cdot 2] = 36$
   h) $[(-3) \cdot (-5)] \cdot [(-7) \cdot (-1)]$ $= +15$ $\cdot +7 = +105$

2. Verify the following:
   a) $2 \cdot 3 \cdot 5 = 5 \cdot 3 \cdot 2$
   b) $(-2) \cdot (-3) \cdot 4 = 24$
   c) $(-3) \cdot (-5) = (-5) \cdot (-3)$
   d) $3 \cdot (-4) \cdot 5 = 5 \cdot 3 \cdot (-4)$
   e) $[2 \cdot (-3)] \cdot (2 \cdot 3) = (2 \cdot 2) \cdot [(-3) \cdot 3]$
   f) $[(-2) \cdot (-3)] \cdot [(-4) \cdot (-5)] = [(-5) \cdot (-2)] \cdot [(-4) \cdot (-3)]$

3. Find and graph the solution set of the following equations.
   a) $2x = -10$
   b) $-3x = 12$
   c) $-2x = -6$
   d) $3x - 5 = 13$
   e) $-5x + 3 = -12$
   f) $(x-1)(x+1) = 0$
   g) $x(x-2) = 0$
   h) $x(x-1)(x+1) = 0$
   i) $5(-x) = 10$

4. Show that $(-1) \cdot a = -a$ for any integer a.

5. Under what conditions will $a + b = a - b$?

6. Mark *true* or *false* (a,b,c,x are any integers).
   a) The set of positive integers is closed under multiplication. True
   b) The set of negative integers is closed under multiplication. False
   c) Multiplication on the set of integers is commutative. T.
   d) Multiplication on the set of integers is associative. T.
   e) $(-1) \cdot a = -a$ True
   f) $(-1) \cdot (-a) = -a$
   g) $(-a) \cdot (-b) = ab$
   h) If $ab = ac$, then $b = c$.
   i) If $3x = 0$, then $x = 0$.
   j) If $a = b$, then $ac = bc$.
   k) If $ab = 0$, then $a = 0$.
   l) If $ab = 0$, then $a = 0$ and $b = 0$.

7. Give a counterexample to each of the false statements in problem 6.

8. State the property of multiplication used in each of the following statements, where each symbol represents an integer.
   a) $x \cdot y$ is an integer   Closure prop of ·
   b) $(-a) \cdot 1 = -a$ Multip. identity · I
   c) If $3x = 3y$, then $x = y$
   d) $3 \cdot (4 \cdot 5) = (3 \cdot 4) \cdot 5$ assoc · I
   e) $x \cdot y = y \cdot x$ comm · I
   f) If $x = y$, then $5x = 5y$
   g) $x \cdot [(-y) \cdot z] = x \cdot [z \cdot (-y)]$
   h) $x \cdot [y \cdot (-z)] = (x \cdot y) \cdot (-z)$ assoc · S
   i) $(ab)(cd) = (cd)(ab)$ comm
   j) $(ab)(cd) = a[b(cd)]$ assoc · S

9. Simplify the following expressions.
   a) $(a+b) - (a-b)$
   b) $(a-b) - (a+b)$
   c) $(a-b) + (a+b)$
   d) $(a+b) + (b-a)$

10. If the integer a is its own additive inverse, show that $a = 0$.

11. Let x and y be integers such that $xy = 0$. If $x \neq 0$, use the cancellation property to prove that $y = 0$.

12. If a and b are integers such that $ab = a$, what do you conclude?

13. If the temperature at 6 a.m. was 3 degrees below zero and went up 4 degrees each hour, what was the temperature at 9 a.m.?

14. If the temperature at 8 p.m. was 5 degrees above zero and went down 5 degrees an hour, what was the temperature at midnight?

15. The temperature at 1 a.m. was 3 degrees below zero. It rose 12 degrees by 6 a.m. If the temperature at 11 a.m. is twice as much as the temperature at 6 a.m., what is the temperature at 11 a.m.?

16. The temperature at 1 a.m. was 2 degrees below zero. It rose 7 degrees by 6 a.m. The temperature at 9 a.m. was twice the temperature at 6 a.m. Find the temperature at 2 p.m. if it is three times the temperature at 9 a.m.?

17. Frank borrowed some money from his friend John in January. He returned 175 dollars in February and 100 dollars in March. If Frank still owes John 4 times the amount of money he returned in March, how many dollars did he borrow from him?

18. Prove each of the following.
  a) $a(-b)c = ca(-b)$
  b) $a(-bc) = (-a)bc$
  c) $(-a)(-b) = ba$
  d) $(ab)(cd) = (ad)(bc)$

19. Define an operation $*$ on the set of integers by $a * b = a + b + ab$.
  a) Is I closed under $*$?
  b) Is $*$ commutative?
  c) Is $*$ associative?
  d) Does I have an identity element with respect to $*$?
  e) Does the integer 0 have an inverse with respect to $*$?

20. Repeat problem 19, with the operation $\odot$ defined by $a \odot b = a + b - ab$.

★21. Using the ordered pair notation, define the operation of multiplication on I by $(a,b) \cdot (c,d) = (ac + bd, ad + bc)$, where $a,b,c,d \in W$.
  a) Evaluate $(3,5) \cdot (2,7)$.
  b) Evaluate $(5,3) \cdot (1,0)$.
  c) For what value of x is $(x,3) \cdot (5,2) = (9,3)$?

4.5 THE DISTRIBUTIVE PROPERTY

So far in our discussion of integers, we have considered only one operation at a time: either addition, subtraction, or multiplication. In the case of whole numbers, we observed that multiplication distributes over both addition and subtraction. Since the set of integers is obtained by extending the set of whole numbers, and these operations on I are obtained by extending those defined on W, it is only proper to ask the following question: Is multiplication distributive over addition and subtraction, in the set of integers?

Example 4.22   Verify that $5 \cdot [7+(-3)] = 5 \cdot 7 + 5 \cdot (-3)$.

Solution:
$$5 \cdot [7+(-3)] = 5(7-3) = 5 \cdot 4 = 20$$
$$5 \cdot 7 + 5 \cdot (-3) = 35 - 15 = 20$$

Thus,
$$5 \cdot [7+(-3)] = 5 \cdot 7 + 5 \cdot (-3)$$

Example 4.23   Verify that $7 \cdot (-8+5) = 7 \cdot (-8) + 7 \cdot 5$.

Solution:
$$7 \cdot (-8+5) = 7 \cdot (-3) = -21$$
$$7 \cdot (-8) + 7 \cdot 5 = -56 + 35 = -21$$

Thus,
$$7 \cdot (-8+5) = 7 \cdot (-8) + 7 \cdot 5$$

These two examples now suggest that the answer could be "yes" to the question posed above.

Theorem 4.17   (***Distributive Property***)   *For any integers a, b, and c,*

$$a \cdot (b+c) = a \cdot b + a \cdot c$$

This theorem states that multiplication distributes over addition. However, as in the case of whole numbers, observe that addition does not distribute over multiplication. For instance, $2 + 3 \cdot 5 \neq (2+3) \cdot (2+5)$. (Why?)

In Section 4.3, we observed that $b - c = b + (-c)$ for any integers b and c (Theorem 4.8). It now follows from Theorem 4.17 that $a(b-c) = ab - ac$. That is, multiplication also distributes over subtraction (but not vice versa).

The distributive property is very useful in our discussion of the system of integers. It helps us to evaluate any product of large numbers, as illustrated by the following examples.

Example 4.24   Use the distributive property to evaluate $8 \cdot 452$.

Solution:

$8 \cdot 452 = 8(450+2)$	*definition of addition of integers*
$= 8 \cdot 450 + 8 \cdot 2$	*distributive property*
$= 8(400+50) + 8 \cdot 2$	*definition of addition of integers*
$= (8 \cdot 400 + 8 \cdot 50) + 8 \cdot 2$	*distributive property*
$= (3200 + 400) + 16$	*definition of multiplication of integers*
$= 3600 + 16$	*definition of addition of integers*
$= 3616$	*definition of addition of integers*

Example 4.25   Use the distributive property to evaluate $7 \cdot 327$.
Solution: $7 \cdot 327$

$= 7(330 - 3)$ *definition of subtraction of integers*

$= 7 \cdot 330 - 7 \cdot 3$ *distributive property*

$= 7(300 + 30) - 7 \cdot 3$ *definition of addition of integers*

$= (7 \cdot 300 + 7 \cdot 30) - 7 \cdot 3$ *distributive property*

$= (2100 + 210) - 21$ *definition of multiplication of integers*

$= 2310 - 21$ *definition of addition of integers*

$= 2289$ *definition of subtraction of integers*

Exercise 4.5

1. Mark *true* or *false* (a,b,c,x,y are any integers).
   a) Addition distributes over multiplication.
   b) Multiplication distributes over addition.
   c) Multiplication distributes over subtraction.
   d) Subtraction distributes over multiplication.
   e) $2(a - 1) = 2a - 1$
   f) $(a + 1)(a - 1) = a^2 - 1$
   g) $[5 + (-3)]x = 5x - 3x$
   h) $(-1)(x - y) = -x - y$
   i) There exist integers a, b, and c such that $a + (b \cdot c) = (a + b) \cdot (a + c)$.
   j) There exist integers a, b, and c such that $a - (b \cdot c) = (a - b) \cdot (a - c)$.
2. Give a counterexample to each of the false statements in problem 1.
3. Use the distributive property to evaluate the following products.
   a) $8 \cdot 123$ b) $5 \cdot 97$ c) $7 \cdot 565$
4. Solve the following equations.
   a) $3x - 8 = 16$ b) $5(x - 7) = 2(x - 13)$
   c) $7(2x - 3) = 3(1 - 2x) - 4$ d) $3(3 + x) = 5x - 1$
   e) $4x + 1 = 3(4x - 5)$ f) $7(x - 3) + 11 = 7x - 10$
5. The temperature at 6 a.m. was 3 degrees below the temperature at 9 a.m. If twice the temperature at 6 a.m. was 18 degrees, what is the temperature at 9 a.m.?
4. If the sides of a square are increased by 3 inches, the area of the new square is 39 square inches more than the area of the original square. Find the length of a side of the original square.
7. If three times the sum of two consecutive integers is 45, find the integers.
8. If four times the sum of three consecutive integers is 144, find the integers.

9. Peter and his brother John are 6 and 14 years of age. In how many years will five times the age of Peter be equal to three times that of John?

10. Use the distributive property to prove that $a \cdot 0 = 0$ for every integer a. [*Hint:* $0 = 0 + 0$.]

11. Prove each of the following, where a, b, c, d, x are all integers. (Recall that x^2 means $x \cdot x$.)

a) $(a+b)c = ac + bc$
b) $a(b-c) = ab - ac$
c) $(a+b)^2 = a^2 + 2ab + b^2$
d) $(a-b)^2 = a^2 - 2ab + b^2$
e) $(a+b)(a-b) = a^2 - b^2$
f) $(a+b)(c+d) = ac + ad + bc + bd$
g) $a(b+c+d) = ab + ac + ad$
h) $a(b-c-d) = ab - ac - ad$

4.6 THE ORDER RELATION

In Section 3.6, we defined the order relation "$<$" on the set of whole numbers and used it to assign a unique position to each whole number on the number line. If $a < b$ for whole numbers, then the point with coordinate a lies to the left of the point with coordinate b, on the number line. In this section we define the order relation on the set of integers by extending the definition of the order relation on W. Recall that if a and b are whole numbers, we defined $a < b$ if and only if there exists a natural number x such that $a + x = b$. Our definition of the order relation on I is motivated by the desire that all the properties of the order relation on W be preserved. Accordingly we make the following definition:

Definition 4.5   Let a and b be any integers. The integer a is said to be *less than* b, written $a < b$, if there exists a positive integer x such that $a + x = b$.

If $a < b$, we also say that b is *greater than* a and write $b > a$. If a is not less than b, we write $a \not< b$.

Example 4.26 $\quad\quad 3 < 8 \quad\quad$ since $\quad\quad 3 + 5 = 8$

$\quad\quad\quad\quad\quad -5 < 3 \quad$ since $(-5) + 8 = 3$

$\quad\quad\quad\quad\quad -7 < -2 \quad$ since $(-7) + 5 = -2$

$\quad\quad\quad\quad\quad -4 \not< -6 \quad$ since there exists no positive integer x such that $(-4) + x = -6$. (Why?)

Definition 4.5 helps us to represent integers on the integer line. If $a < b$ for integers, then the point with coordinate a lies to the left of the point with coordinate b, on the integer line, as shown in Fig. 4.17. Observe that negative integers are represented by points on the left of the origin and the positive integers by points on the right of the origin.

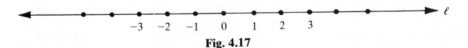

$$-3 \quad -2 \quad -1 \quad 0 \quad 1 \quad 2 \quad 3$$

Fig. 4.17

Notice that the definition of the order relation implies that $a < b$ if and only if $b - a = x$ is a positive integer.

Let's now discuss the properties of the order relation on the set of integers. In the case of whole numbers, we observed that if a and b are any two whole numbers, either $a < b$ or $a = b$ or $b < a$. This property can now be extended to the set of integers.

Theorem 4.18   (**Law of Trichotomy**)   *If a and b are any two integers, then either a < b or a = b or b < a.*

The law of trichotomy states that any two integers either are equal or one integer is less than the other. In particular, if we let $b = 0$, then either $a < 0$ or $a = 0$ or $0 < a$ for any integer a. In other words, if a is any integer, then either a is a negative integer or $a = 0$ or a is a positive integer. This is how the law of trichotomy is sometimes stated.

Theorem 4.19   (**Transitive Property**)   *If a,b,c are integers such that a < b and b < c, then a < c.*

Example 4.27 $2 < 3$ since $2 + 1 = 3$

                         $3 < 5$ since $3 + 2 = 5$

Now, $2 < 5$ since $2 + (1 + 2) = 2 + 3 = 5$

This example is demonstrated geometrically in Fig. 4.18. Observe that the point with coordinate 2 lies to the left of the point with coordinate 5.

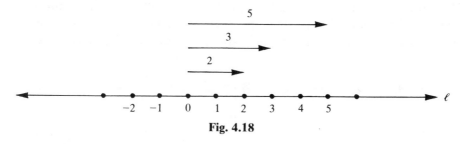

Fig. 4.18

Theorem 4.19 helps us to order the integers in the following way:

$$\cdots < -3 < -2 < -1 < 0 < 1 < 2 < 3 < \cdots$$

Theorem 4.20   *If a and b are integers such that a < b, then a + c < b + c for every integer c.*

This theorem states that the sense of an inequality is preserved under the addition of the same integer to both sides of the inequality, corresponding to the addition property. Observe that the theorem also states that if $a + c < b + c$, then $a < b$ (why?), corresponding to the cancellation property of addition.

Example 4.28 Observe that $-2 < 3$ since $(-2) + 5 = 3$. Let's add the same integer, -5, to both sides of the inequality $-2 < 3$ to get the integers -7 and -2. The numbers -7 and -2 are then related as $-7 < -2$ (why?).

Example 4.29 Find and graph the solution set of the inequality $3x + 5 < 2x + 8$.

Solution:

$3x + 5 < 2x + 8$	*given*
$(-2x) + (3x + 5) < (-2x) + (2x + 8)$	*add $-2x$ to both sides (Theorem 4.20)*
$[(-2x) + 3x] + 5 < [(-2x) + 2x] + 8$	*associative property of addition*
$x + 5 < 0 + 8$	*definition of addition of integers*
$x + 5 < 8$	*0 is the additive identity*
$(x + 5) + (-5) < 8 + (-5)$	*add -5 to both sides (Theorem 4.20)*
$x + [5 + (-5)] < 8 + (-5)$	*associative property of addition*
$x + 0 < 3$	*definition of addition of integers*
$x < 3$	*0 is the additive identity*

Therefore, the solution set is $\{x \in I \mid x < 3\}$. Figure 4.19 illustrates the graph of the solution set on the integer line. The heavy dots on the number line show that the integers corresponding to them are solutions of the given inequality.

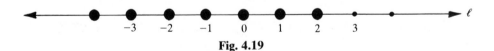

Fig. 4.19

Theorem 4.21 *Let a, b, and c be integers. Then*
1. *$a < b$ if and only if $ac < bc$, if c is a positive integer, and*
2. *$a < b$ if and only if $ac > bc$, if c is a negative integer.*

This theorem states that the sense of an inequality is preserved if both sides are multiplied by the same positive integer or if the same positive integer is cancelled from both sides. The sense of the inequality is reversed if both sides are multiplied by the same negative integer or if the same negative integer is cancelled from both sides. For instance, if $3x < 3y$ then $x < y$, and if $-3x < -3y$ then $x > y$.

It now follows from Theorem 4.21 that $a < b$ if and only if $-a > -b$.

Example 4.30   We know that $-3 < 1$. Multiplying both sides by 2, we have $-6 = (-3) \cdot 2 < 1 \cdot 2 = 2$. Multiplying both sides by -2, we have $6 = (-3) \cdot (-2) > 1 \cdot (-2) = -2$.

Let a and b be any integers. If a is less than or equal to b, we write $a \le b$. Similarly, $a \ge b$ means a is greater than or equal to b. The expression $a < x < b$ means $a < x$ and $x < b$. For example, $A = \{x \in I \mid -3 < x < 2\} = \{-2, -1, 0, 1\}$. Notice that the set A can also be written $\{x \in I \mid -3 < x \le 1\}$ or $\{x \in I \mid -2 \le x \le 1\}$.

We remark that Theorems 4.19, 4.20, and 4.21 all remain true if "$<$" is replaced by "$\le$." In Theorem 4.21, c can very well be zero (why?).

Example 4.31   Find and graph the solution set of the inequality $-2 < -x + 3 \le 5$.

Solution: We ask the reader to justify each of the following steps.

$$-2 < -x + 3 \le 5$$
$$-2 < -(x-3) \le 5$$
$$2 > x - 3 \ge -5$$
$$-5 \le x - 3 < 2$$
$$(-5) + 3 \le (x-3) + 3 < 2 + 3$$
$$(-5) + 3 \le [x + (-3)] + 3 < 2 + 3$$
$$(-5) + 3 \le x + [(-3) + 3] < 2 + 3$$
$$-2 \le x + 0 < 5$$
$$-2 \le x < 5$$

Therefore, the solution set is $\{x \in I \mid -2 \le x < 5\} = \{-2, -1, 0, 1, 2, 3, 4\}$. Figure 4.20 illustrates the graph of the solution set of the given inequality.

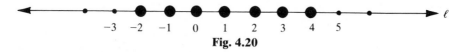

Fig. 4.20

Example 4.32   Find and graph the solution set of the inequality $-9 \le 2x - 3 < 5$.

Solution: We ask the reader to justify each of the following steps.

$$-9 \le 2x - 3 < 5$$
$$-9 \le 2x + (-3) < 5$$
$$(-9) + 3 \le [2x + (-3)] + 3 < 5 + 3$$
$$(-9) + 3 \le 2x + [(-3) + 3] < 5 + 3$$
$$-6 \le 2x < 8$$
$$-3 \le \ x < 4$$

Therefore, the solution set is $\{x \in I \mid -3 \le x < 4\} = \{-3,-2,-1,0,1,2,3\}$. Figure 4.21 exhibits the solution set on the integer line.

Fig. 4.21

In Section 3.6, we proved that if a and b are whole numbers, then $a \le b$ if and only if $a^2 \le b^2$. In other words, if a and b are nonnegative integers, then $a \le b$ if and only if $a^2 \le b^2$. What happens if a or b or both are negative? Is the result still true? Well, observe that $-5 < 3$, but $25 = (-5)^2 > 3^2 = 9$. Also, $-5 < -3$, but $25 = (-5)^2 > (-3)^2 = 9$. Thus, $a^2 \le b^2$ does not mean that $a \le b$! (Why?)

Example 4.33   Find the set of all nonnegative integers x such that $-96 < 4 - 4x^2 < -12$, and graph the solution set.

Solution:
$$-96 < 4 - 4x^2 < -12$$
$$-96 - 4 < -4x^2 < -12 - 4$$
$$-100 < -4x^2 < -16$$
$$25 > x^2 > 4$$
$$5 > x > 2$$

since x is a nonnegative integer. Therefore, the solution set is $\{x \in I \mid 2 < x < 5\} = \{3,4\}$. Its graph is given in Fig. 4.22.

Fig. 4.22

Exercise 4.6

1. Verify the following:
   a) $3 < 7$ **b)** $-8 < 3$ **c)** $-8 < -5$ **d)** $-4 < 0$

2. Find the sets $A \cup B$, $A \cap B$, $A - B$, and $B - A$ if $A = \{x \in I \mid -5 < x < 3\}$ and $B = \{x \in I \mid -2 \le x < 5\}$.

3. If $A = \{x \in I \mid -6 < x < -2\}$, $B = \{x \in I \mid -4 \le x < 1\}$, $C = \{x \in I \mid -2 < x \le 3\}$, and $U = \{x \in I \mid -7 \le x \le 4\}$, find the following sets.
   a) $A \cup B$ **b)** $A \cap B$ **c)** $A - B$
   d) A' **e)** $A \cap B'$ **f)** $A - C'$
   g) $(A \cup B) \cap C'$ **h)** $(A \cup B)' \cap C$ **i)** $A' \cup B' \cup C'$

4. Find and graph the solution set of each of the following inequalities, x being an integer.
   a) $x + 5 < 8$ **b)** $-x + 3 < -2$ **c)** $(x-2) + (-3) < 4$
   d) $-(x-2) < -3$ **e)** $-x - 4 < 1$ **f)** $-2 < x + 3 \le 1$
   g) $-3 < 3 - 2x \le 5$ **h)** $-12 < 4x - 4 < 8$ **i)** $-2 \le 3x + 1 \le 4$

5. Mark *true* or *false* (a,b,c,x,y are any integers).
   a) Every negative integer is less than 0.
   b) If $a<b$ and $b<c$, then $a<c$.
   c) If $a<b$ and $a<c$, then $b<c$.
   d) If $x<y$, then $4+x<4+y$.
   e) If $a<b$, then $a-c<b-c$.
   f) If $x<y$, then $4-x<4-y$.
   g) If $a<b$, then $-a<-b$.
   h) If $a<b$, then $ac<bc$.
   i) If $a>0$ and $b>0$, then $ab>0$.
   j) If $a<0$ and $b<0$, then $ab<0$.
6. Give a counterexample to each of the false statements in problem 5.
7. a) Find the set of all positive integers x such that
   1. $x^2\leq 9$ 2. $-5\leq 2x^2-7<11$
   3. $-24\leq 24-3x^2\leq 24$ 4. $1<x^2-3<13$
   5. $3(x-27)+1\leq 3x+1-x^2\leq 3(x-1)$
   b) Graph each of the sets in (*a*).
8. Dick bought 7 books at the same price. What are the possible values for the price of a book in dollars if the total cost of the books does not exceed 42 dollars?
9. If twice the maximum and twice the minimum temperatures of a certain day were 32 degrees above zero and 6 degrees below zero, respectively, what were the possible temperatures in degrees at various times of the day?
10. If the sides of a square are increased by 3 inches, the area of the new square is less than 49 square inches. Find the possible values in inches for a side of the original square.
11. If 5 times an integer is 18 more than twice that integer, what are the possible values for the integer?
12. Show that $a\leq a$ for every integer a (reflexive property).
★**13.** Prove that if $a\leq b$ and $b\leq c$, then $a\leq c$ (transitive property).
★**14.** Prove that if $a\leq b$ and $b\leq a$, then $a=b$.
★**15.** Prove that if $a\leq b$, then (1) $a+c\leq b+c$ and (2) $a-c\leq b-c$.

4.7 DIVISION OF INTEGERS

Recall that we expanded the set of whole numbers to the set of integers so that all the properties of addition and multiplication of whole numbers are preserved. Now the set of integers has an additional property: it is closed under subtraction. In other words, every equation of the form $a=b+x$ is solvable in the set of integers. Subtraction, we observed, is the inverse operation of addition. Our aim in this section is to define a new operation, division, on the set of integers, which is the inverse operation of multiplication. If two pounds of candy costs 4 dollars, how much does it cost for a pound? Observe that the answer to this problem is given by the solution of the equation $2x = 4$. If Mr. Rogers drove 240 miles in 4 hours, how fast did he drive? The solution of the equation $4y = 240$ gives the speed at which he drove. Even though these concrete situations can be solved using the cancellation property of multiplication, the operation of division might be more handy and useful. The price of one pound of candy is given by $4 \div 2 = 2$ dollars and the speed of the car is given by $240 \div 4 = 60$ miles per hour. Recall that we defined subtraction in terms of addition. Similarly, we define the operation of division in terms of multiplication.

Definition 4.6   Let a and b be any two integers. Then $a \div b = x$ if and only if $a = bx$ and x is unique. The operation we have defined is called *division* of integers. We read $a \div b$ as "a divided by b." Instead of $a \div b$, we also write $\dfrac{a}{b}$.

Observe from our definition that $a \div b$ is the unique solution of the equation $a = bx$.

Example 4.34   Evaluate $6 \div 3$.

Solution: $6 \div 3 = x$ if and only if $6 = 3x$. But $6 = 3x$ if and only if $x = 2$, by the cancellation property of multiplication. Therefore, $6 \div 3 = 2$.

Example 4.35   Evaluate $-10 \div 2$.

Solution: $-10 \div 2 = x$ if and only if $-10 = 2x$. Since $-10 = 2x$ if and only if $x = -5$, thus $-10 \div 2 = -5$.

Example 4.36   Evaluate $-12 \div -3$.

Solution: $-12 \div -3 = x$ if and only if $-12 = -3x$, and $-12 = -3x$ implies $x = 4$. Therefore $-12 \div -3 = 4$.

Let's now discuss division of zero by a nonzero integer a. By Definition 4.6, $0 \div a = x$ if and only if $0 = ax$. Now, $ax = 0$ implies $x = 0$ by the product law. Thus $0 \div a = 0$, where $a \neq 0$.

Let a be a nonzero integer. Then $a \div 0 = x$ if and only if $a = 0 \cdot x$. Thus $a = 0$, which is a contradiction. Therefore $a \div 0$ does not exist and is left undefined.

Now, notice that $0 \div 0 = x$ if and only if $0 = 0 \cdot x$. But this equation has no unique solution by the zero property of multiplication of integers. Hence $0 \div 0$ is undefined.

Let's now summarize our observations: division by zero is undefined and zero divided by any nonzero integer is zero.

Example 4.37   Evaluate $5 \div 3$.

Solution: $5 \div 3 = x$ if and only if $5 = 3x$ has a solution in the set of integers. But the equation $5 = 3x$ is not solvable in the set of integers. Therefore, $5 \div 3$ does not exist in the set of integers.

What do we conclude from this example? The set of integers is not closed under division. Equivalently, not every equation of the form $a = bx$ is solvable in the set of integers. This is a deficiency of the set of integers. Thus, I is not enough to solve every equation of the form $a = bx$.

If $a \div b = x$ exists, we also say that *a is divisible by b* or *b divides a* or *b is a factor* of a.

Exercise 4.7

1. Use the definition of division to evaluate the following:

    a) $12 \div 4$ **b)** $15 \div (-3)$ **c)** $(-18) \div 6$ **d)** $(-21) \div (-7)$

2. Evaluate the following:

    a) $3 + (12 \div 3)$ **b)** $-5 + [6 \div (-3)]$ **c)** $7 - [(-15) \div 3]$

    d) $-2 - [(-6) \div (-2)]$ **e)** $2 \cdot (0 \div 2)$ **f)** $3 \cdot [(-8) \div 4]$

    g) $24 \div (2 \cdot 3)$ **h)** $18 \div [(-9) \cdot 2]$ **i)** $(-6) \div [8 \div (-4)]$

3. Mark *true* or *false* (a,b,c are any integers).

    a) The set of integers is closed under division.

    b) The operation of division is commutative.

    c) The operation of division is associative.

    d) Division distributes over addition.

    e) Subtraction is distributive over division.

    f) Any integer divided by itself is one.

    g) $a \div 0 = 0$

    h) Every equation of the form $a = bx$ is solvable in I.

    i) Every integer has a multiplicative inverse.

    j) $a \div 1 = a$

4. Give a counterexample to each of the false statements in problem 3.

5. If 4 gallons of gasoline cost 2 dollars, how much is it for a gallon?

6. If 10 pounds of onions cost 150 cents, how much does it cost for a pound of onions?

7. How many hours does it take to drive a distance of 275 miles at a rate of 55 miles an hour?

8. How fast does Andy have to drive, if he wants to cover 434 miles in 7 hours?

9. If an hour is divided into 10 equal parts, how many minutes are there in each part?

10. In how many equal parts of 12 minutes each can you divide an hour?

11. Verify the following:

    a) $(a^2 - b^2) \div (a - b) = a + b$ **b)** $(a^3 + b^3) \div (a + b) = a^2 - ab + b^2$

    c) $(a^3 - b^3) \div (a - b) = a^2 + ab + b^2$ **d)** $ab \div ac = b \div c$, where $ac \neq 0$.

12. a) Is the set $\{0,1\}$ closed under division? Why?

    b) Is the set $\{1,2\}$ closed under division? Why?

13. Prove that

    a) a divides a for every nonzero integer a.

    b) If a divides b and b divides c, then a divides c (transitive property).

★14. Using the ordered pair notation for integers, define $(a,b) \div (c,d) = (x,y)$ if and only if $(a,b) = (c,d)(x,y)$, where (x,y) is unique. Use this definition to evaluate

    a) $(16,4) \div (4,1)$ **b)** $(11,3) \div (3,5)$

4.8 SUMMARY AND COMMENTS

We extended the set of whole numbers to a "larger" set, called the set of integers, where we can solve every equation of the form $a = b + x$. We defined the operations of addition, subtraction, and multiplication on the set of integers by extending their definitions on the set of whole numbers.

The set of integers is closed and has a unique identity with respect to addition and multiplication. Both addition and multiplication are commutative and associative. Every integer has a unique additive inverse. Consequently, I is closed under subtraction. Subtraction on the set of integers is neither commutative nor associative. The set I has no identity element under subtraction. Multiplication distributes over both addition and subtraction.

The set of integers satisfies the following additional properties with respect to addition and multiplication:

1. If $a = b$ and $c = d$, then $a + c = b + d$ (*addition property*).

2. If $a + b = a + c$, then $b = c$ (*cancellation property*).

3. If $a = b$ and $c = d$, then $a \cdot c = b \cdot d$ (*multiplication property*).

4. If $ab = ac$ and $a \neq 0$, then $b = c$ (*cancellation property*).

5. If $ab = 0$, then either $a = 0$ or $b = 0$ (*product law*).

6. $a \cdot 0 = 0$ (*zero property of multiplication*).

We extended the order relation on the set of whole numbers to the set of integers, with respect to which we have the following properties:

1. If a and b are any two integers, then either $a < b$ or $a = b$ or $b < a$ (*law of trichotomy*).

2. If $a < b$ and $b < c$, then $a < c$ (*transitive property*).

3. If $a < b$, then $a + c < b + c$ for every integer c.

4. $a < b$ if and only if $\begin{cases} ac < bc \text{ if } c \text{ is a positive integer,} \\ ac > bc \text{ if } c \text{ is a negative integer.} \end{cases}$

5. Let a and b be nonnegative integers. Then $a < b$ if and only if $a^2 < b^2$. These properties remain true if "$<$" is replaced by "$\leq$."

We represented integers by points on a straight line, the collection of all those points being called the integer line. We observed that the integer line is full of "gaps." For example, there are no integers corresponding to the infinitely many points between those with coordinates 2 and 3.

Finally, we defined the new operation of division of integers, with respect to which I is not closed. Also, division is neither commutative nor associative. Since $W \subset I$, we have the following line diagram for our number systems.

Fig. 4.23

Is the "tree" in Fig. 4.23 large enough for our purposes? We observed that not every equation of the form $a = bx$ is solvable in I. Therefore, we have to enlarge our set of integers to a larger set, in order to solve every equation of the type $a = bx$. So we conclude that our "tree" has not "grown" enough.

SUGGESTED READING

Dubisch, R., "A 'Proof' that $(-)(-) = +$," *The Mathematics Teacher*, vol. 64 (Dec. 1971), p. 750.

Eynden, C. V., "How Do You Read '$-x$'?" *School Science and Mathematics*, vol. 74 (Feb. 1974), pp. 134–136.

Henry, B., "Zero, the Troublemaker," *The Arithmetic Teacher*, vol. 16 (May 1969), pp. 365–367.

Sherzer, L., "Adding Integers Using Only the Concepts of One-to-One Correspondence and Counting," *The Arithmetic Teacher*, vol. 16 (May 1969), pp. 360–361.

Wheeler, R. E., *Modern Mathematics: An Elementary Approach* (3rd ed.), Brooks/Cole Publishing Co., Monterey, Calif. (1973), pp. 169–201.

"Mathematics is the queen of the Sciences and arithmetic is the queen of Mathematics."

C. F. GAUSS

5/The Theory of
Numbers

After studying this chapter, you should be able to:
- *find the quotient and the remainder when an integer is divided by a positive integer*
- *use simple properties of even and odd integers*
- *establish if a natural number is a prime*
- *write the prime factorization of a (composite) number*
- *find the gcd and the lcm of two or more natural numbers*
- *apply the Euclidean algorithm to the gcd of two natural numbers*
- *name some special numbers and some related unsolved problems*

5.0 INTRODUCTION

Recall that we developed the system of integers from the system of whole numbers in the preceding chapter. In this chapter, our goal is to introduce the reader to one of the most fascinating areas of mathematics—the theory of numbers. The theory of numbers is the branch of mathematics that deals with properties of numbers.

The history of theory of numbers dates back more than 20 centuries. The ancient Greeks had great interest in number theory. The Pythagorean Brotherhood—followers of the famous Greek mathematician Pythagoras (569–500 B.C.)—regarded odd numbers as male and good, even numbers as female and bad. The number 1 was considered neither male nor female. The number 5, being the sum of the first masculine and feminine numbers, was regarded as a symbol of marriage! Some philosophers, supported by theologians of the early centuries of Christianity, identified the number 1 with God.

To the ancient Greeks and Hindus, puzzles and solutions of interesting problems in number theory were a source of fun and amusement.

The three outstanding mathematicians who contributed considerably in this area of mathematics are Pierre de Fermat (1601–1665), a Frenchman, Leonhard Euler (1707–1783), a Swiss, and Carl Frederick Gauss (1777–1855), a German. Gauss, usually referred to as the prince of mathematics, is reported to have once said, "Mathematics is the queen of the Sciences and arithmetic is the queen of Mathematics," where by arithmetic he meant number theory.

The first recorded proofs of simple properties of numbers were given by Euclid (330?–275 B.C.), especially those related to divisibility.

5.1   THE DIVISION ALGORITHM

In Section 4.7 we defined the operation of division of integers. Recall that the integer a is said to be *divisible* by an integer $b(\neq 0)$ or b is said to *divide a* if and only if the equation $a = bx$ is solvable in I. If b divides a, we also say that b is a *factor (divisor)* of a or a is a *multiple* of b. Observe that ± 1 and $\pm x$ are factors of every integer $x(\neq 1)$. They are called the *trivial factors* of x.

Example 5.1   The number 2 is a factor of 6 since $2 \cdot 3 = 6$; the number -3 is a divisor of 12 since $12 = (-3)(-4)$; and 10 is a multiple of 5 since $10 = 5 \cdot 2$; but 18 is not a multiple of 5 since there exists no integer x such that $18 = 5x$.

Notice that 17 is not a multiple of 5. Consequently, $17 \div 5$ does not exist in the set of integers. However, 17 and 5 are related by the equation $17 = 3 \cdot 5 + 2$. Also, $-7 \div 2$ does not exist in the set of integers. Still they are connected by the equation $-7 = (-4) \cdot 2 + 1$. This now suggests the following important property:

Theorem 5.1   (*Division Algorithm*)   *If a is any integer and b any positive integer, then there exist unique integers q and r such that $a = bq + r$, where $0 \le r < b$.*

The division algorithm simply means that if we divide an integer a, called the *dividend*, by a positive integer b by the long division method taught in elementary school, we get a unique *quotient q* and a unique nonnegative integer r less than b as the *remainder*. If $r = 0$, then $a = bq$, in which case a is divisible by b and $a \div b = q$ is the quotient.

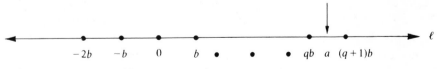

Fig. 5.1

The division algorithm may be interpreted geometrically as follows. On the number line, mark off the points with coordinates $\ldots, -2b, -b, 0, b, 2b, \ldots$, as shown in Fig. 5.1. Then there exists a suitable largest integer q such that the point with coordinate a either coincides with the point with coordinate qb or falls between the points with coordinates qb and $(q+1)b$. The remainder r in the division algorithm is the "distance" between the points with coordinates qb and a; that is, $r = a - qb$.

Example 5.2   Find the integers q and r of the division algorithm if $a = 237$ and $b = 11$.

$$
\begin{array}{r}
21 \\
11 \overline{\smash{\big)}\ 237} \\
\underline{22} \\
17 \\
\underline{11} \\
6
\end{array}
$$

Thus, $q = 21$ and $r = 6$. Observe that $237 = 11 \cdot 21 + 6$ and $0 \le r < 11$.

Example 5.3   Compute the integers q and r of the division algorithm if $a = -2429$ and $b = 15$.

$$
\begin{array}{r}
161 \\
15 \overline{\smash{\big)}\ 2429} \\
\underline{15} \\
92 \\
\underline{90} \\
29 \\
\underline{15} \\
14
\end{array}
$$

Thus,

$$
\begin{aligned}
2429 &= 15 \cdot 161 + 14 \\
-2429 &= 15 \cdot (-161) + (-14) \qquad \text{(recall } 0 \le r < 15\text{)} \\
&= 15 \cdot (-161) + (-15) + 15 + (-14) \qquad \text{(adding } (-15) + 15 = 0\text{)} \\
&= [15 \cdot (-161) + 15 \cdot (-1)] + [15 + (-14)] \\
&= 15 \cdot [(-161) + (-1)] + 1 \\
&= 15 \cdot (-162) + 1
\end{aligned}
$$

Therefore, $q = -162$ and $r = 1$.

If the numbers a and b are small enough, q and r can be found by inspection.

Exercise 5.1
1. List **a)** three positive multiples of 11, 22, 33, 44
    b) five negative multiples of 2 -4,-6,-8, -10
    c) three positive multiples of −3 9 12 15
    d) five negative multiples of −3 -3 -6 -9 -12
2. List **a)** three positive divisors of 27 1, 3, 9
    b) five negative divisors of 24 -1,-2,-3,-4,-6
    c) three positive divisors of −30 2, 3, 5
    d) five negative divisors of −12
3. What is the set of all possible remainders when an integer is divided by
    a) 5   **b)** 7   **c)** 2   **d)** 12
4. Let *a* and *b* be integers such that *a* divides *b* and *b* divides *a*. What is your
    conclusion?
5. Mark *true* or *false* (*a*, *b*, and *c* are any integers).
    a) Every integer is a multiple of 1. true
    b) 1 is a factor of every integer. true
    c) If *a* is a factor of *b*, then *a* is a factor of −*b*.
    d) *a* is a factor of 0.
    e) If *a* is a factor of *b* and *b* is a factor of *a*, then *a* = *b*. F
    f) If *a* is a factor of *b*, then *a* < *b*.
    g) If *a* < *b*, then *a* is a factor of *b*.
    h) If *a* is a factor of *b* and *b* is a factor of *c*, then *a* is a factor of *c*.
    i) If *a* is not a factor of *b*, then *b* is not a factor of *a*.
    j) There is a 1-1 correspondence between the set of days of a week and
        the set {0,1,2,3,4,5,6}.

can't divide by zero

6. Give a counterexample to each of the false statements in problem 5.
7. Compute the integers *q* and *r* of the division algorithm if
    a) *a* = 28 and *b* = 5   **b)** *a* = 78 and *b* = 11
    c) *a* = −57 and *b* = 8   **d)** *a* = −325 and *b* = 13
    e) *a* = −23 and *b* = 25   **f)** *a* = 57 and *b* = 75
★8. Prove each of the following:
    a) If *a* divides *b* and *b* divides *c*, then *a* divides *c*.
    b) If *a* divides *b* and *a* divides *c*, then *a* divides (*b* + *c*).
    c) If *a* divides *b* and *a* divides *c*, then *a* divides (*b* − *c*).

5.2 EVEN AND ODD INTEGERS

Our aim in this section is to divide the set of integers into two disjoint sets, the
sets of even and odd integers, and to discuss some simple properties of these
sets.

Let *a* be any integer. Let's apply the division algorithm with the integers *a*
and 2, treating *a* as the dividend. Then, there exist unique integers *q* and *r* such
that *a* = 2*q* + *r*, where 0 ≤ *r* < 2. Since 0 ≤ *r* < 2, there are only two possible

values for the remainder r: either $r = 0$ or $r = 1$. If $r = 0$, then $a = 2q$ and a is divisible by 2. If $r = 1$, then $a = 2q + 1$. Thus, we observe that every integer a is either of the form $a = 2m$ or $a = 2m + 1$ for some integer m. Accordingly, we make the following definition:

Definition 5.1   An integer a is called an *even integer* if it can be expressed in the form $a = 2m$ for some integer m. An integer a is called an *odd integer* if it can be expressed in the form $a = 2n + 1$ for some integer n.

Example 5.4 $2 = 2 \cdot 1 + 0,$ *an even integer*

$$-2 = 2 \cdot (-1) + 0, \quad \textit{an even integer}$$

$$3 = 2 \cdot 1 + 1, \quad \textit{an odd integer}$$

$$-3 = 2 \cdot (-2) + 1, \quad \textit{an odd integer}$$

$$0 = 2 \cdot 0 + 0, \quad \textit{an even integer}$$

Notice that every even integer is divisible by 2, that is, the remainder is 0 when divided by 2. Every odd integer when divided by 2 leaves the remainder 1. Therefore, no integer is both even and odd. Consequently the sets of even and odd integers are disjoint; also, their union is the entire set of integers.

Motorists with registration plates ending in even numbers will be able to buy gasoline in Massachusetts today, second day of state's 'Oregon plan.'

Fig. 5.2

Today being the 13th of the month —an "odd" day—motorists whose registration plates end in an odd digit may buy gasoline under state's voluntary distribution plan. Owners of vanity plates are also eligible.

Fig. 5.3

The following theorem was known to the followers of Pythagoras.

Theorem 5.2

a) *The sum and product of any two even integers is even.*

b) *The sum of any two odd integers is even.*

c) *The product of any two odd integers is odd.*

d) *The product of any even integer and any odd integer is even.*

e) *The sum of any even integer and any odd integer is odd.*

Proof: We will prove only part (c) and leave the other parts to the reader.

c) Let a and b be any odd integers. Then there exist integers m and n such that $a = 2m + 1$ and $b = 2n + 1$, by Definition 5.1, and

$$ab = (2m + 1)(2n + 1)$$
$$= 2m(2n + 1) + 1 \cdot (2n + 1)$$
$$= 4mn + 2m + 2n + 1$$
$$= 2(2mn + m + n) + 1$$

Therefore, ab is an odd integer.

This theorem may be summarized as follows:

even + even = even;	even · even = even
even + odd = odd;	even · odd = even
odd + odd = even;	odd · odd = odd

It now follows from the theorem that the square of any even integer is even and that of any odd integer is odd. Conversely, we now have the following result:

Theorem 5.3

a) *If the square of an integer is even, then the integer itself is even.*

b) *If the square of an integer is odd, then the integer itself is odd.*

Proof: We will prove only part (*b*) of the theorem and leave the other half to the reader. Let a be any integer such that a^2 is odd. Our aim is to show that then a must be odd. Now assume the contrary, that is, assume a is even. Then a^2 is even, by Theorem 5.2. But this contradicts our assumption that a^2 is odd. Thus, if a^2 is odd, so must be a.

Exercise 5.2

1. Mark *true* or *false*.

    a) 0 is neither even nor odd.

    b) There is no remainder when an even integer is divided by 2.

    c) The set of even integers is closed under addition.

    d) The set of odd integers is closed under addition.

    e) The set of even integers is closed under multiplication.

    f) The set of odd integers is closed under multiplication.

    g) Every even integer is divisible by 2.

    h) Every odd integer is the sum of an even integer and the integer 1.

    ★**i)** There exists a 1-1 correspondence between the sets of even and odd integers.

2. Give a counterexample to each of the false statements in problem 1.

License Plate Plan

Fig. 5.4

3. Identify as even or odd.
   a) 37   b) 64   c) $31 \cdot 8$ d) $21 \cdot 73$
   e) $26 \cdot 62$ f) $21 + 31$ g) $53 - 27$ h) $58 + 61$
   i) $74 - 47$ j) $68 - 39$ k) $51 - 15$ l) $15(51 + 13)$

4. What are the possible values for the remainder when an integer is divided by 3?

5. Prove that the sum of any two even integers is even.

6. Prove that the sum of any two odd integers is even.

7. Prove that the product of any two even integers is even.

8. Prove that if the square of an integer is even, then the integer itself is even. [*Hint*: imitate the proof given in Theorem 5.3.]

9. Show that the product of any two consecutive integers is even.

10. Prove that the sum of two integers of the form $4k + 1$ is even.

★11. Prove that the product of two integers of the form $3k + 1$ is also of the same form.

★12. Prove that every odd integer is of the form $4k + 1$ or $4k + 3$. [*Hint*: use the division algorithm.]

5.3 PRIME AND COMPOSITE NUMBERS

For the rest of the chapter, we will be considering only positive integers. In this section, we split the set of positive integers into three mutually disjoint sets and discuss some of the very fascinating properties of them.

Consider the natural numbers 5 and 6. The divisors of 5 are ± 1 and ± 5, which are the trivial factors of 5. The divisors of 6 are ± 1, ± 2, ± 3 and ± 6. Thus 6 has divisors other than the trivial ones. Accordingly, we like to distinguish those natural numbers that have only the trivial factors from those that have others.

Definition 5.2 A natural number greater than one is called a *prime number* or simply a *prime* if it has only trivial factors. A natural number greater than 1 that is not a prime is called a *composite number* or simply a *composite*.

Example 5.5 The numbers 2, 3, 5, 7, 11, 13, 17, 19, 23, and 29 are the first ten prime numbers, while 4, 6, 8, 9, 10, 12, 14, 15, 16, and 18 are the first ten composite numbers.

Notice that Definition 5.2 excludes the natural number 1 from the set of prime numbers and from the set of composite numbers. Also, the set of natural numbers is the union of three mutually disjoint sets: the sets of primes, the set of composites, and {1}, as shown in Fig. 5.5.

How do we determine if a positive integer x is a prime? We must check if it is divisible by any prime smaller than x. This procedure is difficult if x is a fairly large number. However, if d is a number less than x, but $d^2 > x$, then we need only check if x is divisible by all the primes smaller than d. If x is not divisible by

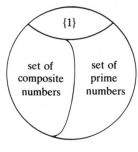

Fig. 5.5

any prime less than *d*, then *x* must be a prime. (Why?) This procedure is illustrated in the following examples.

Example 5.6 Determine whether 71 is a prime or a composite.

Solution: Observe that $8^2 = 64 < 71 < 81 = 9^2$. The prime numbers less than 9 are 2, 3, 5, and 7, and none of these is a divisor of 71. Hence 71 is a prime.

Example 5.7 Determine whether 161 is a prime or not.

Solution: Notice that $13^2 = 169 > 161 > 144 = 12^2$. The prime numbers less than 13 are 2, 3, 5, 7, and 11. The primes 2, 3, and 5 do not divide 161, but 7 does, and $161 \div 7 = 23$. Therefore, 161 is composite.

We now proceed to discuss a fundamental property of natural numbers. Notice that the number 120 can be written as a product of natural numbers in several ways, as shown below:

$$120 = 1 \cdot 120$$
$$= 2 \cdot 60$$
$$= 3 \cdot 40$$
$$= 4 \cdot 30$$
$$= 3 \cdot 4 \cdot 10$$
$$= 2 \cdot 4 \cdot 15$$
$$= 2 \cdot 2 \cdot 2 \cdot 3 \cdot 5$$

Each of these is called a *factorization* of 120. Of these different factorizations, the last one deserves special attention. All the factors in the last factorization of 120 are prime numbers, so it is called a *prime factorization* of 120. Also observe that $120 = 2 \cdot 3 \cdot 2 \cdot 2 \cdot 5 = 3 \cdot 5 \cdot 2 \cdot 2 \cdot 2$, etc., so that the prime factorization is unique except for the order in which the factors are written. This brings about the following result:

Theorem 5.4 (*Prime Factorization Theorem*) *Every composite number can be written as a product of primes, and this product is unique except for the order of the factors.*

The statement of this theorem may seem somewhat simple and obvious. This is one of the cornerstones of number theory. Because of its fundamental nature, the theorem is also called the *fundamental theorem of arithmetic*. The prime factorization theorem states that every composite number can be obtained as a product of primes. Thus, prime numbers are the building blocks of natural numbers and hence of integers.

Definition 5.3 Let a be any integer and n any natural number. Then

$$a^n = \underbrace{a \cdot a \ldots a}_{n \text{ times}}$$

is called the *nth power* of a. The number n is called the *exponent* or *power* of a and a the *base*. Also, if a is any nonzero integer, we define $a^0 = 1$.

For example, $5^3 = 5 \cdot 5 \cdot 5 = 125$ and $3^5 = 3 \cdot 3 \cdot 3 \cdot 3 \cdot 3 = 243$. Notice that $2^3 \cdot 2^2 = 8 \cdot 4 = 32 = 2^5 = 2^{3+2}$. Also, $3 \cdot 3^4 = 3 \cdot 81 = 243 = 3^5 = 3^{1+4}$.

This now suggests that if m and n are any natural numbers, then $a^m \cdot a^n = a^{m+n}$ for any integer a. We will discuss more about exponents in Section 7.11.

In the prime factorization of 120, the prime number 2 appears 3 times, 3 once, and 5 also once. Using the above definition, we can write the factorization as $120 = 2^3 \cdot 3 \cdot 5$.

There are two commonly used techniques of finding the prime factorization of a composite number. The first method involves finding the prime divisors, starting from the smallest prime, 2, and continuing until all the prime factors have been extracted, as illustrated in the following example:

Example 5.8 Find the prime factorization of 2520.

Solution: Since the smallest prime, 2, divides 2520, we have

$$2520 = 2 \cdot 1260$$

Now we observe that 2 is a divisor of 1260. Therefore,

$$2520 = 2 \cdot 2 \cdot 630$$

Also 2 is a factor of 630 and hence we have

$$2520 = 2 \cdot 2 \cdot 2 \cdot 315$$

Since 2 does not divide 315, check if the next prime, 3, is a factor of 315. Indeed, 3 divides 315. Thus,

$$2520 = 2 \cdot 2 \cdot 2 \cdot 3 \cdot 105$$

Since 3 is a factor of 105, we have

$$2520 = 2 \cdot 2 \cdot 2 \cdot 3 \cdot 3 \cdot 35$$

Since 35 is not a prime, we continue with 35. Since 2 and 3 are not divisors of 35, check if 5 is a factor or not: 5 divides 35 and we have

$$2520 = 2 \cdot 2 \cdot 2 \cdot 3 \cdot 3 \cdot 5 \cdot 7$$

where 7 is a prime and the procedure stops here. This is a prime factorization of 2520. Using Definition 5.4, we can now rewrite this as

$$2520 = 2^3 \cdot 3^2 \cdot 5 \cdot 7$$

where the primes 2, 3, 5, and 7 occur in ascending order of magnitude.

The second method of finding the prime factorization of a composite number is not only simple, but more efficient. It involves expressing the given composite number n as a product of two natural numbers r and s and then continuing with r and s until all the prime factors have been obtained, as illustrated in the following example:

Example 5.9   Find the prime factorization of 2520 by the second method.
Solution: Notice that

$$2520 = 40 \cdot 63$$

Now, $40 = 4 \cdot 10$ and $63 = 9 \cdot 7$. Therefore,

$$2520 = 4 \cdot 10 \cdot 9 \cdot 7$$

Observe that 4, 10, and 9 are not primes and hence we continue factorizing:

$$4 = 2 \cdot 2$$
$$10 = 2 \cdot 5$$
$$9 = 3 \cdot 3$$

Thus, $2520 = 2 \cdot 2 \cdot 2 \cdot 5 \cdot 3 \cdot 3 \cdot 7$

This gives a prime factorization of 2520. Rearranging the primes, we have

$$2520 = 2^3 \cdot 3^2 \cdot 5 \cdot 7$$

This method can be illustrated using a tree diagram, usually called a *factor tree*. In such a diagram, if a divides b, we write b just above a and connect them by a straight line. A factor tree of 2520 is illustrated in Fig. 5.6.

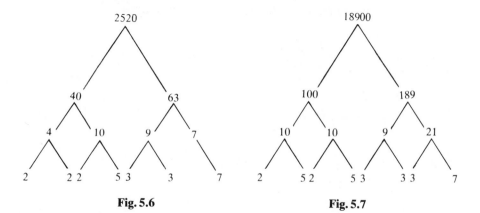

Fig. 5.6   Fig. 5.7

Figure 5.7 demonstrates a factor tree of the composite number 18,900, where $18,900 = 2 \cdot 5 \cdot 2 \cdot 5 \cdot 3 \cdot 3 \cdot 3 \cdot 7$. The prime factorization of 18,900 is given by

$$18900 = 2^2 \cdot 3^3 \cdot 5^2 \cdot 7$$

In Section 1.11, we proved that the set of natural numbers is infinite. Does this mean that the set of primes is also infinite? Do you believe that there are only 25 primes < 100 and they are distributed unevenly over the range? Does this generate a feeling that the primes are scarce? Let's not jump into any easy conclusions. About 300 B.C., Euclid (330?–275 B.C.) proved that the set of primes is infinite.

Theorem 5.5 (*Euclid*)   *The set of prime numbers is infinite.*

Since 2 times a prime is composite, it now follows that the set of composite numbers is also infinite. (Why?)

Exercise 5.3

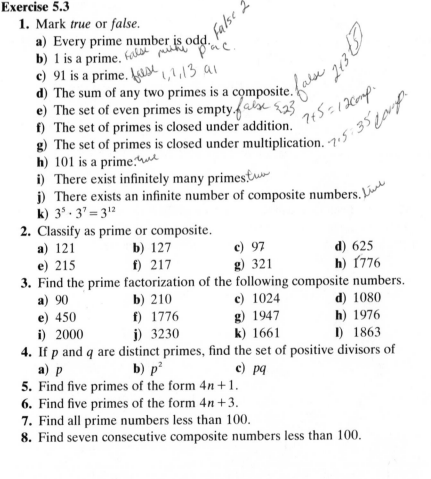

1. Mark *true* or *false*.
    a) Every prime number is odd.
    b) 1 is a prime.
    c) 91 is a prime.
    d) The sum of any two primes is a composite.
    e) The set of even primes is empty.
    f) The set of primes is closed under addition.
    g) The set of primes is closed under multiplication.
    h) 101 is a prime.
    i) There exist infinitely many primes.
    j) There exists an infinite number of composite numbers.
    k) $3^5 \cdot 3^7 = 3^{12}$
2. Classify as prime or composite.
    a) 121   b) 127   c) 97   d) 625
    e) 215   f) 217   g) 321   h) 1776
3. Find the prime factorization of the following composite numbers.
    a) 90   b) 210   c) 1024   d) 1080
    e) 450   f) 1776   g) 1947   h) 1976
    i) 2000   j) 3230   k) 1661   l) 1863
4. If *p* and *q* are distinct primes, find the set of positive divisors of
    a) *p*   b) p^2 c) *pq*
5. Find five primes of the form $4n + 1$.
6. Find five primes of the form $4n + 3$.
7. Find all prime numbers less than 100.
8. Find seven consecutive composite numbers less than 100.

9. Find four consecutive composite numbers. [*Hint*: $(n+1)!+2$, $(n+1)!+3, \ldots, (n+1)!+(n+1)$ are n consecutive composite numbers, where $n!$ (*n factorial*) $= 1 \cdot 2 \cdot 3 \ldots (n-1)n$.]

10. Find five consecutive composite numbers.

★11. Prove that every odd prime is of the form $4n+1$ or $4n+3$.

★12. Prove that 2 and 3 are the only two consecutive primes.

★13. Consider the set of numbers $A = \{3n+1 \mid n \in W\}$. A number $p(\neq 1)$ in A is called a *prime* if its only divisors in A are 1 and p.

    a) Find five primes and five composites in the set A.

    b) Show that the prime factorization theorem fails in this set of numbers. [*Hint*: Consider the number 460.]

★14. Repeat problem 13 with the set of numbers $B = \{4n+1 \mid n \in W\}$. [*Hint*: Consider the number 693.]

5.4   GREATEST COMMON DIVISOR

Let a be any natural number. Recall that d is a divisor of a if and only if there exists an integer x such that $a = dx$. It may happen that d is a divisor of two or more natural numbers. For instance, 3 is a divisor of both 15 and 24. Accordingly, we make the following definition:

Definition 5.4   A natural number d is called a *common divisor* of the natural numbers a and b if it is a positive divisor of both a and b.

Since 1 divides every natural number, we notice that 1 is a common divisor of any two natural numbers. Consequently, the set of common divisors of any two natural numbers is nonempty.

Let A be the set of positive divisors of 18. Then $A = \{1, 2, 3, 6, 9, 18\}$. The set of positive divisors of 24 is given by $B = \{1, 2, 3, 4, 6, 8, 12, 24\}$. Then $A \cap B = \{1, 2, 3, 6\}$ is the set of common divisors of 18 and 24. Among the common divisors 1, 2, 3, and 6 of 18 and 24, the number 6 has a unique place, in that it is the largest of all the common divisors. Also observe that every common divisor of 18 and 24 is also a divisor of 6.

Definition 5.5   The *greatest common divisor* (usually abbreviated gcd) of two positive integers a and b is the largest positive integer that divides both a and b. The gcd of a and b is usually denoted gcd$\{a,b\}$.

From our discussion above, we observe that 6 is the greatest common divisor of 18 and 24, that is, gcd$\{18,24\} = 6$.

There are four methods of finding the gcd of two natural numbers. The first method, which has already been illustrated, is a consequence of the definition. This method is fairly simple if the numbers are small enough.

Example 5.10   Find the gcd of 24 and 36.

Solution: $A = \{1, 2, 3, 4, 6, 8, 12, 24\}$ and $B = \{1, 2, 3, 4, 6, 9, 12, 18, 36\}$ are the sets of positive divisors of 24 and 36, respectively. Then $A \cap B = \{1, 2, 3, 4, 6, 12\}$ is the set of common divisors of 24 and 36. Since the largest member of $A \cap B$ is 12, the gcd of 24 and 36 is 12.

The second method involves finding the prime factorizations of the natural numbers and then taking the product of common prime factors in the factorizations, as illustrated by the following example:

Example 5.11 Find the gcd of 60 and 90.

Solution:
$$60 = 2^2 \cdot 3 \cdot 5$$
$$90 = 2 \cdot 3^2 \cdot 5$$

Notice that 2^2 is a factor of 60 and 2 is a factor of 90. Thus, 2 is a common factor of 60 and 90. Similarly, 3 and 5 are also common factors of 60 and 90. Thus, $2 \cdot 3 \cdot 5 = 30$ is the gcd of 60 and 90.

This method could be quite tedious if the numbers are quite large. Hence we look for a practical and systematic procedure to find the gcd of any two natural numbers. This method, usually called the *Euclidean algorithm* after Euclid, depends on repeated application of the division algorithm.

Example 5.12 (*Euclidean algorithm*) Find the gcd of 105 and 1980.

Solution: We apply the division algorithm, treating 1980 (the larger of the two) as the dividend and 105 as the divisor:

$$1980 = 18 \cdot 105 + 90$$

Repeat this with 105 and 90, using 105 as the dividend:

$$105 = 1 \cdot 90 + 15$$

Apply the division algorithm again with 90 and 15, treating 90 as the dividend. Thus we have

$$90 = 6 \cdot 15 + 0$$

Once we get the remainder 0, the procedure stops. Thus we have

$$1980 = 18 \cdot 105 + 90$$
$$105 = 1 \cdot 90 + 15$$
$$90 = 6 \cdot 15 + 0$$

The last nonzero remainder we get in this procedure is the gcd of 1980 and 105. That is, $\gcd\{1980, 105\} = 15$.

Let's now actually verify this using the second method. Notice that $1980 = 2^2 \cdot 3^2 \cdot 5 \cdot 11$ and $105 = 3 \cdot 5 \cdot 7$. It follows clearly that $\gcd\{1980, 105\} = 15$.

The fourth method to find the gcd of two natural numbers involves division by primes beginning with the smallest prime, as demonstrated in the following example:

Example 5.13 Find the gcd of 36 and 90.
Solution:

2	36	90
3	18	45
3	6	15
	2	5

We observe, by inspection, that 2 is a common divisor of 36 and 90. They leave the quotients 18 and 45 on division by 2. Now divide 18 and 45 by 3 to leave quotients 6 and 15. Continue this procedure until the numbers in the last row have no common prime factors. The gcd of the numbers 36 and 90 is now obtained by taking the product $2 \cdot 3 \cdot 3 = 18$ of all the common divisors on the left.

We can now extend our definition of the gcd of two natural numbers to more than two natural numbers in an obvious manner.

Example 5.14 Find the gcd of 24, 56, and 60.
Solution:
$$24 = 2^3 \cdot 3$$
$$56 = 2^3 \cdot 7$$
$$60 = 2^2 \cdot 3 \cdot 5$$

Clearly, 2^2 is the only common divisor of 24, 56, and 60, other than 1. Thus the gcd of 24, 56, and 60 is given by

$$\gcd\{24,56,60\} = 2^2 = 4$$

Definition 5.6 Two natural numbers are said to be *relatively prime* if their gcd is 1.

For example, 4 and 15 are relatively prime numbers, whereas 6 and 10 are not (why?).

Exercise 5.4

1. Find the gcd of a and b if
   a) $a = 2^2 \cdot 3 \cdot 5$ and $b = 2 \cdot 3 \cdot 5^2$
   b) $a = 3^2 \cdot 5^3 \cdot 7$ and $b = 2 \cdot 5 \cdot 7$
   c) $a = 5 \cdot 7^3 \cdot 11^2$ and $b = 2 \cdot 3^2 \cdot 7 \cdot 11$
   d) $a = 5^{11} \cdot 11^5$ and $b = 3^5 \cdot 5^3$

2. Find the gcd of
   a) 108 and 40   b) 48 and 162   c) 72 and 108
   d) 294 and 450   e) 175 and 192   f) 85 and 525

3. Use the Euclidean algorithm to find the gcd of
   a) 429 and 117   b) 1661 and 77   c) 1947 and 327
   d) 1973 and 379   e) 1962 and 1976   f) 2000 and 1776

4. Find the gcd of
   a) 24, 28, and 36   b) 54, 120, and 210   c) 72, 77, and 87
5. What is the gcd of a and b if
   a) a divides b b) b divides a c) $a = 1$ d) $a = b$ _value_
   e) a and b are consecutive natural numbers $\{5, 6\}$ _Common_
   f) a and b are distinct primes
6. Mark *true* or *false* (a,b,c are any natural numbers, p any prime).
   a) The set of common divisors of any two positive integers is nonempty. _true_
   b) The set of common divisors of any two natural numbers is finite. _true_
   c) If a divides b, gcd$\{a,b\} = b$. _False - a/b   gcd has to be g because it is smaller than divides_
   d) If p does not divide a, gcd$\{p,a\} = 1$. _True_
   e) If a does not divide b, gcd$\{a,b\} = 1$. _false_
   f) Any two consecutive natural numbers are relatively prime.
   g) If d divides ab, then d divides a or d divides b.
   h) gcd$\{a,p\} = 1$
   i) If gcd$\{a,b\} = 1$ and gcd$\{b,c\} = 1$, then gcd$\{a,c\} = 1$.
   j) If gcd$\{a,b\} = 2$ and gcd$\{b,c\} = 2$, then gcd$\{a,c\} = 2$.
7. Give a counterexample to each of the false statements in problem 6.
8. John has two sticks, one 18 inches long and the other 24 inches long. If he wants to cut them up into pieces all of equal length, what are the largest-length pieces he can cut?
★9. Prove that if a prime p does not divide the natural number a, then p and a are relatively prime.
★10. Prove that any two consecutive natural numbers are relatively prime.

5.5   LEAST COMMON MULTIPLE

Closely associated with the notion of the gcd of two natural numbers is their least common multiple. Recall that if a divides b, then b is a multiple of a. For instance, 4 is a multiple of 2, 12 is a multiple of 3.

Let A be the set of all positive multiples of 12. Then $A = \{12, 24, 36, 48,...\}$. The set of all positive multiples of 18 is given by $B = \{18, 36, 54, 72,...\}$. Then $A \cap B = \{36, 72, 108, 144,...\}$ is the set of all common positive multiples of 12 and 18. Since the multiples increase in size, we can not talk about the largest element in $A \cap B$. However, we can certainly talk about the smallest element, 36, in $A \cap B$. The number 36 is called the least common multiple of 12 and 18. It is the smallest positive integer divisible by both 12 and 18.

Definition 5.7   The *least common multiple* (usually abbreviated lcm) of two natural numbers a and b is the smallest natural number that is a multiple of both a and b, denoted lcm$\{a,b\}$.

It follows from the definition that the lcm of two natural numbers a and b is the smallest positive integer divisible by both a and b.

We have already discussed one way of finding the lcm of two natural numbers. The second method uses the prime factorization of the natural numbers, as illustrated in the following example:

Example 5.15 Find the lcm of 15 and 18.

Solution:
$$15 = 3 \cdot 5$$
$$18 = 2 \cdot 3^2$$

The prime factors appearing in the prime factorizations of 15 and 18 are 2, 3, 5. The maximum number of times the primes 2, 3, and 5 appear in either factorization is 1, 2, and 1, respectively. Accordingly, the lcm is obtained by taking the product of 2, 3^2, and 5. Thus, lcm$\{15,18\} = 2 \cdot 3^2 \cdot 5 = 90$.

Example 5.16 Find lcm$\{24,35\}$.

Solution:
$$24 = 2^3 \cdot 3$$
$$35 = 5 \cdot 7$$

The lcm of 24 and 35 is given by $2^3 \cdot 3 \cdot 5 \cdot 7 = 24 \cdot 35 = 840$. Notice that 24 and 35 are relatively prime and their lcm is given by their product.

Example 5.17 Find lcm$\{24,90\}$.

Solution:
$$24 = 2^3 \cdot 3$$
$$90 = 2 \cdot 3^2 \cdot 5$$
$$\text{lcm}\{24,90\} = 2^3 \cdot 3^2 \cdot 5 = 360$$

Observe that gcd$\{24,90\} = 2 \cdot 3 = 6$ and $360 = \dfrac{24 \cdot 90}{6}$. That is,

$$\text{lcm}\{24,90\} = \frac{24 \cdot 90}{\text{gcd}\{24,90\}}$$

This example now suggests the following remarkable result:

Theorem 5.6 *If a and b are any two natural numbers, then*

$$lcm\{a,b\} = \frac{ab}{gcd\{a,b\}}$$

This theorem, which expresses the lcm of two natural numbers in terms of their gcd, gives a third method to compute their lcm. Once we know the gcd of two natural numbers, we can obtain their lcm by dividing the product of the numbers by their gcd.

Example 5.18 Find gcd$\{36,54\}$ and lcm$\{36,54\}$.

Solution:
$$36 = 2^2 \cdot 3^2$$
$$54 = 2 \cdot 3^3$$
$$\text{gcd}\{36,54\} = 2 \cdot 3^2 = 18$$
$$\text{lcm}\{36,54\} = \frac{36 \cdot 54}{18} = 108$$

The fourth method to find the lcm of two natural numbers involves division by primes beginning with the smallest prime, as illustrated in the following example:

Example 5.19   Find the lcm of 18 and 24.

Solution:

$$
\begin{array}{r|cc}
2 & 18 & 24 \\
3 & 9 & 12 \\ \hline
 & 3 & 4
\end{array}
$$

By inspection we observe that 2 is a divisor of both 18 and 24. They leave the quotients 9 and 12 on division by 2. Now divide 9 and 12 by 3 to leave the quotients 3 and 4. Continue this procedure until the last row consists of ones and/or two relatively prime numbers. The lcm of 18 and 24 is obtained by taking the product of the numbers in the last row and all the common divisors on the left. Thus,

$$\text{lcm}\{18,24\} = 2 \cdot 3 \cdot 3 \cdot 4 = 72$$

We now can extend the definition of the lcm of two natural numbers to more than two natural numbers as illustrated by the following example:

Example 5.20   Find the lcm of 12, 18, and 24.

Solution:

$$12 = 2^2 \cdot 3$$
$$18 = 2 \cdot 3^2$$
$$24 = 2^3 \cdot 3$$

Then the lcm of 12, 18, and 24 is given by

$$\text{lcm}\{12,18,24\} = 2^3 \cdot 3^2 = 72$$

The scheme illustrated in example 5.19 to find the lcm of two natural numbers is faster and more convenient for finding the lcm of more than two numbers, especially if they are fairly large.

Example 5.21   Find the lcm of 72, 108, 246, and 316.

Solution:

$$
\begin{array}{r|cccc}
2 & 72 & 108 & 246 & 316 \\
2 & 36 & 54 & 123 & 158 \\
3 & 18 & 27 & 123 & 79 \\
3 & 6 & 9 & 41 & 79 \\ \hline
 & 2 & 3 & 41 & 79
\end{array}
$$

Notice that in the third row of this procedure, since 3 does not divide 79, the 79 is simply brought down to the fourth row. In the fourth row, since 3 does not divide both 41 and 79, they are brought down to the fifth row. Our procedure stops at the fifth row as all the numbers on this row are primes. Thus, the lcm of 72, 108, 246, and 316 is $2 \cdot 2 \cdot 3 \cdot 3 \cdot 2 \cdot 3 \cdot 41 \cdot 79 = 699,624$.

Exercise 5.5

1. Find the lcm of a and b if $2^3 \cdot 3^2 \cdot 5$
   $6 \quad 9 \cdot 5 \quad$ 525

   a) $a = 2^3 \cdot 3 \cdot 5$ and $b = 2 \cdot 3^2 \cdot 5$ **b)** $a = 2 \cdot 3 \cdot 7^2$ and $b = 2 \cdot 3^2 \cdot 7$
   c) $a = 3^5 \cdot 5^3 \cdot 7$ and $b = 3^3 \cdot 5^5 \cdot 7$ **d)** $a = 2^3 \cdot 3^2$ and $b = 5^2 \cdot 7$

2. Find the lcm of
   a) 24 and 30   **b)** 28 and 63   **c)** 96 and 108
   d) 48 and 96   **e)** 110 and 210   **f)** 65 and 66

3. Find the lcm of
   a) 24, 36, and 48   **b)** 30, 40, and 50   **c)** 54, 72, and 96 (in back)
   d) 15, 18, and 24   **e)** 12, 13, and 14   **f)** 132, 123, and 312

4. What is the lcm of a and b if
   a) a divides b **b)** b divides a **c)** $a = 1$ **d)** $a = b$
   e) a and b are distinct primes
   f) a and b are consecutive natural numbers

5. Find lcm$\{a,b\}$ if gcd$\{a,b\} = 3$ and $ab = 693$.
6. Find gcd$\{a,b\}$ if $ab = 192$ and lcm$\{a,b\} = 48$.
7. Find gcd$\{a,b\}$ if $ab = 156$ and lcm$\{a,b\} = 156$.
8. Find ab if gcd$\{a,b\} = 3$ and lcm$\{a,b\} = 53$.
9. Mark *true* or *false* (a,b,c are any natural numbers, p is a prime).
   a) The set of common multiples of any two natural numbers is finite. *False*
   b) The lcm of two primes is their product. *False*
   c) The lcm of two consecutive natural numbers is their product. *True*
   d) The lcm of two distinct primes is their product. *True*
   e) If gcd$\{a,b\} = 1$, then lcm$\{a,b\} = ab$.
   f) If p does not divide a, then lcm$\{p,a\} = pa$.
   g) If lcm$\{a,b\} = 1$, then $a = b = 1$.
   h) If lcm$\{a,b\} = b$, then $a = 1$.
   i) If lcm$\{a,b\} = b$, then a divides b.

10. Give a counterexample to each of the false statements in problem 9.
11. Show that if lcm$\{a,b\} = ab$, then a and b are relatively prime.
12. Evaluate a if lcm$\{a, a+1\} = 132$.
13. Evaluate lcm$\{a,b,c\}$ and $\dfrac{abc}{\gcd\{a,b,c\}}$ if
    a) $a = 20$, $b = 24$, and $c = 30$
    b) $a = 12$, $b = 25$, and $c = 30$
    c) $a = 6$, $b = 35$, and $c = 143$
    d) $a = 45$, $b = 52$, and $c = 77$

14. Fred and his friend Frank can throw a javelin exactly 36 yards and 42 yards, respectively. If they throw their javelins from the same spot and in the same direction, what is the minimum number of throws required by each so that their javelins will have been thrown the same distance? What is this distance?

15. One car can travel at 45 miles per hour and another at 60 miles per hour. If they travel the same route from the same place, what is the minimum number of hours taken by each so that the cars will have traveled the same distance? What is this distance?

5.6 EXOTIC NUMBERS

Are some numbers "luckier" than others? Do some numbers possess some "magic power"? Is 13 an unlucky number? Some people believe that Apollo 13 could not accomplish its mission because 13 is an unlucky number. The number 13 is used 13 times in the Bible. The number 7, a prime, has been considered a lucky number by several civilizations. There are seven days a week and seven days in the Biblical creation. It is used 395 times in the Bible.

The number 6 was once considered to have some marvelous powers. It has a special place in number theory. Recall that God's work of creation was completed over a period of six days. St. Augustin believed that God's work is perfect because the number 6 is "perfect." What are the positive divisors of 6? They are 1, 2, 3, and 6. A *proper divisor* of a number is a positive divisor other than the number. Observe that the sum $1 + 2 + 3$ of the proper divisors of 6 is 6. Such a number is said to be perfect.

Definition 5.8 A natural number is called a *perfect number* if it is equal to the sum of its proper divisors.

Thus 6 is a perfect number. The next perfect number is 28, since $28 = 1 + 2 + 4 + 7 + 14$, where 1, 2, 4, 7, and 14 are the proper divisors of 28.

Perfect numbers have been a center of attention since the ancient Greeks. They have been considered "good" numbers.

The first five perfect numbers are 6, 28, 496, 8,128, and 33,550,336. Observe that these are all of the form $2^{n-1}(2^n - 1)$, where $2^n - 1$ is a prime! How many perfect numbers are there? Are there infinitely many primes? How many of them are even? How many are odd?

No one knows how many perfect numbers there are. However, it is known that every even perfect number must end in 6 or 8. No one knows if there are any odd perfect numbers. In 1968, Bryant Tuckerman proved that an odd perfect number, if it exists, must be larger than 10^{36}.

Primes of the form $2^n - 1$ play an important role in the discussion of even perfect numbers. This created an interest in the study of primes of the form $2^n - 1$. If $n = 2$, then $2^n - 1 = 3$, a prime. For $n = 3$, we have $2^n - 1 = 7$, a prime. For $n = 4$, we get $2^n - 1 = 15$, not a prime. Prime numbers of this form are called *Mersenne primes*, in honor of the Franciscan priest Marin Mersenne (1588–1648). The numbers 3 ($n = 2$), 5 ($n = 3$), 31 ($n = 5$), and 127 ($n = 7$) are the first four Mersenne primes. For $n = 11$, $2^n - 1 = 2^{11} - 1 = 2047 = 23 \cdot 89$, not a prime.

That there are infinitely many Mersenne primes remains an unsolved problem. If this problem can be settled in the affirmative, then it will, in turn, tell us that there is an infinite number of even perfect numbers and hence an infinite number of perfect numbers. The largest known prime, $2^{19937} - 1$, was found in 1971 by Bryant Tuckerman (see Fig. 5.8). Consequently, the largest known even perfect number is $2^{19936}(2^{19937} - 1)$.

 Thomas J. Watson Research Center
P.O. Box 218
Yorktown Heights, New York 10598

Fig. 5.8

Fig. 5.9

There exists a second type of numbers that has been a source of fun and excitement for several investigators. They are the so-called amicable numbers. **Definition 5.9**   Two natural numbers are said to be *amicable* (*friendly*) if each number is the same as the sum of the proper divisors of the other.

Thus 220 and 284, 1184 and 1210, 2620 and 2924, 5020 and 5564, 6232 and 6368 are some examples of amicable pairs. The second amicable pair, 1184 and 1210, was discovered by a 16-year-old boy in 1866. It is not known if there are infinitely many amicable pairs or not.

Fermat, in his search for a formula that would yield all the primes, conjectured that numbers of the form $2^{2^n} + 1$ are primes for every whole number n. This was in about 1640. He was right for $n = 0, 1, 2, 3$, and 4. In 1739, Euler established the falsity of his conjecture by producing a counterexample. He showed that $2^{2^5} + 1 = 641 \cdot 6700417$, a composite. No one knows what happens after $n = 5$. Prime numbers of the form $2^{2^n} + 1$ are called Fermat primes.

Another outstanding problem that remains unsolved is concerned with what are called twin primes. Two primes that differ by 2 are called *twin primes*. For instance, 3 and 5, 5 and 7, 11 and 13, 17 and 19, 29 and 31, 41 and 43, 59 and 61, 71 and 73 are the twin primes less than 100. That there are several twin primes less than 100 might generate a feeling that there are infinitely many primes. But there are no twin primes between 700 and 800, and none between

900 and 1000. The conjecture that there are infinitely many twin primes remains unsettled one way or the other.

Before we close this section, we will mention a kind of number that is associated with geometric figures. Any natural number of the form $n(n+1)/2$ is called a *triangular number*. Thus, 1, 3, 6, 10, and 15 are the first five triangular numbers. The number of dots arranged in the form of an equilateral triangle is a triangular number. If you have visited any bowling alley, you might have observed that the bowling pins are arranged in the shape of an equilateral triangle. Accordingly, the number of bowling pins is a triangular number. The first five triangular numbers are represented geometrically in Fig. 5.10.

Fig. 5.10

Any natural number of the form n^2 is called a *square number*. The first five square numbers are 1, 4, 9, 16, and 25. Each of these is the number of dots that can be arranged in the shape of a square, as shown in Fig. 5.11.

Fig. 5.11

Observe that

$$1 = 1$$
$$4 = 1 + 3$$
$$9 = 3 + 6$$
$$16 = 6 + 10$$
$$25 = 10 + 15$$

This suggests that every square number except the first is the sum of two consecutive triangular numbers.

Exercise 5.6

   1. Mark *true* or *false*.
   a) 6 is a perfect number.
   b) Every perfect number is a prime.
   c) Every perfect number ends in 6.
   d) Every prime number is perfect.
   e) Every number of the form $2^n - 1$ is a prime.
   f) 6 and 28 are amicable numbers.
   g) 9 and 11 are twin primes.
   h) Every square number is the sum of two triangular numbers.
   i) The sum of the first five odd natural numbers is a square number.
   j) The sum of the first five natural numbers is a triangular number.
   2. Verify that 496 is a perfect number.
   3. Verify that 220 and 284, 1184 and 1210 are amicable pairs.
   4. Find the first five Mersenne numbers. Which of them are primes?
   5. Find the first five Fermat numbers. Which of them are primes?
   6. Verify that $n^2 - n + 41$ is prime if $n < 41$. Does it yield a prime if $n = 41$? Why?
   7. Find the first ten triangular numbers.
   8. Find the first ten square numbers.
   9. Express the eighth, ninth, and tenth square numbers as the sum of two consecutive triangular numbers.
   10. A *palindrome* is a natural number that reads the same backward as well as forward. For example, 12321 is a palindrome.
   a) Find all the two-digit palindromes.
   b) Find all the three-digit palindromes less than 120.
 ★11. Prove that the sum of two consecutive triangular numbers is a square number.
 ★12. Show that the sum of the first n natural numbers is $n(n + 1)/2$.
 ★13. Find a number that can be expressed as the sum of two cubes in two different ways. [*Hint*: consider the number 1729.]

5.7 THE RABBIT PROBLEM

   Leonardo Fibonacci (1170–1250), the most outstanding mathematician of the Middle Ages, around 1202 proposed the following problem about rabbits in his *Liber Abaci*:

Suppose we have two newborn rabbits, one male and the other female. How many pairs of rabbits will be produced in a year if rabbits begin to bear two months after their birth and each pair of rabbits produces a mixed pair every month from the second month on?

Since we are concerned with rabbit population over a period of one year, let's assume, for convenience, that the original pair of rabbits was born on January 1. Rabbits are assumed to take a month to become mature and become productive two months after their birth. Therefore, there is still only one pair of rabbits by the end of February. However, on March 1 the original pair, being two months old, produces a new mixed pair. By the end of March, we have two pairs of rabbits. On April 1, the original pair produces a new mixed pair, to have three pairs by the end of April, since the pair born in March are still "babies." On May 1, there are two new pairs, one from the original pair and the other from the pair born in March. We have a total of five pairs of rabbits by the end of May. Continuing like this, we would have eight pairs of rabbits by the end of June, 13 pairs by the end of July, etc.

Table 5.1 gives the rabbit population in pairs for the first nine months.

Table 5.1

	Jan.	Feb.	March	April	May	June	July	August	Sept.
Adults in pairs	0	1	1	2	3	5	8	13	21
Babies in pairs	1	0	1	1	2	3	5	8	13
Total in pairs	1	1	2	3	5	8	13	21	34

We can write the number of pairs of rabbits at the end of each month in a sequence: $1, 1, 2, 3, 5, 8, 13, 21, \ldots$. If we examine these numbers carefully, we observe that each number after the second is the sum of the two preceding numbers. For example, the fifth number, 5, is the sum of the third number, 2, and the fourth number, 3. This fact now helps us to find the number of pairs of rabbits at the end of any number of months. The numbers $1, 1, 2, 3, 5, 8, 13, \ldots$ are called *Fibonacci numbers.* Thus, at the given rate, there will be 144 pairs of rabbits at the end of 12 months.

If F_n denotes the nth Fibonacci number, then $F_{n+2} = F_{n+1} + F_n$ where $F_1 = F_2 = 1$. The rabbit growth can conveniently be illustrated using a tree diagram, as shown in Fig. 5.12.

Exercise 5.7

1. Let F_n denote the nth Fibonacci number. Verify that
   $$F_{n+2} \cdot F_n - F_{n+1}^2 = (-1)^{n+1} \text{ for } n = 1, 2, 3, 4, \text{ and } 5.$$
2. Suppose we start with a mixed pair of rabbits. If they bear a new mixed pair every month after their birth, how many pairs of rabbits will there be at the end of 5 months? 12 months?

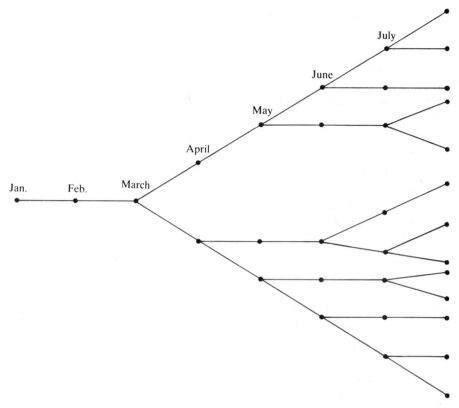

Fig. 5.12

5.8 MAGIC SQUARES

Magic squares were known to the ancient Chinese and Indians. However, it was not until the fifteenth century that they became popular in the West.

A *magic square* is an arrangement of distinct natural numbers in the form of a square such that the sums of the numbers in every row, column, and diagonal are the same. The sum is called the *magic constant* of the square. The magic square is said to be of *order n* if the numbers $1, 2, 3, \ldots, n^2$ are used. For example,

8	1	6
3	5	7
4	9	2

2	25	12
23	13	3
14	1	24

(a)   **Fig. 5.13**   (b)

are both magic squares, the first one being a magic square of order 3. Observe that the sums of numbers in every row, column and diagonal in the square are $15 = 3 \cdot 10/2 = 3(3^2+1)/2$ and 39, respectively. In Fig. 5.14, we give a magic square of order 4. The magic constant of this square is $34 = 4 \cdot 17/2 = 4(4^2+1)/2$.

The magic constants of magic squares of order 3 and 4 now suggest to us the following simple result:

Theorem 5.7 *The magic constant of a magic square of order n is $n(n^2+1)/2$.*

We now illustrate a systematic procedure to construct a magic square of any odd order, say, $n = 5$ (see Fig. 5.15). Draw a square and divide it into 25 subsquares, usually called *cells*. Place the first number, 1, in the middle cell of the top row. Now, starting with 2, write the remaining 24 numbers diagonally from left to right and from bottom to top, subject to the following conditions: (1) If the diagonal leads out of the top of the square, then the next number is written in the bottom cell of the next column, as in the case of the number 2. (2) If the diagonal leads out of the side of the square, then the next number is written in the leftmost cell of the next row, as in the case of the number 4. (3) If the diagonal leads to a cell that has already been occupied, then the next number is written in the cell just below the number entered last, as in the case of the number 6. Observe that the numbers on the rows, columns, and diagonals add up to $65 = 5 \cdot 26/2 = 5(5^2+1)/2$.

1	14	15	4
12	7	6	9
8	11	10	5
13	2	3	16

Fig. 5.14

17	24	1	8	15
23	5	7	14	16
4	6	13	20	22
10	12	19	21	3
11	18	25	2	9

Fig. 5.15

Exercise 5.8

1. Fill in the blank cells to get a magic square with nine entries
   a) with magic constant 39   b) with magic constant 51

5		15
	7	

7		
	11	27

c) with magic constant 33

	19	
13		17

d) with magic constant 135

35	85	
		55

2. Fill in the blank cells to get a magic square

a) of order 4

1		14	4
12		7	
13			16

b) with magic constant 130

			37
29	35	34	
	31	30	
			25

3. Verify that the sums of numbers along the rows, columns, and diagonals in the following square are equal.

3	4	5	1	2
2	3	4	5	1
1	2	3	4	5
5	1	2	3	4
4	5	1	2	3

4. Which of the following are magic squares?

a)

10	1	14
11	9	5
4	15	6

b)

55	1	73
61	43	25
13	85	31

c)

1	2	3	4
4	1	2	3
3	4	1	2
2	3	4	1

d)

1	24	17	15	8
7	5	23	16	14
13	6	4	22	20
19	12	10	3	21
25	18	11	9	2

5. Rearrange the columns so that the following becomes a magic square.

1	8	15	17	24
7	14	16	23	5
13	20	22	4	6
19	21	3	10	12
25	2	9	11	18

6. Fill in the cells of the following diagram with the numbers 1, 2, 3, 4, 5, 6, 7, 8, 9, 10, 11, and 12 such that no two consecutive numbers appear in neighboring cells.

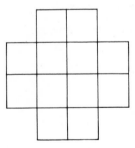

5.9 SUMMARY AND COMMENTS

In this chapter, we discussed several simple and important properties of numbers. We said that a is a divisor of b or b is a multiple of a if there exists an integer c such that $b = ac$. The following important property was discussed: given any integer a and any positive integer b, there exist unique integers q and r such that $a = bq + r$, where $0 \leq r < b$ (the division algorithm).

Using the division algorithm we observed that every integer is either even or odd, but not both. The sum and the product of two even integers is even. The sum of two odd integers is even, whereas the product of two odd integers is odd. The sum of an even and odd integer is odd, and their product is even. These results were known to the Pythagoreans.

We partitioned the set of natural numbers into three mutually disjoint sets: the set of primes, the set of composites, and the set $\{1\}$. Every composite number can be expressed as a product of primes, and this product is unique except for the order of primes (prime factorization theorem). Two methods were discussed to find the prime factorization of a composite number. We observed that there are infinitely many primes.

We discussed four different methods to compute the gcd and the lcm of two natural numbers. The Euclidean algorithm gives us an efficient method to find the gcd of any two natural numbers a and b. If the gcd is 1, then a and b are said to be relatively prime. The $\gcd\{a,b\}$ and the $\text{lcm}\{a,b\}$ of any two natural numbers a and b are connected by the relation $\text{lcm}\{a,b\} = ab/\gcd\{a,b\}$.

Some special types of numbers were discussed: perfect numbers, amicable numbers, twin primes, Mersenne primes, Fermat primes, triangular and square numbers. The Pythagoreans have been given the credit for inventing many of these sets of numbers. The first perfect number, 6, is the area of the right triangle with legs 3 and 4 and hypotenuse 5.

We discussed Fibonacci's well-known rabbit problem and observed that $F_{n+2} = F_{n+1} + F_n$. Since the sum of any two sides of a triangle is greater than the third side, this relation tells us that we can not have a triangle with consecutive Fibonacci numbers as sides. Also, it can be proved that $F_{n-1} \cdot F_{n+1} - F_n^2 = (-1)^n$. Consequently, the area of a rectangle with sides F_{n-1} and F_{n+1} and that of a square with side F_n differ by a square unit.

The following challenging unsolved problems were presented:

1. Are there infinitely many perfect numbers?
2. Are there infinitely many even perfect numbers?
3. Is there any odd perfect number?
4. Are there infinitely many amicable numbers?
5. Are there infinitely many twin primes?

Finally, we presented magic squares, which have been a source of fun for several investigators, including Fermat.

SUGGESTED READING

Barnett, I. A., "The Fascination of Whole Numbers," *The Mathematics Teacher*, vol. 54 (Feb. 1971), pp. 103–108.

Coltharp, F. L., "Determining the LCM and GCF through the Use of Set Theory," *The Arithmetic Teacher*, vol. 13 (April 1966), pp. 282–284.

Dilley, C. A., and W. E. Rucker, "Teaching Division by Two-Digit Numbers," *The Arithmetic Teacher*, vol. 16 (April 1969), pp. 306–308.

Holdan, Gregory, "Prime: A Drill in the Recognition of Prime and Composite Numbers," *The Arithmetic Teacher*, vol. 16 (Feb. 1969), pp. 149–151.

McLean, Robert C., "Estimating Quotient for the New Long Division Algorithm," *The Arithmetic Teacher*, vol. 16 (May 1969), pp. 398–400.

Schloff, Charles E., "A Pictured Approach to an Idea for Division," *The Arithmetic Teacher*, vol. 16 (May 1969), pp. 403–404.

Wheeler, R. E., *Modern Mathematics*: *An Elementary Approach* (3rd ed.), Brooks/Cole Publishing Co., Monterey, Calif. (1973), pp. 203–240.

Zink, M. H., "Greatest Common Divisor and Least Common Multiple," *The Arithmetic Teacher*, vol. 13 (Feb. 1966), pp. 138–140.

"The number of a class is the class of all classes similar to the given class."

B. Russell

6/ The System of Rational Numbers

After studying this chapter, you should be able to:
- *indicate if two rational numbers are equal*
- *find the sum of two rational numbers*
- *identify the properties of addition of rational numbers*
- *subtract one rational number from another rational number*
- *evaluate the product of two rational numbers*
- *identify the properties of multiplication of rational numbers*
- *check if a rational number is less than another rational number*
- *identify the properties of the order relation*
- *divide a rational number by a nonzero rational number*
- *find a rational number between any two distinct rational numbers*

6.0 INTRODUCTION

Recall that in Chapter 4 we discussed a major deficiency of the system of integers: not every equation of the form $a = bx(b \neq 0)$ is solvable in the set of integers. This equation has a solution in the set of integers if and only if b is a factor of a. This motivates us to extend the set of integers to a still richer set, where we can solve all equations of the form $a = bx$ provided $b \neq 0$. This new and familiar set is called the set of "rational numbers" or "fractions."

Do we make use of fractions in our day-to-day life? How important are fractions in our everyday life? Assume Carol bought six apples from a shop for 69 cents. She wants to know the price of one apple. How will you solve this physical situation if we have only integers at our disposal? Observe that her problem amounts to solving the equation $69 = 6x$, and $69 \div 6$ does not exist in the set of integers. From our experience, we are tempted to denote the solution as $\frac{69}{6}$ or 69/6. Assume she cuts one apple into four parts. Clearly, each piece is a part of the whole apple that consists of four pieces ("one-fourth" or 1/4); two pieces constitute two of the four pieces ("two-fourths" or 2/4); three pieces constitute three of the four parts ("three-fourths" or 3/4).

Assume Mrs. Smith cuts a pie equally among her three children. How much of the pie does each one get? Notice that it is impossible to handle this concrete situation with just integers. The problem is that the equation $3x = 1$ does not have a solution in the set of integers. From what we have learned from school, we know that each gets "one-third" or 1/3 of the pie; two children get "two-thirds" or 2/3 of the pie.

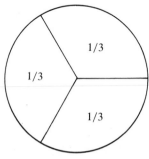

Fig. 6.1

Almost every day, on television or radio, we see or hear statements like the following: four out of five (that is, 4/5) people surveyed like a certain brand of toothpaste; two out of every three (that is, 2/3) people interviewed prefer a certain kind of chewing gum to the other leading brand, etc.

You should now be convinced that numbers other than integers are required to deal with several physical situations that occur in our everyday life.

6.1 RATIONAL NUMBERS

In our above discussion we observed that each problem boils down to solving an equation of the form $a = bx$, where a and b are integers and $b \neq 0$. Intuitively, we know that $x = a \div b$, that is, $x = \frac{a}{b}$ is a solution of the equation; that is, $a = b\left(\frac{a}{b}\right)$. Recall that the symbol $\frac{a}{b}$, which is the *ratio* of the integers a and b, need not represent an integer.

Definition 6.1 The ratio $\frac{a}{b}$ of the integers a and $b(\neq 0)$ is called a *fraction* or a *rational number*. The integers a and b are respectively called *numerator* and *denominator* of the rational number $\frac{a}{b}$.

The fractions $\frac{2}{3}, \frac{1}{4}, \frac{-2}{6}, \frac{13}{-8}, \frac{7}{7}, \frac{-3}{1}$, etc., are all rational numbers. In the fraction $\frac{2}{3}$, the number 2 is the numerator and 3 the denominator, and $\frac{2}{3}$ simply means that if we divide something, say a pie, into three equal parts, we consider two of them together. The rational number $\dfrac{a}{b}$ is also sometimes denoted a/b for convenience. The set of all rational numbers is denoted by the letter Q.

When do we say that two rational numbers a/b and c/d are equal? Let's divide a square region into two equal parts (two one-halves), as in Fig. 6.2a and into four equal parts (four one-fourths), as in Fig. 6.2b. The shaded region in Fig. 6.2a represents $\frac{1}{2}$ of the whole region, and that in Fig. 6.2b represents $\frac{2}{4}$ of the total area. Intuitively, we observe that these two shaded regions represent the same area. Consequently, we feel that the rational numbers $\frac{1}{2}$ and $\frac{2}{4}$ must be the same; that is, $\frac{1}{2} = \frac{2}{4}$; and both are equal to $\frac{4}{8}$ (see Fig. 6.2c). Notice that the statement $\frac{1}{2} = \frac{2}{4}$ is equivalent to the statement $1 \cdot 4 = 2 \cdot 2$. This discussion leads us now to the following definition.

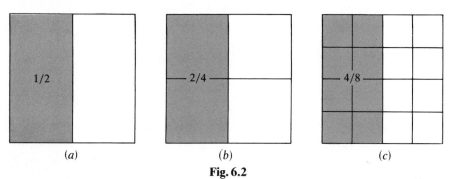

(a)   (b)   (c)

Fig. 6.2

Definition 6.2   Two rational numbers a/b and c/d are said to be *equal* if and only if $ad = bc$; that is, $\dfrac{a}{b} = \dfrac{c}{d}$ if and only if $ad = bc$.

Example 6.1
$$\frac{2}{3} = \frac{4}{6} \quad \text{since} \quad 2 \cdot 6 = 3 \cdot 4$$
$$\frac{3}{5} = \frac{12}{20} \quad \text{since} \quad 3 \cdot 20 = 5 \cdot 12$$
$$\frac{-4}{7} = \frac{12}{-21} \quad \text{since} \quad (-4) \cdot (-21) = 7 \cdot 12$$
$$\frac{3}{-2} = \frac{-3}{2} \quad \text{since} \quad 3 \cdot 2 = (-2) \cdot (-3)$$
$$\frac{0}{3} = \frac{0}{5} \quad \text{since} \quad 0 \cdot 5 = 3 \cdot 0$$
$$\frac{0}{-2} = \frac{0}{7} \quad \text{since} \quad 0 \cdot 7 = (-2) \cdot 0$$

It follows from Definition 6.2 that $\dfrac{a}{-b} = \dfrac{-a}{b}$ (why?); for example, $\frac{-2}{3} = \frac{2}{-3}$, $\frac{1}{-3} = \frac{-1}{3}$, $\frac{-3}{-5} = \frac{3}{5}$, etc. Consequently, it is always possible to rewrite a rational number with positive denominator. A rational number a/b with $b > 0$ is called a *positive rational number* if $a > 0$ and a *negative rational number* if $a < 0$. For example, $\frac{2}{3}$ and $\frac{-3}{-5}$ are positive rational numbers, whereas $\frac{-4}{7}$ and $\frac{3}{-5}$ are negative rational numbers.

Consider the rational numbers $\frac{3}{5}$ and $\frac{24}{40}$. Notice that the gcd (Section 5.4) of the integers 3 and 5 in $\frac{3}{5}$ is 1, whereas that of the integers 24 and 40 in $\frac{24}{40}$ is 8 (why?). However, $\frac{3}{5} = \frac{24}{40}$ (why?). More generally, we observe that the rational numbers $\dfrac{ac}{bc}$ and $\dfrac{a}{b}$ are equal; that is, $\dfrac{ac}{bc} = \dfrac{a}{b}$. (We may cancel c from the numerator ac and the denominator bc of the fraction ac/bc without changing its value; that is, $\dfrac{a\not{c}}{b\not{c}} = \dfrac{a}{b}$.) Also, the value of a fraction does not change even if its numerator and denominator are multiplied by the same nonzero integer. This prompts us to make the following definition.

Definition 6.3 A rational number a/b (assume $b > 0$) is said to be in its *simplest form* if the integers a and b have no common factors other than 1.

Example 6.2 Rewrite the fractions $\frac{36}{96}$, $\frac{-50}{110}$, and $\frac{-30}{-45}$ in their simplest forms.

Solution: We have

$$\frac{36}{96} = \frac{\not{2} \cdot \not{2} \cdot \not{3} \cdot 3}{\not{2} \cdot \not{2} \cdot \not{3} \cdot 8} = \frac{3}{8}$$

$$\frac{-50}{110} = \frac{(-1) \cdot \not{2} \cdot \not{5} \cdot 5}{\not{2} \cdot \not{5} \cdot 11} = \frac{(-1) \cdot 5}{11} = \frac{-5}{11}$$

and

$$\frac{-30}{-45} = \frac{(-\not{1}) \cdot 2 \cdot \not{3} \cdot \not{5}}{(-\not{1}) \cdot \not{3} \cdot 3 \cdot \not{5}} = \frac{2}{3}$$

Observe that there is a 1-1 correspondence between the set I of integers and the set $\left\{ \dfrac{n}{1} \mid n \in I \right\}$, which is a proper subset of the set of rational numbers, as shown below:

$$
\begin{array}{ccccccc}
\cdots & -2 & -1 & 0 & 1 & 2 & \cdots \\
 & \updownarrow & \updownarrow & \updownarrow & \updownarrow & \updownarrow & \\
\cdots & \frac{-2}{1} & \frac{-1}{1} & \frac{0}{1} & \frac{1}{1} & \frac{2}{1} & \cdots
\end{array}
$$

Since there is no difference between the numbers n and $n/1$ for all practical purposes, we identify the integer n with the rational number $n/1$. Thus the set of integers is a proper subset of the set of rational numbers.

Notice that the rational numbers a/b and b/a need not be equal. Consequently, we can very well consider the rational number a/b as an ordered pair (a,b) of integers a and b. Even though it may be easier to work with the ordered pair notation for rational numbers, we will be using only the more familiar notation a/b.

Recall that the integer line (Fig. 6.3) we constructed in Section 4.1 is full of "gaps." For instance, there are no integers corresponding to the points between the points with coordinates 2 and 3 on the number line. How do we

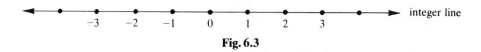

Fig. 6.3

assign a position to every rational number on the number line? Let's discuss how to represent the rational numbers $\frac{2}{7}$ and $\frac{-3}{7}$. Divide each unit length into 7 equal parts; the length of each part then will be $\frac{1}{7}$. The second point of division, P, to the right of the origin 0 corresponds to $\frac{2}{7}$, and the third point of division, Q, to the left of 0 has coordinate $\frac{-3}{7}$.

More generally, to represent the rational numbers a/b and $-c/b$, where we assume a, b, and c are positive integers, we subdivide each unit length into b equal parts. Then the ath point to the right of 0 corresponds to a/b, and the cth point to the left of 0 corresponds to $-c/b$. The points corresponding to rational numbers are called *rational points* and constitute the *rational line* (Fig. 6.4).

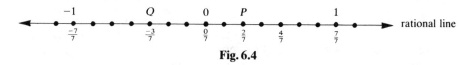

Fig. 6.4

Exercise 6.1

1. Find the numerator and the denominator of the following rational numbers.

   a) $\frac{2}{5}$ **b)** $\frac{-3}{8}$ **c)** $\frac{-4}{-3}$

   d) $\frac{0}{3}$ **e)** 3   **f)** -3

2. Determine which of the following pairs of rational numbers are equal.

   a) $\frac{5}{7}$ and $\frac{15}{21}$ **b)** $\frac{-24}{18}$ and $\frac{20}{-15}$

   c) $\frac{25}{45}$ and $\frac{40}{73}$ **d)** $\frac{-6}{-15}$ and $\frac{-12}{-30}$

3. Name three fractions equal to each of the following.

   a) $\frac{2}{5}$ **b)** $\frac{-3}{7}$ **c)** $\frac{4}{-3}$ **d)** $\frac{-5}{-6}$

4. Solve the following equations.

   a) $\dfrac{x}{3} = \dfrac{2}{6}$ **b)** $\dfrac{2}{x} = \dfrac{3}{2}$ **c)** $\dfrac{x}{-5} = \dfrac{3}{4}$

   d) $\dfrac{x}{4} = \dfrac{-2}{4}$ **e)** $\dfrac{5}{-x} = \dfrac{-7}{3}$ **f)** $\dfrac{2}{-x} = \dfrac{3}{-4}$

5. Rewrite each of the following fractions in its simplest form.

   a) $\frac{10}{15}$ **b)** $\frac{30}{25}$ **c)** $\frac{-27}{66}$ **d)** $\frac{88}{-121}$

   e) $\frac{234}{432}$ **f)** $\frac{121}{1221}$ **g)** $\frac{-123}{231}$ **h)** $\frac{-76}{-1776}$

6. Mark *true* or *false* (a,b,c,d are any nonzero integers).

   a) $a = \dfrac{a}{1}$

   b) $\frac{3}{7} = \frac{-3}{-7}$

   c) $\frac{0}{8} = \frac{0}{-8}$

d) $\dfrac{-a}{b} = \dfrac{a}{-b}$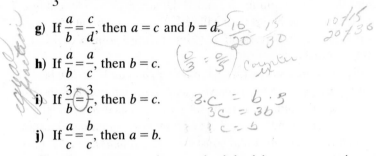

e) $\frac{2}{3}$ is a positive rational number.

f) $\dfrac{a}{3}$ is a positive rational number.

g) If $\dfrac{a}{b} = \dfrac{c}{d}$, then $a = c$ and $b = d$.

h) If $\dfrac{a}{b} = \dfrac{a}{c}$, then $b = c$.

i) If $\dfrac{3}{b} = \dfrac{3}{c}$, then $b = c$.

j) If $\dfrac{a}{c} = \dfrac{b}{c}$, then $a = b$.

7. Give a counterexample to each of the false statements in problem 6.

8. If $\dfrac{a}{b} = 0$ and $b \neq 0$, prove that $a = 0$. In other words, if a fraction is to be zero, then its numerator must be zero.

9. Solve the following equations.

a) $\dfrac{x-2}{5} = 0$ **b)** $\dfrac{x-8}{3+5} = 0$ **c)** $\dfrac{x+3}{x-5} = 0$

10. If $\dfrac{a}{b} = \dfrac{b}{a}$, what can you say about the integers a and b?

6.2 ADDITION OF RATIONAL NUMBERS

Recall that in Section 4.2, we defined addition on the set of integers in such a way that all properties of addition on the set of whole numbers are preserved. In a similar fashion, we define addition on the set of rational numbers without losing any properties of addition of integers. As in the case of addition of integers, the rational line can very well be used as a visual aid to discuss the operation of addition on the set of rational numbers.

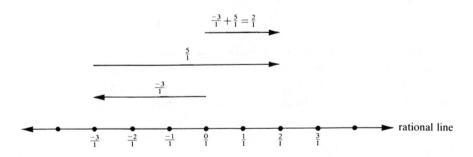

Fig. 6.5

Since we know how to add two integers, we can now easily add rational numbers of the form $a/1$; for example, $\frac{-3}{1} + \frac{5}{1} = -3 + 5 = 2$, as illustrated in Fig. 6.5. More generally,

$$\frac{a}{1} + \frac{b}{1} = a + b = \frac{a+b}{1}$$

Let's now discuss how to find the sum of two rational numbers with the same denominator. For example, how do we evaluate the sum $\frac{2}{3} + \frac{5}{3}$? It is clear from Fig. 6.6 that the arrow $\frac{2}{3}$ units long followed by the arrow $\frac{5}{3}$ units long terminates

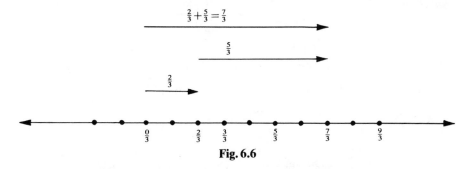

Fig. 6.6

at the rational point $\frac{7}{3}$. From our experience with the integer line, we can say that

$$\frac{2}{3} + \frac{5}{3} = \frac{7}{3} = \frac{2+5}{3}$$

and more generally,

$$\frac{a}{b} + \frac{c}{b} = \frac{a+c}{b}$$

Recall that there are pairs of rational numbers whose denominators are not the same. How do we add such rational numbers? For example, how do we evaluate the sum $\frac{3}{4} + \frac{5}{7}$? Let's make use of the above idea to solve this. We have,

$$\frac{3}{4} = \frac{3 \cdot 7}{4 \cdot 7} \quad \text{and} \quad \frac{5}{7} = \frac{5 \cdot 4}{7 \cdot 4} = \frac{5 \cdot 4}{4 \cdot 7}$$

Since $\dfrac{3 \cdot 7}{4 \cdot 7}$ and $\dfrac{5 \cdot 4}{4 \cdot 7}$ have the same denominator, $4 \cdot 7 = 28$, we have

$$\frac{3}{4} + \frac{5}{7} = \frac{3 \cdot 7}{4 \cdot 7} + \frac{5 \cdot 4}{4 \cdot 7} = \frac{3 \cdot 7 + 5 \cdot 4}{4 \cdot 7}$$

$$= \frac{21 + 20}{28} = \frac{41}{28}$$

These observations lead us to make the following definition.

Definition 6.4 The *sum* of the rational numbers a/b and c/d is defined by

$$\frac{a}{b}+\frac{c}{d}=\frac{ad+bc}{bd}$$

Notice that this definition makes sense, since

$$\frac{a}{b}+\frac{c}{d}=\frac{ad}{bd}+\frac{bc}{bd}=\frac{ad+bc}{bd}$$

Example 6.3 Evaluate the sum $\frac{7}{8}+\frac{5}{11}$.

Solution:
$$\frac{7}{8}+\frac{5}{11}=\frac{7\cdot 11}{8\cdot 11}+\frac{8\cdot 5}{8\cdot 11}=\frac{7\cdot 11+8\cdot 5}{8\cdot 11}$$

$$=\frac{77+40}{88}=\frac{117}{88}$$

Example 6.4 Find the sum of the rational numbers $\frac{-3}{8}$ and $\frac{2}{7}$.

Solution:
$$\frac{-3}{8}+\frac{2}{7}=\frac{(-3)\cdot 7+8\cdot 2}{8\cdot 7}=\frac{-21+16}{56}=\frac{-5}{56}$$

Observe that in order to find the sum $a/b+c/d$, first we have to make the denominators the same by multiplying the numerator and the denominator of the first fraction a/b by d (the denominator of the second fraction) to get ad/bd and, similarly, multiplying those of the second fraction c/d by b to get bc/bd. However, more often we make use of the concept of the lcm (Section 5.5) of two integers as a tool for finding a *common denominator* and then proceed as usual. This method is more efficient, especially if the denominators are fairly large, as illustrated in the following example.

Example 6.5 Evaluate the sum $\frac{5}{18}+\frac{7}{24}$.

Solution:

2	18	24
3	9	12
	3	4

Notice that $\text{lcm}\{18,24\}=2\cdot 3\cdot 3\cdot 4=72$. Also, $72=4\cdot 18=3\cdot 24$. We say that 72 is the *least common denominator* of the rational numbers $\frac{5}{18}$ and $\frac{7}{24}$. Multiply the numerator and denominator of the fractions $\frac{5}{18}$ and $\frac{7}{24}$ by 4 and 3, respectively. We then have

$$\frac{5}{18}+\frac{7}{24}=\frac{5\cdot 4}{18\cdot 4}+\frac{3\cdot 7}{3\cdot 24}=\frac{20}{72}+\frac{21}{72}$$

$$=\frac{41}{72}$$

If we had used the definition, then we would have

$$\frac{5}{18}+\frac{7}{24}=\frac{5\cdot 24+18\cdot 7}{18\cdot 24}=\frac{120+126}{432}$$

$$=\frac{246}{432}=\frac{41}{72}$$

Let's now proceed to discuss the properties of addition on the set of rational numbers. As mentioned earlier, we expect all the properties of addition discussed in Chapter 4 to hold in Q, also.

We defined the sum of the rational numbers $\frac{a}{b}$ and $\frac{c}{d}$ as $\frac{ad+bc}{bd}$. Since a, b, c, and d are integers, it follows by the closure properties of addition and multiplication of integers that $ad+bc$ is also an integer. Since b and d are nonzero integers, so is bd, by the product law. Consequently, we conclude that the sum $\frac{ad+bc}{bd}$ is a unique rational number and addition is a binary operation on the set of rational numbers.

Theorem 6.1   (***Closure Property***)   *The set of rational numbers is closed under addition.*

We observed in Chapter 4 that the order in which we add any two integers is not at all important. Is this true in the case of rational numbers, also? Let's consider the following example before we answer this.

Example 6.6   Evaluate the sums $\frac{-5}{12}+\frac{7}{18}$ and $\frac{7}{18}+\frac{-5}{12}$.

Solution: Observe that lcm$\{12,18\}=36$. Also, $36=3\cdot 12=2\cdot 18$.

$$\frac{-5}{12}+\frac{7}{18}=\frac{(-5)\cdot 3}{12\cdot 3}+\frac{7\cdot 2}{18\cdot 2}=\frac{-15}{36}+\frac{14}{36}$$

$$=\frac{-15+14}{36}=\frac{-1}{36}$$

and

$$\frac{7}{18}+\frac{-5}{12}=\frac{7\cdot 2}{18\cdot 2}+\frac{(-5)\cdot 3}{12\cdot 3}$$

$$=\frac{14+(-15)}{36}=\frac{-1}{36}$$

Thus,

$$\frac{-5}{12}+\frac{7}{18}=\frac{7}{18}+\frac{-5}{12}$$

Theorem 6.2   (***Commutative Property***)   *For any two rational numbers x and y,*

$$x+y=y+x$$

Recall from Section 4.2 that addition on the set of integers is associative; that is, $a + (b + c) = (a + b) + c$ for all integers a, b, and c. Does this property hold on the set of rational numbers?

Example 6.7 Evaluate the sums $\frac{2}{5} + (\frac{-3}{8} + \frac{4}{7})$ and $(\frac{2}{5} + \frac{-3}{8}) + \frac{4}{7}$.

Solution: We have

$$\frac{2}{5} + \left(\frac{-3}{8} + \frac{4}{7}\right) = \frac{2}{5} + \frac{(-3)\cdot 7 + 8 \cdot 4}{8 \cdot 7} = \frac{2}{5} + \frac{-21 + 32}{56}$$

$$= \frac{2}{5} + \frac{11}{56} = \frac{2 \cdot 56 + 5 \cdot 11}{5 \cdot 56}$$

$$= \frac{112 + 55}{280} = \frac{167}{280}$$

and

$$\left(\frac{2}{5} + \frac{-3}{8}\right) + \frac{4}{7} = \frac{2 \cdot 8 + 5 \cdot (-3)}{5 \cdot 8} + \frac{4}{7} = \frac{16 - 15}{40} + \frac{4}{7}$$

$$= \frac{1}{40} + \frac{4}{7} = \frac{1 \cdot 7 + 40 \cdot 4}{40 \cdot 7}$$

$$= \frac{7 + 160}{280} = \frac{176}{280}$$

Thus,

$$\left(\frac{2}{5} + \frac{-3}{8}\right) + \frac{4}{7} = \frac{2}{5} + \left(\frac{-3}{8} + \frac{4}{7}\right)$$

More generally, we now have the following result.

Theorem 6.3 (*Associative Property*) *If x, y, z are any rational numbers, then*

$$x + (y + z) = (x + y) + z$$

Since addition of rational numbers is associative, the expression $x + y + z$ is not ambiguous. Using the commutative and the associative properties, the sum of any finite number of rational numbers can be found in any order we like.

Example 6.8 Evaluate the sum $\frac{2}{7} + \frac{7}{8} + \frac{11}{12}$.

Solution: Since addition is associative, we can evaluate the sum either as $\frac{2}{7} + (\frac{7}{8} + \frac{11}{12})$ or as $(\frac{2}{7} + \frac{7}{8}) + \frac{11}{12}$. Thus,

$$\frac{2}{7} + \frac{7}{8} + \frac{11}{12} = \frac{2}{7} + \left(\frac{7}{8} + \frac{11}{12}\right) = \frac{2}{7} + \frac{7 \cdot 3 + 11 \cdot 2}{24} \qquad \text{[Note: lcm\{8,12\} = 24]}$$

$$= \frac{2}{7} + \frac{21 + 22}{24} = \frac{2}{7} + \frac{43}{24}$$

$$= \frac{2 \cdot 24 + 7 \cdot 43}{7 \cdot 24} = \frac{48 + 301}{168} \qquad \text{[Note: lcm\{7,24\} = 168]}$$

$$= \frac{349}{168}$$

We observed in Chapter 4 that the set of integers contains a very special element, 0, the additive identity. Since addition of rational numbers is an extension of that of integers, it is only proper to ask if the set of rational numbers contains a similar element. This is answered by the following theorem.

Theorem 6.4   (*Existence of a Unique Additive Identity*)   *The set of rational numbers contains a unique additive identity, 0. That is,*

$$x + 0 = x = 0 + x$$

for every rational number x.

Example 6.9   Evaluate the sums $\frac{-3}{5} + 0$ and $0 + \frac{2}{3}$.

Solution:

$$\frac{-3}{5} + 0 = \frac{-3}{5} + \frac{0}{5} = \frac{-3 + 0}{5} = \frac{-3}{5}$$

and

$$0 + \frac{2}{3} = \frac{0}{3} + \frac{2}{3} = \frac{0 + 2}{3} = \frac{2}{3}$$

If x is any rational number, can we find a rational number y such that $x + y = 0$, the additive identity? In other words, does every rational number have an additive inverse?

Example 6.10   Evaluate the sums $\frac{-3}{7} + \frac{3}{7}$ and $\frac{2}{5} + \frac{-2}{5}$.

Solution:

$$\frac{-3}{7} + \frac{3}{7} = \frac{-3 + 3}{7} = \frac{0}{7} = 0$$

$$\frac{2}{5} + \frac{-2}{5} = \frac{2 + (-2)}{5} = \frac{0}{5} = 0$$

Theorem 6.5   (*Existence of a Unique Additive Inverse*)   *Every rational number a/b has a unique additive inverse $-a/b$; that is,*

$$\frac{a}{b} + \frac{-a}{b} = 0$$

Since $\frac{-a}{b} = \frac{a}{-b}$, the additive inverse of $\frac{a}{b}$ can also be denoted as $\frac{a}{-b}$.

However, we usually denote the additive inverse of $\frac{a}{b}$ as $-\frac{a}{b}$. Thus, $\frac{-a}{b} = \frac{a}{-b} = -\frac{a}{b}$. Observe that $-(-x) = x$ for every rational number x. (Why?) For example, $-(-\frac{2}{3}) = \frac{2}{3}$. Also $-(x + y) = (-x) + (-y)$ for any rational numbers x and y. (Why?) This simply means that the additive inverse of a sum is the same as the sum of their additive inverses. For example,

$$-\left(\frac{2}{3} + \frac{1}{4}\right) = -\frac{2 \cdot 4 + 3 \cdot 1}{3 \cdot 4} = -\frac{8 + 3}{12} = -\frac{11}{12}$$

and

$$\left(-\frac{2}{3}\right)+\left(-\frac{1}{4}\right)=\frac{-2}{3}+\frac{-1}{4}=\frac{(-2)\cdot 4+3\cdot(-1)}{3\cdot 4}$$

$$=\frac{-8+(-3)}{12}=\frac{-11}{12}=-\frac{11}{12}$$

Thus,

$$-\left(\frac{2}{3}+\frac{1}{4}\right)=\left(-\frac{2}{3}\right)+\left(-\frac{1}{4}\right)$$

Example 6.11   Observe that $\frac{3}{4}=\frac{15}{20}$ and $\frac{2}{11}=\frac{6}{33}$. Let's now find the sums $\frac{3}{4}+\frac{2}{11}$ and $\frac{15}{20}+\frac{6}{33}$.

$$\frac{3}{4}+\frac{2}{11}=\frac{3\cdot 11+4\cdot 2}{4\cdot 11}=\frac{33+8}{44}=\frac{41}{44}$$

and

$$\frac{15}{20}+\frac{6}{33}=\frac{15\cdot 33+20\cdot 6}{20\cdot 33}=\frac{495+120}{660}$$

$$=\frac{615}{660}=\frac{41}{44}$$

Thus, $\frac{3}{4}=\frac{15}{20}, \frac{2}{11}=\frac{6}{33}$, and $\frac{3}{4}+\frac{2}{11}=\frac{15}{20}+\frac{6}{33}$.

This example now suggests us the following useful property.
Theorem 6.6   (*Addition Property*)   *If w, x, y, z are any rational numbers such that w = x and y = z, then w + y = x + z.*

The addition property simply states that if equal rational numbers are added to equal rational numbers, then the sums are still equal. Conversely, if $x + y = x + z$, then $y = z$, as in the case of addition of integers.
Theorem 6.7   (*Cancellation Property*)   *If x, y, z are any rational numbers such that x + y = x + z, then y = z.*

The addition and cancellation properties are extremely useful in solving problems, as illustrated by the following examples.
Example 6.12   Find the solution set of the equation $x + \frac{2}{3} = \frac{3}{2}$.
Solution: Let's isolate x by getting rid of $\frac{2}{3}$ from the left-hand side of the equation.

$x + \frac{2}{3} = \frac{3}{2}$	*given*
$(x + \frac{2}{3}) + (-\frac{2}{3}) = \frac{3}{2} + (-\frac{2}{3})$	*addition property*
$x + [\frac{2}{3} + (-\frac{2}{3})] = \frac{3}{2} + (-\frac{2}{3})$	*associative property of addition*
$x + 0 = \frac{3}{2} + (\frac{-2}{3})$	*definition of additive inverse*
$x = \frac{3}{2} + (\frac{-2}{3})$	*0 is the additive identity*
$x = \dfrac{3\cdot 3+2\cdot(-2)}{2\cdot 3}$	*definition of addition of rational numbers*

$$x = \frac{9-4}{6} \qquad \text{\textit{definition of multiplication of integers}}$$

$$x = \tfrac{5}{6} \qquad \text{\textit{definition of subtraction of integers}}$$

Let's now verify if $\tfrac{5}{6}$ is the actual solution of the given equation. Substitute $\tfrac{5}{6}$ for x and check if $x + \tfrac{2}{3}$ is the same as $\tfrac{3}{2}$:

$$x + \frac{2}{3} = \frac{5}{6} + \frac{2}{3} = \frac{5 \cdot 1 + 2 \cdot 2}{6} = \frac{5+4}{6} = \frac{9}{6} = \frac{3}{2}$$

Thus, $\tfrac{5}{6}$ is the solution and $\{\tfrac{5}{6}\}$ is the solution set.

Example 6.13   Solve the equation $x + \tfrac{3}{4} + \tfrac{5}{3} = \tfrac{2}{9} + \tfrac{5}{3}$.

Solution: We ask the reader to justify each of the following steps.

$$x + \tfrac{3}{4} + \tfrac{5}{3} = \tfrac{2}{9} + \tfrac{5}{3}$$

$$x + \tfrac{3}{4} = \tfrac{2}{9}$$

$$(x + \tfrac{3}{4}) + (-\tfrac{3}{4}) = \tfrac{2}{9} + (-\tfrac{3}{4})$$

$$x + [\tfrac{3}{4} + (-\tfrac{3}{4})] = \tfrac{2}{9} + (-\tfrac{3}{4})$$

$$x + 0 = \tfrac{2}{9} + (-\tfrac{3}{4})$$

$$x = \tfrac{2}{9} + (-\tfrac{3}{4})$$

$$x = \frac{2 \cdot 4 + (-3) \cdot 9}{9 \cdot 4}$$

$$x = \frac{8-27}{36}$$

$$x = \tfrac{-19}{36}$$

It can now easily be checked that $-\tfrac{19}{36}$ is indeed the solution of the given equation.

Before we conclude this section, let's discuss different types of fractions. We are all familiar with fractions of the form $3\tfrac{5}{9}$. What does it mean? It is simply a notation for $3 + \tfrac{5}{9} = \tfrac{3}{1} + \tfrac{5}{9}$. It consists of two parts: an integral part, 3, and a fractional part, $\tfrac{5}{9}$. Fractions of the form $a\tfrac{b}{c}$ are called *mixed fractions*. Observe that $3\tfrac{5}{9} = 3 + \tfrac{5}{9} = \tfrac{3}{1} + \tfrac{5}{9} = \tfrac{32}{9}$. We warn the reader that $-3\tfrac{5}{9}$ is not the same as $-3 + \tfrac{5}{9} = -\tfrac{22}{9}$, but $-3\tfrac{5}{9} = -(3\tfrac{5}{9}) = -\tfrac{32}{9}$. Notice that $\tfrac{32}{9}$ is a fraction with numerator larger than the denominator. Fractions of the form a/b or $-a/b$ (assume a and b are positive integers) are called *improper fractions* if $a \geq b$ and *proper fractions* if $a < b$. For example, $\tfrac{5}{3}, \tfrac{-5}{3}, \tfrac{8}{7}, -\tfrac{31}{13}$ are improper fractions; $\tfrac{3}{5}, \tfrac{-2}{3}, \tfrac{1}{4}, -\tfrac{8}{11}$ are proper fractions.

As illustrated above, mixed fractions can easily be converted to improper fractions. How to convert improper fractions into mixed fractions is illustrated in the following example.

Example 6.14 Convert the improper fractions $\frac{32}{5}$ and $-\frac{29}{8}$ into mixed fractions.

Solution: Observe that $32 = 5 \cdot 6 + 2$ and $29 = 8 \cdot 3 + 5$, by the division algorithm. Thus

$$\frac{32}{5} = \frac{5 \cdot 6 + 2}{5} = \frac{5 \cdot 6}{5} + \frac{2}{5}$$

$$= \frac{\cancel{5} \cdot 6}{\cancel{5}} + \frac{2}{5} = 6 + \frac{2}{5} = 6\frac{2}{5}$$

and

$$-\frac{29}{8} = -\frac{8 \cdot 3 + 5}{8} = -\left(\frac{8 \cdot 3}{8} + \frac{5}{8}\right)$$

$$= -\left(\frac{\cancel{8} \cdot 3}{\cancel{8}} + \frac{5}{8}\right) = -\left(3 + \frac{5}{8}\right) = -3\frac{5}{8}$$

Exercise 6.2

1. Evaluate each of the following sums.

a) $\frac{1}{2} + \frac{1}{3}$ b) $\frac{5}{7} + \frac{7}{5}$ c) $\frac{-11}{3} + \frac{8}{11}$

d) $\frac{-3}{7} + \frac{-5}{6}$ e) $\frac{2}{3} + \left(\frac{3}{4} + \frac{4}{5}\right)$ f) $\left(\frac{5}{4} + \frac{4}{3}\right) + \frac{3}{2}$

g) $\dfrac{2}{a} + \dfrac{3}{b}$ h) $\dfrac{-3}{a} + \dfrac{6}{5a}$ i) $\dfrac{1}{a} + \dfrac{2}{b} + \dfrac{3}{ab}$

2. Verify each of the following.

a) $\frac{7}{11} + \frac{11}{7} = \frac{11}{7} + \frac{7}{11}$ b) $\frac{-3}{8} + \frac{5}{7} = \frac{5}{7} + \frac{-3}{8}$

c) $\frac{-3}{7} + \frac{-5}{9} = \frac{-5}{9} + \frac{-3}{7}$ d) $\frac{0}{1} + \frac{1}{2} + \frac{-2}{3} = \frac{-2}{3} + \frac{1}{2} + \frac{0}{1}$

e) $\frac{1}{2} + \left(\frac{1}{3} + \frac{1}{4}\right) = \left(\frac{1}{2} + \frac{1}{3}\right) + \frac{1}{4}$ f) $\frac{3}{8} + \left(\frac{-5}{11} + \frac{8}{5}\right) = \left(\frac{3}{8} + \frac{-5}{11}\right) + \frac{8}{5}$

3. Solve the following equations.

a) $x + \frac{1}{3} = 0$ b) $x + \frac{3}{8} = \frac{8}{3}$ c) $-\frac{4}{7} + x = 0$

d) $x + \frac{-5}{7} = \frac{2}{7}$ e) $-\frac{3}{4} + x = -\frac{7}{8}$ f) $x + \frac{3}{5} + \frac{-2}{7} = \frac{3}{10}$

4. Find the integer x if

a) $\dfrac{x}{5} + \dfrac{5}{4} = \dfrac{37}{20}$ b) $\dfrac{x}{2} + \dfrac{1}{3} = \dfrac{5}{6}$

c) $\dfrac{3}{x} + \dfrac{2}{5} = \dfrac{29}{35}$ d) $\dfrac{5}{x} + \dfrac{3}{5} = -\dfrac{13}{20}$

5. If x and y are rational numbers such that $x + y = y$, what is your conclusion?

6. If the rational number x is its own additive inverse, what is your conclusion?

7. Mark *true* or *false* (a, b, c, d are any nonzero integers; x, y, z are any rational numbers).

a) Addition is a binary operation on the set of rational numbers. ~~true~~

 b) $\dfrac{a}{b} + \dfrac{c}{d} = \dfrac{a+c}{bd}$ $= \dfrac{ad}{bd} + \dfrac{bc}{bd}$ $\dfrac{ad+bc}{bd}$

c) If $x = y$, then $x + (-z) = y + (-z)$. *True*

d) If $x + z = y + z$, then $z = 0$. *False*   *why?*

e) $\dfrac{a}{b} + \dfrac{c}{b} = \dfrac{a+c}{b}$ *True*

f) $\dfrac{a}{b+c} = \dfrac{a}{b} + \dfrac{a}{c}$ *False* — *why.* $\dfrac{a}{b} + \dfrac{a}{c} = \dfrac{ac}{bc} + \dfrac{ba}{bc}$

g) $-(x + y) = (-x) + (-y)$ *True*

h) $-(-x) = x$ *True*

i) $2\frac{1}{3} = 2 + \frac{1}{3}$ *true*

j) $-3\frac{2}{5} = -3 + \frac{-2}{5}$ *True*

k) $-a\dfrac{b}{c} = -\left(a + \dfrac{b}{c}\right)$ *False*   *why.* $-a - \dfrac{b}{c}$

8. Give a counterexample to each of the false statements in problem 7.
9. State the property of addition used in each of the following statements, each symbol representing a rational number.
   a) $x + y$ is a rational number.
   b) $(-x) + 0 = -x$
   c) If $x = y$, then $x + \frac{2}{3} = y + \frac{2}{3}$.
   d) If $x + (-z) = y + (-z)$, then $x = y$.
   e) $\frac{3}{4} + \frac{4}{5} = \frac{4}{5} + \frac{3}{4}$
   f) $(w + x) + (y + z) = (y + z) + (w + x)$
   g) $x + [(-y) + z] = [x + (-y)] + z$
   h) $x + [y + (-z)] = x + [(-z) + y]$
   i) If $-\frac{2}{3} + x = -\frac{2}{3} + y$, then $x = y$.
   j) $\frac{3}{4} + 0 = \frac{3}{4}$
10. Which of the following fractions are mixed? proper? improper?
   a) $\frac{2}{5}$ *proper*   **b)** $3\frac{2}{7}$ *mixed*   **c)** $\frac{-3}{8}$ *proper*   **d)** $\frac{11}{7}$ *improper*
   e) $\frac{-18}{7}$ *improper*   **f)** $-8\frac{2}{3}$ *mixed*   **g)** $-2\frac{3}{4}$ *mixed*   **h)** $\frac{4}{5}$ *proper*
11. Convert the following mixed fractions into improper fractions.
   a) $3\frac{4}{5}$ **b)** $5\frac{2}{7}$ **c)** $-4\frac{1}{11}$ **d)** $-4\frac{5}{6}$
12. Convert the following improper fractions into mixed fractions.
   a) $\frac{17}{5}$ **b)** $\frac{-21}{8}$ **c)** $\frac{29}{11}$ **d)** $-\frac{54}{7}$
13. Find a set of integers a, b, and c to show that $\dfrac{a}{b+c} \neq \dfrac{a}{b} + \dfrac{a}{c}$ in general.
14. If Frank did one-half of his homework one day and one-third of the homework the next day, how much of his homework did he do in two days?
15. Paul went for a trip by car. If he traveled $\frac{1}{5}$ of the total distance in two days and $\frac{1}{3}$ of the total distance in the next three days, how much of the total distance did he travel in five days? How far away was he then from his destination?

16. If John and Jane painted two-fifths and one-third of their house, respectively, how much of the house did they paint? How much is left now to be painted?

17. Susan wants to buy a car. If her father promised to pay two-fifths, her mother one-sixth and her uncle two-thirteenths of the price of the car, how much of the price of the car did she receive as promised?

18. Prove each of the following, where w, x, y, z are any rational numbers.

a) $x + y + z = z + x + y$

b) $(w + x) + (y + z) = (w + z) + (x + y)$

c) $(x + y) + [(-x) + (-y)] = 0$

d) $\dfrac{a}{b} + \dfrac{b}{a} = \dfrac{a^2 + b^2}{ab}$

e) $\dfrac{a}{b} + \dfrac{2a}{3b} = \dfrac{5a}{3b}$

f) $\dfrac{a}{b} + \dfrac{-b}{a} = \dfrac{a^2 - b^2}{ab}$

19. Define an operation $*$ on the set of rational numbers Q by $x * y = x + y + \frac{1}{2}$.

a) Is Q closed under $*$?

b) Is $*$ commutative?

c) Is $*$ associative?

d) Does Q have an identity element with respect to $*$?

e) Does every element in Q have an inverse with respect to $*$?

★20. Using the ordered pair notation (a,b) for rational numbers, define addition as $(a,b) + (c,d) = (ad + bc, bd)$ where a,c are any integers and b,c are any nonzero integers. Also, $(a,b) = (c,d)$ if and only if $ad = bc$.

a) Evaluate the sum $(3,7) + (11,13)$.

b) Evaluate the sum $(-7,13) + (5,8)$.

c) Find the integer x such that $(x,3) + (4,9) = (-2,9)$.

d) Find the integer x such that $(4,x) + (3,5) = (1,1)$.

6.3 SUBTRACTION OF RATIONAL NUMBERS

If John mowed one-half of his lawn, how much of the lawn is left to be mowed? The answer is given by the solution of the equation $x + \frac{1}{2} = 1$. If Ron taught two-thirds of his course, how much more of the course does he have to teach to complete it? The solution of the equation $x + \frac{2}{3} = 1$ provides the answer to this problem. Assume Tom, Dick, and Harry were teaching different parts of a course. If Tom taught one-half of the course and Dick taught one-third of the course, how much of the course did Harry teach? This problem boils down to solving the equation $x + \frac{1}{2} + \frac{1}{3} = 1$.

Even though we have enough mathematical tools to handle these physical situations, it will be easier to discuss these cases with the operation of subtraction. We now define the operation of subtraction of rational numbers in terms of addition, by extending the definition of subtraction of integers.

Definition 6.5   Let a and b be any two rational numbers. Then $a - b$ is defined as the solution of the equation $a = b + x$. The operation "$-$" we defined is called *subtraction*.

Example 6.15   Evaluate $\frac{2}{3} - \frac{3}{7}$.

Solution: $\frac{2}{3} - \frac{3}{7} = x$ if and only if $\frac{2}{3} = \frac{3}{7} + x$. Let's now solve this last equation

$$x + \frac{3}{7} = \frac{2}{3}$$

$$x + \frac{3}{7} + \left(\frac{-3}{7}\right) = \frac{2}{3} + \left(\frac{-3}{7}\right)$$

$$x + 0 = \frac{14 - 9}{21}$$

$$x = \frac{5}{21}$$

Thus, $\frac{2}{3} - \frac{3}{7} = \frac{5}{21}$.

Example 6.16   Evaluate $\frac{7}{15} - \frac{3}{11}$ and $\frac{2}{13} - \frac{5}{7}$.

Solution:
$$\frac{7}{15} - \frac{3}{11} = \frac{32}{165} \quad \text{since} \quad \frac{7}{15} = \frac{3}{11} + \frac{32}{165}$$

and
$$\frac{2}{13} - \frac{5}{7} = -\frac{51}{91} \quad \text{since} \quad \frac{2}{13} = \frac{5}{7} + \left(-\frac{51}{91}\right)$$

In Section 4.3 we observed that subtracting an integer b from an integer a is the same as adding the additive inverse $-b$ of b to a; that is, $a - b = a + (-b)$. Is this true in the case of rational numbers? This is answered in the affirmative by the following theorem.

Theorem 6.8   *If x and y are any two rational numbers, then $x - y = x + (-y)$.*

Example 6.17   Evaluate $\frac{3}{4} - \frac{5}{7}$ and $\frac{3}{4} + (-\frac{5}{7})$.

Solution: $\frac{3}{4} - \frac{5}{7} = z$ if and only if $\frac{3}{4} = \frac{5}{7} + z$. Let's now solve the equation $z + \frac{5}{7} = \frac{3}{4}$.

$$z + \frac{5}{7} + \left(-\frac{5}{7}\right) = \frac{3}{4} + \left(\frac{-5}{7}\right)$$

$$z + 0 = \frac{3 \cdot 7 + 4 \cdot (-5)}{4 \cdot 7} = \frac{21 - 20}{28}$$

$$z = \frac{1}{28}$$

We have
$$\frac{3}{4} + \left(-\frac{5}{7}\right) = \frac{3 \cdot 7 + 4 \cdot (-5)}{4 \cdot 7}$$

$$= \frac{21 - 20}{28} = \frac{1}{28}$$

Thus,
$$\frac{3}{4}-\frac{5}{7}=\frac{3}{4}+\left(-\frac{5}{7}\right)$$

The operation of subtraction of rational numbers can be illustrated on the rational line using Theorem 6.8, however it is relatively difficult. Let's evaluate $\frac{5}{7}-\frac{2}{7}$ (same denominator for convenience) using the rational line. Notice that $\frac{5}{7}-\frac{2}{7}=\frac{5}{7}+(-\frac{2}{7})$. Draw an arrow from 0 to terminate at the point $\frac{5}{7}$ and then an arrow $\frac{2}{7}$ units long in the opposite direction to terminate at the point labeled $\frac{3}{7}$. Thus, $\frac{5}{7}-\frac{2}{7}=\frac{5}{7}+(-\frac{2}{7})=\frac{3}{7}$.

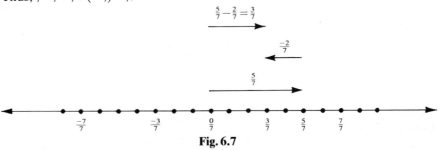

Fig. 6.7

Theorem 6.8 is a very useful and a remarkable property; it expresses subtraction directly in terms of addition. It follows by the theorem that $x-(-y)=x+[-(-y)]=x+y$ and $-y+x=(-y)+x=x+(-y)=x-y$ for any rational numbers x and y. Since the set of rational numbers is closed under addition, it now follows that the set is also closed under subtraction.

Example 6.18 If Ted waxed $\frac{1}{5}$ of a car, Ned waxed $\frac{3}{13}$ of the car, and Ed did the rest, how much of the car did Ed wax?

Solution: Let x be the part of the car waxed by Ed. Since Ted and Ned waxed one-fifth and three-thirteenths of the car, respectively, we have

$$x+\frac{1}{5}+\frac{3}{13}=1$$

$$x+\frac{13+15}{65}=1$$

$$x+\frac{28}{65}=1$$

$$x=1-\frac{28}{65}$$

$$=\frac{65-28}{65}$$

$$x=\frac{37}{65}$$

Thus, Ed waxed $\frac{37}{65}$ of the car.

Exercise 6.3

1. Evaluate each of the following.

   a) $\frac{1}{2}-\frac{1}{3}$ **b)** $\frac{2}{3}-\frac{3}{2}$ **c)** $-\frac{5}{8}-\frac{3}{7}$

   d) $\frac{1}{2}-(\frac{1}{3}-\frac{1}{4})$ **e)** $(\frac{1}{2}-\frac{1}{3})-\frac{1}{4}$ **f)** $(\frac{4}{9}+\frac{1}{6})-\frac{4}{5}$

   g) $-\frac{3}{4}-(-\frac{5}{6})$ **h)** $\frac{2}{5}-(\frac{5}{3}+\frac{-1}{5})$ **i)** $\frac{2}{3}-(\frac{3}{4}-\frac{-4}{5})$

2. Solve the following equations.

   a) $x-\frac{2}{3}=\frac{2}{3}$ **b)** $x-\frac{8}{11}=\frac{3}{5}+\frac{1}{2}$

   c) $-\frac{2}{9}+\frac{-7}{3}=\frac{8}{24}-x$ **d)** $\frac{5}{6}-\frac{7}{12}=-x+\frac{3}{7}$

3. Find the integer x if

   a) $\dfrac{x}{3}-\dfrac{5}{7}=\dfrac{13}{21}$ **b)** $\dfrac{-2}{3}-\dfrac{x}{2}=\dfrac{5}{6}$

   c) $\dfrac{2}{x}=\dfrac{1}{3}-\dfrac{2}{7}$ **d)** $\dfrac{4}{x}-\dfrac{5}{6}=\dfrac{3}{4}-\dfrac{85}{84}$

4. Verify each of the following.

   a) $\frac{2}{5}-\frac{7}{8}\neq\frac{7}{8}-\frac{2}{5}$ **b)** $\frac{3}{4}-\frac{4}{3}\neq\frac{4}{3}-\frac{3}{4}$

   c) $\frac{3}{8}-(\frac{2}{3}-\frac{1}{5})\neq(\frac{3}{8}-\frac{2}{3})-\frac{1}{5}$ **d)** $\frac{1}{3}-(\frac{1}{5}-\frac{1}{7})\neq(\frac{1}{3}-\frac{1}{5})-\frac{1}{7}$

5. Mark *true* or *false* (x, y, z are any rational numbers).

   a) The set of rational numbers is closed under subtraction.

   b) Subtraction is commutative on the set of rational numbers.

   c) The set of rational numbers contains an identity under subtraction.

   d) $x-y=x+(-y)$

   e) $-(x+y)=-x+y$

   f) $x-(-y)=x+y$

   g) If $x=y$, then $x-z=y-z$.

   h) If $x-z=y-z$, then $x=y$.

   i) If $x+\frac{3}{4}=\frac{4}{3}$, then $x=\frac{4}{3}-\frac{3}{4}$.

   j) There are rational numbers x, y, z such that $x-(y-z)=(x-y)-z$.

6. Give a counterexample to each of the false statements in problem 5.

7. Simplify each of the following.

   a) $\dfrac{1}{a}-\dfrac{1}{a+1}$ **b)** $\dfrac{a}{a+1}-\dfrac{b}{b+1}$ **c)** $\dfrac{a+1}{a}-\dfrac{b+1}{b}$

   d) $\dfrac{a}{b}-\dfrac{b}{a}$ **e)** $\dfrac{1}{a-1}-\dfrac{1}{a+1}$ **f)** $\dfrac{a}{a-1}-\dfrac{b}{b-1}$

8. If Linda did $\frac{5}{13}$ of her homework, how much is left to be done now?

9. If Ron and Don raked $\frac{3}{4}$ and $\frac{1}{11}$ of the lawn, respectively, and Bob did the rest, how much of the lawn did Bob rake?

10. Joe borrowed 1248 dollars from his friend John. If he returned $\frac{5}{8}$ and later $\frac{3}{13}$ of the amount, how much of the amount does he still owe to John?

11. Paul, Peter, and Porter together did a project. If Paul did $\frac{2}{7}$ of the project and Porter $\frac{5}{12}$ of the project, how much of the project did Peter do?

12. Find two fractions whose sum is $\frac{13}{6}$ and difference is $\frac{5}{6}$.

★**13.** Using the ordered pair notation for rational numbers, define subtraction by $(a,b)-(c,d)=(x,y)$ if and only if $(a,b)=(c,d)+(x,y)$.
   a) Evaluate $(3,7)-(4,3)$.
   b) Evaluate $(-2,3)-(-5,-7)$.
   c) Find the integer x such that $(3,5)-(2,4)=(x,10)$.

6.4 MULTIPLICATION OF RATIONAL NUMBERS

   Recall that we defined the operation of multiplication on the set of integers by extending the operation on the set of whole numbers in such a way that all the properties of multiplication of whole numbers are preserved. Our definition of multiplication of rational numbers will be motivated by exactly similar considerations, but first a few physical examples.

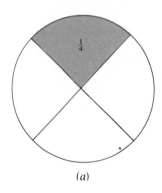
(a)

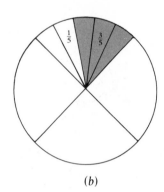
(b)

Fig. 6.8

   Assume Carol gave $\frac{1}{4}$ of an apple pie to her son Rob and Rob gave $\frac{3}{5}$ of that to his friend Bob. How much of the pie did he give Bob? Let's denote $\frac{3}{5}$ of $\frac{1}{4}$ by $\frac{3}{5} \cdot \frac{1}{4}$. Let's now divide the original pie into 20 equal parts, by dividing each one-fourth into five equal parts (Fig. 6.8b). Then the shaded region in Fig. 6.8b, which is the part of the pie Rob gave Bob, represents $\frac{1}{20}+\frac{1}{20}+\frac{1}{20}=\frac{3}{20}$ of the pie. Thus, we observe that

$$\frac{3}{5} \cdot \frac{1}{4}=\frac{3}{20}=\frac{3 \cdot 1}{5 \cdot 4}$$

Similarly, if Linda bought a dozen pencils and gave $\frac{3}{4}$ of them to her daughter, how many pencils did she give her daughter? From the previous example, we feel that the answer must be

$$\frac{3}{4} \cdot 12 =\frac{3}{4} \cdot \frac{12}{1}=\frac{3 \cdot 12}{4 \cdot 1}=9$$

Also observe that

$$3 \cdot 5 =\frac{3}{1} \cdot \frac{5}{1}=\frac{3 \cdot 5}{1 \cdot 1}=\frac{15}{1}=15$$

These discussions now motivate us to make the following definition of multiplication of rational numbers.

Definition 6.6   The *product* of any two rational numbers a/b and c/d is defined as

$$\frac{a}{b} \cdot \frac{c}{d} = \frac{a \cdot c}{b \cdot d}$$

Example 6.19   Evaluate the products $\frac{3}{4} \cdot \frac{5}{7}$, $\frac{-5}{8} \cdot \frac{6}{11}$, and $\frac{-3}{14} \cdot \frac{-21}{12}$.

Solution:

$$\frac{3}{4} \cdot \frac{5}{7} = \frac{3 \cdot 5}{4 \cdot 7} = \frac{15}{28}$$

$$\frac{-5}{8} \cdot \frac{6}{11} = \frac{(-5) \cdot 6}{8 \cdot 11} = \frac{(-5) \cdot \cancel{2} \cdot 3}{\cancel{2} \cdot 4 \cdot 11} = \frac{-15}{44}$$

$$\frac{-3}{14} \cdot \frac{-21}{12} = \frac{(-3) \cdot (-21)}{14 \cdot 12} = \frac{(-1) \cdot \cancel{3} \cdot (-1) \cdot 3 \cdot \cancel{7}}{2 \cdot \cancel{7} \cdot \cancel{3} \cdot 4} = \frac{3}{8}$$

Example 6.20   Mrs. Smith has to drive a distance of 840 miles for a trip. If she drove $\frac{1}{7}$ of the total distance in 2 hours, how far did she drive during that time? How far away is she now from her destination?

Solution: Since Mrs. Smith drove $\frac{1}{7}$ of 840 miles in two hours, the distance she traveled during that time is given by

$$\frac{1}{7} \cdot 840 = \frac{1}{7} \cdot \frac{840}{1} = \frac{1 \cdot 840}{7 \cdot 1}$$

$$= \frac{840}{7} = \frac{\cancel{7} \cdot 120}{\cancel{7} \cdot 1}$$

$$= \frac{120}{1} = 120$$

She is now $840 - 120 = 720$ miles away from her destination.

Example 6.21   The number 18 is $\frac{3}{8}$ of what integer?

Solution: Let x be the integer $\frac{3}{8}$ of which is 18. But $\frac{3}{8}$ of x is given by $\frac{3}{8} \cdot x$. Thus,

$$\frac{3}{8} \cdot x = 18$$

$$\frac{3}{8} \cdot \frac{x}{1} = \frac{18}{1}$$

$$\frac{3x}{8} = \frac{18}{1}$$

$$1 \cdot 3x = 8 \cdot 18$$

$$3x = 8 \cdot 3 \cdot 6$$

$$x = 8 \cdot 6 = 48$$

Thus, the desired integer is 48. Observe that $\frac{3}{8} \cdot 48 = 18$.

Notice from these examples that the effect of multiplying an integer by a fraction is the same as multiplying the integer by the numerator of the fraction and keeping the same denominator. For example,

$$3 \cdot \frac{5}{7} = \frac{3 \cdot 5}{7} = \frac{15}{7}; \qquad 2 \cdot \frac{-3}{5} = \frac{2 \cdot (-3)}{5} = \frac{-6}{5}$$

and, more generally,

$$a \cdot \frac{b}{c} = \frac{a \cdot b}{c} = \frac{ab}{c}$$

At this point we advise the reader not to confuse $a \cdot \dfrac{b}{c}$ with $a\dfrac{b}{c}$ (why?).

The rational line can theoretically be used as a geometric tool to discuss the operation of multiplication of rational numbers, however it is a difficult task. That $3 \cdot \frac{2}{5} = \frac{6}{5}$ is illustrated geometrically in Fig. 6.9.

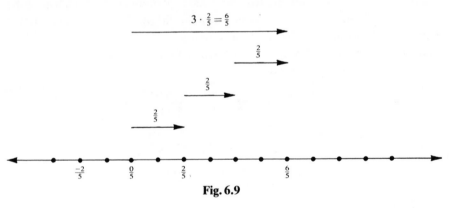

Fig. 6.9

Let's now proceed to discuss the fundamental properties of multiplication of rational numbers. Since the set of integers is closed under multiplication, it now follows from the definition $\dfrac{a}{b} \cdot \dfrac{c}{d} = \dfrac{ac}{bd}$ that the set of rational numbers is also closed under multiplication. In other words, multiplication is a binary operation on the set of rational numbers.

Theorem 6.9 (*Closure Property*)　*The set of rational numbers is closed under multiplication.*

Example 6.22　Evaluate the products $\frac{-3}{5} \cdot \frac{7}{8}$ and $\frac{7}{8} \cdot \frac{-3}{5}$.

Solution:
$$\frac{-3}{5} \cdot \frac{7}{8} = \frac{(-3) \cdot 7}{5 \cdot 8} = \frac{-21}{40}$$

and
$$\frac{7}{8} \cdot \frac{-3}{5} = \frac{7 \cdot (-3)}{8 \cdot 5} = \frac{-21}{40}$$

Thus,
$$\frac{-3}{5} \cdot \frac{7}{8} = \frac{7}{8} \cdot \frac{-3}{5}$$

More generally, as in the case of multiplication of integers, we have the following result, which tells us that two rational numbers can be multiplied in any order we like.

Theorem 6.10   (***Commutative Property***)   *For any two rational numbers x and y,*

$$x \cdot y = y \cdot x$$

Example 6.23   Evaluate the products $\frac{2}{5} \cdot (\frac{7}{3} \cdot \frac{6}{11})$ and $(\frac{2}{5} \cdot \frac{7}{3}) \cdot \frac{6}{11}$.

Solution:

$$\frac{2}{5} \cdot \left(\frac{7}{3} \cdot \frac{6}{11}\right) = \frac{2}{5} \cdot \left(\frac{7 \cdot 6}{3 \cdot 11}\right) = \frac{2}{5} \cdot \frac{42}{33} = \frac{2 \cdot 42}{5 \cdot 33} = \frac{84}{165}$$

and

$$\left(\frac{2}{5} \cdot \frac{7}{3}\right) \cdot \frac{6}{11} = \left(\frac{2 \cdot 7}{5 \cdot 3}\right) \cdot \frac{6}{11} = \frac{14}{15} \cdot \frac{6}{11} = \frac{14 \cdot 6}{15 \cdot 11} = \frac{84}{165}$$

Thus,

$$\frac{2}{5} \cdot \left(\frac{7}{3} \cdot \frac{6}{11}\right) = \left(\frac{2}{5} \cdot \frac{7}{3}\right) \cdot \frac{6}{11}$$

That this is true for any three rational numbers is given by the following theorem. Consequently, in a product of any finite number of rational numbers, we can group them in any way we wish to.

Theorem 6.11   (***Associative Property***)   *For any three rational numbers x, y, and z,*

$$x \cdot (y \cdot z) = (x \cdot y) \cdot z$$

Example 6.24   Evaluate the products $1 \cdot \frac{3}{5}$ and $\frac{-8}{3} \cdot 1$.

Solution:

$$1 \cdot \frac{3}{5} = \frac{1 \cdot 3}{5} = \frac{3}{5}$$

and

$$\frac{-8}{3} \cdot 1 = \frac{(-8) \cdot 1}{3} = \frac{-8}{3}$$

We observed in Chapter 4 that the set of integers possesses a unique multiplicative identity, namely, the integer 1. Does the set of rational numbers contain a multiplicative identity? Is it unique? These are answered by the following theorem.

Theorem 6.12   (***Existence of a Unique Multiplicative Identity***)   *The set of rational numbers contains a unique multiplicative identity* 1. *That is, x · 1 = x = 1 · x for every rational number x.*

To illustrate another important property, let's examine the following example:

Example 6.25   Evaluate the products $\frac{3}{5} \cdot \frac{5}{3}$ and $\frac{-2}{3} \cdot \frac{-3}{2}$.

Solution:

$$\frac{3}{5} \cdot \frac{5}{3} = \frac{3 \cdot 5}{5 \cdot 3} = \frac{15}{15} = 1$$

and

$$\frac{-2}{3} \cdot \frac{-3}{2} = \frac{(-2) \cdot (-3)}{3 \cdot 2} = \frac{6}{6} = 1$$

What do we conclude from this example? If $x = \frac{3}{5}$, then there is a $y = \frac{5}{3}$ such that $x \cdot y = \frac{3}{5} \cdot \frac{5}{3} = 1$. If $x = \frac{-2}{3}$, we can find a $y = \frac{-3}{2}$ such that $x \cdot y = \frac{-2}{3} \cdot \frac{-3}{2} = 1$. For every rational number x, does there exist a rational number y such that $x \cdot y = 1$, the multiplicative identity? The answer is yes, except when $x = 0$. If $x \cdot y = 1$, then y is called the *multiplicative inverse (reciprocal)* of x, denoted by x^{-1}. Observe that $x^{-1} = 1/x$.

Theorem 6.13   (*Existence of a Unique Multiplicative Inverse*)   *Every non-zero rational number x has a unique multiplicative inverse $x^{-1} = 1/x$. That is, $x \cdot x^{-1} = 1 = x^{-1} \cdot x$ for every nonzero rational number x.*

For example, the multiplicative inverse of $\frac{3}{5}$ is $\frac{5}{3}$, and that of $\frac{-2}{3}$ is $\frac{-3}{2}$, as seen from example 6.24. That every nonzero element of Q has a multiplicative inverse is a fundamental property possessed by no other previous number systems. Consequently, it is a major step in the development of the number systems.

Example 6.26   Evaluate the products $\frac{2}{3} \cdot \frac{5}{7}$ and $\frac{6}{9} \cdot \frac{25}{35}$.

Solution:

$$\frac{2}{3} \cdot \frac{5}{7} = \frac{2 \cdot 5}{3 \cdot 7} = \frac{10}{21}$$

and

$$\frac{6}{9} \cdot \frac{25}{35} = \frac{6 \cdot 25}{9 \cdot 35} = \frac{150}{315} = \frac{15 \cdot 10}{15 \cdot 21} = \frac{10}{21}$$

Observe that $\frac{2}{3} = \frac{6}{9}$, $\frac{5}{7} = \frac{25}{35}$, and $\frac{2}{3} \cdot \frac{5}{7} = \frac{6}{9} \cdot \frac{25}{35}$.

More generally, we now have the following result.

Theorem 6.14   (*Multiplication Property*)   *If w, x, y, z are any rational numbers such that $w = x$ and $y = z$, then $w \cdot y = x \cdot z$.*

In other words, products of equal rational numbers by equal rational numbers are equal. In the case of multiplication of integers, we observed that if $ab = ac$ and $a \neq 0$, then $b = c$. Is there a parallel result in the rational number system? This is answered by the following theorem.

Theorem 6.15   (*Cancellation Property*)   *If x, y, z are rational numbers such that $xy = xz$ and $x \neq 0$, then $y = z$.*

We warn the reader that the condition that x be nonzero in the theorem is crucial. For example, $0 \cdot \frac{2}{3} = 0 \cdot \frac{5}{7}$, however $\frac{2}{3} \neq \frac{5}{7}$. The cancellation property plays a significant role in the solution of equations, as shown by the following examples.

Example 6.27   Solve the equation $\frac{x}{3} = \frac{4}{9}$.

Solution: We must isolate the x on the left-hand side of the equation.

$$\frac{x}{3} = \frac{4}{9} \qquad \textit{given}$$

$$\frac{1}{3} \cdot \frac{x}{1} = \frac{1}{3} \cdot \frac{4}{3} \qquad \textit{definition of multiplication}$$

$$\frac{x}{1} = \frac{4}{3} \qquad \textit{cancellation property of multiplication}$$

$$x = \frac{4}{3} \qquad \frac{x}{1} = x$$

It can now be verified that $\frac{4}{3}$ is indeed the solution by substituting $\frac{4}{3}$ for x in the given equation.

Example 6.28   Solve the equation $\dfrac{x+2}{14} = \dfrac{3}{35}$.

Solution:

$$\frac{x+2}{14} = \frac{3}{35}$$

$$\frac{x+2}{2 \cdot 7} = \frac{3}{5 \cdot 7}$$

$$\frac{x+2}{2} = \frac{3}{5}$$

$$5(x+2) = 2 \cdot 3$$

$$5x + 10 = 6$$

$$5x = -4$$

$$x = -\frac{4}{5}$$

Thus, $-\frac{4}{5}$ is the desired solution.

In Section 4.4, we observed that the product of two integers is zero if and only if at least one of them is zero. This property can now be extended to include rational numbers.

Theorem 6.16   (***Product Law***)   *If x and y are any two rational numbers such that $x \cdot y = 0$, then either $x = 0$ or $y = 0$. Equivalently, the product of any two nonzero rational numbers is nonzero.*

Example 6.29   Solve the equation $(x + \frac{2}{3})(x - \frac{3}{2}) = 0$.

Solution: Let's make use of the product law to solve the given equation. Since $(x + \frac{2}{3})(x - \frac{3}{2}) = 0$, we have

$$x + \tfrac{2}{3} = 0 \qquad \text{or} \qquad x - \tfrac{3}{2} = 0$$

$$x = -\tfrac{2}{3} \qquad \text{or} \qquad x = \tfrac{3}{2}$$

Thus, $-\frac{2}{3}$ and $\frac{3}{2}$ are the solutions of the equations.

Recall that the integer 0 possesses a unique place under multiplication: the product of any integer by zero is always zero. This is true in the set of rational numbers, too.

Theorem 6.17   (*Zero Property of Multiplication*)   *For any rational number x,*

$$x \cdot 0 = 0 = 0 \cdot x$$

We conclude this section by discussing the distributive property of multiplication over addition in the rational number system.

Example 6.30   Verify that $\frac{2}{5} \cdot (\frac{3}{4} + \frac{-2}{7}) = \frac{2}{5} \cdot \frac{3}{4} + \frac{2}{5} \cdot \frac{-2}{7}$.

Solution: $\dfrac{2}{5} \cdot \left(\dfrac{3}{4} + \dfrac{-2}{7} \right) = \dfrac{2}{5} \cdot \left(\dfrac{3 \cdot 7 + 4 \cdot (-2)}{4 \cdot 7} \right) = \dfrac{2}{5} \cdot \left(\dfrac{21 - 8}{28} \right)$

$$= \frac{2}{5} \cdot \frac{13}{28} = \frac{26}{140} = \frac{13}{70}$$

$$\frac{2}{5} \cdot \frac{3}{4} + \frac{2}{5} \cdot \frac{-2}{7} = \frac{2 \cdot 3}{5 \cdot 4} + \frac{2 \cdot (-2)}{5 \cdot 7} = \frac{6}{20} + \frac{-4}{35}$$

$$= \frac{7 \cdot 6 + 4 \cdot (-4)}{140} = \frac{42 - 16}{140}$$

$$= \frac{26}{140} = \frac{13}{70}$$

Thus, $\dfrac{2}{5} \cdot \left(\dfrac{3}{4} + \dfrac{-2}{7} \right) = \dfrac{2}{5} \cdot \dfrac{3}{4} + \dfrac{2}{5} \cdot \dfrac{-2}{7}$

More generally, we have the following result.

Theorem 6.18   (*Distributive Property*)   *For any rational numbers x, y, and z,*

$$x(y + z) = xy + xz$$

The distributive property tells us that multiplication is distributive over addition, as in the previous number systems. It now follows that multiplication is distributive over subtraction, too. (Why?)

Exercise 6.4

   1. Evaluate each of the following.

   a) $\frac{-3}{10} \cdot \frac{15}{7}$ 　　　　 **b)** $\frac{-8}{13} \cdot \frac{26}{48}$ 　　　　 **c)** $-3(\frac{2}{3} + \frac{7}{6})$

d) $\frac{2}{3}(\frac{3}{4}+\frac{4}{5})$ **e)** $\frac{-5}{6}\cdot\frac{3}{7}\cdot\frac{8}{9}$ **f)** $\frac{-5}{11}\cdot\frac{3}{7}\cdot\frac{-7}{6}$

g) $\frac{-4}{3}\cdot\frac{5}{8}+\frac{2}{5}\cdot\frac{-4}{7}$ **h)** $-\frac{1}{3}\cdot\frac{3}{4}-\frac{5}{6}\cdot\frac{-6}{7}$ **i)** $(\frac{1}{2}-\frac{1}{3})(\frac{1}{2}+\frac{1}{3})$

2. Verify each of the following.

a) $\frac{3}{4}\cdot\frac{-5}{9}=\frac{-5}{9}\cdot\frac{3}{4}$ **b)** $\frac{-3}{7}\cdot\frac{-8}{9}=\frac{-8}{9}\cdot\frac{-3}{7}$

c) $\frac{1}{2}\cdot(\frac{1}{3}\cdot\frac{1}{4})=(\frac{1}{2}\cdot\frac{1}{3})\cdot\frac{1}{4}$ **d)** $\frac{3}{8}\cdot(\frac{5}{6}\cdot\frac{4}{11})=(\frac{3}{8}\cdot\frac{5}{6})\cdot\frac{4}{11}$

e) $\frac{4}{7}(\frac{3}{5}+\frac{2}{3})=\frac{4}{7}\cdot\frac{3}{5}+\frac{4}{7}\cdot\frac{2}{3}$ **f)** $\frac{-5}{3}(\frac{4}{5}-\frac{7}{11})=\frac{-5}{3}\cdot\frac{4}{5}-\frac{-5}{3}\cdot\frac{7}{11}$

3. Find the multiplicative inverse of

a) $\frac{5}{7}$ **b)** $\frac{-8}{11}$ **c)** $\frac{7}{-10}$ **d)** $5\frac{2}{3}$

4. What is the difference between $3\frac{1}{5}$ and $3\cdot\frac{1}{5}$?

5. Find a set of rational numbers x, y, z such that

   a) $x+(y\cdot z)=(x+y)\cdot(x+z)$

   b) $x-(y\cdot z)=(x-y)\cdot(x-z)$

6. Solve the following equations.

a) $\dfrac{2}{3}\cdot x=\dfrac{10}{21}$ **b)** $\dfrac{5x-8}{8x-5}=0$

c) $\dfrac{3}{4}\left(\dfrac{x}{3}+\dfrac{2}{5}\right)=\dfrac{27}{40}$ **d)** $\dfrac{4}{5}\left(x+\dfrac{1}{4}\right)+\dfrac{5}{4}=\dfrac{5}{4}+\dfrac{4}{5}$

7. Simplify each of the following.

a) $\dfrac{a+b}{b}\cdot\dfrac{a}{a+b}$ **b)** $\dfrac{a-ab}{a+ab}$ **c)** $\dfrac{a^2-1}{a+1}$

d) $\dfrac{a^2-b^2}{a-b}$ **e)** $\left(\dfrac{1}{a}+\dfrac{1}{b}\right)\left(\dfrac{1}{a}-\dfrac{1}{b}\right)$ **f)** $\left(\dfrac{a}{b}+\dfrac{b}{a}\right)\left(\dfrac{a}{b}-\dfrac{b}{a}\right)$

8. If $\dfrac{a}{b}\cdot\dfrac{c}{d}=0$, what is your conclusion?

9. Use the distributive property to evaluate each of the following.

   a) $3\frac{2}{5}\cdot4\frac{3}{5}$ **b)** $2\frac{3}{4}\cdot5\frac{6}{7}$ **c)** $-5\frac{2}{7}\cdot3\frac{9}{11}$

   [*Hint:* $2\frac{5}{7}\cdot3\frac{4}{5}=(2+\frac{5}{7})(3+\frac{4}{5})$]

10. Find a set of integers a, b, c, d such that $\dfrac{a}{b}\cdot\dfrac{c}{d}=\dfrac{ad}{bc}$.

11. If $\dfrac{a}{b}\cdot\dfrac{c}{d}=\dfrac{ad}{bc}$, prove that $c=\pm d$, assuming all integers are nonzero.

12. Mark *true* or *false* (a, b, c are any nonzero integers; x, y, z are any rational numbers).

a) $\dfrac{a}{b}=a\cdot\dfrac{1}{b}$

b) The set of negative fractions is closed under multiplication.

c) $\dfrac{2}{3}\cdot x=\dfrac{2x}{3}$

d) $(-1)\cdot x=-x$

e) If $xy=xz$, then $y=z$.

f) If x is the multiplicative inverse of y, then y is the multiplicative inverse of x.

g) The multiplicative inverse of a/b is b/a.

h) The multiplicative inverse of $\dfrac{0}{5}$ does not exist.

i) If x is a rational number, so is $1/x$.

j) Multiplication is distributive over subtraction.

k) $x(y-z)=xy-xz$

l) $a\left(\dfrac{1}{b}+\dfrac{1}{c}\right)=\dfrac{a}{b}+\dfrac{a}{c}$

13. Give a counterexample to each of the false statements in problem 12.

14. If the rational number x is its own additive inverse, prove that $x=0$.

15. If the rational number x is its own multiplicative inverse, what is your conclusion?

16. Prove that $a\cdot\dfrac{b}{c}=\dfrac{ab}{c}$. [*Hint*: use the definition of multiplication and $a=a/1$.]

17. Prove that $(-1)\cdot\dfrac{a}{b}=-\dfrac{a}{b}$. [*Hint*: use problem 16.]

18. If $\dfrac{a}{be}=\dfrac{c}{de}$, prove that $\dfrac{a}{b}=\dfrac{c}{d}$ (assume a, b, c, d, e are nonzero integers).

19. What is $\frac{1}{3}$ of a yard?

20. How many minutes are there in $\frac{4}{15}$ of an hour?

21. What part of an hour is 5 minutes?

22. What part of a dollar is 45 cents?

23. How many cents has Andy left if he spent $\frac{2}{3}$ of a dollar on candy?

24. Sandy planned to drive a distance of 760 miles for a trip. She drove 240 miles in 3 hours. How much of the total distance did she travel during that time?

25. If $\frac{2}{5}$ of $\frac{3}{8}$ of an integer is 30, find the integer.

26. Mr. Whipple spends $\frac{3}{14}$ of his monthly income for rent and $\frac{2}{11}$ of what is left on food. If he has 540 dollars left, what is his monthly income?

27. What is the price of $5\frac{1}{4}$ pounds of beef if it costs $1\frac{3}{5}$ dollars a pound?

★28. Prove each of the following, where w, x, y, z are any rational numbers.

    a) $(-x)\cdot y=x\cdot(-y)=-(xy)$ **b)** $(-x)\cdot(-y)=xy$

★29. Use the ordered pair notation for rational numbers to define multiplication by $(a,b)\cdot(c,d)=(ac,bd)$. Recall that $(a,b)=(c,d)$ if and only if $ad=bc$.

    a) Evaluate $(3,5)\cdot(4,3)$.

    b) Evaluate $(-5,7)\cdot(4,-11)$.

    c) Find the integer x such that $(5,6)\cdot(x,4)=(5,8)$.

    d) Find the rational number (a,b) if $(-3,4)\cdot(a,b)=(-1,6)$.

6.5 DIVISION OF RATIONAL NUMBERS

Assume that Mr. Murphy has a $\frac{3}{4}$ acre plot of land. In how many equal parts can he divide the area if each is to be $\frac{3}{20}$ of an acre? Observe that the problem is to evaluate $\frac{3}{4} \div \frac{3}{20}$ or to solve the equation $\frac{3}{4} = \frac{3}{20} \cdot x$. Multiplying both sides of the equation by $\frac{20}{3}$ (the multiplicative inverse of $\frac{3}{20}$), we get

$$\frac{20}{3} \cdot \frac{3}{4} = \frac{20}{3} \cdot \frac{3}{20} \cdot x$$

$$5 = x$$

Therefore, he can divide his plot of land into five equal parts of $\frac{3}{20}$ of an acre. Thus $\frac{3}{4} \div \frac{3}{20} = 5$, which can also be written as $\dfrac{3 \cdot 20}{4 \cdot 3}$.

In Section 4.7, we defined the operation of division on the set of integers by $a \div b = c$ if and only if $a = bc$, assuming $b \neq 0$. We now define the operation of division of rational numbers, keeping in mind that it must be consistent with the definition of division of integers. In light of our experience with division of integers,

$$\frac{a}{b} \div \frac{c}{d} = x \qquad \text{if and only if} \qquad \frac{a}{b} = \frac{c}{d} \cdot x \qquad \text{where } \frac{c}{d} \neq 0$$

Let's now solve the equation $\dfrac{a}{b} = \dfrac{c}{d} \cdot x$. Multiplying both sides by $\dfrac{d}{c}$:

$$\frac{d}{c} \cdot \frac{a}{b} = \frac{d}{c} \cdot \left(\frac{c}{d} \cdot x \right)$$

$$\frac{da}{cb} = \left(\frac{d}{c} \cdot \frac{c}{d} \right) \cdot x$$

$$\frac{ad}{bc} = 1 \cdot x = x$$

Thus,

$$\frac{a}{b} \div \frac{c}{d} = \frac{ad}{bc}$$

This discussion now leads us to make the following definition.

Definition 6.7   Let $\dfrac{a}{b}$ and $\dfrac{c}{d}$ be any two rational numbers with $\dfrac{c}{d} \neq 0$. Then $\dfrac{a}{b} \div \dfrac{c}{d}$ is defined as $\dfrac{a}{b} \div \dfrac{c}{d} = \dfrac{ad}{bc}$. The operation "$\div$" we defined is called *division*.

Observe that $\dfrac{a}{b} \div \dfrac{c}{d} = \dfrac{ad}{bc} = \dfrac{a}{b} \cdot \dfrac{d}{c}$, where $\dfrac{d}{c}$ is the multiplicative inverse of $\dfrac{c}{d}$. Thus, in order to divide $\dfrac{a}{b}$ by $\dfrac{c}{d}$, it is enough to multiply $\dfrac{a}{b}$ by the reciprocal $\dfrac{d}{c}$ of

$\frac{c}{d}$. It now follows that the set of nonzero rational numbers is closed under division, a property the system of integers does not possess.

Example 6.31 Evaluate $\frac{3}{5} \div \frac{9}{10}$ and $\frac{-10}{7} \div \frac{15}{28}$.

Solution:
$$\frac{3}{5} \div \frac{9}{10} = \frac{3 \cdot 10}{5 \cdot 9} = \frac{\cancel{3} \cdot \cancel{5} \cdot 2}{\cancel{5} \cdot \cancel{3} \cdot 3} = \frac{2}{3}$$

and
$$\frac{-10}{7} \div \frac{15}{28} = \frac{-10 \cdot 28}{7 \cdot 15} = \frac{-2 \cdot \cancel{5} \cdot \cancel{7} \cdot 4}{\cancel{7} \cdot \cancel{5} \cdot 3} = \frac{-2 \cdot 4}{3} = \frac{-8}{3}$$

Example 6.32 Susan has $\frac{5}{8}$ of a pie left from yesterday. If she divides it into three equal pieces, how much of the pie will each piece be?

Solution: Notice that it suffices to solve the equation $3x = \frac{5}{8}$. Therefore,
$$x = \frac{5}{8} \div 3 = \frac{5}{8} \div \frac{3}{1}$$
$$= \frac{5 \cdot 1}{8 \cdot 3} = \frac{5}{24}$$

Thus, each piece is $\frac{5}{24}$ of the whole pie.

Exercise 6.5

1. Evaluate each of the following.

a) $\dfrac{7}{8} \div \dfrac{5}{8}$

b) $\dfrac{-3}{16} \div \dfrac{-9}{24}$

c) $\dfrac{\frac{9}{10}}{\frac{2}{5} - \frac{3}{2}}$

d) $\dfrac{\frac{2}{3} - \frac{5}{6}}{\frac{3}{8}}$

e) $\dfrac{\frac{1}{2} - \frac{1}{3}}{\frac{1}{2} + \frac{1}{3}}$

f) $\dfrac{\frac{2}{3} - \frac{4}{5}}{\frac{5}{6} - \frac{7}{8}}$

g) $\dfrac{\frac{2}{3} - \frac{3}{2}}{\frac{3}{4} - \frac{4}{3}}$

h) $\left(\dfrac{-3}{4} \div \dfrac{5}{6} \right) + \dfrac{11}{2}$

i) $\dfrac{5}{11} - \left(\dfrac{-6}{5} \div \dfrac{2}{3} \right)$

2. Solve the following equations.

a) $\dfrac{3x}{5} = \dfrac{4}{7}$

b) $\dfrac{-3x}{2} = \dfrac{2}{3} + \dfrac{1}{4}$

c) $\dfrac{4}{5} \cdot \dfrac{x}{3} = \dfrac{3}{4} \div \dfrac{12}{5}$

d) $\dfrac{1}{3}\left(x - \dfrac{1}{3} \right) + \dfrac{3}{2} = \dfrac{3}{2}$

e) $x + \dfrac{3x}{7} = 2x$

f) $\dfrac{-3x}{4} + \dfrac{1}{2} = x + \dfrac{1}{3}$

3. Verify each of the following.

a) $\frac{2}{3} \div \frac{4}{5} \neq \frac{4}{5} \div \frac{2}{3}$

b) $\frac{3}{4} \div \left(\frac{-4}{5} \div \frac{5}{8} \right) \neq \left(\frac{3}{4} \div \frac{-4}{5} \right) \div \frac{5}{8}$

c) $\frac{1}{2} + \left(\frac{1}{3} \div \frac{1}{4} \right) \neq \left(\frac{1}{2} + \frac{1}{3} \right) \div \left(\frac{1}{2} + \frac{1}{4} \right)$

d) $\frac{2}{3} \div \left(\frac{3}{4} + \frac{4}{5} \right) \neq \left(\frac{2}{3} \div \frac{3}{4} \right) + \left(\frac{2}{3} \div \frac{4}{5} \right)$

4. Is $\dfrac{a}{b} \div c = \dfrac{a}{bc}$?

5. Is $\dfrac{a}{b} \div \dfrac{1}{b} = a$?

6. Simplify each of the following.

a) $\dfrac{1/b - 1/a}{1/b + 1/a}$

b) $\dfrac{1/a - 1/b}{1/ab}$

c) $\dfrac{a/b - b/a}{a/b + b/a}$

d) $\dfrac{a(a-b)}{a+b} \div \dfrac{b(a-b)}{a+b}$ e) $\dfrac{a+b}{ab} \div \dfrac{a-b}{ab}$ f) $\dfrac{a^2b^2}{a^2-b^2} \div \dfrac{ab}{a-b}$

7. Mark *true* or *false* (a, b, c are any nonzero integers).

   a) The set of rational numbers is closed under division.

   b) $\dfrac{a}{c} \div \dfrac{b}{c} = a \div b$

   c) $\frac{3}{5} \div \frac{4}{5} = \frac{3}{4}$

   d) $\frac{3}{4} \div 3 = \frac{1}{4}$

   e) $\frac{3}{4} \div \frac{1}{4} = 3$

8. Give a counterexample to each of the false statements in problem 7.

9. If $\frac{3}{5}$ of a pie is cut into 6 equal parts, how much of the pie will each piece be?

10. June has $6\frac{3}{4}$ dollars. If a can of soup costs $\frac{3}{16}$ of a dollar, how many cans of soup can she buy?

11. Into how many pieces, $\frac{1}{20}$ of a yard each, can you divide a string of $\frac{1}{4}$ of a yard?

12. Mr. Rogers has $8\frac{1}{6}$ gallons of gasoline in the tank of his car. If the car uses $\frac{7}{72}$ of a gallon for a mile, how many miles can be drive with that much gasoline?

13. Harry bought $3\frac{2}{5}$ pounds of one kind of candy and Mary bought $1\frac{1}{2}$ pounds of another kind of candy for a party. If they mix their candy together and give $\frac{7}{30}$ of a pound to each of their friends at the party, how many people are present at the party?

14. Find a set of integers a, b, c, d such that $\dfrac{a}{b} \div \dfrac{c}{d} = \dfrac{ac}{bd}$.

15. If $\dfrac{a}{b} \div \dfrac{c}{d} = \dfrac{ac}{bd}$, prove that $c = \pm d$ (assume $a \neq 0$).

★16. Use the facts that $\dfrac{ac}{bc} = \dfrac{a}{b}$ and $\dfrac{a}{1} = a$ to prove that $\dfrac{a}{b} \div \dfrac{c}{d} = \dfrac{ad}{bc}$.

★17. Using the ordered pair notation for rational numbers, define division by $(a,b) \div (c,d) = (ad,bc)$.

   a) Evaluate $(3,5) \div (7,6)$.

   b) Evaluate $(-7,8) \div (3,-4)$.

   c) Find the integer x if $(-3,4) \div (x,2) = (-1,6)$.

6.6   THE ORDER RELATION

Assume that Peter has a plot of land of $\frac{3}{5}$ of an acre and John has $\frac{2}{7}$ of an acre. Since $\frac{3}{5} \neq \frac{2}{7}$ (why?), we want to know who has more land? Notice that we can find a rational number $x = \frac{11}{35}$ such that $\frac{2}{7} + \frac{11}{35} = \frac{3}{5}$. From our experiences with integers, we conclude that Peter has more land than John. Observe that $\frac{3}{5} - \frac{2}{7} = \frac{11}{35}$, a positive rational number. In mathematical language, we say that $\frac{2}{7}$ is less than $\frac{3}{5}$.

If Susan has $\frac{2}{3}$ of a pound of candy and Sandy has $\frac{3}{4}$ of a pound of candy, who has more candy? Since $\frac{3}{4} - \frac{2}{3} = \frac{1}{12}$, a positive fraction, we conclude that Sandy has more candy.

In Chapter 4, we observed that if a and b are any integers, then a is less than b if and only if $b - a$ is a positive integer. We now extend this property to the set of rational numbers as follows.

Definition 6.8   Let x and y be any two rational numbers. Then x is said to be *less than* y, denoted by $x < y$, if $y - x$ is a positive rational number. If x is less than y, then we also say that y is *greater than* x. The expression $x \leq y$ simply means $x < y$ or $x = y$.

Example 6.33   Verify that $\frac{5}{8} < \frac{7}{11}$ and $\frac{-5}{24} < \frac{-3}{16}$.

Solution: Let's evaluate $\frac{7}{11} - \frac{5}{8}$ and $\frac{-3}{16} - \frac{-5}{24}$ and check that each is a positive rational number.

$$\frac{7}{11} - \frac{5}{8} = \frac{7 \cdot 8 - 11 \cdot 5}{11 \cdot 8} = \frac{56 - 55}{88} = \frac{1}{88}$$

which is a positive rational number. Therefore, $\frac{5}{8} < \frac{7}{11}$. Next,

$$\frac{-3}{16} - \frac{-5}{24} = \frac{(-3) \cdot 3 - (-5) \cdot 2}{48} = \frac{-9 - (-10)}{48}$$

$$= \frac{-9 + 10}{48} = \frac{1}{48}$$

a positive rational number. Therefore, $\frac{-5}{24} < \frac{-3}{16}$.

Let a/b and c/d be any two rational numbers with positive denominators. To decide if $\frac{a}{b} < \frac{c}{d}$, we must check if $\frac{c}{d} - \frac{a}{b} = \frac{bc - ad}{bd}$ is a positive rational number. Since $\frac{bc - ad}{bd} = \frac{-(ad - bc)}{bd}$, where $bd > 0$, we need only check if $ad - bc < 0$. Thus, if $b > 0$ and $d > 0$,

$$\frac{a}{b} < \frac{c}{d} \qquad \text{if and only if} \qquad ad < bc$$

Example 6.34   Determine if $-\frac{2}{3} < -\frac{3}{5}$.

Solution: $-\frac{2}{3} < -\frac{3}{5}$ if and only if $(-2) \cdot 5 < 3 \cdot (-3)$ if and only if $-10 < -9$, which is true. Therefore, $-\frac{2}{3} < -\frac{3}{5}$. We now ask the reader to verify that $-\frac{3}{5} - (-\frac{2}{3})$ is a positive rational number.

That $\frac{a}{b} < \frac{c}{d}$ geometrically means that the point with coordinate a/b lies to the *left* of the point with coordinate c/d on the number line. Since our definition of the less-than relation is consistent with that on the set of integers, we expect all the properties of the order relation discussed in Section 4.6 to remain true here also. In the case of integers we observed that if a and b are any two integers, then either $a < b$ or $a = b$ or $a > b$. That this holds in the set of rational numbers is given in the following theorem.

Theorem 6.19 (*Law of Trichotomy*)   *If x and y are any two rational numbers, then they are related in exactly one of the following ways: either*

$$(1) \ x < y \qquad or \qquad (2) \ x = y \qquad or \qquad (3) \ y < x$$

Equivalently, the law of trichotomy can also be stated as follows: if x is any rational number, then either $x < 0$ or $x = 0$ or $x > 0$. Consequently, the set of rational numbers can be partitioned as in Fig. 6.10.

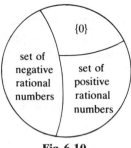

Fig. 6.10

Theorem 6.20 (*Transitive Property*)   *If x, y, z are any rational numbers such that $x < y$ and $y < z$, then $x < z$.*

The transitive property geometrically means that if x lies to the left of y and y lies to the left of z on the rational line, then x lies to the left of z. This is illustrated for $x = -\frac{1}{2}$, $y = \frac{3}{4}$, and $z = \frac{3}{2}$ in Fig. 6.11.

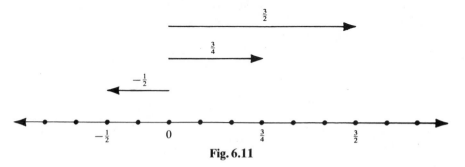

Fig. 6.11

Example 6.35   Observe that $\frac{3}{5} < \frac{7}{8}$. Adding the same rational number $-\frac{3}{11}$ to both sides of the inequality, we get $\frac{3}{5} + (-\frac{3}{11}) = \frac{18}{55}$ and $\frac{7}{8} + (-\frac{3}{11}) = \frac{53}{88}$, where $\frac{18}{55} < \frac{53}{88}$ since $18 \cdot 88 < 55 \cdot 53$. Multiplying both sides of $\frac{3}{5} < \frac{7}{8}$ by $\frac{3}{11}$, we get $\frac{3}{5} \cdot \frac{3}{11} = \frac{9}{55}$ and $\frac{7}{8} \cdot \frac{3}{11} = \frac{21}{88}$, where $\frac{9}{55} < \frac{21}{88}$ (why?). Multiplying both sides of $\frac{3}{5} < \frac{7}{8}$ by $-\frac{3}{11}$, we have

$\frac{3}{5} \cdot (-\frac{3}{11}) = -\frac{9}{55}$ and $\frac{7}{8} \cdot (-\frac{3}{11}) = -\frac{21}{88}$ where $-\frac{9}{55} > -\frac{21}{88}$ (why?). Thus, $\frac{3}{5} < \frac{7}{8}$ implies that

a) $\frac{3}{5} + (-\frac{3}{11}) < \frac{7}{8} + (-\frac{3}{11})$

b) $\frac{3}{5} \cdot \frac{3}{11} < \frac{7}{8} \cdot \frac{3}{11}$

c) $\frac{3}{5} \cdot (-\frac{3}{11}) > \frac{7}{8} \cdot (-\frac{3}{11})$

More generally, we now have the following result.

Theorem 6.21 *If x and y are any two rational numbers such that $x < y$, then*

1. $x + z < y + z$ *for every rational number z,*
2. $xz < yz$ *if z is a positive rational number,*
3. $xz > yz$ *if z is a negative rational number.*

In other words, the theorem states that the same rational number z can be added or subtracted from both sides of an inequality without changing its type; an inequality remains of the same type if both sides are multiplied by the same positive rational number z; the inequality reverses if z is negative. It now follows that $x < y$ if and only if $-x > -y$.

Example 6.36 Find and graph the solution set of the inequality

$$\frac{2x}{3} + \frac{3}{4} < \frac{3x}{5} - \frac{1}{2}$$

Solution:

$$\frac{2x}{3} + \frac{3}{4} < \frac{3x}{5} - \frac{1}{2}$$

$$\frac{2x}{3} + \frac{3}{4} - \frac{3}{4} < \frac{3x}{5} - \frac{1}{2} - \frac{3}{4}$$

$$\frac{2x}{3} < \frac{3x}{5} - \frac{5}{4}$$

$$\frac{2x}{3} - \frac{3x}{5} < \frac{3x}{5} - \frac{3x}{5} - \frac{5}{4}$$

$$\frac{x}{15} < -\frac{5}{4}$$

$$x < -\frac{5}{4} \cdot 15$$

$$x < -\frac{75}{4}$$

Thus, the solution set of the given inequality is $\{x \in Q \mid x < -\frac{75}{4}\}$ and is represented geometrically in Fig. 6.12.

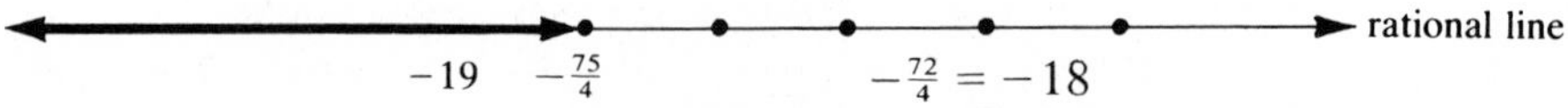

-19 $-\frac{75}{4}$ $-\frac{72}{4} = -18$ rational line

Fig. 6.12

Example 6.37 Find the solution set of

$$\frac{-3}{8}+\frac{1}{6}<\frac{5x}{4}+\frac{3}{2}\le\frac{5}{12}-\frac{1}{4}$$

Solution:

$$\frac{-3}{8}+\frac{1}{6}<\frac{5x}{4}+\frac{3}{2}\le\frac{5}{12}-\frac{1}{4}$$

$$\frac{-9+4}{24}<\frac{5x+6}{4}\le\frac{5-3}{12}$$

$$\frac{-5}{24}<\frac{6(5x+6)}{24}\le\frac{2\cdot2}{24}$$

$$-5<30x+36\le4$$

$$-5-36<30x\le4-36$$

$$-41<30x\le-32$$

$$-\tfrac{41}{30}<x\le-\tfrac{32}{30}$$

Thus, the solution set is $\{x\in Q\mid-\tfrac{41}{30}<x\le-\tfrac{32}{30}\}$.

Example 6.38 If four times a fraction does not exceed $5\tfrac{1}{3}$, what are the possible values of the fraction?

Solution: Let x be a fraction such that $4x$ does not exceed $5\tfrac{1}{3}=\tfrac{16}{3}$.

$$4x\le\tfrac{16}{3}$$

$$x\le\tfrac{4}{3}$$

Thus, any member of the set $\{x\in Q\mid x\le\tfrac{4}{3}\}$ is a candidate for the fraction.

Before we close this section, let's discuss a very remarkable and fascinating property possessed by the system of rational numbers and by no other

THICKLY SETTLED

Fig. 6.13

previously discussed number systems. If a and b are distinct integers, is it always possible to find an integer c between them, that is, an integer c such that $a < c < b$? Clearly not; for example, there are no integers between 2 and 3. However, to our surprise, it is always possible to find a rational number z between any two distinct rational numbers x and y. This is the property known as *denseness* (see Fig. 6.13). Equivalently, we also say that the set of rational numbers is *dense*. For example, how do we find a rational number z such that $x = \frac{2}{7} < z < \frac{3}{8} = y$? A value of z can be chosen such that

$$z = \frac{x+y}{2} = \frac{1}{2}\left(\frac{2}{7} + \frac{3}{8}\right)$$

$$= \frac{1}{2} \cdot \frac{16+21}{56} = \frac{37}{112}$$

The number $z = \dfrac{x+y}{2}$ is called the *arithmetic mean (average)* of the numbers x and y (see Fig. 6.14).

Fig. 6.14

Now notice that

$$\frac{2}{7} < \frac{37}{112} < \frac{3}{8}$$

for $\frac{2}{7} < \frac{37}{112}$ since $2 \cdot 112 = 224 < 7 \cdot 37 = 259$ and $\frac{37}{112} < \frac{3}{8}$ since $37 \cdot 8 = 296 < 112 \cdot 3 = 336$. Thus, by finding the average of the numbers x and y where $x < y$, we can find a number between them. Now by taking the average of the numbers $\frac{2}{7}$ and $\frac{37}{112}$, and of the numbers $\frac{37}{112}$ and $\frac{3}{8}$, we can find a number between each pair, namely,

$$\frac{1}{2}\left(\frac{2}{7} + \frac{37}{112}\right) = \frac{69}{224} \qquad \text{and} \qquad \frac{1}{2}\left(\frac{37}{112} + \frac{3}{8}\right) = \frac{79}{224}$$

Thus we have

$$\frac{2}{7} < \frac{3}{8}$$

$$\frac{2}{7} < \frac{37}{112} < \frac{3}{8}$$

$$\frac{2}{7} < \frac{69}{224} < \frac{79}{224} < \frac{3}{8}$$

Obviously, this procedure can be continued indefinitely, yielding infinitely many rational numbers between any two distinct rational numbers, no matter how close they are!

Exercise 6.6

1. Determine if $x < y$ in each of the following cases.

   a) $x = \frac{3}{4}, y = \frac{4}{3}$ b) $x = \frac{2}{3}, y = \frac{3}{5}$ c) $x = \frac{3}{7}, y = \frac{7}{11}$

   d) $x = \frac{-5}{6}, y = \frac{5}{7}$ e) $x = -\frac{7}{8}, y = -\frac{8}{9}$ f) $x = \frac{-3}{11}, y = \frac{-4}{13}$

2. Verify that $x < y$ in the following cases. In which of them is $\frac{1}{x} > \frac{1}{y}$?

   a) $x = 2, y = 5$ b) $x = \frac{-3}{5}, y = \frac{-2}{7}$ c) $x = \frac{-10}{7}, y = \frac{3}{11}$

3. Identify the sets $A \cup B$, $A \cap B$, $A - B$ and $B - A$ in the set-builder notation if $A = \{x \in Q \mid \frac{-3}{2} \leq x < \frac{3}{5}\}$ and $B = \{x \in Q \mid \frac{3}{8} < x \leq \frac{5}{6}\}$.

4. Find the solution set of the following.

   a) $2x < \frac{3}{5}$ b) $-3x \leq \frac{4}{5}$

   c) $-\frac{5}{7} < 2x + \frac{2}{3} < \frac{5}{7}$ d) $-3x + \frac{5}{8} \leq -2x + \frac{5}{12}$

   e) $\frac{2}{15} < 2x + \frac{4}{5} \leq \frac{9}{5}$ f) $\frac{2}{3} - \frac{13}{2} \leq 4x + \frac{5}{6} \leq -\frac{47}{30}$

5. If $\frac{19}{30}$ is a rational number between $\frac{3}{5}$ and $\frac{2}{3}$, find a rational number between $\frac{3}{2}$ and $\frac{5}{3}$.

6. Find a rational number between each of the following pairs of rational numbers.

   a) $\frac{7}{10}$ and $\frac{3}{4}$ b) $\frac{4}{11}$ and $\frac{8}{13}$ c) $-\frac{5}{6}$ and $\frac{3}{2}$

   d) $-\frac{5}{2}$ and $-\frac{4}{5}$ e) $\frac{3}{5}$ and $\frac{5}{3}$ f) $-\frac{2}{3}$ and $-\frac{3}{4}$

7. Use the fact that if $\frac{a}{b} < \frac{c}{d}$, then $\frac{a}{b} < \frac{a+c}{b+d} < \frac{c}{d}$ to find a rational number between the following pairs of rational numbers.

   a) $\frac{5}{6}$ and $\frac{6}{7}$ b) $\frac{5}{9}$ and $\frac{10}{13}$ c) $\frac{-8}{11}$ and $\frac{-11}{13}$

8. Find three rational numbers between $\frac{1}{3}$ and $\frac{1}{2}$.

9. Are $\frac{1}{4}$ and $\frac{1}{2}$ the rational numbers closest to $\frac{1}{3}$? Why?

10. How many rational numbers exist between $\frac{1}{3}$ and $\frac{1}{2}$? Identify them as a set.

11. Mark *true* or *false* (a, b, c, d are any nonzero integers; x, y, z are any rational numbers).

    a) If $\frac{a}{b} < \frac{c}{d}$, then $-\frac{a}{b} < -\frac{c}{d}$.

    b) If $\frac{a}{b} < \frac{c}{d}$, then $\frac{1}{2} \frac{a}{b} < \frac{1}{2} \frac{c}{d}$.

    c) If $\frac{a}{b} < \frac{c}{d}$, then $\frac{2}{3} + \frac{a}{b} < \frac{2}{3} + \frac{c}{d}$.

    d) If $x < y$, then $xz < yz$.

    e) If $xy < xz$, then $y < z$.

    f) If $x + y < x + z$, then $y < z$.

    g) If $x < y$ and $y < z$, then $x < z$.

    h) If $x \leq y$ and $y \leq x$, then $x = y = 0$.

    i) The set of rational numbers is dense.

    ★j) If $0 < x < y$, then $0 < \dfrac{1}{y} < \dfrac{1}{x}$.

12. Give a counterexample to each of the false statements in problem 11.

13. If three times a fraction is at least as much as $\frac{2}{3}$ more than twice the fraction, find the set of possible values of the fraction.

14. If five times a fraction does not exceed $\frac{2}{3}$ less than three times the fraction, what is the set of all possible values of the fraction?

15. Kathy went to a bookstore with 45 dollars to buy 5 books, all at the same price. If she finds that the total price of the books does not exceed 54 dollars, but 45 dollars is not enough, what are the possible values in fractions for the price of each book?

16. If three times the maximum and the minimum temperatures of a certain day were $112\frac{1}{5}$ degrees above zero and $11\frac{1}{4}$ degrees below zero, what was the set of possible values in fractions at various times of the day?

★17. If x is any rational number, prove that $x \leq x$.

★18. If x, y, z are rational numbers such that $x < y$ and $y < z$, then prove that $x < z$.

★19. If x and y are rational numbers such that $x \leq y$ and $y \leq x$, prove that $x = y$.

★20. Using the ordered pair notation for rational numbers, define $(a,b) < (c,d)$ if and only if $ad < bc$, assuming $b > 0$ and $d > 0$.

    a) Verify that $(3,8) < (2,5)$.

    b) Verify that $(-8,11) < (-4,7)$.

    c) Find the set of all integers x such that $(-3,4) < (x,3)$.

6.7 A DEFICIENCY OF THE SET OF RATIONAL NUMBERS

Recall that we could not solve every equation of the form $a = b + x$ in the set of whole numbers. We overcame this difficulty by extending the set of whole numbers to the set of integers. However, in Section 4.7, we observed that not every equation of the form $a = bx$ is solvable in the set of integers. To remedy this situation, we "enlarged" the set of integers to the set of rational numbers by introducing numbers of the form a/b, where a and b are integers ($b \neq 0$).

We now focus our attention on discussing a deficiency of the set of rational numbers. Is there a rational number whose square is 2? In other words, can we solve the equation $x^2 = 2$ in the set of rational numbers? We now proceed to prove indirectly that this equation cannot be solved in the set of rational numbers. Assume that there is a rational number $x = a/b$ such that $x^2 = 2$. We can very well assume that the integers a and b have no common factors other than 1 (why?). We have

$$\left(\frac{a}{b}\right)^2 = 2$$

$$\frac{a^2}{b^2} = 2$$

$$a^2 = 2b^2$$

This shows that a^2 is an even integer. It now follows, by Theorem 5.3, that the integer a is also even. Therefore, $a = 2c$ for some integer c. We now have

$$(2c)^2 = 2b^2$$

$$4c^2 = 2b^2$$

$$b^2 = 2c^2$$

Consequently b^2 and hence b are even integers. Since a and b are both even integers, they have a common factor, 2. But, this contradicts our assumption that a and b have no common factors other than 1. Thus we conclude that our assumption is wrong and there exists no rational number x such that $x^2 = 2$.

Let's denote the solutions of the equation $x^2 = 2$ by $\pm\sqrt{2}$ (*square root of* 2). We now warn the reader that $\sqrt{2}$ and $-\sqrt{2}$ are not rational numbers, consequently, they cannot be expressed as the ratio of two integers. Numbers having this property are called *irrational numbers*. For example, $\sqrt{2}$, $-\sqrt{2}$, $\sqrt{3}$, $-\sqrt{3}$, etc., are all irrational numbers. More generally, $\pm\sqrt{a}$, where a is any positive rational number that is not a perfect square, are irrational numbers.

One of the irrational numbers that plays a very significant role in mathematics is π (pi), which denotes the ratio of the circumference of a circle, $2\pi r$, to its diameter, $2r$. The symbol π was first used by the English mathematician William Oughtred in 1647 to denote the circumference of a circle. It was the English writer William Jones who in 1706 for the first time used π to name the ratio of the circumference of a circle to its diameter. However, it did not become popular until Euler used it in 1737.

Exercise 6.7

1. Mark *true* or *false*.

   a) The equation $x^2 = 2$ has no solutions in the set of rational numbers.
   b) The equation $x^2 = 2$ has exactly two solutions.
   c) $\sqrt{3}$ is not a rational number.
   d) The equation $x^2 = a$, where a is any positive integer, has no rational solution.
   e) The set of rational numbers is closed under the operation "taking square root of."
   f) $\pi = \frac{22}{7}$.

2. Give a counterexample (whenever possible) to each of the false statements in problem 1.

3. Prove that $\sqrt{3}$ is not a rational number. [*Hint*: imitate the proof in Section 6.7.]

4. Prove that the equation $x^2 = 5$ has no rational solution. [This will mean that $\sqrt{5}$ is not a rational number.]

★**5.** Prove that there is no rational number x such that $x^3 = 2$. [This will prove that $\sqrt[3]{2}$ (*cube root* of 2) is not a rational number.]

6.8 SUMMARY AND COMMENTS

In this chapter, we developed the system of rational numbers from that of the integers. We now can solve every equation of the form $a = bx$ where a and $b (\neq 0)$ are integers. A rational number (fraction) is the ratio a/b of two integers a and $b (\neq 0)$. The number a is the numerator and b the denominator of the fraction a/b. Every integer is also a rational number. Rational numbers a/b and c/d are equal if and only if $ad = bc$.

We defined the operations of addition, subtraction, multiplication, and division (by nonzero elements) on the set of rational numbers. The set of rational numbers is closed under the first three operations. Addition and multiplication are both commutative and associative. Subtraction is neither commutative nor associative. The same is the case with division. The set of rational numbers contains a unique additive identity, 0, and a unique multiplicative identity, 1. Every rational number a/b has a unique additive inverse,

$$-\frac{a}{b} = \frac{-a}{b} = \frac{a}{-b},$$ while every nonzero element a/b has a unique multiplicative

inverse, b/a. As in the case of integers, we observed that $x - y = x + (-y)$ for any two rational numbers x and y. Multiplication is distributive over both addition and subtraction.

The following properties of addition and multiplication were discussed in the preceding sections:

1. If $w = x$ and $y = z$, then $w + y = x + z$ (*addition property*).

2. If $x + y = x + z$, then $y = z$ (*cancellation property of addition*).

3. If $w = x$ and $y = z$, then $w \cdot y = x \cdot z$ (*multiplication property*).

4. If $xy = xz$ and $x \neq 0$, then $y = z$ (*cancellation property of multiplication*).

5. If $xy = 0$, then either $x = 0$ or $y = 0$ (*product law*).

6. $x \cdot 0 = 0$ (*zero property of multiplication*).

The order relation on the set of rational numbers has the following basic properties:

1. If x and y are any two rational numbers, then either $x < y$ or $x = y$ or $x > y$ (*law of trichotomy*).

2. If $x < y$ and $y < z$, then $x < z$ (*transitive property*).

3. If $x < y$, then $x + z < y + z$ for every rational number z.

4. If $x < y$, then $\begin{cases} xz < yz & \text{if } z \text{ is a positive rational number,} \\ xz > yz & \text{if } z \text{ is a negative rational number.} \end{cases}$

One of the fascinating properties possessed by the system of rational numbers is the property of denseness. This simply means that there are infinitely many rational numbers between any two distinct rational numbers. Recall that neither the whole number system nor the integer system satisfies this remarkable property. The concept of averaging was used to find a rational number between any two distinct rational numbers. The fact that if $\dfrac{a}{b} < \dfrac{c}{d}$, then $\dfrac{a}{b} < \dfrac{a+c}{b+d} < \dfrac{c}{d}$ can also be used to find a rational number between a/b and c/d.

The set of all points on a straight line corresponding to rational numbers constitutes the rational line. We discussed a serious deficiency of the set of rational numbers: not every equation of the form $x^2 = a$, where a is any positive rational number, is solvable in the set of rational numbers. This motivated us to introduce irrational numbers, which will correspond to the infinitely many "gaps" or "holes" on the rational line. The Pythagoreans have been given the credit for discovering these new numbers.

We now have the line diagram as given in Fig. 6.15 for our number systems. Since not all equations of the form $x^2 = a$ are solvable in Q, we conclude that the "tree" in Fig. 6.15 is not yet "large enough" for all our purposes.

Fig. 6.15

SUGGESTED READING

Bray, C. J., "To Invert or not to Invert," *The Arithmetic Teacher*, vol. 10 (May 1963), pp. 274–276.

Chabe, A. M., "Rationalizing 'inverting and multiplying'," *The Arithmetic Teacher*, vol. 10 (May 1963), pp. 272–273.

Jordon, A. E., "One," *The Arithmetic Teacher*, vol. 14 (Oct. 1967), pp. 498–499.

Kolesnik, T. S., "Illustrating the Multiplication and Division of Common Fractions," *The Arithmetic Teacher*, vol. 10 (May 1963), pp. 268–271.

Mielke, P. T., "Rational Points on the Number Line," *The Mathematics Teacher*, vol. 63 (Oct. 1970), pp. 475–479.

Ohmer, M. M., C. V. Aucoin, and M. J. Cortez, *Elementary Contemporary Mathematics* (2nd ed.), Xerox Publishing Co., Lexington, Mass. (1972), pp. 202–251.

Rappolee, W. E., "Illustrating the Division of Fractions," *The Arithmetic Teacher*, vol. 10 (May 1963), p. 292.

Wheeler, R. E., *Modern Mathematics: An Elementary Approach* (3rd ed.), Brooks/Cole Publishing Co., Monterey, Calif. (1973), pp. 241–293.

"The miraculous powers of modern calculation are due to three inventions: the Arabic Notation, Decimal Fractions and Logarithms."

F. CAJORI

7/The System of Real Numbers

After studying this chapter, you should be able to:
- *convert fractions into decimal fractions, whenever possible*
- *perform the arithmetic operations on the set of decimal fractions*
- *do simple problems involving percent*
- *be familiar with the metric units of length, weight, and volume*
- *find the decimal representation of rational numbers*
- *identify rational and irrational numbers*
- *identify the properties of operations and the order relation*
- *find the square root of a positive real number*
- *find the absolute value of a real number*
- *evaluate the distance between any two points on the real line*

7.0 INTRODUCTION

Recall that in Section 6.7, we discussed a shortcoming of the set of rational numbers: not every equation of the form $x^2 = a$ is solvable in the set of rational numbers. This equation has a solution in the set of rational numbers if and only if a is a perfect square. For example, $x^2 = 4$ is solvable with rational solutions $x = \pm 2$; so is $x^2 = \frac{4}{9}$ with solutions $x = \pm \frac{2}{3}$; but $x^2 = 2$, $x^2 = \frac{2}{3}$, etc., are not solvable in Q. This motivates us to extend the set of rational numbers to a still larger set by introducing new numbers called *irrational numbers*. The resulting set is the set of *real numbers*.

How often do we come across irrational numbers in our daily life? If there is a square of area 15 square units, what is the length of a side of the square? Observe that the length of a side is given by the positive solution of the equation

$x^2 = 15$, namely, $\sqrt{15}$. If 2 and 3 are the lengths of the legs of a right triangle, then by the Pythagorean theorem (see Chapter 13), the length of its hypotenuse is given by $\sqrt{2^2 + 3^2} = \sqrt{4+9} = \sqrt{13}$. We shall introduce irrational numbers more formally in Section 7.5.

7.1   DECIMAL FRACTIONS

Fractions were introduced in the previous chapter, where we observed that in order to perform the arithmetical operations of addition and subtraction of two fractions, we first have to rewrite them with the same denominator. Fractions become more handy if their denominators are integral powers of 10. The first work on the theory of decimal fractions was published by Simon Stevin (1548–1620), who realized the need for developing new techniques for faster and more accurate computations. He was an influential mathematician of the Low Countries in the sixteenth century and a quartermaster general of the Dutch army.

Definition 7.1   A fraction whose denominator is an integral power of 10 is called a *decimal fraction*.

For example, $\dfrac{365}{100} = \dfrac{365}{10^2}$, $\dfrac{12345}{1000} = \dfrac{12345}{10^3}$ are decimal fractions.

Since every fraction is of the form a/b $(b \neq 0)$ with a,b in I, every decimal fraction can be expressed in the form $\dfrac{x}{10^k}$ with x in I and k in N.

Notice that $\frac{1}{2}$ is not a decimal fraction. However, it can be expressed as $\frac{5}{10}$, which is a decimal fraction. The student is warned that not every fraction can be converted to a decimal fraction. For example, $\frac{1}{3}, \frac{2}{3}, \frac{2}{7}$, and $\frac{4}{15}$ cannot be expressed as decimal fractions. Is there a device that can be used to decide if a given fraction can be converted to a decimal fraction? The following theorem gives a partial answer to this question.

Theorem 7.1   *If the denominator b of a fraction a/b (in simplest form) has no prime factors other than 2 and 5, then a/b can be expressed as a decimal fraction. Conversely, if the fraction a/b (in simplest form) can be expressed as a decimal fraction, then the prime factors of the denominator b must belong to the set $\{2,5\}$.*

Example 7.1   Convert the fraction $\frac{21}{250}$ into a decimal fraction.
Solution: We have

$$\frac{21}{250} = \frac{21}{2 \cdot 5^3} = \frac{2^2 \cdot 21}{2^2 \cdot 2 \cdot 5^3}$$

$$= \frac{4 \cdot 21}{2^3 \cdot 5^3} = \frac{84}{(2 \cdot 5)^3}$$

Thus,
$$\frac{21}{250} = \frac{84}{10^3}$$

Example 7.2 Convert the fraction $\frac{33}{60}$ into a decimal fraction, if possible.
Solution: Since $\frac{33}{60} = \frac{11}{20}$ and $20 = 2^2 \cdot 5$, by Theorem 7.1, $\frac{33}{60}$ can be converted to a decimal fraction:

$$\frac{33}{60} = \frac{11}{2^2 \cdot 5} = \frac{11 \cdot 5}{2^2 \cdot 5^2}$$

$$= \frac{55}{(2 \cdot 5)^2} = \frac{55}{10^2}$$

A convenient alternate way of writing a decimal fraction is by inserting a <u>dot</u>, called *decimal point,* between two digits of the numerator so that the number of digits on the right of the decimal point is the same as the exponent of the ten in the denominator. If the power of the ten in the denominator of a decimal fraction is more than the number of digits in the numerator, then we write a sufficient number of zeros between the decimal point and the digits in the numerator so that the number of digits to the right of the decimal point corresponds to the exponent of the ten.

For example, the decimal fraction $\frac{12345}{1000} = \frac{12345}{10^3}$ can be written, using a decimal point, as 12.345; also, $\frac{3443}{10} = 344.3$, $\frac{25}{1000} = \frac{25}{10^3} = 0.025$, $\frac{32}{10000} = \frac{32}{10^4} = 0.0032$, etc.

Notice that 2.57 is the same as 2.570 or 2.5700, etc., since

$$2.57 = \frac{257}{100} = \frac{2570}{1000} = \frac{25700}{10000}, \text{etc.}$$

Recall that only fractions whose denominator contains only 2 or 5 or both as prime factors are convertible to decimal fractions. A representation of the form $a_1 a_2 \ldots a_m . b_1 b_2 \ldots b_n$ is called a *decimal representation.* We now illustrate an algorithm to find decimal representations of fractions that can be converted to decimal fractions.

Example 7.3 Find the decimal representation of $\frac{3}{4}$.
Solution:

$$
\begin{array}{r}
0.75 \\
4\overline{\smash{\big)}\,3.00} \\
2\,8 \\
\hline
20 \\
20 \\
\hline
0
\end{array}
$$

Since $3 = 3.00$, we divide the number 300 by 4 in the ordinary way and then place a decimal point in the quotient just above the decimal point of the dividend. Thus $\frac{3}{4} = 0.75$. This procedure is justified since,

$$\frac{3}{4} = \frac{3}{4} \cdot \frac{100}{100}$$

$$= \frac{300}{4} \cdot \frac{1}{100}$$

$$= 75 \cdot \frac{1}{100}$$

$$= \frac{75}{100}$$

$$= 0.75$$

Example 7.4   Find the decimal representation of $\frac{237}{40}$.

Solution:

$$
\begin{array}{r}
5.925 \\
40\overline{\smash{)}237.000} \\
200 \\
\hline
370 \\
360 \\
\hline
100 \\
80 \\
\hline
200 \\
200 \\
\hline
0
\end{array}
$$

Thus, $\frac{237}{40} = 5.925$.

The rules of the arithmetical operations of addition, subtraction, multiplication, and division of decimal fractions, taught in elementary schools, are illustrated in the following examples.

Example 7.5   Find the sum of 953.6, 3.58, and 71.438.

Solution: First, we write the numbers in such a way that the decimal points in all numbers appear under one another:

$$
\begin{array}{r}
953.600 \\
3.580 \\
71.438
\end{array}
$$

Now we add the numbers as usual and then insert a decimal point in the sum, just below the decimal points:

$$
\begin{array}{r}
953.600 \\
3.580 \\
\underline{71.438} \\
1028.618
\end{array}
$$

We justify this procedure as follows.

$$953.6 + 3.58 + 71.438 = \frac{9536}{10} + \frac{358}{100} + \frac{71438}{1000}$$

$$= \frac{953600}{1000} + \frac{3580}{1000} + \frac{71438}{1000}$$

$$= \frac{1028618}{1000}$$

$$= 1028.618$$

Example 7.6   Subtract 28.57 from 63.2.
Solution:

$$
\begin{array}{r}
63.20 \\
28.57 \\
\hline
34.63
\end{array}
$$

Subtract 28.57 from 63.20 disregarding the decimal points, and then put a decimal point in the resulting number so that all the numbers have the same number of digits to the right of the decimal point, as in addition. Thus, $63.2 - 28.57 = 34.63$. This procedure is justified, since

$$63.20 - 28.57 = \frac{6320}{100} - \frac{2857}{100}$$

$$= \frac{6320 - 2857}{100}$$

$$= \frac{3463}{100}$$

$$= 34.63$$

Example 7.7   Find the product of 128.3 and 35.87.
Solution:

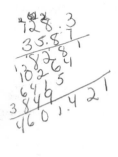

$$
\begin{array}{r}
128.3 \\
35.87 \\
\hline
8\ 981 \\
102\ 64 \\
641\ 5 \\
3849 \\
\hline
4602.121
\end{array}
$$

We multiply the numbers as usual, disregarding the decimal points. Insert a decimal point in the resulting number so that the number of digits to the right of the decimal point equals the sum of the number of digits to the right of the decimal points of the two given numbers. This method makes sense, since

$$(128.3)(35.87) = \frac{1283}{10} \cdot \frac{3587}{100} = \frac{4602121}{1000} = 4602.121$$

Example 7.8   Divide 196.52 by 8.5.

Solution: Since the denominator, 8.5, of $\dfrac{196.52}{8.5}$ is not a natural number, let's

multiply the numerator and the denominator by 10 to get a natural number in the denominator, namely, $(8.5)10 = 85$.

$$
\begin{array}{r}
23.12 \\
85\overline{\smash{)}\,1965.20} \\
170 \\
\hline
265 \\
255 \\
\hline
102 \\
85 \\
\hline
170 \\
170 \\
\hline
0
\end{array}
$$

Thus, $\dfrac{196.52}{8.5} = \dfrac{1965.2}{85} = 23.12.$

Exercise 7.1

1. Mark *true* or *false*.
   a) The additive inverse of a decimal fraction is a decimal fraction.
   b) The product of two decimal fractions is a decimal fraction.
   c) The multiplicative inverse of a decimal fraction is a decimal fraction.
2. Give a counterexample to each of the false statements in problem 1.
3. Convert the following fractions to decimal fractions.
   a) $\frac{3}{4}$ b) $\frac{4}{5}$ c) $\frac{23}{20}$ d) $\frac{31}{50}$ e) $\frac{123}{100}$
   f) $\frac{35}{200}$ g) $\frac{27}{60}$ h) $\frac{126}{1400}$ i) $3\frac{4}{5}$ j) $15\frac{1}{2}$
4. Convert the following fractions to decimals.
   a) $\frac{1}{4}$ b) $\frac{7}{20}$ c) $\frac{13}{40}$ d) $\frac{83}{200}$ e) $\frac{12}{75}$
   f) $\frac{15}{24}$ g) $\frac{24}{300}$ h) $\frac{21}{700}$ i) $\frac{5}{16}$ j) $\frac{17}{25}$
5. Convert the following decimals to fractions.
   a) 0.5   b) 1.25   c) 2.75   d) 0.12   e) 0.012
   f) 0.123   g) 1.11   h) 4.203   i) 2.003   j) 5.0505
6. Perform the indicated operations.
   a) $37.583 + 73.708$ b) $23.18 + 82.7 + 5.038$
   c) $73.51 - 37.15$ d) $23.105 - 36.034$
   e) $(2.387)(57.201)$ f) $(51.15)(262.02)$
   g) $84.7632 \div 3.2$ h) $5.6457 \div 0.41$

7.2   PERCENT

So far in our discussion of the theory of fractions, we have discussed two ways of representing fractions: as the ratio of two integers and then as decimals. Closely allied to decimal fractions is the concept of *percent*. We often use the word percent (%) in our daily life. Statements like, "The interest rate at the savings bank is 5%," "The prime lending rate is $7\frac{1}{4}\%$," "John is in the 20% tax bracket," "There is a 70% chance for rain today," "10% of the students failed in mathematics 101," "The cost of food went up two-tenths of 1% this month," "There is a 20% discount on all winter coats" are not new to us.

If John has \$400 in a savings bank that pays 5% interest annually, then at the end of the year, he receives $\frac{5}{100} \cdot 400 = 20$ dollars as interest. The statement that the bank pays 5% interest annually simply means that for every 100 dollars, it pays 5 dollars at the end of the year. If John has x dollars in the bank, then at this rate, he receives $\frac{5}{100} \cdot x$ dollars as interest at the end of the year.

If we have a fraction $x/100$ whose denominator is 100, then we say x *percent* ($x\%$). That is, $x\% = x/100$. Percent simply means per hundred.

For example, $25\% = \frac{25}{100}$, $41\% = \frac{41}{100}$, $69\% = \frac{69}{100}$. Notice that $25\% = 0.25$, $41\% = 0.41$, and $69\% = 0.69$.

Example 7.9   Convert 39%, 8.7%, and 9% to decimals.

Solution:
$$39\% = \frac{39}{100} = 0.39$$

$$8.7\% = \frac{8.7}{100} = \frac{87}{1000} = 0.087$$

$$9\% = \frac{9}{100} = 0.09$$

Notice that in order to convert percents to decimals, we drop the percent sign and place the decimal point two places to the left.

Example 7.10   Convert 0.23, 0.3, and 0.057 to percents.

Solution:
$$0.23 = \frac{23}{100} = 23\%$$

$$0.3 = \frac{3}{10} = \frac{30}{100} = 30\%$$

$$0.057 = \frac{57}{1000} = \frac{5.7}{100} = 5.7\%$$

To convert decimals to percents, we need only move the decimal point two places to the right and then append the percent sign.

Example 7.11   Convert $\frac{3}{5}$, $\frac{2}{8}$, and $\frac{31}{25}$ to percents.

Solution:
$$\frac{3}{5} = \frac{6}{10} = \frac{60}{100} = 60\%$$

$$\frac{2}{8} = \frac{1}{4} = \frac{25}{100} = 25\%$$

$$\frac{31}{25} = \frac{124}{100} = 124\%$$

Observe that to convert fractions to percents, first we have to convert them to fractions with denominator 100.

Example 7.12   Find 24% of 625.

Solution: Since $24\% = \frac{24}{100}$, 24% of 625 is $\frac{24}{100} \cdot 625 = 150$.

Example 7.13   What percent of 350 is 56?

Solution: Let $x\%$ of 350 be 56. Then,

$$\frac{x}{100} \cdot 350 = 56$$

$$x = \frac{56 \cdot 100}{350}$$

$$x = 16$$

Thus, 16% of 350 is 56.

Example 7.14   Ron spends 4% of his monthly earnings on gasoline for his car. If his total bill for gasoline was 36 dollars, what is his monthly income?

Solution: Let Ron's monthly income be x dollars. Then 4% of his monthly earnings $= \frac{4}{100} \cdot x = 36$, so

$$x = \frac{36 \cdot 100}{4} = 900$$

Thus, Ron's monthly income is $900.

Stores advertise that certain items are sold this week at a certain *discount,* say 15%, that is, 15% off. That there is a 15% discount on a certain item simply means that the customer need only pay $100\% - 15\% = 85\%$ of the retail price. For example, if the original price was $200, then he need only pay $200 \cdot \frac{85}{100} = 170$ dollars, a saving of $200 - 170 = 30$ dollars.

Example 7.15   There is a 20% discount on all lawnmowers at a store. If a lawnmower's original price was $145, what is its price now?

Solution: Since there is a 20% discount, the customer need only pay 80% of $145, which is $145 \cdot \frac{80}{100} = 116$ dollars.

Exercise 7.2

1. Convert the following percents to decimals.

   a) 6%   **b)** 8.35%   **c)** 12.3%

   d) 90%   **e)** 120%   **f)** 0.12%

2. Convert the following decimals to percents.

   a) 2.7   **b)** 1.45   **c)** 3.04

   d) 0.85   **e)** 0.03   **f)** 0.045

3. Convert the following fractions to percents.

   a) $\frac{7}{20}$ **b)** $\frac{13}{40}$ **c)** $\frac{11}{25}$

   d) $\frac{3}{4}$ **e)** $\frac{4}{15}$ **f)** $\frac{17}{45}$

4. a) What is 20% of 80?

   b) What is $12\frac{1}{2}$% of 72?

   c) Find 3.12% of 240.

   d) Find 0.7% of 360.

   e) What percent of 360 is 54?

   f) What percent of 640 is 147.2?

   g) Find the number of which 15% is 78.

   h) Find the number of which 6% is 44.4.

   i) Find the number of which 0.5% is 18.75.

5. If Frank deposited $125 in a bank at the rate of 6% per year, how much money does he receive as interest after one year? What is his balance then? What is his balance after 2 years?

6. Mr. Murphy deposited $450 in a bank that pays 5% interest annually. If he withdraws 20% of his balance at the bank at the end of the year, what will be his balance at the end of the second year?

7. Of the 425 freshmen at Adams University, 8% got grade A and 12% failed in Math. 100. How many freshmen got grade A in Math. 100? How many failed in Math. 100?

8. If 2250 students applied for admission to a certain medical school, but only 270 of them were given admission, what percent of the applicants was granted admission?

9. John pays $172 a month for rent. What is his monthly salary if his rent is 20% of it?

10. Mrs. Smith pays $15 a month for a place to park her car. What percent of her salary is spent for the parking space if she earns $750 a month?

11. Dick wants to buy a car, the price of which has been marked down 10%. His father has agreed to pay 45% of the cost and his mother 25%. If he now has to pay only $688.50 himself, what was the original price of the car?

12. If a refrigerator that costs 530 dollars is sold at a discount of 12%, how much does the customer have to pay?

13. A winter coat now costs 25% off the retail price. If its present price is $150, what was its original price?

14. If a typewriter was advertised to be sold at a discount of 10%, then was reduced another 15%, and later discounted still another 20%, what is now the actual discount from the original price? If the typewriter initially cost $250, what is its present price?

7.3 THE METRIC SYSTEM

We have basically two types of measuring units: units to measure distance (length) and weight (mass). In the *British system (imperial system)*, which we are currently using in the United States, distance is measured in terms of inches, feet, yards, and miles, while weight is measured in terms of ounces, pounds, hundredweights, and tons. Most of us have some kind of difficulty in converting from one unit to another in the imperial system. This problem would have been considerably minimized if the measuring units in the British system were powers of 10. The imperial units of length and weight, used in the United States, are given in Table 7.1, for convenience. It is to be remembered that in Great Britain, 112 pounds equals one hundredweight and 2240 pounds equals one (long) ton, whereas in the United States, one hundredweight contains 100 pounds and a (short) ton consists of 2000 pounds.

Table 7.1   British System of Units of Length and Weight

Units of Length	Units of Weight
12 inches = 1 foot	16 drams = 1 ounce
3 feet = 1 yard	16 ounces = 1 pound
1760 yards = 1 mile	100 pounds = 1 hundredweight
	20 hundredweights = 1 ton

How many feet are there in a mile? The answer is $1760 \times 3 = 5280$ feet (why?). A square mile contains $5280 \times 5280 = 27,878,400$ square feet, which is divided into 640 equal parts, each being called an *acre*. Consequently, an acre consists of $\frac{27878400}{640} = 43560$ square feet.

The *rod* is a unit of linear measure, established in Germany in the sixteenth century, that is not in common usage. What is a rod? It is interesting to note that a sample of 16 men were selected at random from people coming out of the church and they stood toe-to-heel along a line. The total length formed this way was accepted as a rod. A rod is approximately 16.5 feet. There was a time when 36 barleycorns were placed one after the other to measure one foot. The yard was defined by Henry I as the distance from the tip of his nose to the tip of his middle finger, when the arm was outstretched.

There was a time when every town and every province in France had its own system of measurements. As time passed, the need for standardized units for all measurements became evident to avoid confusion and problems in buying and selling things in neighboring communities.

Since multiplication and division of numbers by 10 and its powers are fairly easy to handle, the logical choice is a system of weights and measures based on the number ten. Simon Stevin, who developed the theory of decimal fractions, was the first person to suggest a system based on ten. In 1670, Gabriel Mouton, a French priest, proposed such a system in the French Academy of Sciences. Neither Stevin nor Mouton had any success in promoting such a system.

A committee was appointed in 1790 by the French National Assembly to study the possibilities of a logical and convenient system of measurement. The committee proposed a new system. This system introduced a unit of length called *metre* and a new unit of weight called *gram*. A metre is about 39.36996 inches, while a gram is approximately 0.03527 ounce. Latin prefixes—deci, centi, milli, deka, hecto, and kilo—are used to denote decimal portions and multiples of these units (metre and gram). For example, one-tenth ($\frac{1}{10}$), one-hundredth ($\frac{1}{100}$), and one-thousandth ($\frac{1}{1000}$) of a metre are called decimetre, centimetre, and millimetre, respectively, while 10 metres, 100 metres, and 1000 metres constitute a dekametre, hectometre, and a kilometre, respectively. A kilometre (= 1000 metres) is approximately five-eighths ($\frac{5}{8}$) of a mile, while a kilogram (= 1000 grams) is about 2.2 pounds.

In Table 7.2, we give the units of length and weight in the metric system.

Table 7.2   Metric System

Units of Length	Units of Weight
10 millimetres = 1 centimetre	10 milligrams = 1 centigram
10 centimetres = 1 decimetre	10 centigrams = 1 decigram
10 decimetres = 1 metre	10 decigrams = 1 gram
10 metres = 1 dekametre	10 grams = 1 dekagram
10 dekametres = 1 hectometre	10 dekagrams = 1 hectogram
10 hectometres = 1 kilometre	10 hectograms = 1 kilogram

The basic unit of volume is the *litre*, which is one thousand cubic centimetres or one-thousandth of a cubic metre. As before, Latin prefixes mentioned above are used to designate decimal portions and multiples of a litre.

Metric equivalents of the most widely used imperial units and the equivalents of the important metric units in the imperial system are given in Table 7.3, correct to three decimal places.

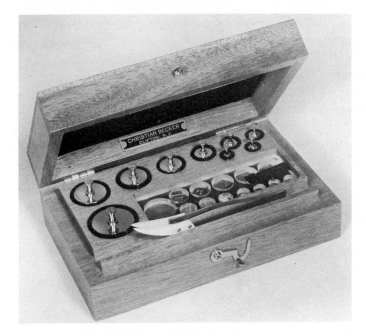

Fig. 7.1

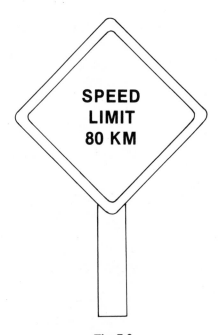

Fig. 7.2

Table 7.3   Conversion Table

Imperial to Metric	Metric to Imperial
1 inch = 2.540 centimetres 1 foot = 0.305 metre 1 mile = 1.609 kilometres	1 millimetre = 0.039 inch 1 metre = 3.281 feet 1 kilometre = 0.621 mile
1 pound = 0.454 kilogram 1 U.S. ton = 907.276 kilograms	1 gram = 0.035 ounce 1 kilogram = 2.205 pounds
1 fluid ounce = 0.030 litre 1 U.S. quart = 0.946 litre 1 U.S. gallon = 3.785 litres	1 millilitre = 0.338 fluid ounce 1 litre = 0.264 U.S. gallon 1 kilolitre = 264.201 U.S. gallons
1 cubic inch = 16.387 cubic centimetres 1 cubic foot = 0.028 cubic metre	1 cubic centimetre = 0.061 cubic inch 1 cubic metre = 35.315 cubic feet

The metric system went into effect in France in 1799. However, it did not become popular until 1840. In the United States, both Thomas Jefferson (1743–1826) and John Quincy Adams (1767–1848) advocated its adoption, but in vain. It has been legal in the United States since 1866, and the imperial system will be replaced by the more natural metric system within the next few years. Have you noticed that weights of pharmaceutical products are already shown on their labels in the metric system? Have you paid any attention to the fact that film sizes are usually given in millimetres? At present, more than 150 countries around the world have changed to the metric system. Scientists, in particular physicists and chemists, all over the world have been using the metric system for a long time.

The metric system has two significant advantages: (1) it is a decimal system, a system based on the number ten, and (2) every unit in the metric system is related to the fundamental linear measure, namely the metre. Consequently, the unit of metre acts like a unifying thread in the metric system and every type of unit can be derived from another type.

Example 7.16   How many square feet are there in 5 square yards?

Solution: Since 1 yard = 3 feet, there are $3 \times 3 = 9$ square feet in a square yard. Therefore,

$$5 \text{ square yards} = 5 \times 9 \text{ square feet}$$

$$= 45 \text{ square feet}$$

Example 7.17   The distance between Boston and New York City is about 216 miles. Express this distance in kilometres.

Solution: distance = 216 miles

$$= 216 \times 1.609 \text{ kilometres} \quad (1 \text{ mile} = 1.609 \text{ kilometres})$$

$$= 347.544 \text{ kilometres}$$

Example 7.18   Find the area of a rectangle 20 centimetres (cm) long and 15 centimetres wide.

Solution: area = (20 cm)(15 cm) = 20 × 15 square centimetres

$$= 300 \text{ square centimetres or } 300 \text{ cm}^2$$

Example 7.19   Find the volume of a cube of edge 5 cm.

Solution: volume = (5 cm)(5 cm)(5 cm)

$$= 5 \times 5 \times 5 \text{ cubic centimetres}$$

$$= 125 \text{ cubic centimetres or } 125 \text{ cm}^3$$

Example 7.20   Find the area in square metres of a rectangular plot of land that is 300 feet long and 200 feet wide.

Solution: 1 foot = 0.305 metre   (see Table 7.3)

$$300 \text{ feet} = 300 \times 0.305 = 91.5 \text{ metres}$$

$$200 \text{ feet} = 200 \times 0.305 = 61 \text{ metres}$$

Therefore, area = 91.5 × 61 = 5581.5 square metres or 5581.5 m².

Example 7.21   Express 120 pounds in kilograms.

Solution: 1 pound = 0.454 kilogram   (see Table 7.3)

Therefore,   120 pounds = 120 × 0.454 kilograms

$$= 54.48 \text{ kilograms or } 54.48 \text{ Kg}$$

Example 7.22   How many grams of salt are there in 2 pounds of salt?

Solution: 1 pound = 0.454 kilogram

$$= 454 \text{ grams}$$

Therefore,   2 pounds = 2 × 454 = 908 grams or 908 gm

Exercise 7.3

1. Express the following in metres.
   a) 1 yard   b) 1 mile   c) 3 miles and 125 yards
2. Express the following in grams.
   a) 1 pound   b) 1 ton   c) 2 tons and 13 pounds
3. Express the following in litres.
   a) 1 pint   b) 3 quarts   c) $\frac{1}{2}$ gallon and 2 quarts
4. Convert 1500 grams to kilograms.
5. Convert 2.34 kilograms to grams.
6. How many feet are there in 100 centimetres?

7. How many centimetres are there in 100 feet?
8. Find the sum of the following quantities:
   a) 25.203 metres, 123.02 centimetres, and 13.37 kilometres
   b) 43 centigrams, 74.3 grams, and 23.532 kilograms
   c) 38.54 decilitres, 54.27 litres, 68.234 dekalitres
9. The distance between Chicago and Montreal is about 850 miles. Express this distance in kilometres.
10. The radius of the earth is about 4000 miles. Express this distance in kilometres.
11. The moon is approximately 2.4×10^5 miles away from the earth. Rewrite this distance in kilometres.
12. The distance of the sun from the earth is about 9.3×10^7 miles. Rewrite this distance in kilometres.
13. If a *light year* (a unit of distance used by astronomers) is approximately 5.88×10^{12} miles, how much is it in kilometres?
14. The net weight of a can of soup is 10.75 ounces. Express this weight in grams. [*Hint:* 1 ounce = 28.350 grams.]
15. Find the volume of water in a vessel that is 40 cm long, 30 cm wide, and 20 cm high. (Neglect the thickness of the vessel.) [*Hint:* 1000 cubic centimetres = 1 litre.]
16. How many litres of gasoline are needed to fill a gas tank with the capacity of 16 gallons? 20 gallons?
17. Which is greater in the following pairs?
    a) 100 metres and 100 yards   b) 100 kilometres and 100 miles
    c) 100 litres and 100 gallons   d) 100 kilolitres and 100 gallons
18. The speed of sound is about 1100 feet per second. Express this in metres per second.
19. A boat can travel in still water at 8 miles per hour.
    a) What is its speed in still water in kilometres per hour?
    b) What will be its speed in kilometres per hour when traveling up the stream with river current 2 miles an hour?
    c) What will be its speed in kilometres per hour when traveling down the stream?
20. The speed of light is approximately 186,400 miles per second. Express this speed in:
    a) kilometres per second   b) kilometres per hour
21. Express 12,345 square millimetres (mm^2) in square metres (m^2).
22. The area of Texas is about 2.67×10^5 square miles. Express this in square kilometres.
23. How many 30 cm × 30 cm tiles are needed to cover the floor of a 3.6 metre × 3.6 metre room?
24. How many cubic centimetres (cm^3) are there in one cubic metre (m^3)?

7.4 DECIMAL REPRESENTATION OF RATIONAL NUMBERS

In Section 7.1, we discussed the decimal representation of a special class of rational numbers, namely, decimal fractions. We observed that only those rational numbers whose denominators contain as factors powers of 2 or 5 or both can be converted to decimal fractions. These decimals contain only a finite number of digits to the right of the decimal point. Consequently, the decimal expansion of every rational number that can be converted to decimal fractions contains only a finite number of digits after the decimal point. Such decimals are called *terminating decimals*. For example, $\frac{1}{2}=0.5$, $\frac{3}{4}=0.75$, $\frac{7}{8}=0.875$, $\frac{234}{500}=0.468$ are all terminating decimals.

Let's now try to find the decimal expansion of $\frac{1}{3}$ and $\frac{2}{7}$ by the familiar long division method. We find that $\frac{1}{3}=0.3333\ldots$, and $\frac{2}{7}=0.285714285714\ldots$.

$$
\begin{array}{r}
0.3333\ldots \\
3\overline{\smash{\big)}\,1.0000} \\
9 \\
\hline
10 \\
9 \\
\hline
10 \\
9 \\
\hline
10 \\
9 \\
\hline
1 \\
\cdot \\
\cdot
\end{array}
\qquad
\begin{array}{r}
0.285714285714\ldots \\
7\overline{\smash{\big)}\,2.000000} \\
1\,4 \\
\hline
60 \\
56 \\
\hline
40 \\
35 \\
\hline
50 \\
49 \\
\hline
10 \\
7 \\
\hline
30 \\
28 \\
\hline
2 \\
\cdot \\
\cdot \\
\cdot
\end{array}
$$

These are examples of *nonterminating decimals*. An interesting and surprising property of these decimals is that a block containing a finite number of digits is repeated in each decimal representation. For instance, the repeating block in the decimal representation of $\frac{1}{3}$ contains only the digit 3, whereas it contains the digits 2, 8, 5, 7, 1, 4 in the decimal expansion of $\frac{2}{7}$. Such decimals are called *periodic or repeating decimals*. The *period* of a repeating decimal is the number of digits in the smallest repeating block in the expansion. The period of the decimal expansion of $\frac{1}{3}$ is 1 and that of $\frac{2}{7}$ is 6. Observe that the terminating decimals 0.5, 0.75 and 0.875 can also be written as repeating decimals, as $0.5000\ldots$, $0.75000\ldots$, and $0.875000\ldots$ respectively.

For convenience, let's draw a bar just above the smallest repeating block in the case of repeating decimals. For example, $\frac{1}{3} = 0.\overline{3}$, $\frac{2}{7} = 0.\overline{285714}$, $\frac{41}{333} = 0.\overline{123}$, $\frac{1}{2} = 0.5 = 0.5\overline{0}$, etc.

Thus, we observe that the decimal expansions of some rational numbers are terminating and those of some rational numbers are repeating. Now, is there any rational number that does not fall into one of these two categories? It turns out that if a rational number has a nonterminating decimal representation, then it should be repeating!

Obviously, every terminating decimal represents a rational number. Does every repeating decimal also represent a rational number? Before we answer this question, let's look at the following example.

Example 7.23   Does $0.\overline{23}$ represent a rational number?

Solution: Let $x = 0.\overline{23}$. Then

$$100x = 23.\overline{23}$$

$$100x - x = 23.\overline{23} - 0.\overline{23} = 23.00$$

$$99x = 23$$

$$x = \frac{23}{99}$$

Thus, $0.\overline{23}$ represents the rational number $\frac{23}{99}$.

This example corroborates our feeling that every repeating decimal represents a rational number. In fact, this is true!

Theorem 7.2   *Every repeating decimal represents a rational number.*

Example 7.24   Find the rational number represented by $12.\overline{25}$.

Solution: Let $x = 12.\overline{25} = 12.2525 \ldots$. Multiply both sides by a suitable power of 10 ($10^2 = 100$ in this case) so that the first repeating block moves to the left of the decimal point. We then have

$$100x = 1225.\overline{25}$$

$$x = 12.\overline{25}$$

Subtracting,

$$99x = 1213$$

$$x = \frac{1213}{99}$$

Thus,

$$12.\overline{25} = \frac{1213}{99}$$

We now ask the reader to verify this by dividing 1213 by 99, applying the long division method.

Example 7.25   Find the rational number represented by $2.3\overline{45}$.

Solution: Let $x = 2.3\overline{45}$. Let's now multiply both sides by $10^3 = 1000$ (why?).

$$1000x = 2345.\overline{45}$$
$$10x = 23.\overline{45}$$

Subtracting,

$$990x = 2322$$
$$x = \frac{2322}{990} = \frac{129}{55}$$

Thus,

$$2.3\overline{45} = \frac{129}{55}$$

Example 7.26   Find the rational number represented by $0.\overline{9}$.
Solution: Let $x = 0.\overline{9}$. Then

$$10x = 9.\overline{9}$$
$$x = 0.\overline{9}$$

Subtracting,

$$9x = 9$$
$$x = 1$$

Thus,

$$0.\overline{9} = 1$$

Exercise 7.4
1. Mark *true* or *false*.
   a) Every terminating decimal represents a rational number.
   b) Every repeating decimal represents a rational number.
   c) Every rational number has a terminating decimal expansion.
   d) The sum of two terminating decimals is a terminating decimal.
   e) The difference of two terminating decimals is a terminating decimal.
2. Give a counterexample to each of the false statements in problem 1.
3. Find the decimal representation of
   a) $\frac{4}{333}$ b) $\frac{5}{9}$ c) $\frac{3}{8}$ d) $\frac{2}{17}$
   e) $\frac{40}{13}$ f) $\frac{21}{23}$ g) $\frac{557}{495}$ h) $\frac{2}{31}$
4. Find the period of the decimal expansion of
   a) $\frac{5}{9}$ b) $\frac{21}{23}$ c) $\frac{40}{13}$ d) $\frac{1}{19}$ e) $\frac{1}{23}$
5. Identify as rational or nonrational numbers.
   a) 2.35 b) $0.999\ldots$ c) $2.134133413334\ldots$
   d) $2.12323\ldots$ e) $1.414213\ldots$ f) $0.1234234\ldots$
6. Find the rational number represented by
   a) 0.25 b) $8.75\overline{6}$ c) $1.2\overline{34}$ d) $0.\overline{34}$
   e) $1.\overline{56}$ f) $2.2\overline{34}$ g) $10.\overline{010}$ h) $1.00\overline{1}$
   i) $8.7\overline{654}$ j) $3.31\overline{256}$ k) $0.\overline{076923}$ l) $0.0\overline{9}$
7. Use the fact that $\frac{1}{7} = 0.\overline{142857}$ to find the decimal representations of $\frac{2}{7}, \frac{3}{7}, \frac{4}{7}$, $\frac{5}{7}$, and $\frac{6}{7}$.

8. What is the maximum possible period in a decimal representation of the rational number a/b? Can you find a rational number with the maximum period?

★**9.** Prove that $\frac{1}{3}$ cannot have a terminating decimal representation. [*Hint:* assume it does and get a contradiction.]

7.5   IRRATIONAL NUMBERS AND REAL NUMBERS

Thus far, we have been discussing decimal representation of rational numbers. We observed that every rational number has either a terminating or a repeating decimal representation. Also, every terminating or repeating decimal represents a rational number. We are now naturally tempted to ask: how about a decimal that is neither terminating nor repeating? For example, consider the decimal $12.343443444\ldots$, which is clearly nonterminating. It is nonrepeating, too, since each 3, except the first, is followed by one more 4 than the preceding 3. On the basis of our discussion, it cannot represent a rational number. Consequently, it represents a number that is not rational, called an *irrational number.* Conversely, every irrational number has a nonterminating and nonrepeating decimal expansion. For instance, $2.34518632\ldots$, $0.325325532555\ldots$, etc., represent irrational numbers. In Section 6.7, we observed that $\pm\sqrt{2}$, $\pm\sqrt{3}$, $\pm\sqrt{5}$, $\pm\pi$, etc., are irrational numbers. Hence the decimal expansion of these numbers will be both nonterminating and nonrepeating. We denote the set of irrational numbers by Q'.

Clearly there are infinitely many irrational numbers. Even though there are infinitely many rational numbers, it turns out that there are more irrational numbers.

We shall not attempt to start a thorough discussion of the system of irrational numbers. However, we point out a few properties that might come as a real surprise. The set of irrational numbers is closed neither under addition nor under multiplication. For example, we know that $\sqrt{2}$ and $-\sqrt{2}$ are irrational numbers; however, their sum, $\sqrt{2}+(-\sqrt{2})=0$, is a rational number. Similarly, $(\sqrt{2})(\sqrt{2})=2$, again a rational number.

What can we say about the sum and the product of a rational number and an irrational number? The sum $0+\sqrt{2}=\sqrt{2}$ is clearly an irrational number, whereas $0\cdot\sqrt{2}=0$ is a rational number. We will now prove indirectly that $3+\sqrt{2}$ is an irrational number. Assume that $3+\sqrt{2}$ is a rational number. Since -3 is a rational number and the set of rational numbers is closed under addition (Theorem 6.1), $-3+(3+\sqrt{2})=(-3+3)+\sqrt{2}=0+\sqrt{2}=\sqrt{2}$ must be a rational number. This is clearly a contradiction (why?). Consequently, our assumption that $3+\sqrt{2}$ is a rational number is wrong and hence it is an irrational number. It can similarly be shown that $3\sqrt{2}$ is also an irrational number.

We observed in Section 6.6 that the set of rational numbers is dense. That is, between any two distinct rational numbers, we can always find a third rational number. Is there a similar result in the case of irrational numbers? Indeed, that is the case! There is at least one irrational number between any two unequal irrational numbers. For example, 1.242332333 ... is an irrational number between the irrational numbers 1.232332333 ... and 1.242442444 .... Also, between any two distinct irrational numbers, we can always find a rational number. For instance, 2.14 is a rational number between the irrational numbers 2.131331333 ... and 2.141441444 .... As a result, it is always possible to find a rational number as close as we please to an irrational number. In other words, any irrational number can be approximated by a rational number to any desired degree of accuracy. For example, recall that π, which is an irrational number, is usually approximated by $\frac{22}{7}$ or 3.142. Finally, we can find an irrational number between two different rational numbers. For example, 1.34010110111 ... is an irrational number between the rational numbers 1.34 and 1.35. In short:

---

1. There exists a rational number between any two distinct rational numbers.
2. There exists a rational number between any two distinct irrational numbers.
3. There exists an irrational number between any two distinct rational numbers.
4. There exists an irrational number between any two distinct irrational numbers.

The union of the set Q of rational numbers and the set Q' of irrational numbers is called the set of *real numbers*, denoted by the letter R; thus, $R = Q \cup Q'$ (see Fig. 7.3). Thus, a real number is either a rational number or an irrational number, but not both; it has either a terminating or repeating decimal expansion, or a nonterminating and nonrepeating decimal expansion, but not both.

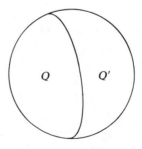

Set R of real numbers

Fig. 7.3

We do not intend to discuss formally the arithmetic operations or the order relation on the set of real numbers. However, we now list all the fundamental properties associated with them.

Addition

1. The set of real numbers is closed under addition.

2. Addition on the set of real numbers is commutative.

3. Addition on the set of real numbers is associative.

4. The set of real numbers contains a unique additive identity, 0.

5. Every real number a has a unique additive inverse, $-a$.

6. If $a = b$ and $c = d$, then $a + c = b + d$ (*addition property*).

7. If $a + c = b + c$, then $a = b$ (*cancellation property*).

Subtraction

1. The set of real numbers is closed under subtraction.

2. $a - b = a + (-b)$ for any real numbers a and b.

Multiplication

1. The set of real numbers is closed under multiplication.

2. Multiplication on the set of real numbers is commutative.

3. Multiplication on the set of real numbers is associative.

4. The set of real numbers contains a unique multiplicative identity, 1.

5. Every nonzero real number a has a unique multiplicative inverse, $a^{-1} = 1/a$.

6. If $a = b$ and $c = d$, then $a \cdot c = b \cdot d$ (*multiplicative property*).

7. If $ab = ac$ and $a \neq 0$, then $b = c$ (*cancellation property*).

8. If $ab = 0$, then either $a = 0$ or $b = 0$ (*product law*).

9. $a \cdot 0 = 0$ for every real number a (*zero property of multiplication*).

Division

1. The set of nonzero real numbers is closed under division.

2. $a \div b = a \cdot b^{-1}$, where b^{-1} denotes the multiplicative inverse of b $(\neq 0)$.

Order relation

1. If a and b are any two real numbers, then either $a < b$ or $a = b$ or $a > b$ (*law of trichotomy*).

2. If $a < b$ and $b < c$, then $a < c$ (*transitive property*).

3. If $a < b$, then $a + c < b + c$ for every real number c.

4. If $a < b$, then $\begin{cases} ac < bc & \text{if } c > 0, \\ ac > bc & \text{if } c < 0. \end{cases}$

5. The set of real numbers is dense; that is, there exists at least one real number between any two unequal real numbers.

Exercise 7.5

1. Identify the following numbers as rational or irrational.

a) $\sqrt{5}$ b) $\sqrt{4}$ c) $\sqrt{1}$

d) π e) $0.123123123\ldots$ f) $1.123000\ldots$

g) $123.122112211221\ldots$ h) $25.24318597\ldots$ i) $\sqrt{\dfrac{9}{4}}$

2. The value of π is $\frac{22}{7}$, true or false?

3. If a and b are real numbers such that $a < b$, how are the following related?

a) $3 + a$ and $3 + b$ **b)** $2a$ and $2b$ **c)** $-a$ and $-b$
d) $-5a$ and $-5b$ **e)** $4 - a$ and $4 - b$ **f)** $2 - 3a$ and $2 - 3b$

4. Solve the following equations.

a) $x^2 = 4$ **b)** $x^2 = 5$ **c)** $x^2 = 16$ **d)** $x^2 = 6$
e) $x^2 = 1$ **f)** $x^2 = 0$ **g)** $x^2 = 7$ **h)** $x^2 = 81$

5. Mark *true* or *false*.

a) The set of irrational numbers is closed under addition.
b) The set of irrational numbers is closed under multiplication.
c) 0 is a rational number.
d) Every irrational number has a nonterminating and repeating decimal representation.
e) Every irrational number has a nonterminating and nonrepeating decimal representation.
f) Every real number has a terminating or repeating decimal expansion.
g) π is an irrational number.
h) There exists an irrational number between any two distinct irrational numbers.
i) The sets of rational and irrational numbers are disjoint.
j) The set of irrational numbers is dense.
k) The set of real numbers is dense.

6. Give a counterexample to each of the false statements in problem 5, whenever possible.

7. Find a real number between each of the following pairs.

a) 2 and 3   **b)** 1.23 and 1.24   **c)** $2.\overline{45}$ and $3.\overline{54}$

8. Show that $-1 + \sqrt{3}$ is not a rational number. [*Hint*: Assume it is a rational number and get a contradiction.]

9. Show that $2\sqrt{3}$ is not a rational number. [*Hint*: Prove by contradiction.]

10. Show that $-1 + 2\sqrt{3}$ is not a rational number.

★11. Show that

a) $\dfrac{3}{\sqrt{3}} = \sqrt{3}$ **b)** $\dfrac{2}{\sqrt{6}} = \dfrac{\sqrt{6}}{3}$ **c)** $\dfrac{\sqrt{3}}{2\sqrt{3}} = \dfrac{1}{2}$

★12. Show that $\dfrac{1}{1 + \sqrt{2}} = \sqrt{2} - 1$. [*Hint*: multiply the numerator and the denominator of $\dfrac{1}{1 + \sqrt{2}}$ by $1 - \sqrt{2}$.]

★13. Find a rational number between the following pairs of irrational numbers.

a) 2.1232332333 ... and 2.1323223222 ...
b) 5.123451689 ... and 5.1233475 ...

★14. Find an irrational number between the irrational pairs in problem 13.
★15. Find an irrational number between the following pairs of rational numbers.

a) 1.23 and 1.24   b) $1.\overline{23}$ and $1.\overline{24}$

7.6 THE REAL LINE

In the previous chapter, we represented rational numbers by points on the rational line. There existed no points on the line to represent the solutions of the equation $x^2 = 2$. Consequently, the rational line still has "holes" or "gaps" in it. The number line we studied in Chapter 6 is given in Fig. 7.4. How do we represent irrational numbers on the number line? Unfortunately, there is no single systematic procedure that can be applied to represent every irrational number. Each case has to be treated individually. For convenience, we will confine ourselves to the discussion of the representation of the irrational numbers $\sqrt{2}$ and $\sqrt{3}$.

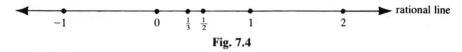

Fig. 7.4

Let Q be the point on the line perpendicular to ℓ at the point P such that $PQ = 1$. Draw a semicircle with center O and radius OQ to cut the line at points A and B. Observe that $OA = OB = OQ$ (radius). Since OPQ is a right triangle, by the Pythagorean theorem (see Section 13.6), $OQ^2 = OP^2 + PQ^2 = 1 + 1 = 2$ and hence $OQ = \sqrt{2}$. Therefore, $OA = OB = \sqrt{2}$. Consequently, the coordinate of A is $-\sqrt{2}$ and that of B is $\sqrt{2}$.

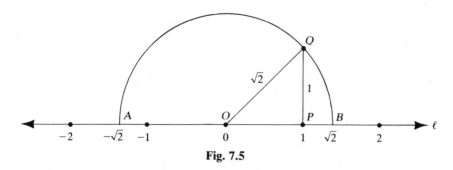

Fig. 7.5

To represent the irrational number $\sqrt{3}$ on the number line, we choose the point R on the perpendicular to the line ℓ at B where $BR = 1$. Let the semicircle with center O and radius OR cut the line ℓ at C and D. Using the Pythagorean theorem, as before, we have $OR = OC = OD = \sqrt{3}$. Thus, the points C and D represent the irrational numbers $-\sqrt{3}$ and $\sqrt{3}$, respectively. Continuing like this, any (irrational) numbers $\pm\sqrt{x}$, with x a positive integer, can be represented on the number line.

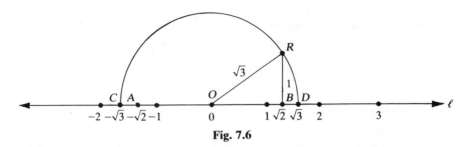

Fig. 7.6

Let a be any rational number. Irrational numbers of the form $a \pm \sqrt{2}$ and $a \pm \sqrt{3}$ can now be represented on the number line by simply moving segments of length $\sqrt{2}$ and $\sqrt{3}$ through the "rational" distance a from the origin.

Corresponding to every real number, there is a unique point on the number line, and conversely, every point on the line corresponds to a unique real number. The irrational numbers "plug" the "holes" in the rational line. The number line that exhibits a 1-1 correspondence between the set of real numbers and the set of points on a straight line is usually called the *real line*. The real line has no "holes" or "gaps" in it.

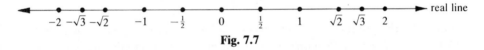

Fig. 7.7

7.7   INTERVALS

So far in our discussion of the development of the system of real numbers, we have come across several subsets of the set of real numbers; for example, the set of natural numbers, the set of whole numbers, the set of even integers, the set of positive rational numbers, etc. Recall that we used number lines to picture some subsets of the set of whole numbers, integers, etc.

If a and b are real numbers such that $a < b$, then the point with coordinate a lies on the left of the point with coordinate b on the real line. Let

$$(a,b) = \{x \in R \mid a < x < b\}$$
$$(a,b] = \{x \in R \mid a < x \le b\}$$
$$[a,b) = \{x \in R \mid a \le x < b\}$$
$$[a,b] = \{x \in R \mid a \le x \le b\}$$

Any subset of R in any one of these forms is called an *interval*. These intervals are called *open, open-closed, closed-open,* and *closed* intervals, respectively. The real numbers a and b are called the *endpoints* of the intervals.

For example, $(1,2)$ is an open interval; $(1,2]$ is an open-closed interval; $[1,2)$ is a closed-open interval; and $[1,2]$ is a closed interval. The thick portions in Figs. 7.8–7.11 illustrate the graphs of these sets on the real line. The parentheses at 1 and 2 indicate that those two numbers do not belong to the interval (1,2).

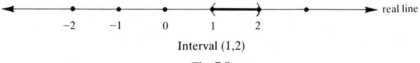

Interval (1,2)

Fig. 7.8

The parenthesis at 1 simply means that 1 is not included in the set, whereas the bracket at 2 indicates that it *is* a member of the set.

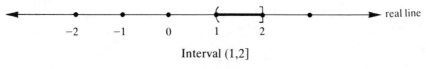

Interval (1,2]

Fig. 7.9

The bracket at 1 shows that 1 is included in the set, whereas the parenthesis at 2 shows that it does not belong to the set.

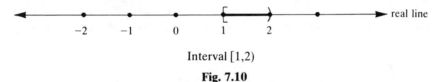

Interval [1,2)

Fig. 7.10

The brackets at 1 and 2 indicate that both numbers belong to the interval [1,2].

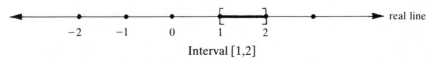

Interval [1,2]

Fig. 7.11

Exercise 7.7

1. Mark *true* or *false*.
   a) [1,2] is an infinite set.
   b) (1,2) is a finite set.
   c) [1,2] = (1,2) ∪ {1,2}.
   d) (2,2) is a finite set.
   e) The union of two intervals is an interval.
   f) The intersection of two intervals is an interval.
2. What is the cardinality of the following sets?
   a) (1,1) **b)** (1,1] **c)** [0,0) **d)** [1,1]

3. Draw the graph of each of the following sets on the real line.

    a) $(-1,3]$ **b)** $[-5,-2)$ **c)** $[-3,5]$ **d)** $(-2,2)$

4. If $A = [-2,3)$, $B = (1,5)$, and $U = R$, find the following sets:

    a) $A \cup B$ **b)** $A \cap B$ **c)** $A - B$ **d)** A'

    e) B' **f)** $A' \cup B'$ **g)** $A' \cap B'$ **h)** $A - B'$

    i) $(A - B)'$ **j)** $A' - B'$

Graph these sets on the real line.

7.8 APPROXIMATION OF $\sqrt{2}$

Recall that since $\sqrt{2}$ is an irrational number, the decimal expansion of $\sqrt{2}$ is both nonterminating and nonrepeating. However, since we can find rational numbers as close as we please to any irrational number, $\sqrt{2}$ can be approximated by rational numbers. Recall that if a and b are positive real numbers, then $a < b$ if and only if $a^2 < b^2$. We will make use of this fact to find an approximate value of $\sqrt{2}$.

Recall that $\sqrt{2}$ is the positive number x such that $x^2 = 2$. Since $1^2 = 1 < 2 < 2^2 = 4$, it follows that $1 < \sqrt{2} < 2$, as a first approximation. We proceed now by trial and error. Since $(1.4)^2 = 1.96 < 2 < (1.5)^2 = 2.25$, we have $1.4 < \sqrt{2} < 1.5$, as a second approximation. We now observe that $(1.41)^2 = 1.9881 < 2 < (1.42)^2 = 2.0164$. Therefore, $1.41 < \sqrt{2} < 1.42$, as a third approximation. This procedure can be carried out until the desired accuracy of approximation is achieved.

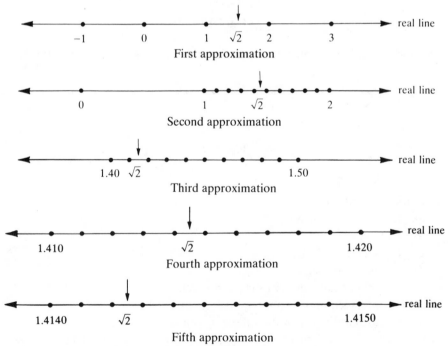

Fig. 7.12

Thus we have,

$$1<\sqrt{2}<2 \qquad \text{since} \qquad 1<2<4$$
$$1.4<\sqrt{2}<1.5 \qquad \text{since} \qquad (1.4)^2<2<(1.5)^2$$
$$1.41<\sqrt{2}<1.42 \qquad \text{since} \qquad (1.41)^2<2<(1.42)^2$$
$$1.414<\sqrt{2}<1.415 \qquad \text{since} \qquad (1.414)^2<2<(1.415)^2$$
$$1.4142<\sqrt{2}<1.4143 \qquad \text{since} \qquad (1.4142)^2<2<(1.4143)^2$$
$$1.41421<\sqrt{2}<1.41423 \qquad \text{since} \qquad (1.41421)^2<2<(1.41423)^2$$

Continuing this procedure, which is never ending, we find that $\sqrt{2} = 1.414213\ldots$. Thus $\sqrt{2}$ can be approximated as 1.41421 correct to five significant decimal places.

Exercise 7.8

1. Is it true that $\sqrt{2} = 1.41421$?

2. Is it true that there cannot be more than 1 million digits in the decimal representation of $\sqrt{2}$? Why?

3. Using the method given in this section, find the approximate value, correct to three significant decimal places, of:

a) $\sqrt{3}$ **b)** $\sqrt{5}$ **c)** $\sqrt{6}$ **d)** $\sqrt{7}$

4. Consider the subset $F = \{a + b\sqrt{2} \mid a,b \in Q\}$ of R. Show each of the following:

a) F is closed under addition.

b) F is closed under multiplication.

c) 0 belongs to F.

d) 1 belongs to F.

e) Addition is associative.

f) Multiplication is associative.

g) Addition is commutative.

h) Multiplication is commutative.

i) Every element of F has an additive inverse.

5. If a,b,c,d are rational numbers such that $a + b\sqrt{2} = c + d\sqrt{2}$, show that $a = c$ and $b = d$.

★6. Show that $a + b\sqrt{2} = 0$ if and only if $a - b\sqrt{2} = 0$, where a and b are rational numbers.

★7. Show that $\sqrt{\sqrt{2}}$ cannot be expressed in the form $a + b\sqrt{2}$ with a,b in Q. [*Hint:* Prove by contradiction.]

7.9   THE SQUARE ROOT ALGORITHM

In the preceding section, we illustrated a method to find an approximate value for the square root of a positive real number. We now proceed to demonstrate a method that is much faster and more efficient than this.

Example 7.27 Find the square root of 523.456 correct to three significant decimal places.

Solution:

Step 1 First, starting at the decimal point of the given number, we group the digits in pairs in both directions. If there is an odd number of digits to the right of the decimal point, we append a zero at the end. If there is an odd number of digits on the left of the decimal point, we group them in pairs as many times as possible, leaving the digit on the extreme left by itself. After the grouping, the number looks like this: 5 23. 45 60.

Step 2 Start with the first pair on the left (5 in this case). Find the largest integer whose square is less than or equal to this pair. This number is 2; we write it just above the digit 5. Since $2^2 = 4$, we write 4 under the first pair and subtract it. The remainder is 1 and we now bring down the next pair, 23.

$$
\begin{array}{r}
2 \\
\overline{\smash{\big|}\ 5\ 23.\,45\ \,60} \\
4 \\
\hline
1\ 23 \\
\end{array}
$$

Step 3 Doubling the 2 (above the bar) gives us 4. By trial and error, we want to find the largest digit x such that $4x$ times x is less than or equal to 123. Observe that $(43)(3) = 129$, so the largest digit we can choose for x is 2, where $(42)(2) = 84$. We write this 2 above the pair 23. Subtract 84 from 123 to leave the remainder 39. Now we bring down the next pair, 45, which lies on the right of the decimal point. Accordingly, we place a decimal point after the 2, above the bar.

$$
\begin{array}{r}
2\ 2 \\
\overline{\smash{\big|}\ 5\ 23.\,45\ \,60} \\
4 \\
\hline
1\ 23 \\
42 \qquad 84 \\
\hline
39\ 45 \\
\end{array}
$$

Step 4 Double the 22 (above the bar) to get 44. Find the largest digit x such that $44x$ times x is contained in 3945. By trial and error, we find that $x = 8$ and $(448)(8) = 3584$. We write this 8 just above the pair 45. Subtract 3584 from 3945 to leave the remainder 361.

$$
\begin{array}{r}
2\ \ 2.\ \ 8 \\
\overline{\smash{\big|}\ 5\ 23.\,45\ \,60} \\
4 \\
\hline
1\ 23 \\
42 \qquad 84 \\
\hline
39\ 45 \\
448 \qquad 35\ 84 \\
\hline
3\ 61 \\
\end{array}
$$

Step 5 Bring down the next pair, 60. Continuing as in Step 4, we get

```
               2  2. 8  7
            | 5 23. 45 60
              4
              ‾‾‾‾
              1 23
     42        84
              ‾‾‾‾‾
              39 45
    448       35 84
              ‾‾‾‾‾‾
               3 61 60
   4567        3 19 69
              ‾‾‾‾‾‾‾
                 41 91
```

Step 6 Since no more pairs remain on the right of the decimal point, we write 00 as the next pair and continue as before. Since the fourth digit after the decimal point is less than 5, we write $\sqrt{523.456}$ correct to three significant decimal places as 22.879.

```
               2  2. 8  7  9  1
            | 5 23. 45 60 00 00
              4
              ‾‾‾‾
              1 23
     42        84
              ‾‾‾‾‾
              39 45
    448       35 84
              ‾‾‾‾‾‾
               3 61 60
   4567        3 19 69
              ‾‾‾‾‾‾‾
                 41 91 00
  45749          41 17 41
              ‾‾‾‾‾‾‾‾‾
                 73 59 00
 457581          45 75 81
              ‾‾‾‾‾‾‾‾‾
                 27 84 19
```

Example 7.28   Find $\sqrt{2}$ correct to four significant decimal places.
Solution:

$$
\begin{array}{r|l}
 & 1.\ \ 4\ \ \ 1\ \ \ 4\ \ \ 2\ \ \ 1 \\
 & \overline{2.00\ \ 00\ \ 00\ \ 00\ \ 00} \\
 & 1 \\
 & \overline{1\ 00} \\
24 & \ \ 96 \\
 & \overline{4\ 00} \\
281 & 2\ 81 \\
 & \overline{1\ 19\ 00} \\
2824 & 1\ 12\ 96 \\
 & \overline{6\ 04\ 00} \\
28282 & 5\ 65\ 64 \\
 & \overline{38\ 36\ 00} \\
282841 & 28\ 28\ 41 \\
 & \overline{10\ 08\ 59}
\end{array}
$$

Thus $\sqrt{2} = 1.4142$ correct to four significant decimal places. Observe that this agrees with the answer we obtained in the previous section.

Exercise 7.9

1. Mark *true* or *false* (a and b are any positive real numbers).

   a) $\sqrt{a} + \sqrt{b} = \sqrt{(a+b)}$

   b) $\sqrt{a} - \sqrt{b} = \sqrt{(a-b)}$

   c) $a = b$ if and only if $\sqrt{a} = \sqrt{b}$.

   d) If $a < b$, then $\sqrt{a} < \sqrt{b}$.

   e) If $\sqrt{a} < \sqrt{b}$, then $a < b$.

   f) Every positive real number has a positive square root.

   g) Every positive real number has a unique positive square root.

   h) Every rational number has a positive square root in Q.

   i) $\sqrt{a} \cdot \sqrt{a} = a$

  ★j) $\sqrt{a} \cdot \sqrt{b} = \sqrt{ab}$

  ★k) $\dfrac{\sqrt{a}}{\sqrt{b}} = \sqrt{\dfrac{a}{b}}$

2. Give a counterexample to each of the false statements in problem 1.

3. Find the square root of each of the following numbers correct to three significant decimal places.

a) 3	**b)** 5	**c)** 7	**d)** 81.81
e) 35.21	**f)** 256.256	**g)** 1976	**h)** 23.123
i) 1776.1776	**j)** 0.05	**k)** 2000	**l)** 0.007

4. Show that $\dfrac{1}{2+\sqrt{3}} = 2 - \sqrt{3}$.

★5. Prove that if $a > b > 0$, then $\dfrac{a+b}{2} > \sqrt{ab}$. [*Hint:* $\sqrt{a} - \sqrt{b} > 0$.]

7.10   ABSOLUTE VALUE

Recall that, by the law of trichotomy, if a is any real number, then either $a < 0$ or $a = 0$ or $a > 0$. Also, $a < 0$ if and only if $-a > 0$. We now introduce a new concept that is very closely associated with the law of trichotomy.

Definition 7.2   The *absolute value* of a real number a, denoted by $|a|$, is defined as follows:

$$|a| = \begin{cases} a & \text{if} \quad a \geq 0 \\ -a & \text{if} \quad a < 0 \end{cases}$$

Example 7.29

$|5| = 5$	since $\quad 5 \geq 0$
$|-5| = -(-5) = 5$	since $\quad -5 < 0$
$\left|\frac{2}{3}\right| = \frac{2}{3}$	since $\quad \frac{2}{3} \geq 0$
$|0| = 0$	since $\quad 0 \geq 0$
$|-1.2| = -(-1.2) = 1.2$	since $\quad -1.2 < 0$
$|\pi| = \pi$	since $\quad \pi \geq 0$

Roughly speaking, when we take the absolute value of a real number, we are interested only in its *magnitude* and not in its sign. The concept of the absolute value of a real number can be used as an aid to study and understand the number line better.

What can we say about the absolute value of a number a? Is it always positive? negative? Can it be zero for any value of a? Clearly, if $a = 0$, then $|a| = |0| = 0$. Conversely, if $|a| = 0$, then $a = 0$. Recall that if $a < 0$, then $-a > 0$. It now follows from the definition that whether $a > 0$ or $a < 0$, $|a| > 0$ always. Thus $|a| \geq 0$ for any real number a.

How many real numbers are there with the same absolute value 5? Clearly there are two, namely, 5 and -5; that is, $|5| = |-5| = 5$. More generally, if a is any real number, then $|a| = |-a|$. In other words, if $a > 0$, then the equation $|x| = a$ has exactly two solutions, $\pm a$. For example, the solutions of the equation $|x| = \frac{2}{3}$ are $\pm\frac{2}{3}$.

Example 7.30   Verify that $|a \cdot b| = |a| \cdot |b|$ in the following cases:

a) $a = 2, b = 3$ b) $a = -2, b = 3$

c) $a = 2, b = -3$ d) $a = -2, b = -3$

Solution:

a) $|a \cdot b| = |2 \cdot 3| = 6 = 2 \cdot 3 = |2| \cdot |3| = |a| \cdot |b|$

b) $|a \cdot b| = |(-2) \cdot 3| = 6 = 2 \cdot 3 = |-2| \cdot |3| = |a| \cdot |b|$

c) $|a \cdot b| = |2 \cdot (-3)| = 6 = 2 \cdot 3 = |2| \cdot |-3| = |a| \cdot |b|$

d) $|a \cdot b| = |(-2) \cdot (-3)| = 6 = 2 \cdot 3 = |-2| \cdot |-3| = |a| \cdot |b|$

We now summarize all the important properties of the absolute value in the following theorem.

Theorem 7.3   *Let x, a, and b be any three real numbers. Then*

1. $|a| \geq 0$
2. $|a| = 0$ *if and only if* $a = 0$.
3. $|-a| = |a|$
4. $a \leq |a|$
5. $|a \cdot b| = |a| \cdot |b|$
6. $|a + b| \leq |a| + |b|$ *(triangle inequality)*
7. If $|x| = a$, *then either* $x = a$ *or* $x = -a$, *provided* $a \geq 0$.
8. $|x| \leq a$ *if and only if* $-a \leq x \leq a$, *provided* $a \geq 0$.
9. $|x| \geq a$ *if and only if either* $x \geq a$ *or* $x \leq -a$, *provided* $a \geq 0$.

Example 7.31   Solve the equation $|x - 2| = \frac{2}{3}$.

Solution: By part (7) of Theorem 7.3, $|x - 2| = \frac{2}{3}$ implies that either

$$x - 2 = \tfrac{2}{3} \qquad \text{or} \qquad x - 2 = -\tfrac{2}{3}$$

$$x = 2 + \tfrac{2}{3} \qquad \text{or} \qquad x = 2 - \tfrac{2}{3}$$

$$x = \tfrac{8}{3} \qquad \text{or} \qquad x = \tfrac{4}{3}$$

Thus, the solutions of the given equation are $\frac{8}{3}$ and $\frac{4}{3}$. We now ask the reader to verify that these are indeed solutions of the equation.

Example 7.32 Find and graph the solution set of the inequality $|x-\tfrac{2}{5}|\le\tfrac{3}{2}$.
Solution: By part (8) of Theorem 7.3, $|x-\tfrac{2}{5}|\le\tfrac{3}{2}$ if and only if

$$-\tfrac{3}{2}\le x-\tfrac{2}{5}\le\tfrac{3}{2}$$

$$-\tfrac{3}{2}+\tfrac{2}{5}\le x\le\tfrac{3}{2}+\tfrac{2}{5}$$

$$-\tfrac{11}{10}\le x\le\tfrac{19}{10}$$

(Why?) Thus the solution set is $\{x\in R\mid-1.1\le x\le1.9\}$. Its graph is exhibited in Fig. 7.13.

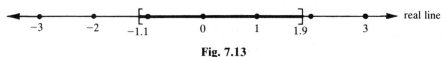

Fig. 7.13

Example 7.33 Find and graph the solution set of the inequality $|x+\tfrac{3}{4}|\ge\tfrac{1}{2}$.
Solution: By part (9) of Theorem 7.3, $|x+\tfrac{3}{4}|\ge\tfrac{1}{2}$ implies that either

$$x+\tfrac{3}{4}\ge\tfrac{1}{2}\qquad\text{or}\qquad x+\tfrac{3}{4}\le-\tfrac{1}{2}$$

$$x\ge\tfrac{1}{2}-\tfrac{3}{4}\qquad\text{or}\qquad x\le-\tfrac{1}{2}-\tfrac{3}{4}$$

$$x\ge-\tfrac{1}{4}\qquad\text{or}\qquad x\le-\tfrac{5}{4}$$

Therefore, the solution set is $\{x\in R\mid x\le-\tfrac{5}{4}\text{ or }x\ge-\tfrac{1}{4}\}$. Figure 7.14 exhibits the graph of the solution set.

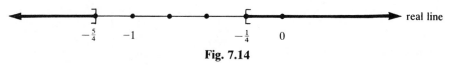

Fig. 7.14

Closely related to the concept of absolute value is that of the distance between two points on the real line. Consider the points A, B, C, and D on the real line, with coordinates $\tfrac{8}{5}$, $\tfrac{2}{3}$, $-\tfrac{2}{5}$, and $-\tfrac{3}{2}$, respectively. The distance from B to A is given by the difference

$$\frac{8}{5}-\frac{2}{3}=\frac{24-10}{15}=\frac{14}{15}$$

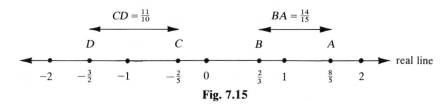

Fig. 7.15

What is the distance from C to D? Is it the following difference?

$$-\frac{3}{2}-\left(-\frac{2}{5}\right)=-\frac{3}{2}+\frac{2}{5}=\frac{-15+4}{10}=-\frac{11}{10}$$

Well, we do not say that the distance from C to D is less than zero. This awkward situation can be avoided by taking the absolute value of $-\frac{11}{10}$, namely, $\left|-\frac{11}{10}\right| = \frac{11}{10}$. Thus,

$$BA = \frac{8}{5} - \frac{2}{3} = \left|\frac{8}{5} - \frac{2}{3}\right| = \frac{14}{15}$$

and
$$CD = \left|-\frac{3}{2} - \left(-\frac{2}{5}\right)\right| = \left|-\frac{11}{10}\right| = \frac{11}{10}$$

These observations now motivate us to make the following definition.

Definition 7.3   Let a and b be any two real numbers. Then the *distance* from a to b, denoted by $d(a,b)$, is defined by

$$d(a,b) = |b - a|$$

$$\longleftarrow d(a,b) \longrightarrow$$

real line

a b

Fig. 7.16

From our discussion above, $d(\frac{2}{3}, \frac{8}{5}) = \frac{14}{15}$ and $d(-\frac{2}{3}, -\frac{3}{2}) = \frac{11}{10}$.

Example 7.34 $d(-3,0) = |0 - (-3)| = |0 + 3| = |3| = 3$

$$d(3,-3) = |-3 - 3| = |-6| = 6$$
$$d(3,3) = |3 - 3| = |0| = 0$$
$$d(-3,-2) = |-2 - (-3)| = |-2 + 3| = |1| = 1$$
$$d\left(\frac{1}{2}, \frac{1}{3}\right) = \left|\frac{1}{3} - \frac{1}{2}\right| = \left|\frac{2 - 3}{6}\right| = \left|-\frac{1}{6}\right| = \frac{1}{6}$$

The important properties of $d(a,b)$ are summarized in the following theorem.

Theorem 7.4   *Let a, b, and c be any three real numbers. Then*
1. $d(a,b) \geq 0$
2. $d(a,b) = 0$ *if and only if* $a = b$
3. $d(a,b) = d(b,a)$
4. $d(a,c) \leq d(a,b) + d(b,c)$ *(triangle inequality)*

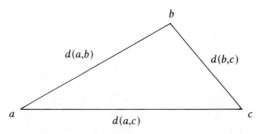

Fig. 7.17

The triangle inequality is so called since the sum of the lengths of any two sides of a triangle is at least as large as the length of its third side (see Fig. 7.17).

Exercise 7.10

1. Evaluate each of the following.
   a) $|5|$ b) $|-2|$ c) $|(-3)\cdot 2|$ d) $|7+3|$
   e) $|7|+|3|$ f) $|5+(-2)|$ g) $|11-3|$ h) $|11|-|3|$

2. Let $A = \{x \in I \mid |x| \leq 2\}$ and $B = \{x \in I \mid x^2 \leq 1\}$. Find the following sets in the roster method.
   a) $A \cup B$ b) $A \cap B$ c) $A - B$ d) $B - A$

3. Evaluate each of the following.
   a) $d(3,0)$ b) $d(0,5)$ c) $d(-4,0)$ d) $d(0,-7)$
   e) $d(3,5)$ f) $d(5,3)$ g) $d(2,5)+d(5,9)$ h) $d(2,9)$

4. Find the value of $x/|x|$ if
   a) $x = 3$ b) $x = -3$ c) $x = \frac{2}{3}$ d) $x = -\frac{2}{3}$

5. What is the difference between $|-x|$ and $-|x|$?

6. If x is a real number such that $|-x| = -|x|$, what is your conclusion?

7. Find a set of values for a and b such that
   a) $|a+b| = |a|+|b|$ b) $|a+b| < |a|+|b|$

8. Mark *true* or *false* (a and b are any two real numbers).
   a) $|a| > 0$
   b) $|a| = a$
   c) If $|a| = 0$, then $a = 0$.
   d) $|-a| = -|a|$
   e) If $|a| = |b|$, then $a = b$.
   f) If $|a| = a$, then $a > 0$.
   g) $|a+b| = |a|+|b|$
   h) $d(a,0) = |a|$
   i) $d(b,a) = -d(a,b)$
   j) If $d(a,b) = 0$, then $a = b = 0$.

9. Give a counterexample to each of the false statements in problem 8.

10. When can $d(x,y) = 0$?

11. Find the set of real numbers x such that
    a) $d(x,7) = 2$ b) $d(-3,x) = 5$ c) $d(x,-7) = 2$
    d) $d(x,-3) = -3$ e) $d(x,0) = 4$ f) $d(0,-x) = 1$

12. Find the set of all integers x such that
    a) $|2x-7| = 11$ b) $|5x-4| = 3$ c) $|2x| \leq 6$
    d) $|3-x| < 3$ e) $|2x-5| \leq 7$ f) $|3x| \geq 9$

13. Find the solution set of the following equations.
    a) $|3x-6| = 0$ b) $|2x+5| = 3$ c) $3+|x-3| = 5$
    d) $|2x| = 6$ e) $|4-3x| = 2$ f) $|3x-5| = -4$
    g) $\left|\dfrac{2x-3}{3x+2}\right| = 0$ h) $\left|\dfrac{5x-3}{2x-4}\right| = 2$ i) $|2x| = |3x-5|$

14. Find the solution set of each of the following.
   a) $d(x,3) \leq 2$ **b)** $d(-2,x) \leq \frac{2}{3}$ **c)** $d(1,-x) \leq \frac{1}{2}$
15. Prove that $|x^2| = |x|^2$ for every real number x.
16. Prove that $|x^3| = |x|^3$ for every real number x.
★**17.** Prove that if $d(a,b) = 0$, then $a = b$.
★**18.** Prove that $d(a,b) = d(b,a)$.
★**19.** Prove that $d(a,b) = d(a-b,0)$.

7.11   EXPONENTS

Since we will be coming across products of numbers in which a number is repeated more than once, especially in Chapter 8, it is only appropriate that we discuss the properties of exponents, introduced in Chapter 5. Recall, from Section 5.3, that if a is any integer and n any positive integer, we defined a^n as the product $a \cdot a \ldots a$ a total of n times. The numbers a and n are the base and exponent of a^n, respectively. We now extend this definition to include real bases and integral exponents.

Definition 7.4   Let a be any real number and n any positive integer. Then a^n denotes the product

$$\underbrace{a \cdot a \ldots a}_{n \text{ times}}$$

the number a is the *base* and n the *exponent* of a^n. A number, when expressed in the form a^n, is said to be in *exponential form*. We read a^n as *a raised to n* or the *nth power of a*.

For example,

$$2^3 = 2 \cdot 2 \cdot 2 = 8$$

$$(-2)^3 = (-2) \cdot (-2) \cdot (-2) = -8$$

$$(\sqrt{2})^4 = \sqrt{2} \cdot \sqrt{2} \cdot \sqrt{2} \cdot \sqrt{2} = 2 \cdot 2 = 4$$

$$(\tfrac{1}{3})^2 = \tfrac{1}{3} \cdot \tfrac{1}{3} = \tfrac{1}{9}$$

$$10^5 = 10 \cdot 10 \cdot 10 \cdot 10 \cdot 10 = 100,000$$

The exponential notation, which is very useful in shortening an expression containing products, was introduced by René Descartes. We will extend our above definition to include zero and negative integers as exponents, but first a few examples. Let's evaluate $7^5 \div 7^3$.

$$7^5 \div 7^3 = \frac{7^5}{7^3} = \frac{7 \cdot 7 \cdot 7 \cdot 7 \cdot 7}{7 \cdot 7 \cdot 7} = 7 \cdot 7 = 7^2$$

Observe now that

$$\frac{7^5}{7^3} = 7^2 = 7^{5-3}$$

More generally, if $m > n$, then

$$\frac{a^m}{a^n} = a^{m-n} \qquad \text{provided } a \neq 0$$

How is this result affected if we let $m = n$? If $m = n$, then

$$\frac{a^m}{a^m} = a^{m-m} = a^0$$

But from our experiences, we know that $a^m/a^m = 1$. This now forces us to define $a^0 = 1$ for every nonzero real number a.

Definition 7.5   If a is any nonzero real number, then $a^0 = 1$ and $a^{-m} = 1/a^m$ for every integer m. The quantity a^0 is undefined if $a = 0$.

For example, $2^0 = 1$, $(-3)^0 = 1$, $(\sqrt{2})^0 = 1$, $-(5^0) = -1$. Also,

$$2^{-3} = \frac{1}{2^3} = \frac{1}{8}$$

$$5^{-1} = \frac{1}{5^1} = \frac{1}{5}$$

$$(2^0)^{-3} = 1^{-3} = \frac{1}{1^3} = \frac{1}{1} = 1$$

Observe that

$$3^2 \cdot 3^4 = (3 \cdot 3)(3 \cdot 3 \cdot 3 \cdot 3) = 3 \cdot 3 \cdot 3 \cdot 3 \cdot 3 \cdot 3 = 3^6 = 3^{2+4}$$

$$2^3 \cdot 2^1 = (2 \cdot 2 \cdot 2)(2) = 2 \cdot 2 \cdot 2 \cdot 2 = 2^4 = 2^{3+1}$$

and $\qquad 5^2 \cdot 5^3 = (5 \cdot 5)(5 \cdot 5 \cdot 5) = 5 \cdot 5 \cdot 5 \cdot 5 \cdot 5 = 5^5 = 5^{2+3}$

More generally, $a^m \cdot a^n = a^{m+n}$. Thus,

$$3^5 \cdot 3^8 = 3^{5+8} = 3^{13}$$

and $\qquad\qquad\qquad 7^4 \cdot 7^6 = 7^{4+6} = 7^{10}$

Let m and n be positive integers. How should $(a^m)^n$ be defined? For example,

$$(2^3)^4 = 2^3 \cdot 2^3 \cdot 2^3 \cdot 2^3 = 8 \cdot 8 \cdot 8 \cdot 8 = 4096$$

and $\qquad\qquad\qquad 2^{3 \cdot 4} = 2^{12} = \underbrace{2 \cdot 2 \dots 2}_{\text{12 times}} = 4096$

Thus, $(2^3)^4 = 2^{3 \cdot 4}$. More generally, $(a^m)^n = a^{mn}$. Therefore,

$$(5^3)^2 = 5^{3 \cdot 2} = 5^6 = 15625$$

and $\qquad\qquad\qquad 3^8 = (3^4)^2 = 81^2 = 6561$

Finally, how is $(ab)^m$ defined? We have,

$$(2 \cdot 3)^5 = 6^5 = 7776$$

$$2^5 = 32$$

$$3^5 = 243$$

$$2^5 \cdot 3^5 = 32 \cdot 243 = 7776$$

Thus $(2 \cdot 3)^5 = 2^5 \cdot 3^5$. Using the commutative and associative properties of multiplication of real numbers, it can be proved that $(ab)^m = a^m \cdot b^m$. We now summarize the properties of exponents in the following theorem.

Theorem 7.5 (**Laws of Exponents**) *Let a and b be any two real numbers, m and n any integers. Then:*

1. $a^m \cdot a^n = a^{m+n}$

2. $(ab)^m = a^m \cdot b^m$

3. $a^0 = 1$ *provided $a \neq 0$*

4. $a^{-m} = \dfrac{1}{a^m}$ *provided $a \neq 0$*

5. $\dfrac{a^m}{a^n} = a^{m-n}$ *provided $a \neq 0$*

6. $\left(\dfrac{a}{b}\right)^m = \dfrac{a^m}{b^m}$ *provided $b \neq 0$*

Example 7.35   Evaluate $2^3/2^5$.

Solution:

$$\frac{2^3}{2^5} = 2^{3-5} = 2^{-2} = \frac{1}{2^2} = \frac{1}{4}$$

Example 7.36   Evaluate $3^8/6^8$.

Solution:

$$\frac{3^8}{6^8} = \left(\frac{3}{6}\right)^8 = \left(\frac{1}{2}\right)^8 = \frac{1^8}{2^8} = \frac{1}{256}$$

Example 7.37   Evaluate $(-2)^{-3}$ and $(2^{-3})^{-4}$.

Solution:

$$(-2)^{-3} = \frac{1}{(-2)^3} = \frac{1}{(-2)(-2)(-2)} = \frac{1}{-8} = -\frac{1}{8}$$

$$(2^{-3})^{-4} = 2^{(-3)(-4)} = 2^{12} = 4096$$

Example 7.38   Simplify $\dfrac{3^5 \cdot 3^3 \cdot 3^0}{3^4 \cdot 3^2}$.

Solution:

$$\frac{3^5 \cdot 3^3 \cdot 3^0}{3^4 \cdot 3^2} = \frac{3^{5+3+0}}{3^{4+2}} = \frac{3^8}{3^6} = 3^{8-6} = 3^2 = 9$$

Example 7.39   Simplify $\dfrac{2^5 \cdot 3^4 \cdot 5}{2^4 \cdot 5^3 \cdot 3^5}$.

Solution:

$$\frac{2^5 \cdot 3^4 \cdot 5}{2^4 \cdot 5^3 \cdot 3^5} = \frac{2^5}{2^4} \cdot \frac{3^4}{3^5} \cdot \frac{5}{5^3} = 2^{5-4} \cdot 3^{4-5} \cdot 5^{1-3}$$

$$= 2^1 \cdot 3^{-1} \cdot 5^{-2} = 2 \cdot \frac{1}{3} \cdot \frac{1}{5^2}$$

$$= \frac{2}{3 \cdot 5^2} = \frac{2}{3 \cdot 25} = \frac{2}{75}$$

Example 7.40   Simplify $\dfrac{7^{-2} \cdot 5^4 \cdot 2^8}{7^{-3} \cdot 10^3 \cdot 3^2}$.

Solution: $\dfrac{7^{-2} \cdot 5^4 \cdot 2^8}{7^{-3} \cdot 10^3 \cdot 3^2} = \dfrac{7^{-2} \cdot 5^4 \cdot 2^8}{7^{-3} \cdot 5^3 \cdot 2^3 \cdot 3^2}$ $[10^3 = (5 \cdot 2)^3 = 5^3 \cdot 2^3]$

$$= \frac{7^{-2}}{7^{-3}} \cdot \frac{5^4}{5^3} \cdot \frac{2^8}{2^3} \cdot \frac{1}{3^2} = 7^{-2+3} \cdot 5^{4-3} \cdot 2^{8-3} \cdot \frac{1}{3^2}$$

$$= 7^1 \cdot 5^1 \cdot 2^5 \cdot \frac{1}{3^2} = \frac{7 \cdot 5 \cdot 2^5}{3^2}$$

$$= \frac{7 \cdot 5 \cdot 32}{9} = \frac{1120}{9}$$

Exercise 7.11

1. Mark *true* or *false* (*a* and *b* are any real numbers, *m* and *n* are any positive integers).

a) $3^0 = 1$

b) $(-4)^0 = -1$

c) $(2^0)^4 = 16$

d) $0^{-3} = -3$

e) $a^0 = 1$

f) $(a^m)^n = (a^n)^m$

g) If $a \neq 0$, then $a^{-m} = \dfrac{1}{a^m}$.

h) If $a \neq 0$ and $b \neq 0$, then $\left(\dfrac{a}{b}\right)^{-1} = \dfrac{b}{a}$.

i) $a^m + b^m = (a + b)^m$

j) $a^{m+n} = a^m + a^n$

2. Give a counterexample (whenever possible) to each of the false statements in problem 1.

3. Write the following numbers in exponential form.

a) 8	b) 49	c) 81
d) 256	e) 243	f) 1024

4. Simplify each of the following.

a) $2a + 3a$

b) $a + 3b + 2c + b + 5a + 7c$

c) $1 + b + b^2 + 3b^2 + 2b + 4$

d) $a^2 + 2a^2 + 5a^2 + b^3 + 4b^3$

e) $3x^3 + 2y^2 + 5y^2 + x^2$

f) $3a^4 + b^2 + 5 + 4a^4$

5. Evaluate each of the following.

a) $2^3 \times 2^7 \times 2^0$

b) $(2 \times 3)^4$

c) $\dfrac{2^5}{2^3}$

d) 2^{-3}

e) $\dfrac{3^5}{3^{-2}}$

f) $(4^2)^3$

g) $5^7 \times 5^{-3}$

h) 10^0

i) $(-10)^0$

j) $4^3 \times 2^5$

k) $2^4 + 3^4$

l) $(2 + 3)^4$

6. Simplify each of the following.

a) $\dfrac{5^3 \times 7^4}{5^4 \times 7^3}$

b) $\dfrac{(3 \times 5)^4}{3^5 \times 5^2}$

c) $\dfrac{2^0 \times (-3)^0}{(-4^0) \times 5^0}$

d) $\dfrac{2^3 \times 3^4 \times 4^5}{4^3 \times 3^2 \times 2}$

e) $\dfrac{5^7 \times 3^{-2} \times 2^6}{10^5 \times 3^{-5}}$

f) $\dfrac{7 \times 8^3 \times 5^{-2}}{4^3 \times 5 \times 7^2}$

7. Simplify:

a) $a^5 \times a^3 \times a^{-7}$

b) $\dfrac{a^6 \times b^5}{a^3 \times b^7}$

c) $\dfrac{a^8}{a^{-10}}$

d) $\dfrac{b^{-8}}{b^{-5}}$

e) $(a^2 \times a^3)^4$

f) $(b^5 \times b^0)^{-1}$

g) $(a^{-3} \times a^5)^2$

h) $\dfrac{a^2 \times b^3 \times c^4}{a^4 \times b^3 \times c^2}$

i) $\dfrac{a^{-3} \times b^5 \times c^7}{a^4 \times b^{-2} \times c^7}$

8. Solve the following equations.

a) $3^{4x} = 3^{20}$

b) $2^{3x} = 1$

c) $(5^x)^2 = 625$

d) $2^{-x} = \frac{1}{2}$

e) $3^{2+x} = 27$

f) $(3x)^2 = 36$

g) $(5^4)^x = (5^3)^4$

h) $4^{-3x} = 2^{18}$

i) $(7^x)^2 = (2^0)^7$

7.12   SUMMARY AND COMMENTS

A special type of fractions called the decimal fraction was introduced. A fraction in its simplest form can be converted to a decimal fraction if and only if the prime factors of its denominator belong to the set {2,5}. We discussed the four basic arithmetic operations on the set of decimal fractions. The metric system, which is in practice in most countries, makes use of decimal fractions. Closely related to decimal fractions is percent, a concept that is used often in our daily life.

Every rational number, we observed, has either a terminating or a periodic decimal expansion. Also, every terminating or repeating decimal represents a rational number. This motivated a discussion of decimal representations that are neither terminating nor periodic. They give rise to the set of irrational numbers. Consequently, nonterminating and nonperiodic decimals represent numbers that cannot be written as the ratio of two integers. The square root of every rational number that is not a perfect square is an irrational number. However, these are not the only irrational numbers.

The set of irrational numbers, to our big surprise, is closed under neither addition nor multiplication. However, as in the case of rational numbers, the set of irrational numbers is dense; that is, there exist infinitely many irrational numbers between any two unequal irrational numbers. Also, there is an infinite number of rational numbers between any two distinct irrational numbers. Using that result, irrational numbers can always be approximated by rational numbers to any desired accuracy.

The union of the sets of rational and irrational numbers constitutes the set of real numbers. We did not formally define the arithmetic operations and the order relation on the set of real numbers, since they are somewhat complicated. However, in Section 7.5, we listed all the basic properties. The set of real numbers, like the sets of rational and irrational numbers, is also dense. We now can solve equations of the form $x^2 = a$, for any positive real number a.

We illustrated the square root algorithm: a systematic procedure to find the square root of any positive real number.

The concepts of the absolute value of a real number and the distance between two real numbers were introduced.

The following laws of exponents were discussed:

1. $a^m \cdot a^n = a^{m+n}$

2. $(ab)^m = a^m \cdot b^m$

3. $a^0 = 1$ provided $a \neq 0$

4. $\dfrac{a^m}{a^n} = a^{m-n}$ if $a \neq 0$

5. $a^{-m} = \dfrac{1}{a^m}$ provided $a \neq 0$

6. $\left(\dfrac{a}{b}\right)^m = \dfrac{a^m}{b^m}$ if $b \neq 0$

The real line establishes a 1-1 correspondence between the set of real numbers and the set of points on a straight line, both being infinite sets.

The notion of intervals was introduced. There are four different types of intervals: open, (a,b); open-closed, $(a,b]$; closed-open, $[a,b)$; and closed, $[a,b]$.

Unlike the previous number systems, the set R of real numbers is *complete*: roughly speaking, this means that the real line has no "holes" or "gaps" in it.

Finally, we observed that, even though the real number system is richer than any other previously discussed number system, it is still not the richest number system. Not every equation of the form $x^2 = a$, where a is any real number, can be solved in the set of real numbers. For example, the equation $x^2 = -1$ has no solution in R, since there is no real number x such that x^2 is negative. This deficiency of the set of real numbers motivates us to extend it to a still larger set. This is the topic of discussion of Chapter 9.

The line diagram of the number systems discussed thus far is given in Fig. 7.18. The "tree" in Fig. 7.18 is still growing!

Fig. 7.18

SUGGESTED READING

Boomer, L. W., "An Intuitive Approach to Square Root," *The Arithmetic Teacher*, vol. 16 (Oct. 1969), pp. 463–464.

Frederiksen, J. G., "Square Root+," *The Arithmetic Teacher*, vol. 16 (Nov. 1969), pp. 549–555.

Freitag, H. T., and A. H. Freitag, "Non-Terminating Periodically Repeating Decimals," *School Science and Mathematics*, vol. 69 (March 1969), pp. 226–240.

Hall, L. T., Jr., "Persuasive Arguments: .999 ... = 1," *The Mathematics Teacher*, vol. 64 (Dec. 1971), pp. 749–750.

Nelson, J., "Percent: A Rational Number or a Ratio," *The Arithmetic Teacher*, vol. 16 (Feb. 1969), pp. 105–109.

Steen, L. A., "New Models of the Real-Number Line," *Scientific American*, vol. 225 (Aug. 1971), pp. 92–99.

Wheeler, R. E., *Modern Mathematics*: *An Elementary Approach* (3rd ed.), Brooks/Cole Publishing Co., Monterey, Calif. (1973), pp. 295–332.

Wilson, P., D. Mundt, and F. Porter, "A Different Look at Decimal Fractions," *The Arithmetic Teacher*, vol. 16 (Feb. 1969), pp. 95–98.

"The grandest achievement of the Hindus and the one which, of all mathematical investigations, has contributed most to the general progress of intelligence, is the invention of the principle of position in writing numbers."

F. Cajori

8/Systems of Numeration

After studying this chapter, you should be able to:
- *represent a counting number in Egyptian, Babylonian, Roman, and Mayan numerals*
- *perform fundamental algorithms in the Hindu-Arabic system*
- *convert a nondecimal numeral to a decimal numeral*
- *express a decimal numeral in any base*
- *perform the fundamental operations in any nondecimal system*

8.0 INTRODUCTION

Thus far, we have been mainly concerned with a logical but informal development of the various number systems. However, we paid only little attention to the symbols used to name numbers and to the rules for combining them to label new numbers. Most of us, knowingly or unknowingly, use the words "number" and "numeral" interchangeably. Recall from Section 3.1 that *number* is an abstract property common to all equivalent sets, whereas a *numeral* is just a symbol used to label a number. For example, 5, V, and ||||| are a few numerals symbolizing the same number, five. As a result, the same number can have more than one symbolic representation.

What is a numeration system? It consists of a set of symbols denoting numbers and a set of rules that allow us to combine them to represent additional numbers.

In this chapter, we outline a few ancient numeration systems so that we may better understand and realize the merits of the numeration system we are all using in our everyday life, namely the *Hindu-Arabic numeration system.* We will also discuss the algorithms of the fundamental arithmetic operations in the Hindu-Arabic system and in a few other systems.

One of the oldest, most primitive numeration systems is the *tally system*. In this system, a stroke or a vertical line segment "|" is used to designate the number one. This symbol is used as many times as needed to denote larger numbers (*repetitive principle*), making use of the concept of 1-1 correspondence introduced in Section 1.10. For instance, the number seven is denoted as |||||||. The concept of the tally system was used by primitive man while keeping a count of his flock of sheep, by drawing marks on sand or making scratches on a piece of wood. The tally system with a slight modification is used even today. We group the strokes in sets of five or more. For example, five and twelve could be represented by |||| and |||| |||| ||, respectively.

Observe that the tally system becomes tedious and cumbersome as the number gets larger and larger.

8.1   THE EGYPTIAN SYSTEM

The *Egyptian numeration system*, developed by the ancient Egyptians, dates back to around 3400 B.C. The Egyptians used picture symbols (*hieroglyphics*) to label numbers. They used a stroke | to represent the number one (1), a heel bone ∩ for ten (10), a scroll ၅ for one hundred (100), a lotus flower ⚇ for one thousand (1,000), a pointed finger ∫ for ten thousand (10,000), a burbot fish ⌒ for one hundred thousand (100,000), and an astonished man ⵊ for one million (1,000,000). Observe that the Egyptian system contains a unique symbol for each of the first few powers of ten and is based on the number ten.

How is a number represented in Egyptian numerals? First, find how many times the largest possible power $n (\leq 6)$ of ten is contained in the number; continue this procedure with the remainder and the next largest power, $n - 1$, of ten, until we get a remainder less than 10. For example, since

$$254 = 200 + 50 + 4 = 2(10^2) + 5(10) + 4(1)$$

we can represent 254 as ၅၅∩∩∩∩∩|||| in Egyptian numerals. Similarly,

$$2,001 = 2(10^3) + 1(1)$$
$$= ⚇⚇|$$
$$10,312 = 1(10^4) + 3(10^2) + 1(10) + 2(1)$$
$$= ∫၅၅၅∩||$$
$$1,123,214 = 1(10^6) + 1(10^5) + 2(10^4) + 3(10^3) + 2(10^2) + 1(10) + 4(1)$$
$$= ⵊ⌒∫∫⚇⚇⚇၅၅∩||||$$

Example 8.1   What number does the Egyptian numeral ∫∫⚇၅၅၅∩∩||||||| represent?

Solution: ∫∫⚇၅၅၅∩∩||||||| = 20,000 + 1,000 + 300 + 20 + 7
$$= 21,327$$

Notice that the same number 123 can be represented in different ways:

$$123 = ?\cap\cap||| = \cap?\cap||| = |\cap?\cap||, \text{etc.}$$

Consequently, the order of the symbols in $?\cap\cap|||$ is immaterial. In the system with which we are familiar, we know that 256 is not the same as 625 or 265, etc., even though each of these contains the same three digits 2, 5, and 6. In 256, for example, 2 has a special position; so do 5 and 6. We usually say that each of the digits 2, 5, and 6 has a *place value* in the number represented by 256. Since the position of a symbol in an Egyptian numeral is unimportant, it is not a *place-value system*. The Egyptian system is based on the number ten, which is called the *base* of the system. Also, the system contains no symbol for zero.

Exercise 8.1

1. Find the numbers represented by the following Egyptian numerals.
   a) $?\cap\cap|$ b) $£ £$ c) $\mathcal{C}|\cap?\cap$
   d) $||\bigcirc|$ e) $\mathcal{X}\cap\cap$ f) $?\cap£|||$
2. Represent the following numbers in Egyptian numerals.
   a) 121   b) 1,331   c) 2,586   d) 196

8.2 THE BABYLONIAN SYSTEM

The early Babylonians of around 3000 B.C. had a numeration system of their own. They employed only two symbols to designate a number, namely, Υ for one and $\blacktriangleleft$ for ten. Numbers 1 through 59 were represented using these two symbols, using the repetitive principle. For example,

$$24 = 2(10) + 4 = \blacktriangleleft\blacktriangleleft\Upsilon\Upsilon\Upsilon\Upsilon$$

$$39 = 3(10) + 9 = \blacktriangleleft\blacktriangleleft\blacktriangleleft\begin{matrix}\Upsilon\Upsilon\Upsilon\\\Upsilon\Upsilon\Upsilon\\\Upsilon\Upsilon\Upsilon\end{matrix}$$

The number 60 was also denoted by the same symbol Υ that represented one. To avoid the possible confusion between the two usages, enough space was left between symbols, using the concept of place value. Consequently, each set of symbols from right to left would be multiplied by $60^0 = 1, 60^1 = 60, 60^2 = 3,600$, $60^3 = 216,000$, etc. For example,

$$\Upsilon\Upsilon\Upsilon \quad \blacktriangleleft \quad \Upsilon\Upsilon = 3(60^2) + 10(60) + 2(1)$$
$$= 10,800 + 600 + 2$$
$$= 11,402$$

$$\blacktriangleleft \quad \Upsilon\Upsilon \quad \blacktriangleleft\Upsilon \quad \blacktriangleleft = 10(60^3) + 2(60^2) + 11(60) + 10$$
$$= 10(216,000) + 2(3,600) + 11(60) + 10$$
$$= 2,160,000 + 7,200 + 660 + 10$$
$$= 2,167,870$$

Example 8.2 Convert 2,532 to a Babylonian numeral.
Solution: Since the largest integral power of 60 contained in 2,532 is the first power of 60, namely, $60^1 = 60$, we have

$$2,532 = 42(60) + 12$$

by the division algorithm. Now the largest integral power of 60 contained in the remainder, 12, is $60^0 = 1$ and $12 = 12(60^0)$. Thus, there are forty-two 60's and twelve units in 2,532; that is,

$$2532 = 42(60) + 12(60^0)$$
$$= \text{◀◀◀◀} \text{YY} \; \text{◀} \text{YY}$$

It is clear from our discussion that sixty is the base of the Babylonian system. It is a place-value system; however, the system did not have a symbol for zero. One can easily misread the positions of the symbols, unless they are properly spaced; the system did not have enough tools to handle situations where one or more powers of 60 were missing.

In passing, we remark that the Greek numeration system, developed around 450 B.C. employed the letters α, β, γ, etc. of the Greek alphabet to represent the numbers one, two, three, etc. The Greek system had no symbol for zero and was not a place-value system.

Exercise 8.2
1. What are the numbers represented by the following Babylonian numerals?

a) ◀   b) Y◀   c) YY ◀◀

d) Y ◀ Y   e) ◀ Y ◀   f) Y YY Y

2. Convert the following to Babylonian numerals.

a) 99   b) 123   c) 256   d) 1024

8.3 THE ROMAN SYSTEM
One of the early numeration systems used even today, though on a limited scale, is the Roman system. Most of us are more familiar with this system than any of the previously discussed numeration systems. Haven't you seen Roman numerals on dials of watches and clocks and on cornerstones of buildings? The Romans used five distinct symbols to represent numbers: I for one, V for five, X for ten, L for fifty, C for one hundred, D for five hundred and M for one thousand.

The Roman system also makes use of the repetitive principle to label numbers. For example,

$$XXIIII = 10 + 10 + 1 + 1 + 1 + 1 = 24$$
$$DCXVIII = 500 + 100 + 10 + 5 + 1 + 1 + 1 = 618$$
$$MDCII = 1000 + 500 + 100 + 1 + 1 = 1,602$$
$$MDLXXVI = 1000 + 500 + 50 + 10 + 10 + 5 + 1 = 1,576$$

However, the Romans were clever enough to use the *subtractive principle* to reduce the number of symbols in the representation of a number: if a symbol of lesser value appears on the left of a symbol of higher value, then the number represented by the first symbol is to be subtracted from that represented by the second symbol. For instance,

$$IV = 5 - 1 = 4$$
$$XXIV = 10 + 10 + (5 - 1) = 24$$
$$DLCXI = 500 + (100 - 50) + 10 + 1 = 561$$
$$MCMLXXVI = 1000 + (1000 - 100) + 50 + 10 + 10 + 5 + 1 = 1,976$$

Using the subtractive principle, no symbol need be used more than three times in a Roman numeral. We now warn the reader that whenever the subtractive principle is used in a numeral, then subtraction must preceed addition.

One of the distinctive features of the Roman system is that it makes use of a bar ($\overline{}$) above a symbol to show that its value is to be multiplied by 1,000. For example,

$$\overline{VI} = (5 + 1)(1,000) = 6(1,000) = 6,000$$
$$\overline{IV} = (5 - 1)(1,000) = 4(1,000) = 4,000$$
$$\overline{\overline{XII}} = (10 + 1 + 1)(1,000)(1,000) = 12(1,000)(1,000) = 12,000,000$$
$$M\overline{DC}XVI = 1,000 + (500 + 100)(1000) + 10 + 5 + 1 = 601,016$$

Observe that the Roman system has base ten; it is not a place-value system.

Exercise 8.3

1. What numbers do the following Roman numerals represent?
a) CD   b) DC   c) XIVII
d) DMXV   e) CDLXVI   f) MMI

2. Rewrite the following numbers in Roman numerals.
a) 59   b) 234   c) 891   d) 1,234

★**3.** Perform the indicated operations.
a) CXVII + DLXIII   b) CDLXII + CLIX
c) CMDXC + LXIV   d) LXIII − XXII
e) XCIV − XXXII   f) CCXXXI − CXXIII

8.4   THE MAYAN SYSTEM

The Mayan numeration system flourished among the Mayan Indians of Central America around 300 A.D. Its base, unlike that of the other systems, is twenty. It is a place-value system; however, the third group of symbols from the left is multiplied by $18 \cdot 20 = 360$, instead of by $20^2 = 400$. That the Mayan calendar had 360 days is said to be one of the reasons for this strange situation. The Mayans used the repetitive principle. The symbols they used are: $\ominus$ for zero, a thick dot • for one, and a horizontal line segment − for five. For example,

$$\underset{=}{\bullet\bullet\bullet} = 5+5+1+1+1+1 = 14$$

$$\underset{\equiv}{\bullet\bullet\bullet} = 5+5+5+5+1+1+1 = 23$$

Like the Chinese and the Japanese numerals, the Mayan numerals are also written vertically. The groups of symbols from bottom are multiplied by $20^0 = 1$, $20^1 = 20$, $18 \cdot 20 = 360$, $20 \cdot 360 = 7200$, $20 \cdot 7200 = 144,000$, etc., in succession. For instance,

$$\bullet\bullet \rightarrow 7 \cdot 7200 = 50,400$$
$$- \rightarrow 5 \cdot 360 \;\; = \;\; 1,800$$
$$\bullet\bullet \rightarrow 2 \cdot 20 \;\;\; = \;\;\;\;\; 40$$
$$\underset{-}{\bullet} \rightarrow 6 \cdot 1 \;\;\;\; = \;\;\;\;\;\; 6$$
$$\text{sum} = \overline{52,246}$$

represents the number 52,246.

Example 8.3 Convert 1,967 to a Mayan numeral.

Solution: We have

$$1,967 = 5 \cdot 360 + 167$$
$$167 = 8 \cdot 20 + 7$$
$$7 = 7 \cdot 1 + 0$$

Thus, there are five 360's, eight 20's and seven 1's in 1967:

$$1967 = 5(360) + 8(20) + 7(1)$$

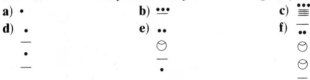

Exercise 8.4

1. Find the numbers represented by the following Mayan numerals.

a) •

b) •••

c) (symbols)

d) (symbols)

e) (symbols)

f) (symbols)

2. Express the following numbers in Mayan numerals.

   a) 23   **b)** 89   **c)** 234   **d)** 1,976

8.5 THE HINDU-ARABIC SYSTEM

Let's now discuss the numeration system we are all familiar with and use in our everyday life, the *Hindu-Arabic numeration system.* It is so called since it was invented by the Hindus of India around the third century B.C. and carried to Europe by the Arabs, who had a trade relationship with India. It is not known if they had any symbol for zero during the early stages. The first evidence that the Hindus used a symbol for zero appears in a book written around 825 A.D. by the Arab mathematician Muhammad ibn Musa al-Khowarizmi. We now remark that the English word algorithm has its origin from the title of its Latin translation, "Algorithmi de numero Indorum."

The Hindu-Arabic numerals were introduced in Spain by the Arabs who invaded the country in 711 A.D. It took several centuries for the Hindu-Arabic numerals to become popular in Europe, where the Roman system was widely used. The invention of the printing press has played a prominent role in this revolutionary change.

As we all know, the Hindu-Arabic system makes use of the ten symbols 0, 1, 2, . . . , 9 to represent any number. These symbols are called *digits*. Since the base of the system is ten, it is also called the *decimal system* from the Latin word "decem," meaning ten. The principal reason for this choice is that the early men, as some people do even today, used their ten fingers for counting and computing. For example, to count the apples in a basket, they grouped them in tens; if there are 23 apples, then there are two groups of ten apples each and three left: $23 = 2(10) + 3$. Unlike the previous numeration systems, our system contains a unique representation for every number. Also, it is a place-value system: every digit in the representation of a number has a unique place value. Each digit is to be multiplied by a suitable power of 10, which is determined by the location of the digit with respect to a reference point, called the *decimal point*. The decimal point is usually omitted when writing integers. The first digit to the left of the decimal point is multiplied by $10^0 = 1$, the second digit by $10^1 = 10$, the third digit by $10^2 = 100$, etc. Similarly, the digits to the right are multiplied by the negative powers $10^{-1} = \frac{1}{10}$, $10^{-2} = \frac{1}{100}$, $10^{-3} = \frac{1}{1000}$, etc. (We now remark that the concept of decimal fractions, and hence the writing of digits to the right of the decimal point, did not become an added feature of the Hindu-Arabic system until Simon Stevin.) For example,

$$256 = 2(10^2) + 5(10^1) + 6(10^0)$$
$$= 2(100) + 5(10) + 6(1)$$

The place value of 2 in the numeral 256 is 100, that of 5 is 10, and that of 6 is 1. In other words, the numeral 256 contains 2 hundreds, 5 tens, and 6 units; hence the name two hundred and fifty-six.

$$12,345 = 1(10^4) + 2(10^3) + 3(10^2) + 4(10^1) + 5(10^0)$$
$$= 1(10,000) + 2(1,000) + 3(100) + 4(10) + 5(1)$$

In the numeral 12,345, the digits 1, 2, 3, 4, and 5 have place values 10,000; 1,000; 100; 10; and 1, respectively. The digits 5, 4, 3, etc., are called the *units digit*, the *tens digit*, the *hundreds digit*, etc.

Example 8.4 $378 = 3(10^2) + 7(10^1) + 8(10^0)$ (1)

$$= 3(100) + 7(10) + 8(1)$$

$7801.4 = 7(10^3) + 8(10^2) + 0(10^1) + 1(10^0) + 4(10^{-1})$ (2)

$$= 7(1,000) + 8(100) + 0(10) + 1(1) + 4(\tfrac{1}{10})$$

$305.07 = 3(10^2) + 0(10^1) + 5(10^0) + 0(10^{-1}) + 7(10^{-2})$ (3)

$$= 3(100) + 0(10) + 5(1) + 0(\tfrac{1}{10}) + 7(\tfrac{1}{100})$$

The right-hand side of each of the equations (1)–(3) is called the *expanded form* of the numbers 378, 7801.4, and 305.07, respectively; each number on the left-hand side is the *decimal representation* of what is on the right-hand side. For example, the decimal representation of $2(10^2) + 1(10^1) + 0(10^0) + 3(10^{-1}) + 8(10^{-2})$ is 210.38.

Exercise 8.5

1. Write the expanded form of each of the following numerals.
   a) 13   b) 123   c) 23.58
   d) 101.24   e) 1221.003   f) 3005.04
2. Write the decimal representation of the following numbers.
   a) $2(10) + 5$ b) $3(10^2) + 5(10) + 7(10^0)$
   c) $4(10^3) + 7(10^2) + 8(10^0)$ d) $3(10^2) + 7(10) + 5(10^{-1})$
   e) $2(10^2) + 4(10^{-1}) + 3(10^{-2})$ f) $4(10^{-1}) + 9(10^{-2})$
3. Find the place value of the digit 3 in the following numerals.
   a) 23   b) 123   c) 231
   d) 2.357   e) 1030.27   f) 256.03
4. Find the units digit in each of the following.
   a) 29   b) 28.3   c) 324.05
5. Find the hundreds digit in each of the following.
   a) 234   b) 1024   c) 25
6. Write the number that contains
   a) 2 hundreds, 5 tens, and 7 units.
   b) 5 thousands, 3 tens, 2 units, and no hundreds.
   c) 6 units, no tens, no hundreds, and 3 thousands.

8.6 ADDITION ALGORITHM

After having mentioned the origin and the advantages of our numeration system, we now turn our attention to the four arithmetic algorithms of the system. The concept of place value in the system has contributed significantly to ease of computation. The commutative, associative, and distributive properties, discussed in the preceding chapters, play a vital role in arithmetic manipulations.

Example 8.5 Find the sum of the numbers 23 and 45.

Solution: We have

$$23 = 2(10^1) + 3(10^0) = 2(10) + 3$$
$$\underline{45 = 4(10^1) + 5(10^0) = 4(10) + 5}$$
$$23 + 45 = 2(10) + 3 + 4(10) + 5$$
$$= 2(10) + 4(10) + 3 + 5$$
$$= (2 + 4)(10) + 8$$
$$= 6(10) + 8(10^0)$$
$$= 68$$

Let's now summarize these steps as follows:

$$23 = 2(10) + 3(10^0)$$
$$\underline{45 = 4(10) + 5(10^0)}$$
$$\text{sum} = 23 + 45 = (2 + 4)(10) + (3 + 5)(10^0)$$
$$= 6(10) + 8(10^0)$$
$$= 68$$

If we look at the various steps in this example carefully, it is clear that in order to find the sum $23 + 45$, it was enough to do the following: add the units digits to get $3 + 5 = 8$ (Step 1 below) and write the 8 in the units place; now add the tens digits to get 6 as the tens digit of the answer (Step 2). Thus the sum is 68.

Step 1	*Step* 2
23	23
+45	+45
8	68

Example 8.6   Evaluate the sum $29 + 65$.
Solution: We have

$$29 = 2(10) + 9(10^0)$$
$$\underline{65 = 6(10) + 5(10^0)}$$

Adding,

$$29 + 65 = (2 + 6)(10) + (9 + 5)(10^0)$$
$$= 8(10) + 14(10^0)$$
$$= 8(10) + (10 + 4)(10^0)$$
$$= 8(10) + 1(10) + 4(10^0)$$
$$= (8 + 1)(10) + 4(10^0)$$
$$= 9(10) + 4(10^0)$$
$$= 94$$

If we add the units digits 9 and 5, we get $9 + 5 = 14 = 1(10) + 4$; now "carry" this 1 (one ten) to the tens place to get $8 + 1 = 9$ as the tens digit in the answer. This is summarized as follows:

Step 1	*Step 2*
①⤸carry	①⤸carry
29	29
+65	+65
—	—
4	94

Example 8.7   Add 128 and 75.

Solution:

$$128 = 1(10^2) + 2(10) + 8$$

$$75 = \qquad\qquad 7(10) + 5$$

$$\text{sum} = 128 + 75 = 1(10^2) + (2+7)(10) + (8+5)$$

$$= 1(10^2) + (2+7)(10) + 1(10) + 3$$

$$= 1(10^2) + (2+7+1)(10) + 3$$

$$= 1(10^2) + 10(10) + 3$$

$$= 1(10^2) + 1(10^2) + 3$$

$$= (1+1)(10^2) + 3$$

$$= 2(10^2) + 0(10) + 3$$

$$= 203$$

This procedure may be shortened as follows:

$$\begin{array}{r} ①① \\ 1\ 2\ 8 \\ +\quad 7\ 5 \\ \hline \\ 2\ 0\ 3 \end{array}$$

Our procedure can easily be extended to include numbers with fractional parts. We now remark that the above examples are illustrations of the algorithm for addition. They do not, by any means, constitute a proof of the validity of the algorithm.

Exercise 8.6

1. Find the sum of the following pairs of numbers in the expanded form.
  a) $7(10) + 4(10^0)$ and $2(10) + 5(10^0)$
  b) $2(10^2) + 3(10) + 5(10^0)$ and $4(10^2) + 5(10) + 1(10^0)$
  c) $5(10^3) + 7(10) + 8(10^0) + 3(10^{-1})$ and $3(10^2) + 4(10^0) + 5(10^{-1})$
  d) $2(10^0) + 3(10^{-1}) + 4(10^{-2})$ and $7(10) + 2(10^{-1})$

2. Use the expanded notation to add the following pairs of numbers.
  a) 29 and 73   **b)** 46 and 64
  c) 13.5 and 5.7   **d)** 21.03 and 0.54

8.7   MULTIPLICATION ALGORITHMS

Let's now take a look at the algorithms for multiplication. The distributive property of multiplication over addition plays a significant part in our discussion:

$$a(b+c) = ab + ac, \qquad (a+b)c = ac + bc$$

$$a(b+c+d) = ab + ac + ad, \qquad \text{etc.}$$

Notice that $(a+b)(c+d) = a(c+d) + b(c+d) = (ac + ad) + (bc + bd)$

$$= ac + ad + bc + bd$$

The products of single-digit whole numbers can easily be found by using the definition of multiplication of whole numbers (Definition 3.4). Once this is known, products of other numbers can be evaluated, as illustrated by the examples to follow.

We all know that the effect of multiplying an integer by 10 is the same as annexing a zero to the right of its digits. For example, $25 \times 10 = 250$. Also,

$$10 \times 28.31 = 10[2(10) + 8(10^0) + 3(10^{-1}) + 1(10^{-2})]$$

$$= 10[2(10)] + 10[8(10^0)] + 10[3(10^{-1})] + 10[1(10^{-2})]$$

$$= 2(10^2) + 8(10) + 3(10^0) + 1(10^{-1})$$

$$= 283.1$$

More generally, given any number in its decimal representation, in order to multiply it by 10^n where n is a positive integer, it is enough to move the decimal point by n digits to the right. For example,

$$10^3(1.28) = 1280 \qquad \text{and} \qquad 10^5(0.0123) = 1230$$

Example 8.8   Evaluate the product 5×128.

Solution: $5 \times 128 = 5[1(10^2) + 2(10) + 8(10^0)]$

$$= 5[1(10^2)] + 5[2(10)] + 5[8(10^0)]$$

$$= (5 \times 1)(10^2) + (5 \times 2)(10) + (5 \times 8)(10^0)$$

$$= 5(10^2) + 10(10) + 40(10^0)$$

$$= 5(10^2) + 1(10^2) + 4(10)$$

$$= (5+1)(10^2) + 4(10)$$

$$= 6(10^2) + 4(10) + 0(10^0)$$

$$= 640$$

These steps may now be shortened to the familiar algorithm shown in Step 4 below as follows. First, we write the numbers in such a way that their units digits, tens digits, etc., are aligned in columns (Step 1); multiply the digits 8, 2, and 1 of 128 by 5, one by one, in that order. Recall that $8 \times 5 = 40 = 4(10) + 0(10^0)$. When 40 is divided by 10, we get a quotient 4 and a remainder 0;

write this 0 in the units place and carry the 4 to the tens column (Step 2). Add this 4 to the product $2 \times 5 = 10$ to get $10 + 4 = 14$. Since $14 = 1(10) + 4$, we write the remainder 4 as the tens digit in our answer and carry the 1 to the hundreds column (Step 3). Add this 1 to the product $1 \times 5 = 5$ to get $5 + 1 = 6$. Since 6 is less than 10, we write this 6 as the hundreds digit in the answer (Step 4). There are no more digits left to the left of 1 in 128, so our procedure stops and the final answer is 640.

Step 1	*Step 2*	*Step 3*	*Step 4*
	④	① ④	① ④
1 2 8	1 2 8	1 2 8	1 2 8
$\times 5$	$\times 5$	$\times 5$	$\times 5$
	0	4 0	6 4 0

Example 8.9   Find the product of 25 and 36.

Solution: Let's use the expanded forms of 25 and 36 to evaluate the product of 25 and 36.

$$25 = 2(10) + 5$$
$$\times 36 = 3(10) + 6$$
$$\overline{ 12(10) + 30} \qquad \text{[multiply } 2(10) + 5 \text{ by 6]}$$
$$6(10^2) + 15(10) \qquad \text{[multiply } 2(10) + 5 \text{ by } 3(10)]$$
$$25 \times 36 = \overline{6(10^2) + 27(10) + 30} \qquad \text{[add]}$$
$$= 6(10^2) + [2(10) + 7](10) + 3(10)$$
$$= 6(10^2) + 2(10^2) + 7(10) + 3(10)$$
$$= (6 + 2)(10^2) + (7 + 3)(10)$$
$$= 8(10^2) + 10(10)$$
$$= 8(10^2) + 1(10^2)$$
$$= (8 + 1)(10^2)$$
$$= 9(10^2) + 0(10) + 0(10^0)$$
$$= 900$$

These steps now justify our familiar algorithm, summarized as follows:

Step 1	*Step 2*	*Step 3*	*Step 4*	*Step 5*
		①	①	①
	①③	①③	①③	①③
2 5	2 5	2 5	2 5	2 5
$\times 3$ 6	$\times 3$ 6	$\times 3$ 6	$\times 3$ 6	$\times 3$ 6
	1 5 0	1 5 0	1 5 0	1 5 0
		7 5 0	7 5 0	7 5
			9 0 0	9 0 0

A closer look at the above steps reveals that we first multiplied 25 by 6 (units digit in 36) to get 150 (Step 2); then multiplied 25 by 3 (tens digit in 36) to get 750 (Step 3) and added 150 and 750 to yield 900 (Step 4). We now remark that Step 5 is usually written for Step 4.

The algorithm we illustrated in the above examples holds even if there are digits to the right of the decimal point.

In order to show that this algorithm for multiplication is not unique, we now illustrate an entirely different algorithm for multiplication, called the *lattice method of multiplication*. Knowing the product of single-digit numbers, we can evaluate the product of any two numbers.

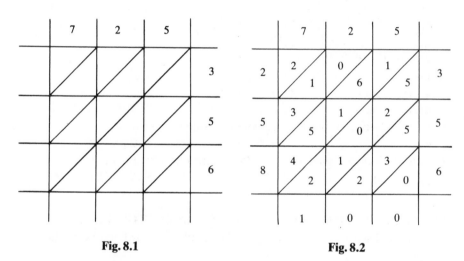

Fig. 8.1   Fig. 8.2

Let's illustrate this method by finding the product of 725 and 356. Since both numerals contain three digits each, divide a square into $3 \times 3 = 9$ cells. Divide these cells now by drawing diagonals from left to right (see Fig. 8.1). Now place the numeral 725 at the top of the square and the numeral 356 on its right side, as in Fig. 8.1. Each of the cells is now filled with two digits, one above and the other below the diagonal, as follows: to find the digits in the lower right-hand corner, we multiply 5 by 6 to get 30; write the units digit, 0, below the diagonal and the tens digit, 3, above the diagonal, as in Fig. 8.2. The other entries in various cells are found similarly. If the tens digit is missing in a product, we account for that by putting a zero in the upper half of the cell. The completed square is exhibited in Fig. 8.2.

Let's now add the digits diagonally from left to right, starting at the lower right-hand corner. Since there is only one digit, 0, in the first diagonal, we write 0 at the left of the diagonal and below the square. The sum of the numbers along the next diagonal is $2 + 3 + 5 = 10$. So we place the units digit, 0, of 10 at the bottom of the diagonal and carry the tens digit, 1, to the sum of the numbers along the next diagonal. This procedure is continued until all the diagonals have been exhausted. The answer is now obtained by reading down the digits on the left side of the square and across the bottom of the square: 258,100.

There is nothing sacred about our choice of these two numbers. If there are more digits in one numeral than the other, then instead of a square, we will have a rectangle. Also, this algorithm works even if there are digits to the right of the decimal point, provided we put the decimal point at the right place in the answer.

Exercise 8.7
1. Find xy in the expanded form if
    a) $x = 2(10) + 5(10^0)$ and $y = 4(10^0)$
    b) $x = 3(10) + 8(10^0)$ and $y = 4(10) + 7(10^0)$
    c) $x = 5(10^2) + 2(10) + 3(10^0)$ and $y = 2(10) + 9(10^0)$
    d) $x = 4(10) + 3(10^0)$ and $y = 8(10^0) + 3(10^{-1})$
2. Use the expanded form to multiply the following pairs.
    a) 19 and 23   **b)** 123 and 43   **c)** 256 and 31
    d) 1.23 and 13   **e)** 27.03 and 1.52   **f)** 0.003 and 0.05
3. Use the lattice method to evaluate the following products.
    a) 16×75 **b)** 123×321 **c)** 25×128
4. Find the value of each of the following products by any method.
    a) 23×101 **b)** 175×1001 **c)** 47×1010

8.8 SUBTRACTION AND DIVISION ALGORITHMS

Let's first discuss the familiar algorithm for subtraction of numbers. The facts

$$(a + b) - c = a + (b - c)$$

and $$(a + b) - (c + d) = (a - c) + (b - d)$$

are useful in our discussion. For example,

$$(15 + 8) - 10 = (15 - 10) + 8$$
$$(28 + 5) - (20 + 3) = (28 - 20) + (5 - 3)$$
$$[3(10) + 8(10^0)] - [2(10) + 3(10^0)] = (3 - 2)(10) + (8 - 3)(10^0)$$

Example 8.10   Evaluate $78 - 35$.
Solution:
$$78 = 7(10) + 8(10^0)$$
$$-35 = 3(10) + 5(10^0)$$
$$\overline{78 - 35 = (7 - 3)(10) + (8 - 5)(10^0)}$$
$$= 4(10) + 3(10^0)$$
$$= 43$$

Let's now write the numerals 78 and 25 columnwise (Step 1). If we observe the above steps carefully, it is clear that we could have obtained our answer by subtracting 5 from 8 to yield 3 in the units position (Step 2) and by subtracting 3 from 7 to get 4 in the tens place (Step 3).

Step 1	*Step* 2	*Step* 3
78	78	78
-35	-35	-35
	3	43

This example does not employ the concept of *borrowing*, which is illustrated in the following example:

Example 8.11   Subtract 28 from 95.

Solution:

$$95 = 9(10) + 5(10^0)$$

$$\frac{-28 = 2(10) + 8(10^0)}{95 - 28 = (9-2)(10) + (5-8)(10^0)}$$

Since $5 - 8$ is not one of the ten digits, we rewrite the expanded form of 95 as

$$8(10) + 1(10) + 5(10^0) = 8(10) + 10(10^0) + 5(10^0)$$

$$= 8(10) + (10+5)(10^0)$$

Thus we have

$$95 = 8(10) + (15)(10^0)$$

$$\frac{-28 = 2(10) + 8(10^0)}{95 - 28 = (8-2)(10) + (15-8)(10^0)}$$

$$= 6(10) + 7(10^0)$$

$$= 67$$

Since the units digit of 95 is less than that of 28, we borrowed one ten from the tens position to get $10 + 5 = 15$ in the units place and then subtracted 8 from 15 to yield 7 as the units digit in the answer. Now there is only 8 left in the tens position of the minuend, and $8 - 2$ gives 6 in the tens place in our answer. This is summarized in Steps 1–3.

Step 1	*Step* 2	*Step* 3
	⑧ ⑮	⑧ ⑮
95	9̸ 5̸	9̸ 5̸
-28	-2 8	-2 8
	7	6 7

The familiar algorithm for division, usually called the *long division method*, makes use of the repeated application of the division algorithm (Theorem 5.1): if a is any integer and b any positive integer, then there exist unique integers q and r such that $a = bq + r$, where $0 \le r < b$.

Example 8.12   Compute $1,976 \div 38$.

Solution: Computing $1,976 \div 38$ is equivalent to finding the number of sets, containing 38 apples each, that we can make from a pile of 1976 apples!

Observe that $38 \times 100 = 3,800$ is much larger than 1,976, whereas $10 \times 38 = 380 < 1,976$. Find, by trial and error, how many 380's are contained in 1976. The answer is 5; that is, $5 \times 10 \times 38 = 50 \times 38 = 1900$. Subtracting 1900 from 1976, we get $1976 - 1900 = 76$. Again, by trial and error, find how many 38's are contained in 76. Clearly it is 2, since $2 \times 38 = 76$. Subtracting 76 from 76, we get a zero remainder and the procedure stops. Let's now summarize this discussion in more traditional lines as follows:

$$
\begin{array}{r|l}
38 \,\lfloor\, 1976 & \\
\underline{1900} & \leftarrow 50 \times 38 \\
76 & \\
\underline{76} & \leftarrow \ \ 2 \times 38 \\
0 & \\
& \overline{52 \times 38} = \text{sum}
\end{array}
$$

Observe that we subtracted $50 + 2 = 52$ times 38 from 1,976 and $38 \times 52 = 1,976$. Consequently $1,976 \div 38 = 52$.

We now justify this procedure using the expanded forms of 1,976 and 38.

$$
\begin{array}{r}
5(10) + 2(10^0) \\
3(10) + 8(10^0) \,\lfloor\, \overline{1(10^3) + 9(10^2) + 7(10) + 6(10^0)} \\
\underline{1(10^3) + 9(10^2) + 0(10)} \\
7(10) + 6(10^0) \\
7(10) + 6(10^0)
\end{array}
$$

This is now abbreviated, line by line, in accordance with our usual understanding of the long division method, as

$$
\begin{array}{r}
5\,2 \\
38 \,\lfloor\, \overline{1976} \\
\underline{190} \\
76 \\
\underline{76}
\end{array}
$$

Before closing this section, let's discuss the algorithms of multiplication and division, developed by the early Egyptians. The fact that every natural number can be uniquely expressed as a sum of powers of 2 forms the basis of these fascinating algorithms. For example,

$$3 = 1 + 2$$

$$5 = 1 + 4$$

$$13 = 1 + 4 + 8$$

$$28 = 4 + 8 + 16$$

Suppose we wish to evaluate the product 23×45. Since $23 = 1 + 2 + 4 + 16$,

$$23 \times 45 = (1 + 2 + 4 + 16)45$$
$$= 1 \times 45 + 2 \times 45 + 4 \times 45 + 16 \times 45$$

Consequently, 23×45 is obtained by adding multiples of 45 by suitable powers of 2. Let's now construct a table (Table 8.1) consisting of two columns, one headed by 1 and the other by 45; each row of the table, except the first, is obtained by doubling the entries in the preceding row.

Table 8.1

1	45*
2	90*
4	180*
8	360
16	720*

Table 8.2

1	23*	
2	46*	
4	92	
8	184*	← 256
16	368	

Now, the value of the product 23×45 is obtained by adding the starred (*) numbers in the second column of Table 8.1. That is,

$$23 \times 45 = 45 + 90 + 180 + 720 = 1035$$

An algorithm very similar to the one just illustrated was developed by the Egyptians to find the quotient and the remainder when one natural number is divided by another.

Suppose we want to find the quotient and remainder when 256 is divided by 23. As before, we make a table (Table 8.2), the first column headed by 1 and the second row by the divisor, 23; each row, after the first, is obtained by doubling the numbers in the preceding row; the procedure is continued until the number in the second column exceeds the dividend, 256. We now express 256 as a sum using numbers from the second column of Table 8.2:

$$256 = 184 + 72 = 184 + 46 + 26$$
$$= 184 + 46 + 23 + \boxed{3}$$

Since 8, 2, and 1 are the numbers in the first column of Table 8.2 across from the starred numbers 184, 46, and 23, respectively, the quotient, when 256 is divided by 23, is given by $8 + 2 + 1 = 11$ and the remainder is 3. Notice now that $256 = 11 \times 23 + 3$.

Exercise 8.8

1. Find the difference $x - y$ in the expanded form if
   a) $x = 5(10^2) + 7(10) + 4(10^0)$ and $y = 3(10^2) + 6(10) + 2(10^0)$
   b) $x = 6(10) + 5(10^0)$ and $y = 2(10) + 3(10^0)$

   c) $x = 9(10^2) + 3(10^0)$ and $y = 4(10^2) + 7(10) + 1(10^0)$
   d) $x = 3(10^3) + 4(10) + 5(10^{-1})$ and $y = 8(10) + 3(10^0) + 4(10^{-1})$

2. Use the expanded form to evaluate $x - y$ if
   a) $x = 67$ and $y = 34$ **b)** $x = 128$ and $y = 69$
   c) $x = 13.8$ and $y = 11.5$ **d)** $x = 27.72$ and $y = 16.61$

3. Use the expanded form to find the quotient q and the remainder r when a is divided by b if
   a) $a = 65$ and $b = 13$ **b)** $a = 112$ and $b = 16$
   c) $a = 284$ and $b = 23$ **d)** $a = 1234$ and $b = 43$

4. Express the following numbers as a sum of powers of 2.
   a) 11   **b)** 23   **c)** 31   **d)** 97

5. Use the Egyptian method of doubling to evaluate each of the following products.
   a) 8×56 **b)** 19×91 **c)** 37×63

6. Use the Egyptian method to evaluate the quotient q and the remainder r when a is divided by b, where
   a) $a = 97, b = 15$ **b)** $a = 135, b = 31$
   c) $a = 328, b = 41$ **d)** $a = 1001, b = 13$

8.9 NONDECIMAL SYSTEMS

Primitive man's use of his ten fingers for counting could have been the reason for the choice of ten as the base of the decimal system. If man had twelve or twenty fingers, it is conceivable that twelve or twenty would have been a natural choice for the base of the system. The modern digital computers employ numeration systems other than the more familiar decimal system.

Is it feasible to develop a place-value system based on a number other than ten? This is answered in the affirmative. Let b be a counting number. Then any numeral in the system with base b can be written using the symbols $0, 1, 2, \ldots$, $b - 2$, and $b - 1$. For example, if the base is 5, then the base five system uses the symbols 0, 1, 2, 3, and 4 to represent any number; in the *binary system* (base two), we use the symbols 0 and 1. In a nondecimal system, the decimal point is replaced by a similar reference point. In base five system, the place values are powers of five: $5^0, 5^1, 5^2$, etc., to the left and $5^{-1}, 5^{-2}, 5^{-3}$, etc., to the right of the reference point. For example, consider the numeral 123 in base five system; the place values of 3, 2, and 1 are $5^0, 5^1$, and 5^2, respectively. The place values of the digits 3, 0, 2, and 4 in the base five numeral 42.03 are $5^{-2}, 5^{-1}, 5^0$, and 5^1, respectively. The numeral 123 in base b is denoted as $(123)_b$, using the subscript b. If a numeral appears without any subscript, it is assumed that the base is ten. For example, $123 = (123)_{10}$.

Recall that in the decimal system, we group in tens. Similarly, in base five system, we group in fives. For instance, $(23)_5$ means 2 sets of five and a set of 3 units: $(23)_5 = 2(5^1) + 3(5^0)$; $(123)_7$ denotes one group of $7^2 = 49$, two groups of $7^1 = 7$, and a group of 3 units: $(123)_7 = 1(7^2) + 2(7^1) + 3(7^0)$.

Example 8.13   Rewrite $(11011)_2$ in base ten.

Solution: Let's first write the expanded form of $(11011)_2$:

$$(11011)_2 = 1(2^4) + 1(2^3) + 0(2^2) + 1(2^1) + 1(2^0)$$
$$= 1(16) + 1(8) + 0(4) + 1(2) + 1(1)$$
$$= 16 + 8 + 0 + 2 + 1$$
$$= 27$$

We now remark that 27 and $(11011)_2$ are only different representations of the same number. Whether a numeral is written in base ten or base five or Roman numerals, it does not affect the value of the numeral.

Example 8.14   Express $(230.24)_5$ as a base ten numeral.

Solution: $(230.24)_5 = 2(5^2) + 3(5^1) + 0(5^0) + 2(5^{-1}) + 4(5^{-2})$

$$= 2(25) + 3(5) + 0(1) + 2(\tfrac{1}{5}) + 4(\tfrac{1}{25})$$
$$= 2(25) + 3(5) + 0(1) + 2(0.2) + 4(0.04)$$
$$= 50 + 15 + 0 + 0.4 + 0.16$$
$$= 65.56$$

If the base b of a system is larger than 10, then we need to introduce new symbols for $10, 11, \ldots, (b-1)$. For example, we might be tempted to use the symbols 0, 1, 2, 3, 4, 5, 6, 7, 8, 9, 10, and 11 in base twelve system, which is usually referred to as *duodecimal system*. However, this could create some confusion because of the symbols 10 and 11; it would not be clear as to whether 10 should be treated as a single symbol or as two symbols, 1 and 0; the same holds for the symbol 11. To avoid this problem, we use the symbols t for 10 and e for 11. Consequently, the symbols used in base twelve system are $0, 1, 2, \ldots,$ 9, t, and e.

Example 8.15   Convert $(2t5e)_{12}$ to base ten.

Solution: $(2t5e)_{12} = 2(12^3) + t(12^2) + 5(12^1) + e(12^0)$

$$= 2(1728) + 10(144) + 5(12) + 11(1)$$
$$= 3456 + 1440 + 60 + 11$$
$$= 4967$$

Thus far in this section, we have been illustrating how to convert a nondecimal numeral to a decimal numeral. How do we now express a base ten numeral in a different base? This procedure, which makes use of the repeated application of the division algorithm, is illustrated in the following examples.

Example 8.16   Convert 478 to a base five numeral.

Solution: Since we wish to express 478 in base five, let's first compute the first few powers of five for reference, so that the last number is greater than 478.

$$5^0 = 1$$
$$5^1 = 5 \qquad \leftarrow 3$$
$$5^2 = 25$$
$$5^3 = 125 \qquad \leftarrow 103$$
$$5^4 = 625 \qquad \leftarrow 478$$

Since 478 lies between 125 and 625, the highest power of five that is less than or equal to 478 is the three namely, $5^3 = 125$. How many 125's are contained in 478? The answer is found by applying the division algorithm:

$$478 = 3(125) + 103$$

Observe now that the remainder, 103, lies between 25 and 125. The highest power of five contained in 103 is the two namely, $5^2 = 25$. By the division algorithm, there are four 25's in 103:

$$103 = 4(25) + 3$$

Since the new remainder, 3, is less than 5, the procedure stops and we have

$$478 = 3(125) + 103$$
$$= 3(125) + 4(25) + 3$$
$$= 3(5^3) + 4(5^2) + 3(5^0) \qquad [\textit{Note}: \text{the term in } 5^1 \text{ is missing}]$$
$$= 3(5^3) + 4(5^2) + 0(5^1) + 3(5^0)$$
$$= (3403)_5$$

This procedure is summarized in Fig. 8.3. To obtain the answer, we need only read the quotients 3, 4, 0, 3 down the right-hand side of Fig. 8.3.

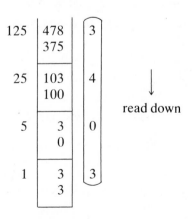

Fig. 8.3

Example 8.17 Express 676 in base seven.

Solution: As before, we first compute a list of powers of 7 so that 676 is less than the highest power.

$$7^0 = 1$$
$$7^1 = 7 \qquad \leftarrow 4$$
$$7^2 = 49 \qquad \leftarrow 39$$
$$7^3 = 343 \qquad \leftarrow 333$$
$$7^4 = 2401 \qquad \leftarrow 676$$

Since 676 lies between 343 and 2401, we check how many 343's are contained in 676. The answer is 1, since $676 = 1(343) + 333$. Therefore, the left-most digit in the base five representation of 676 is 1. Since $49 < 333 < 343$, find how many 49's are in 333. There are six 49's in 333, leaving a remainder of 39. Consequently, 6 is the next digit in the representation. Since there are five 7's and four units in 39, 5 and 4 are the next two digits on the right of 6 in our answer. Thus $676 = (1654)_7$.

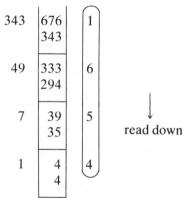

Fig. 8.4

The procedure we illustrated could get difficult, especially if the base is a fairly large number. The algorithm we now illustrate is not much different from the above steps; however, it looks much simpler. Let's now demonstrate this algorithm for the above example.

Divide 676 by 7; we find that the quotient is 96 and the remainder is 4:

$$676 = 7 \cdot 96 + 4$$

Now divide 96 by 7 to get quotient 13 and remainder 5:

$$96 = 7 \cdot 13 + 5$$

When the new quotient, 13, is divided by 7, we find 1 as the quotient and 6 as the remainder:

$$13 = 7 \cdot 1 + 6$$

When the last quotient, 1, is divided by 7, we get quotient 0 and remainder 1:

$$1 = 7 \cdot 0 + 1$$

The procedure stops once a zero quotient is reached. Thus we have

$$\begin{aligned} 676 &= 7 \cdot 96 + \boxed{4} \\ 96 &= 7 \cdot 13 + \boxed{5} \\ 13 &= 7 \cdot 1 \ + \boxed{6} \\ 1 &= 7 \cdot 0 \ + \boxed{1} \end{aligned} \quad \text{read up} \ \uparrow$$

Now the base five numeral of 676 is obtained by writing the remainders in the reverse order. That is, $676 = (1654)_5$.

Example 8.18   Rewrite 17,327 in base twelve.

Solution: Dividing 17,327 and its successive quotients by 12, we have

$$\begin{aligned} 17327 &= 12 \cdot 1443 + \boxed{11} \\ 1443 &= 12 \cdot 120 \ + \boxed{3} \\ 120 &= 12 \cdot 10 \ \ + \boxed{0} \\ 10 &= 12 \cdot 0 \ \ \ + \boxed{10} \end{aligned} \quad \text{read up} \ \uparrow$$

Recalling that the symbols 10 and 11 are denoted by t and e, respectively, and writing the remainders in the reverse order,

$$17{,}327 = (t03e)_{12}$$

Exercise 8.9

1. Write the expanded form of each of the following.
   a) $(256)_7$ b) $(1011)_2$ c) $(t0e)_{12}$
   d) $(10201)_3$ e) $(2030)_5$ f) $(121)_8$

2. Use the shorthand form to rewrite the following.
   a) $1(5^4) + 3(5^2) + 0(5) + 4$ b) $2(3^2) + 1(3) + 2(3^0)$
   c) $3(7^4) + 5(7^2) + 4(7)$ d) $10(12^3) + 8(12^2) + 5(12) + 11$
   e) $1(2^4) + 1(2^3) + 1(2^0)$ f) $7(11^4) + 3(11^3) + 5(11^2) + 2(11)$

3. Convert the following to decimal numerals.
   a) $(123)_5$ b) $(24101)_5$ c) $(1188)_{12}$
   d) $(2102210)_3$ e) $(5555)_7$ f) $(101101)_2$

4. Express the following numerals as required.
   a) $1976 = (\quad)_2$ b) $1976 = (\quad)_3$
   c) $1976 = (\quad)_7$ d) $23 = (\quad)_2$
   e) $101 = (\quad)_{12}$ f) $1776 = (\quad)_{12}$
   g) $256 = (\quad)_3$ h) $1947 = (\quad)_5$

5. Show that $(121)_b = (b+1)^2$.

6. Which represents a larger number in each of the following pairs of numerals?

a) $(10)_5$ and $(11)_5$ b) $(123)_7$ and $(321)_7$
c) $(110)_2$ and $(101)_2$ d) $(10)_5$ and $(10)_7$
e) $(121)_3$ and $(1101011)_2$ f) $(276)_{11}$ and $(267)_{12}$

7. Arrange the following numerals in increasing order of magnitude.
   a) $(1234)_5$ b) $(2431)_5$ c) $(1432)_5$ d) $(3214)_5$

8. Arrange the following numerals in increasing order of value.
   a) $(10011)_2$ b) $(1022)_3$ c) $(231)_5$ d) $(1011)_7$

9. Rewrite the following numerals as required.
   a) $(101011)_2 = ($ $)_3$ b) $(12021)_3 = ($ $)_5$
   c) $(t0e)_{12} = ($ $)_2$ d) $(234)_7 = ($ $)_5$

10. Find the value of b if:
    a) $(54)_b = 64$ b) $(1001)_b = 9$ c) $(1001)_b = 126$
    d) $(105)_b = 41$ e) $(121)_b = 25$ f) $(144)_b = 49$

11. Construct a table showing the representation of the first twelve natural numbers in base ten, base two, base three, and base five.

12. Assume an integer is represented in base two. What can you say about its units digit if it is (a) an even integer? (b) an odd integer?

8.10   ADDITION AND SUBTRACTION IN NONDECIMAL SYSTEMS

In this section, we illustrate the fundamental algorithms for addition and subtraction in nondecimal systems, in the hope that we may better understand the system we are using for everyday purposes. It might seem easier to perform these operations in base ten for obvious reasons; however, since the algorithms for addition and subtraction in any base are analogous to those in the decimal system, they need not sound complicated to us.

In order to add two numbers in base ten, we must know the sum of any two digits. Addition in base ten was easy since we memorized the sum of all single-digit numbers. Likewise, if we know the sum of all single digits in any base, it will definitely speed up our computation.

Since 0 and 1 are the digits used in the binary system, we must know each of the following in base two: $0+0$, $1+0$, $0+1$, and $1+1$. Clearly, $0+0=0$ and $1+0=0+1=1$ in any system. Since $1+1=2$ and there is no such numeral as 2 in base two, we must express 2 in base two numerals. Since there exist one 2 and zero units in the number 2, observe that $1+1=2=(10)_2$. This is summarized in Table 8.3.

Table 8.3.   Addition
Table in Base Two

+	0	1
0	0	1
1	1	10

Table 8.4   Addition Table in Base Five

+	0	1	2	3	4
0	0	1	2	3	4
1	1	2	3	4	10
2	2	3	4	10	11
3	3	4	10	11	12
4	4	10	11	12	13

The sum of two numbers in any base can be found by using the same concepts and ideas we used in the case of the decimal system.

Example 8.19   Evaluate $(123)_5 + (241)_5$.

Solution:

$$(123)_5 = 1(5^2) + 2(5) + 3(5^0)$$

$$+ (241)_5 = 2(5^2) + 4(5) + 1(5^0)$$

$$\text{sum} = (1+2)(5^2) + (2+4)(5) + (3+1)(5^0)$$

$$= 3(5^2) + 6(5) + 4(5^0)$$

$$= 3(5^2) + (5+1)(5) + 4(5^0)$$

$$= 3(5^2) + 1(5^2) + 1(5) + 4(5^0)$$

$$= (3+1)(5^2) + 1(5) + 4(5^0)$$

$$= 4(5^2) + 1(5) + 4(5^0)$$

$$= (414)_5$$

As in the case of addition of numbers in the decimal system, a careful study of the above steps shows that we can shorten our steps as in Fig. 8.5.

$$
\begin{array}{ll}
\text{①} & \text{①} \\
(1\ 2\ 3)_5 & = 3\ 8 \\
+(2\ 4\ 1)_5 & = 7\ 1 \\
\hline
\text{sum} = (4\ 1\ 4)_5 & = 1\ 0\ 9 \\
\end{array}
$$

Fig. 8.5   **Fig. 8.6**

To find the sum, we first add the units digits 3 and 1 to get $3 + 1 = 4$. Since $4 < 5$, we simply write 4 as the units digit in the answer. Adding the digits 2 and 4 in the fives column, we get $2 + 4 = 6$. Since $6 = 1(5) + 1$, we write the remainder, 1, in the fives column and carry the quotient, 1, to the twenty-fives column. Since $1 + 1 + 2 = 4$ is less than 5, we write 4 in the twenty-fives column. Since there are no more digits left to the left of the twenty-fives column, our procedure stops and the answer is $(414)_5$.

We can check our computation by converting the numerals to base ten, adding them in base ten, and converting the answer back to base five.

$$(123)_5 = 1(5^2) + 2(5) + 3(5^0) = 38$$
$$(241)_5 = 2(5^2) + 4(5) + 1(5^0) = 71$$

We have $38 + 71 = 109$ (see Fig. 8.6) and $109 = (414)_5$.

Example 8.20   Find the sum $(101011)_2 + (11001)_2$.

Solution: Recalling that $(1)_2 + (1)_2 = (10)_2$ and employing the familiar short-hand algorithm for addition, we have

$$
\begin{array}{ccccccccl}
 & ① & ① & ① & & ① & ① & & \\
(& 1 & 0 & 1 & 0 & 1 & & 1)_2 &= 43 \\
+(& & 1 & 1 & 0 & 0 & & 1)_2 &= 25 \\
\hline
\text{sum} = (1 & 0 & 0 & 0 & 1 & 0 & & 0)_2 &= 68
\end{array}
$$

Example 8.21   Add $(t58e)_{12}$ and $(9t3)_{12}$.

Solution:

$$
\begin{array}{crcccccccc}
 & & ① & ① & ① & & & & & \\
 & (t & 5 & 8 & e)_{12} &= 1 & 8 & 1 & 0 & 7 \\
 & +(& 9 & t & 3)_{12} &= & 1 & 4 & 1 & 9 \\
\hline
\text{sum} = & (e & 3 & 7 & 2)_{12} &= 1 & 9 & 5 & 2 & 6
\end{array}
$$

How do we subtract one number from another in any base? The procedure remains exactly the same as that in the decimal system, as illustrated by the following examples:

Example 8.22   Evaluate $(312)_5 - (123)_5$.

Solution:

$$
\begin{array}{crccll}
 & & & ⑩ & & \\
 & & ② & ⑩ & ⑫ & \\
 & (\cancel{3} & \cancel{1} & 2)_5 &= 82 \\
 & -(1 & 2 & 3)_5 &= 38 \\
\hline
\text{difference} = & (1 & 3 & 4)_5 &= 44
\end{array}
$$

First, write the numerals in such a way that the corresponding digits in the numerals appear one below the other. Beginning at the units column, since 2 in $(312)_5$ is less than 3 in $(123)_5$, we borrow one five from the fives column of $(312)_5$ to replace 2 by 12 (recall, all in base five), leaving a zero there. Subtracting 3 from 12, we get 4 in the units column of the answer. Proceed now to the fives column. Since $0 < 2$, we borrow one 25 from the twenty-fives column to replace 0 by 10, leaving a 2 in the twenty-fives column. Since $10 - 2 = 3$, we get 3 as the fives digit in our answer. Since $2 > 1$ in the twenty-fives column, we write $2 - 1 = 1$ in the twenty-fives column. Thus,

$$(312)_5 - (123)_5 = (134)_5$$

We now ask the reader to verify that $(123)_5 + (134)_5 = (312)_5$.

Example 8.23   Subtract $(10110)_2$ from $(11011)_2$.

Solution:

$$
\begin{array}{r}
①\;\;⑩ \\
(1 \quad \cancel{1} \quad \cancel{0} \quad 1 \quad 1)_2 = 27 \\
-(1 \quad 0 \quad 1 \quad 1 \quad 0)_2 = 22 \\
\hline
\text{difference} = (\qquad\quad 1 \quad 0 \quad 1)_2 = \;\;5
\end{array}
$$

Thus, $(11011)_2 - (10110)_2 = (101)_2$.

Example 8.24   Evaluate $(t74)_{12} - (39e)_{12}$.

Solution:

$$
\begin{array}{r}
⑯ \\
⑨ \;\; ⑥ \;\; ⑭ \\
(\cancel{t} \quad \cancel{7} \quad \cancel{4})_{12} = 1528 \\
-(3 \quad 9 \quad e)_{12} = \;\;551 \\
\hline
\text{difference} = (6 \quad 9 \quad 5)_{12} = \;\;977
\end{array}
$$

Exercise 8.10

1. Perform the following operations.
   a) $(1)_3 + (2)_3$ b) $(2)_5 + (3)_5$ c) $(4)_7 + (6)_7$
   d) $(2)_5 + (4)_5$ e) $(7)_{12} + (5)_{12}$ f) $(t)_{12} + (e)_{12}$

2. Construct an addition table for base three.

3. Construct an addition table for base six.

4. Construct an addition table for base seven.

5. Find the value of $x + 1$ if
   a) $x = (123)_5$ b) $x = (101)_2$ c) $x = (121)_3$
   d) $x = (25t)_{12}$ e) $x = (27e)_{12}$ f) $x = (666)_7$

6. Evaluate the following sums.
   a) $(102)_3 + (211)_3$ b) $(123)_5 + (231)_5$ c) $(101)_2 + (11)_2$
   d) $(11011)_2 + (1011)_2$ e) $(1234)_5 + (1202)_5$ f) $(89e)_{12} + (5t6)_{12}$

7. Evaluate each of the following differences.
   a) $(3)_5 - (2)_5$ b) $(e)_{12} - (t)_{12}$ c) $(e)_{12} - (7)_{12}$

8. Find the value of $x - 1$ if
   a) $x = (123)_5$ b) $x = (111)_2$ c) $x = (100)_2$
   d) $x = (37e)_{12}$ e) $x = (101)_3$ f) $x = (210)_7$

9. Perform the following operations.
   a) $(110)_2 - (11)_2$ b) $(201)_3 - (12)_3$ c) $(214)_5 - (132)_5$
   d) $(10110)_2 - (1001)_2$ e) $(2000)_7 - (1336)_7$ f) $(89e)_{12} - (5t6)_{12}$

10. Simplify each of the following.
    a) $(110)_2 + (101)_2 + (1101)_2$ b) $(1221)_3 + (200)_3 - (1121)_3$
    c) $(1234)_5 - (142)_5 - (241)_5$ d) $(753)_{12} + (537)_{12} - (375)_{12}$

★11. Find the set of possible values less than 8 for the base b if $(120)_b + (211)_b = (331)_b$.

★12. If $T + F = T = F + T$, $F + F = F$ and $T + T = TF$, evaluate each of the following.
    a) $TFT + TT$ b) $TFTF + TFFT$ c) $TFFT + TTT$

8.11   MULTIPLICATION AND DIVISION IN NONDECIMAL SYSTEMS

The ideas and concepts used to illustrate the algorithms for multiplication and division in the decimal system can easily be extended to discuss corresponding algorithms in any nondecimal system. As in the case of addition and subtraction, it will be helpful to know the products of any two single digits in a base for multiplication and division in that base. For example, the multiplication tables in base three and base five are exhibited in Tables 8.5 and 8.6, respectively.

Table 8.5   Multiplication Table in Base Three

·	0	1	2
0	0	0	0
1	0	1	2
2	0	2	11

Table 8.6   Multiplication Table in Base Five

·	0	1	2	3	4
0	0	0	0	0	0
1	0	1	2	3	4
2	0	2	4	11	13
3	0	3	11	14	22
4	0	4	13	22	31

We advise the reader to check the entries of these tables. For instance, what is $2 \cdot 2$ in base three? Since $2 \cdot 2 = 4 = 1(3) + 1$, we have $(2)_3 \cdot (2)_3 = (11)_3$.

Example 8.25   Evaluate the product $(23)_5 \cdot (34)_5$.

Solution:

$$
\begin{array}{rl}
(23)_5 = & 2(5) + 3 \\
\times (34)_5 = & 3(5) + 4 \\
\hline
& 8(5) + 12 \qquad \text{[multiply } 2(5) + 3 \text{ by 4]} \\
6(5^2) + 9(5) & \qquad \text{[multiply } 2(5) + 3 \text{ by } 3(5)] \\
\hline
\text{product} = 6(5^2) + 17(5) + 12 & \qquad \text{[add]}
\end{array}
$$

$$= 6(5^2) + 17(5) + [2(5) + 2]$$

$$= 6(5^2) + (17 + 2)(5) + 2$$

$$= 6(5^2) + 19(5) + 2$$

$$= 6(5^2) + [3(5) + 4](5) + 2$$

$$= 6(5^2) + 3(5^2) + 4(5) + 2$$

$$= 9(5^2) + 4(5) + 2$$

$$= [1(5) + 4](5^2) + 4(5) + 2$$

$$= 1(5^3) + 4(5^2) + 4(5) + 2$$

$$= (1442)_5$$

As in the case of multiplication in base ten, we may shorten this discussion as follows:

$$
\begin{array}{ccccccc}
① & ① & & & & & \\
② & ② & & & & & \\
(& & 2 & 3)_5 = & 1 & 3 \\
\times (& & 3 & 4)_5 = & 1 & 9 \\
\hline
& 2 & 0 & 2 & 1 & 1 & 7 \\
1 & 2 & 4 & & 1 & 3 \\
\hline
\text{product} = (1 & 4 & 4 & 2)_5 = 2 & 4 & 7
\end{array}
$$

Our computation can easily be checked by transferring the numerals to base ten. Since

$$(23)_5 = 2(5) + 3 = 13$$

and

$$(34)_5 = 3(5) + 4 = 19$$

we have

$$(23)_5 \cdot (34)_5 = 13 \cdot 19 = 247 = (1442)_5$$

Example 8.26 Multiply $(1011)_2$ by $(101)_2$.

Solution:

$$
\begin{array}{l}
(1\,0\,1\,1)_2 = 1\,1 \\
\times (1\,0\,1)_2 = \ \ 5 \\
\hline
1\,0\,1\,1 \\
0\,0\,0\,0 \\
1\,0\,1\,1 \\
\hline
\text{product} = (1\,1\,0\,1\,1\,1)_2 = 5\,5
\end{array}
$$

Thus, $(1011)_2 \cdot (101)_2 = (110111)_2$.

Example 8.27 Compute $(256)_{12} \cdot (29)_{12}$.

Solution:

$$
\begin{array}{ccccccccc}
& & ① & & & & & & \\
& ① & ④ & ④ & & & & & \\
(& 2 & 5 & 6)_{12} = & & & 3 & 5 & 4 \\
\times (& & 2 & 9)_{12} = & & & & 3 & 3 \\
\hline
& 1 & t & 1 & 6 & & 1 & 0 & 6 & 2 \\
& 4 & e & 0 & & 1 & 0 & 6 & 2 \\
\hline
\text{product} = (6 & 9 & 1 & 6)_{12} = 1 & 1 & 6 & 8 & 2
\end{array}
$$

Let's now discuss how we divide a number by another in an arbitrary base. From experience, we know that division is more difficult to carry out than multiplication; however, the procedure remains exactly the same as in base ten.

Example 8.28 Divide $(110111)_2$ by $(101)_2$.

Solution: Just for convenience, let's use the familiar shorthand technique to find the quotient and the remainder when $(110111)_2$ is divided by $(101)_2$.

$$
\begin{array}{r}
1\,0\,1\,1 \\
1\,0\,1\,\overline{)\,1\,1\,0\,1\,1\,1} \\
1\,0\,1 \\
\overline{1\,1} \\
0 \\
\overline{1\,1\,1} \\
1\,0\,1 \\
\overline{1\,0\,1} \\
1\,0\,1 \\
\overline{0}
\end{array}
$$

Thus, $\qquad (1\,1\,0\,1\,1\,1)_2 \div (1\,0\,1)_2 = (1\,0\,1\,1)_2$

We now ask the reader to verify that $(101)_2 \cdot (1011)_2 = (110111)_2$.

Example 8.29 Find the quotient and the remainder when $(1230)_5$ is divided by $(31)_5$.

Solution:

$$
\begin{array}{r}
2\,1 \\
3\,1\,\overline{)\,1\,2\,3\,0} \\
1\,1\,2 \\
\overline{1\,1\,0} \\
3\,1 \\
\overline{2\,4}
\end{array}
$$

Thus, the quotient is $(21)_5$ and the remainder is $(24)_5$. It may now be verified that $(1230)_5 = (31)_5 \cdot (21)_5 + (24)_5$.

Example 8.30 Find the quotient q and the remainder r of the division algorithm when $(1976)_{12}$ is divided by $(23)_{12}$.

Solution:

$$
\begin{array}{r}
9\,7 \\
2\,3\,\overline{)\,1\,9\,7\,6} \\
1\,8\,3 \\
\overline{1\,4\,6} \\
1\,3\,9 \\
\overline{9}
\end{array}
$$

Thus, $q = 97$ and $r = 9$; verify now that

$$
(1976)_{12} = (23)_{12} \cdot (97)_{12} + (9)_{12}
$$

Exercise 8.11

1. Evaluate each of the following products.

  a) $(2)_5 \cdot (3)_5$ **b)** $(2)_7 \cdot (3)_7$ **c)** $(2)_3 \cdot (2)_3$
  d) $(5)_7 \cdot (6)_7$ **e)** $(8)_{12} \cdot (5)_{12}$ **f)** $(t)_{12} \cdot (e)_{12}$

2. Construct a multiplication table for base two.

3. Construct a multiplication table for base six.

4. Construct a multiplication table for base seven.

5. Simplify each of the following products.

  a) $(110)_2 \cdot (101)_2$ **b)** $(123)_5 \cdot (123)_5$ **c)** $(102)_3 \cdot (211)_3$

  d) $(256)_7 \cdot (135)_7$ **e)** $(37)_{12} \cdot (73)_{12}$ **f)** $(5e)_{12} \cdot (1t0)_{12}$

6. Use the lattice method of multiplication (Section 8.7) to evaluate the following products.

  a) $(101)_2 \cdot (11)_2$ **b)** $(102)_3 \cdot (12)_3$ **c)** $(24)_5 \cdot (31)_5$

  d) $(103)_5 \cdot (203)_5$ **e)** $(13)_7 \cdot (235)_7$ **f)** $(29)_{12} \cdot (43)_{12}$

7. Find the quotient q and the remainder r when a is divided by b, if

  a) $a = (23)_5$, $b = (10)_5$ **b)** $a = (11011)_2$, $b = (110)_2$

  c) $a = (11011)_3$, $b = (110)_3$ **d)** $a = (123)_7$, $b = (43)_7$

  e) $a = (275)_{12}$, $b = (17)_{12}$ **f)** $a = (2000)_{12}$, $b = (31)_{12}$

8. Evaluate $x(x + 1)$ if

  a) $x = (110)_2$ **b)** $x = (102)_3$ **c)** $x = (243)_5$

9. Evaluate $x(x - 1)$ if

  a) $x = (101)_2$ **b)** $x = (121)_3$ **c)** $x = (325)_7$

8.12 SUMMARY AND COMMENTS

One of the primary objectives of this chapter was to make the reader aware of the existence of numeration systems other than the one we are all familiar with and use in everyday life. We discussed briefly the tally system, Egyptian system, Babylonian system, Roman system, and Mayan system. We pointed out the merits of the Hindu-Arabic system over the others: it has base ten; it is a place-value system; it contains a symbol for zero; every number has a unique representation. The Hindu-Arabic system is perhaps one of the greatest contributions of India to mankind.

We illustrated the fundamental algorithms for addition, subtraction, multiplication, and division. In order to show that an algorithm for an operation need not be unique, we illustrated the lattice and the ancient Egyptian methods of multiplication and the Egyptian method of division. The Egyptian methods are based on the fundamental property that every natural number can be uniquely written as a sum of powers of two.

Even though ten seems to be a natural choice for a base in the Hindu-Arabic system, it is certainly possible to develop a place-value system with any base b. In a base b system, the symbols used to represent a number are $0, 1, 2, \ldots,$ $(b-2)$, and $(b-1)$. The digits to the left of the reference point in this system have place values b^0, b^1, b^2, etc., successively, while those on its right have place values b^{-1}, b^{-2}, b^{-3}, etc., successively. The advent of computers has necessitated the need for nondecimal systems. More about the binary system will be discussed in Section 16.4.

The algorithms for addition, subtraction, multiplication, and division in any nondecimal system are based on the same principle as those in the Hindu-Arabic system.

Finally, we are already familiar with some nondecimal systems: the base seven system for counting days of a week, the base twenty-four system for counting hours of a day, the base sixty system for counting seconds and minutes in an hour, etc.

SUGGESTED READING

Bernstein, A. L., "Use of Manipulative Devices in Teaching Mathematics," *The Arithmetic Teacher*, vol. 10 (May 1963), pp. 280–283.

Ikeda, H., and M. Ando, "Introduction to the Numeration of Two-Place Numbers," *The Arithmetic Teacher*, vol. 16 (April 1969), pp. 249–251.

Miller, C. D., and V. E. Heeren, *Mathematical Ideas, an Introduction* (2nd ed.), Scott, Foresman and Co., Glenview, Ill. (1973), pp. 1–42.

Shurlow, H. J., "The Game of Five," *The Arithmetic Teacher*, vol. 10 (May 1963), pp. 290–291.

Wessel, G., "The Base Minus-Ten Numeration System," *School Science and Mathematics*, vol. 68 (Nov. 1968), pp. 701–706.

Wolfers, E. P., "The Original Counting Systems of Papua and New Guinea," *The Arithmetic Teacher*, vol. 18 (Feb. 1971), pp. 77–83.

"The imaginary numbers are a wonderful flight of God's spirit; they are almost an amphibian between being and not being."

G. LEIBNIZ

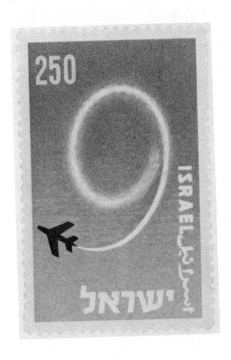

9/The System of Complex Numbers

After studying this chapter, you should be able to:

● *write a complex number in the standard form*
● *find the real and imaginary parts of a complex number*
● *find the conjugate of a complex number*
● *solve equations of the form $x^2 + a = 0$, where a is any real number*
● *represent a complex number on the complex plane*
● *evaluate the sum, the difference, the product, and the quotient of two complex numbers*
● *find the multiplicative inverse of a nonzero complex number*

9.0 INTRODUCTION

Toward the end of Section 7.11, we observed a deficiency of the system of real numbers: if a is a positive real number, then the equation $x^2 = -a$ does not have a solution in the set of real numbers. For example, there is no real number x such that $x^2 = -1$, since the square of no real number—positive, zero, or negative—can be negative. Consequently, in order to solve the equation $x^2 = -1$, we must introduce new numbers. From our experience with irrational numbers, we are almost inclined to write the solutions as $x = \pm\sqrt{-1}$, where $\sqrt{-1}$ is not a real number. We now introduce, for convenience, a new symbol, i (i for imaginary) to denote $\sqrt{-1}$, which was so designated by Euler in 1777. Accordingly, the solutions of the equation $x^2 = -1$ are $\pm i$. Therefore, $i^2 = -1 = (-i)^2$. With the advent of the new number i, we now can solve every equation of the form $x^2 = -a$. For example, the solutions of the equation

$x^2 = -4$ are given by

$$x = \pm\sqrt{-4} = \pm\sqrt{4 \cdot (-1)} = \pm\sqrt{4} \cdot \sqrt{-1} = \pm 2i$$

The solutions of the equation $x^2 = -5$ are

$$x = \pm\sqrt{-5} = \pm\sqrt{5 \cdot (-1)} = \pm\sqrt{5} \cdot \sqrt{-1} = \pm\sqrt{5}\, i$$

Similarly, if $x^2 = -2$, then $x = \pm\sqrt{-2} = \pm\sqrt{2 \cdot (-1)} = \pm\sqrt{2}\, i$; if $x^2 = -16$, then $x = \pm\sqrt{-16} = \pm\sqrt{16 \cdot (-1)} = \pm 4i$; if $x^2 = -\frac{4}{9}$, then $x = \pm\sqrt{-\frac{4}{9}} = \pm\sqrt{\frac{4}{9}} \cdot (-1) = \pm\frac{2}{3}i$; and if $x^2 = -32$, then $x = \pm\sqrt{-32} = \pm\sqrt{16 \cdot 2 \cdot (-1)} = \pm 4\sqrt{2}\, i$.

9.1   COMPLEX NUMBERS

In our above discussion, we came across numbers of the form bi, where b is a real number; for example, $2i$, $-2i$, $\sqrt{5}\, i$, $-\sqrt{3}\, i$, etc. Numbers of the form bi are called *pure imaginary numbers*. For example, $3i$, i, $-5i$, $\sqrt{2}\, i$, πi are all *pure imaginary numbers*. The sum $a + bi$ of a real number a and a pure imaginary number bi is called a *complex number*. The term "complex" was coined by the brilliant German mathematician Gauss. This notation for complex numbers was also introduced by Gauss. Thus, $1 + 2i$, $-3 + \sqrt{5}\, i$, $0 - \sqrt{2}\, i$, $0 + 4i$ are all examples of complex numbers. Recall that in all our discussions, i stands for $\sqrt{-1}$. The set of all complex numbers is denoted by the letter C.

The real numbers a and b are respectively called the *real part* and the *imaginary part* of the complex number $a + bi$. For example, the real part of the complex number $2 + \sqrt{5}\, i$ is 2, and its imaginary part is $\sqrt{5}$; the real part of $3i = 0 + 3i$ is 0, and its imaginary part is 3.

Observe that $2 + \sqrt{-3}\, i$ is a complex number. But it is not in the *standard form* $a + bi$ where a and b are real numbers, since $\sqrt{-3}$ is not a real number. However, we can rewrite $2 + \sqrt{-3}\, i$:

$$2 + \sqrt{-3}\, i = 2 + \sqrt{3 \cdot (-1)}\, i = 2 + \sqrt{3}\, i \cdot i$$

$$= 2 + \sqrt{3}\, i^2 = 2 + \sqrt{3} \cdot (-1) \quad \text{[recall that } i^2 = -1\text{]}$$

Thus, $\quad 2 + \sqrt{-3}\, i = 2 - \sqrt{3} = (2 - \sqrt{3}) + 0i$

Hence the real part of $2 + \sqrt{-3}\, i$ is $2 - \sqrt{3}$, and its imaginary part is 0.

Consider the complex number $a + bi$. If the real part is $a = 0$, then $a + bi = 0 + bi = bi$, a pure imaginary number unless $b = 0$. Similarly, if the imaginary part is zero, then the complex number $a + bi$ is the same as the real number a.

When do we say that the complex numbers $a + bi$ and $c + di$ are the same? We are intuitively tempted to say that two complex numbers are equal if and only if their real and imaginary parts are separately equal. Indeed, this is exactly how we define the equality of complex numbers.

Definition 9.1   Two complex numbers $a + bi$ and $c + di$ are *equal* if and only if $a = c$ and $b = d$.

For example, $2 + 3i = 3i + 2$ and $-1 + 4i = 4i - 1$. If $a + bi = 2 + 3i$, then $a = 2$ and $b = 3$; if $a + bi = 1 - 2i$, then $a = 1$ and $b = -2$ (why?).

Example 9.1   Find the values of a and b if $a + bi = \sqrt{-4} + 3i$.

Solution: Since $\sqrt{-4} + 3i$ is not in the standard form (why?), we first convert it to the standard form:

$$a + bi = \sqrt{-4} + 3i = \sqrt{4 \cdot (-1)} + 3i$$

$$= 2i + 3i = 5i$$

Thus, $\qquad\qquad a + bi = 0 + 5i$

Equating the real and the imaginary parts, we get $a = 0$ and $b = 5$.

Example 9.2   Find the values of a and b if $a - bi = \sqrt{-3}\, i + 2i$.

Solution: Let's first convert $\sqrt{-3}\, i + 2i$ to the standard form:

$$a - bi = \sqrt{-3}\, i + 2i = \sqrt{3}\, i \cdot i + 2i$$

$$= \sqrt{3}\, i^2 + 2i = -\sqrt{3} + 2i$$

Equating the real and the imaginary parts, we get $a = -\sqrt{3}$ and $-b = 2$. Thus, $a = -\sqrt{3}$ and $b = -2$.

From a complex number $z = a + bi$, we can easily obtain another complex number, $a - bi$, by changing i to $-i$. This complex number, $a - bi$, denoted by $\bar{z}$ (z bar), is called the *complex conjugate* of z. Thus, if $z = a + bi$, then $\bar{z} = a - bi$. For example, if $z = 1 + \sqrt{3}\, i$, then $\bar{z} = 1 - \sqrt{3}\, i$; if $z = -2 - \sqrt{5}\, i$, then $\bar{z} = -2 + \sqrt{5}\, i$, etc. The concept of complex conjugates is very useful in simplifications, as will be seen in Section 9.3. If the complex number $z = a + bi$ is real, then $b = 0$ and $\bar{z} = a - bi = a - 0i = a$. Thus, the complex conjugate of a real number is itself. On the other hand, if $z = \bar{z}$, then z must be a real number (why?).

Finally, observe that a complex number $a + bi$ need not be the same as $b + ai$; for example, $2 + \sqrt{3}\, i \neq \sqrt{3} + 2i$. With the result, a complex number $a + bi$ can very well be considered an ordered pair (a, b) of the real numbers a and b. That is, $a + bi = (a, b)$ and $a - bi = (a, -b)$.

Exercise 9.1

1. Rewrite the following complex numbers in the standard form.

a) $5 + \sqrt{-9}$ b) $\sqrt{-4} + \sqrt{4}$ c) $2 + \sqrt{-9}\, i$

d) $2i + \sqrt{-4}\, i$ e) $1 + \sqrt{-4}\, i$ f) $\sqrt{3}\, i + \sqrt{-3}\, i$

2. Find the complex conjugate of the following complex numbers.

a) $1+3i$ b) $2-5i$ c) $1-i$

d) $\sqrt{3}\,i-5$ e) $\sqrt{2}$ f) $\overrightarrow{\sqrt{-3}}+\sqrt{3}\,i$

3. Solve the following equations.

a) $x^2=-9$ b) $x^2+25=0$ c) $2x^2+18=0$

d) $x^2+3=0$ e) $4x^2+9=0$ f) $3x^2+21=0$

4. Find the values of a and b if

a) $a+bi=2+3i$ b) $a+bi=1-2i$

c) $a-bi=1+i$ d) $a+bi=-1+\sqrt{4}\,i$

e) $a+bi=1+\sqrt{-1}\,i$ f) $a-bi=\sqrt{9}+\sqrt{-9}$

5. Mark *true* or *false*.

a) Every real number is also a complex number.

b) i is an irrational number.

c) i is an imaginary number.

d) π is a complex number.

e) If $z=2$, then $\bar{z}=-2$.

f) The complex conjugate of $1+i$ is $1-i$.

g) If $z=a-bi$, then $\bar{z}=-a-bi$.

h) If $z=\bar{z}$, then z is a pure imaginary number.

i) The equation $x^2+1=0$ is solvable in the set of complex numbers.

6. Give a counterexample, whenever possible, to each of the false statements in problem 5.

9.2 GEOMETRIC REPRESENTATION OF COMPLEX NUMBERS

In Section 7.6, we observed that every real number can be represented by a point on a straight line and vice versa. In other words, there is a 1-1 correspondence between the set of real numbers and the set of points on a straight line. We now proceed to show that every complex number can be represented by a point on a plane and every point on the plane corresponds to a complex number. Gauss was one of the first few mathematicians to represent complex numbers by points on a plane.

Consider a horizontal straight line and a vertical straight line on a plane, intersecting at a point O. The point O is called the *origin*. These two lines divide the whole plane into four equal regions, I, II, III, and IV, each being called a *quadrant*. Let's now represent real numbers by points on the horizontal and the vertical lines, with positive numbers along the directions OX and OY (see Fig. 9.1). The horizontal line is called the *real axis*. We now relabel each point with coordinate b on the vertical line by bi, as shown in Fig. 9.1. Observe that every point on the vertical line corresponds to exactly one pure imaginary number and vice versa. The vertical line is now called the *imaginary axis*, for obvious reasons.

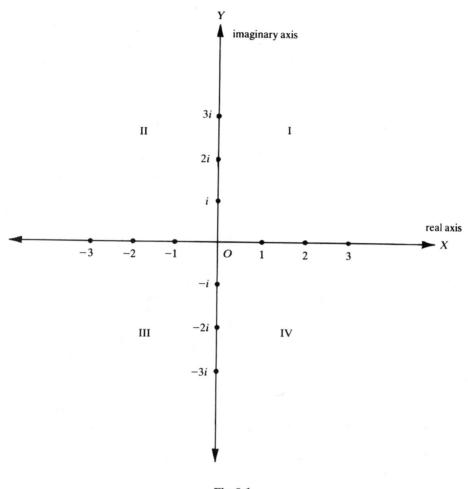

Fig. 9.1

How do we now represent a complex number $z = a + bi$ on the plane? Since the real part of z is a, locate the point P with coordinate a on the real axis. Now mark the point Q with coordinate bi on the imaginary axis. Draw straight lines through P and Q parallel to the imaginary axis and the real axis, respectively, to meet at the point R. This point R represents the complex number $a + bi$, as shown in Fig. 9.2. We have represented the numbers $2 + 3i, 2 - 3i, -2 + 3i$, and $-2 - 3i$ by points in Fig. 9.2.

Conversely, to find the complex number represented by any point R on the plane, draw straight lines through R parallel to the imaginary and the real axes and meeting the axes at P and Q, respectively (see Fig. 9.2). Let a and bi be the coordinates of the points P and Q, respectively. Then the complex number associated with the point R is $a + bi$.

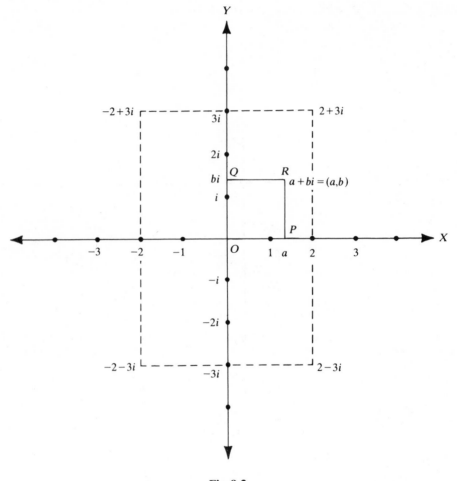

Fig. 9.2

Thus, every complex number corresponds to a point on a plane and every point on the plane corresponds to a complex number. This establishes a 1-1 correspondence between the set of complex numbers and the sets of points on a plane. This plane is usually referred to as the *complex plane*.

9.3   OPERATIONS ON THE SET OF COMPLEX NUMBERS

We shall not attempt to give a formal discussion of the properties of various operations—addition, subtraction, multiplication, and division by nonzero complex numbers—on the set of complex numbers. The fundamental properties are listed in Section 9.4. However, we shall discuss how to evaluate the sum, the difference, the product, and the quotient of two complex numbers. We define these operations as we naturally expect them to be defined.

Definition 9.2   The *sum* $x+y$ of the complex numbers $x=a+bi$ and $y=c+di$ is defined by

$$x+y=(a+bi)+(c+di)=(a+c)+(b+d)i$$

the *difference* $x-y$ is given by

$$x-y=(a+bi)-(c+di)=(a-c)+(b-d)i$$

It is clear from this definition that in order to add two complex numbers, we need only add the real and the imaginary parts separately.

Example 9.3   Find the sum $x+y$ and the difference $x-y$ of the complex numbers $x=2+3i$ and $y=-5+7i$.

Solution:

$$x=2+3i$$

Adding,

$$y=-5+7i$$
$$x+y=(2-5)+(3+7)i$$
$$=-3+10i$$

$$x=2+3i$$
$$y=-5+7i$$

Subtracting,

$$x-y=(2+5)+(3-7)i$$
$$=7-4i$$

Thus, $x+y=-3+10i$ and $x-y=7-4i$.

Example 9.4   Evaluate $x+y$ and $x-y$ if $x=2\sqrt{3}+\sqrt{5}\,i$ and $y=\sqrt{3}-2\sqrt{5}\,i$.

Solution:

$$x=2\sqrt{3}+\sqrt{5}\,i$$

$$y=\sqrt{3}-2\sqrt{5}\,i$$

Adding,

$$x+y=(2\sqrt{3}+\sqrt{3})+(\sqrt{5}-2\sqrt{5})i$$
$$=3\sqrt{3}-\sqrt{5}\,i$$

$$x=2\sqrt{3}+\sqrt{5}\,i$$

$$y=\sqrt{3}-2\sqrt{5}\,i$$

Subtracting,

$$x-y=(2\sqrt{3}-\sqrt{3})+(\sqrt{5}+2\sqrt{5})i$$
$$=\sqrt{3}+3\sqrt{5}\,i$$

Thus, $x+y=3\sqrt{3}-\sqrt{5}\,i$ and $x-y=\sqrt{3}+3\sqrt{5}\,i$.

Consider the complex numbers $x = a + bi$ and $y = c + di$. How should $x \cdot y$ be defined? We have

$$x \cdot y = (a + bi)(c + di)$$
$$= a(c + di) + bi(c + di)$$
$$= ac + adi + bci + bdi^2$$
$$= (ac - bd) + (ad + bc)i \qquad \text{(recall that } i^2 = -1\text{)}$$

Thus, $(a + bi)(c + di) = (ac - bd) + (ad + bc)i$.

Definition 9.3   The *product* of the complex numbers $x = a + bi$ and $y = c + di$ is defined by

$$x \cdot y = (a + bi)(c + di) = (ac - bd) + (ad + bc)i$$

Before we proceed any further, we advise the reader not to try to memorize this formula, since it is easier to evaluate a product by simply following the above steps, as in the case of simplifying the product $(a + b)(c + d)$.

Example 9.5   Evaluate the product $(2 + 3i)(3 - 5i)$.

Solution: $(2 + 3i)(3 - 5i) = 2(3 - 5i) + 3i(3 + 5i)$

$$= 2 \cdot 3 - 2 \cdot 5i + 3i \cdot 3 + 3i \cdot 5i$$
$$= 6 - 10i + 9i - 15 \qquad \text{(recall that } i^2 = -1\text{)}$$
$$= (6 - 15) + (-10 + 9)i = -9 - i$$

Thus, $(2 + 3i)(3 - 5i) = -9 - i$.

Example 9.6   Simplify the product $(-1 + \sqrt{3}\, i)(2 - \sqrt{5}\, i)$.

Solution:

$$(-1 + \sqrt{3}\, i)(2 - \sqrt{5}\, i) = (-1) \cdot 2 + (-1) \cdot (-\sqrt{5}\, i) + \sqrt{3}\, i \cdot 2 + (\sqrt{3}\, i) \cdot (-\sqrt{5}\, i)$$
$$= -2 + \sqrt{5}\, i + 2\sqrt{3}\, i - \sqrt{3} \cdot \sqrt{5}\, i^2$$
$$= -2 + \sqrt{5}i + 2\sqrt{3}i + \sqrt{15}$$
$$= (\sqrt{15} - 2) + (\sqrt{5} + 2\sqrt{3})i$$

Observe that if $z = 1 + 2i$, then

$$z\bar{z} = (1 + 2i)(1 - 2i)$$
$$= 5 = 1 + 4$$
$$= 1^2 + 2^2$$

More generally, if $z = a + bi$, then $z\bar{z} = (a + bi)(a - bi) = a^2 + b^2$. (Why?) This fact is very useful in the simplification of quotients of complex numbers, as shown in the following examples.

Example 9.7   Simplify the quotient $\dfrac{4 - 3i}{5 + 2i}$.

Solution: Our aim is to rewrite $\dfrac{4-3i}{5+2i}$ in the standard form $a+bi$. Since the denominator of the quotient is $5+2i$, we multiply both the numerator and the denominator by the conjugate $5-2i$.

$$\frac{4-3i}{5+2i}=\frac{(4-3i)}{(5+2i)}\cdot\frac{(5-2i)}{(5-2i)}=\frac{(4-3i)(5-2i)}{(5+2i)(5-2i)}$$

$$=\frac{20-8i-15i+6i^{2}}{5^{2}+2^{2}}=\frac{20-23i-6}{25+4}$$

$$=\frac{14-23i}{29}$$

Thus,

$$\frac{4-3i}{5+2i}=\frac{14}{29}-\frac{23}{29}i$$

Example 9.8   Simplify $\dfrac{-1+\sqrt{2}\,i}{3-\sqrt{2}\,i}$.

Solution: Multiply the numerator and the denominator by the conjugate, $3+\sqrt{2}\,i$, of the denominator, $3-\sqrt{2}\,i$.

$$\frac{-1+\sqrt{2}\,i}{3-\sqrt{2}\,i}=\frac{-1+\sqrt{2}\,i}{3-\sqrt{2}\,i}\cdot\frac{(3+\sqrt{2}\,i)}{(3+\sqrt{2}\,i)}=\frac{(-1+\sqrt{2}\,i)(3+\sqrt{2}\,i)}{(3-\sqrt{2}\,i)(3+\sqrt{2}\,i)}$$

$$=\frac{-3-\sqrt{2}\,i+3\sqrt{2}\,i-2}{3^{2}+(\sqrt{2})^{2}}=\frac{-5+2\sqrt{2}\,i}{9+2}$$

$$=-\frac{5}{11}+\frac{2\sqrt{2}\,i}{11}$$

In the system of real numbers, we observed that every nonzero real number x has a multiplicative inverse $1/x$. Likewise, every nonzero complex number z has a multiplicative inverse $1/z$. For example, the multiplicative inverse of $z=2+3i$ is given by

$$\frac{1}{z}=\frac{1}{2+3i}$$

which now can be put in the standard form as before:

$$\frac{1}{z}=\frac{1}{2+3i}=\frac{2-3i}{(2+3i)(2-3i)}$$

$$=\frac{2-3i}{4+9}=\frac{2-3i}{13}$$

$$=\frac{2}{13}-\frac{3}{13}i$$

Thus, the multiplicative inverse of $2+3i$ is $\frac{2}{13}-\frac{3}{13}i$. We now ask the reader to verify that $(2+3i)(\frac{2}{13}-\frac{3}{13}i)=1$.

More generally, if $z=a+bi$ is a nonzero complex number, can you guess what $1/z$ would be?

Exercise 9.3

1. Simplify each of the following as required.
   a) $(1-2i)+(3+5i)$ b) $(2-3i)+(4-5i)$
   c) $(-5+2\sqrt{3}\,i)+(8-\sqrt{3}\,i)$ d) $(5+8i)+(5-8i)$
   e) $(1+i)+(1-i)$ f) $(\sqrt{2}+3\sqrt{2}\,i)+(3\sqrt{2}-\sqrt{2}\,i)$
   g) $(4+5i)-(3+2i)$ h) $(1+i)-(1-i)$
   i) $(2+5i)-(2-5i)$ j) $(2+3\sqrt{5}\,i)-(-1+2\sqrt{5}\,i)$

2. Evaluate each of the products.
   a) $(2+5i)(3-4i)$ b) $(1+i)(1-i)$
   c) $(3+i)(3-i)$ d) $(1+3i)(1-3i)$
   e) $(1+i)^2$ f) $(1+3i)^2$
   g) $(2+\sqrt{3}\,i)(1-\sqrt{3}\,i)$ h) $(-1+\sqrt{2}\,i)(3-\sqrt{2}\,i)$
   i) $(1-i)^3$

3. Simplify each of the following.

   a) $\dfrac{3-4i}{5+2i}$ b) $\dfrac{1+2i}{1-2i}$ c) $\dfrac{\sqrt{2}\,i}{1+\sqrt{2}\,i}$

   d) $\dfrac{1+i}{1-i}$ e) $\dfrac{1-i}{1+i}$ f) $\dfrac{1-2i}{3-4i}$

   g) i^3 h) i^4 i) i^{101}

                                            [*Hint:* $101=4\cdot 25+1$]

4. Find the multiplicative inverse of each of the following complex numbers.
   a) $1+2i$ b) $2-3i$ c) $1+i$

   d) $i-1$ e) $2+\sqrt{3}\,i$ f) $\dfrac{1-3i}{2+5i}$

5. If $z=a+bi$, show that $\bar{\bar{z}}=z$.
6. If $z=a+bi$, show that $z\bar{z}=a^2+b^2$.
7. Solve the following equations.
   a) $x^3-x=0$ b) $x^3+x=0$
   c) $x^4-1=0$ d) $x^3+x=x^2+1$
   e) $x(x^2+1)=i(x^2+1)$ f) $x^3+x^2+x+1=0$
8. If $z=\bar{z}$, prove that z must be a real number.
9. If $z+\bar{z}=0$ and $z\neq 0$, prove that z must be a pure imaginary number.
10. Mark *true* or *false* (z is any complex number).
    a) If $z=1+2i$, then $\bar{z}=1-2i$.

b) $\bar{z} = -z$

c) If $\bar{z} = z$, then z is an imaginary number.

d) $z\bar{z}$ is always a real number.

e) $z + \bar{z}$ is an imaginary number.

f) $z - \bar{z}$ is a real number.

11. Give a counterexample to each of the false statements in problem 12.

12. a) Is the set $\{i, -i\}$ closed under addition? Under multiplication? Why?

b) Same as (*a*) with the set $\{\pm 1, \pm i\}$.

13. Evaluate $\sqrt{b^2 - 4ac}$ in each of the following cases.

a) $a = 1, b = 5, c = 6$ **b)** $a = 1, b = -2, c = 1$

c) $a = 1, b = 1, c = 1$ **d)** $a = 1, b = 0, c = 4$

14. The solutions of the *quadratic equation* $ax^2 + bx + c = 0$ are given by the *quadratic formula,* $x = \dfrac{-b \pm \sqrt{b^2 - 4ac}}{2a}$. Use this formula to solve the following equations.

a) $x^2 + 5x + 6 = 0$ **b)** $x^2 - 2x + 1 = 0$

c) $x^2 + x + 1 = 0$ **d)** $x^2 + 4 = 0$

e) $x^2 - 2x + 2 = 0$ **f)** $6x^2 - x - 1 = 0$

★15. The *absolute value* $|z|$ of a complex number $z = a + bi$ is given by $|z| = \sqrt{a^2 + b^2}$. Find the absolute value of the following complex numbers.

a) $2 + 3i$ **b)** $1 - 3i$ **c)** $7 + \sqrt{2}\, i$

d) i **e)** $1 + i$ **f)** $\sqrt{3} + i$

★16. Prove that $|z| = |\bar{z}|$ for every complex number z.

★17. The *distance* between two complex numbers $x = a + bi$ and $y = c + di$ is given by the formula $d(x, y) = \sqrt{(a - c)^2 + (b - d)^2}$. Find the distance between the following pairs of complex numbers.

a) $1 + 2i$ and $2 + 5i$ **b)** $-2 + 3i$ and $1 - 4i$

c) $1 + i$ and $1 - i$ **d)** i and $-i$

e) 2 and 5 **f)** $1 - 2i$ and $-3 + 2\sqrt{2}\, i$

★18. If $x = a + bi$ and $y = c + di$, prove that $d(x, y) = |x - y|$.

★19. If $x = a + bi$ and $y = c + di$, prove that $d(x, y) = d(x - y, 0)$.

★20. Using the ordered pair notation for complex numbers, define addition by $(a, b) + (c, d) = (a + c, b + d)$ and multiplication by $(a, b) \cdot (c, d) = (ac - bd, ad + bc)$. Evaluate each of the following:

a) $(2, 3) + (1, -5)$ **b)** $(-2, \sqrt{2}) + (3, -3\sqrt{2})$

c) $(1, -\sqrt{2}) \cdot (-2, \sqrt{2})$ **d)** $(1, -1) \cdot (-1, 1)$

9.4 SUMMARY AND COMMENTS

Observing that the real number system is not "large" enough to solve all equations of the form $x^2 + a = 0$, we extended it to a still "larger system," namely, the complex number system. We achieved this by introducing a new symbol, $i = \sqrt{-1}$, where $i^2 = -1$. As a result, every equation of the form $x^2 + a = 0$, where a is any real number, can be solved.

A complex number is of the form $a + bi$, where both a and b are real numbers. It consists of two parts: the real part, a, and the imaginary part, b. Complex numbers whose imaginary parts are zero are clearly real numbers. If the real part of a nonzero complex number is zero, then the complex number is said to be pure imaginary. A complex number $a + bi$ can also be considered as an ordered pair (a,b).

The complex conjugate $\bar{z} = a - bi$ of a complex number $z = a + bi$ is obtained by changing i to $-i$ in z. The conjugate of a complex number equals itself if and only if it is a real number.

A one-to-one correspondence between the set of complex numbers and the set of points on a plane was established.

We now enumerate the fundamental properties of the complex number system:

a) The set of complex numbers is closed under addition.

b) Addition on the set of complex numbers is commutative.

c) Addition on the set of complex numbers is associative.

d) The set of complex numbers contains a unique additive identity, 0.

e) Every complex number z has a unique additive inverse, $-z$.

f) The set of complex numbers is closed under multiplication.

g) Multiplication on the set of complex numbers is commutative.

h) Multiplication on the set of complex numbers is associative.

i) The set of complex numbers possesses a unique multiplicative identity, 1.

j) Every nonzero complex number has a unique multiplicative inverse, $1/z$.

k) Multiplication is distributive over addition.

In our development of number systems, we observed that each number system, except the whole number system, is richer than the previous system. The system of complex numbers is not only richer than that of real numbers, it turns out that it is the richest possible number system. This remarkable property is a consequence of one of the most outstanding principles in mathematics, namely, the *fundamental theorem of algebra*, proved by Gauss at the age of 20, in his doctoral dissertation. It states that every equation of the form $a_0 x^n + a_1 x^{n-1} + \cdots + a_{n-1} x + a_n = 0$ with complex coefficients a_i has at least one complex solution.

There is an area of mathematics that deals with complex numbers, called complex analysis. Complex numbers play a significant role in higher mathematics, the theory of electrical circuits, aerodynamics, and hydrodynamics.

Since our present number system is an extension of the real number system, the line diagram for the number systems looks as in Fig. 9.3. Since the complex number system is the richest number system, we conclude that our "tree" in Fig. 9.3 has attained its full growth.

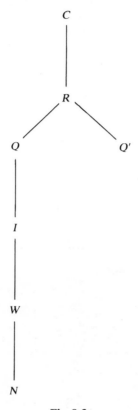

Fig. 9.3

Here ends the discussion of the development of number systems.

SUGGESTED READING

Brady, W. G., "Complex Roots of a Quadratic Equation Graphically," *The Mathematics Teacher*, vol. 63 (March 1970), p. 229.

Graham, M., *Mathematics, Liberal Arts Approach*, Harcourt, Brace, Jovanovich, Inc., New York (1973), pp. 132–136.

Peterson, J. M., *Basic Concepts of Elementary Mathematics*, Prindle, Weber, and Schmidt, Inc., Boston, Mass. (1971), pp. 313–320.

"What science can there be more noble, more excellent, more useful for men, more admirably high and demonstrative than this of the Mathematics."

BENJAMIN FRANKLIN

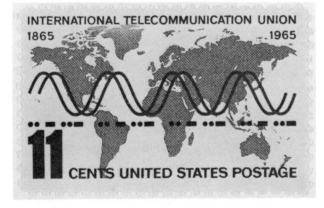

10/Relations and Functions

After studying this chapter, you should be able to:
- *find the domain and the range of a relation*
- *graph a relation*
- *find the inverse of a relation*
- *identify the properties of a relation*
- *decide if a relation is a function*
- *verify if a function is a 1-1 correspondence*
- *decide if a function is a permutation*
- *graph the region defined by linear inequalities*
- *find the coordinates of the corner points of such a region*
- *solve simple linear programming problems*

10.0 INTRODUCTION

Expressions like "teacher-student relation," "husband-wife relation," "doctor-patient relation," "attorney-client relation," "parent-child relation," are very common in our everyday life. What does *relation* mean in the mathematical world?

The concept of relation has already been intuitively introduced when we discussed the equality relation, the order relation, relations between sets—is a subset of, is equal to, is equivalent to, etc.—logical equivalence of statements in logic, etc.

The goal of this chapter is to discuss relations and their various properties with more care and in more depth. We ask the reader to review Section 1.12 before he proceeds any further.

10.1 RELATIONS

Let A and B be any two sets. Recall from Section 1.12 that the cartesian product $A \times B$ is the totality of all ordered pairs (a,b) with a in A and b in B.

Consider the sets

$$A = \{\text{Tom,Dick,Harry}\} \quad \text{and} \quad B = \{\text{Kathy,Nancy,Mary,Cindy}\}$$

Let's form a collection of ordered pairs whose first elements come from A and second elements come from B as follows. Assume that each member of A is married to some member of B. To be specific, let's assume that Tom is married to Nancy, Dick is married to Kathy, and Harry is married to Cindy. Let R denote the set of ordered pairs (Tom,Nancy), (Dick,Kathy) and (Harry,Cindy); that is,

$$R = \{(\text{Tom,Nancy}),(\text{Dick,Kathy}),(\text{Harry,Cindy})\}$$

where in each ordered pair, the first element, which belongs to A, is related to the corresponding second element, which belongs to B, via the relationship of marriage. Notice that R can also be written

$$R = \{(a,b) \in A \times B \mid a \text{ is married to } b\}$$

and that it is the relation given by the phrase "is married to" that helps us to obtain the set R. Notice that R is a subset of $A \times B$. Now we are in a position to make the following definition:

Definition 10.1   A *relation* R from a set A to a set B is a subset of $A \times B$, that is, $R \subseteq A \times B$.

That $(x,y) \in R$ simply means that x has a certain relation R to y. For example, x may be a brother of y, or x may be the father of y, or x may be less than y, etc. If $(x,y) \in R$, we also write xRy; and if $(x,y) \notin R$, we write $x\bar{R}y$. In our above example, (Dick,Kathy) $\in R$, that is, Dick R Kathy, whereas (Dick,Nancy) $\notin R$, since Dick is not married to Nancy.

The relation R in our above example can pictorially be represented as in Fig. 10.1, where an arrow from x to y means that x is related to y by the relation R.

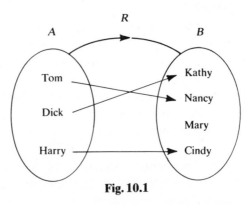

Fig. 10.1

Example 10.1 Consider the sets $C = \{a,b,c,d\}$ and $D = \{x,y,z\}$. Then

$$R_1 = \{(a,x),(b,y),(c,z)\}$$
$$R_2 = \{(b,z),(c,y),(d,x)\}$$
$$R_3 = \{(a,y),(b,y),(c,y),(d,y)\}$$
$$R_4 = \{(a,x),(a,y),(a,z)\}$$

are all relations from C to D, since they are all subsets of $C \times D$. These relations are represented geometrically in Fig. 10.2.

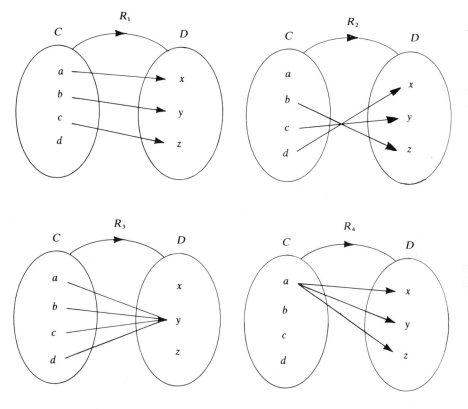

Fig. 10.2

Example 10.2 Let $A = \{Bill,Mary\}$ be the set of parents in a certain family and $B = \{Fred,Carol,Peter,Paul\}$ the set of their children. Then

$$S_1 = \{(Bill,Fred),(Bill,Carol),(Bill,Peter),(Bill,Paul)\}$$
$$S_2 = \{(Mary,Fred),(Mary,Carol),(Mary,Peter),(Mary,Paul)\}$$
$$S_3 = \{(Bill,Fred),(Bill,Paul),(Mary,Carol)\}$$

are all relations from A to B, whereas

$$T_1 = \{(\text{Carol},\text{Bill}),(\text{Carol},\text{Mary})\}$$
$$T_2 = \{(\text{Fred},\text{Bill}),(\text{Peter},\text{Bill}),(\text{Paul},\text{Bill})\}$$
$$T_3 = \{(\text{Carol},\text{Bill})\}$$
$$T_4 = \{(\text{Carol},\text{Bill}),(\text{Fred},\text{Mary})\}$$

are all relations from B to A.

Let R be a relation from A to B. By taking the collection of all first elements in the various ordered pairs in R, we obtain a new set, called the *domain* of the relation R, denoted by dom(R). Likewise, the set of all second elements in the various ordered pairs in R constitutes the *range* of the relation R, denoted by range(R). Observe that dom(R) $\subseteq A$ and range(R) $\subseteq B$.

Example 10.3   Let $A = \{0,1,2,3\}$ and $B = \{2,3,4\}$. Then

$$r_1 = \{(1,2),(2,2),(3,4)\}$$
$$r_2 = \{(0,3),(2,4),(0,2)\}$$
$$r_3 = \{(0,4),(1,3),(2,3),(3,4)\}$$
$$r_4 = \{(1,3),(2,3),(3,3)\}$$

are all relations from A to B, with

dom(r_1) = {1,2,3},	range(r_1) = {2,4}
dom(r_2) = {0,2},	range(r_2) = {2,3,4}
dom(r_3) = {0,1,2,3},	range(r_3) = {3,4}
dom(r_4) = {1,2,3},	range(r_4) = {3}

The graphs of the product set $A \times B$ and the relations r_1, r_2, r_3, and r_4 are given in Fig. 10.3, where the dots correspond to members of $A \times B$ and the thick dots correspond to members of the relations.

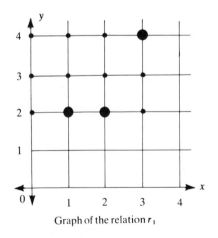

Graph of the relation r_1

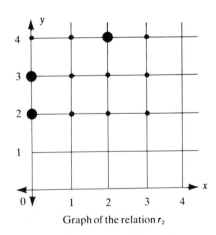

Graph of the relation r_2

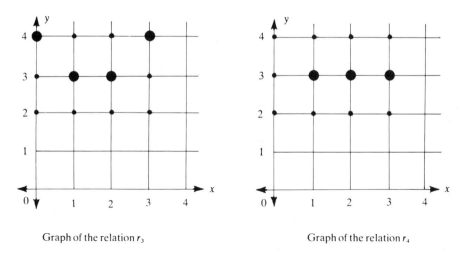

Graph of the relation r_3 Graph of the relation r_4

Fig. 10.3

When we defined a relation from A to B, we did not assume that A and B are distinct; A and B could very well be the same set. Accordingly, we can have relations from a set A to itself. Such a relation is called a *relation on A*. Thus, R is a relation on A if and only if $R \subseteq A \times A$.

Example 10.4   If $A = \{x,y,z\}$, then

$$t_1 = \{(x,x),(y,y),(z,z)\}$$
$$t_2 = \{(x,y),(y,x)\}$$
$$t_3 = \{(x,y),(y,z),(z,x)\}$$

are examples of relations on A.

Phrases like "is a brother of," "is a sister of," "is the father of," "has the same color hair as," "is taller than," "has the same profession as," "is the same as," "is a divisor of" can be used to define relations.

Example 10.5   Let $A = \{\text{Bill,Frank,Linda,Sue}\}$ be the set of all children in a certain family. Let t be the relation defined on A by $t = \{(x,y) \in A \times A \mid x$ is a brother of $y\}$. Notice that

$$t = \{(\text{Bill,Frank}),(\text{Bill,Linda}),(\text{Bill,Sue}),(\text{Frank,Bill}),$$
$$(\text{Frank,Linda}),(\text{Frank,Sue})\}$$

Notice that $(\text{Linda,Frank}) \notin t$ since Linda is not a brother of Frank. This relation may be represented pictorially by a tree diagram, as in Fig. 10.4.

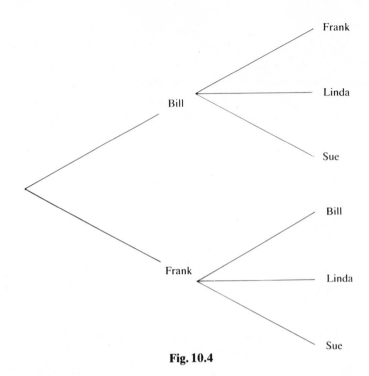

Fig. 10.4

Example 10.6 Graph the relation R defined on the set $B = \{1,2,3,4\}$ by
$$R = \{(x,y) \in B \times B \mid x + y < 5\}$$

Observe that $R = \{(1,1),(1,2),(1,3),(2,1),(2,2),(3,1)\}$. The graph of this relation is displayed by thick dots in Fig. 10.5.

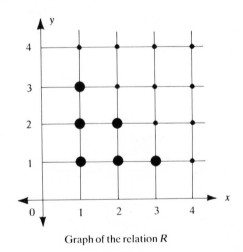

Graph of the relation R

Fig. 10.5

Exercise 10.1

1. Which of the following are relations from $A = \{a,b,c\}$ to $B = \{a,x,y,z\}$?
   a) $\{(a,x),(a,y),(b,z)\}$ b) $\varnothing$
   c) $\{(a,b),(a,c)\}$ d) $\{(a,x),(x,b),(b,y)\}$
   e) $\{(a,x),(x,a)\}$ f) $\{(a,b),(a,a)\}$

2. How many relations exist from $A = \{0\}$ to $B = \{1,2,3\}$? Find them.

3. Find the domain and range of the following relations from $A = \{a,b,c,d\}$
   to $B = \{1,2,3,4,5\}$.
   a) $R_1 = \{(a,4),(b,2),(a,1),(c,1)\}$ b) $R_2 = \{(a,2),(b,2),(c,2)\}$
   c) $R_3 = \{(a,5),(d,1),(c,2),(b,3)\}$ d) $R_4 = \{(b,1),(b,2),(b,3)\}$
   e) $R_5 = \varnothing$ f) $R_6 = A \times B$

4. Rewrite the following relations from $A = \{1,2,3,4\}$ to $B = \{2,4,5\}$ in the
   roster method. Graph each of them.
   a) $S_1 = \{(x,y) \in A \times B \mid x < y\}$ b) $S_2 = \{(x,y) \in A \times B \mid x = y\}$
   c) $S_3 = \{(x,y) \in A \times B \mid x \leq y\}$ d) $S_4 = \{(x,y) \in A \times B \mid y = x + 1\}$
   e) $S_5 = \{(x,y) \in A \times B \mid x \text{ divides } y\}$ f) $S_6 = \{(x,y) \in A \times B \mid y = x^2\}$

5. Mark *true* or *false* (A and B are any two sets and R is an arbitrary
   relation from A to B).
   a) $A \times B$ is a relation from A to B.
   b) $\varnothing$ is a relation from A to B.
   c) $\text{Dom}(R) = A$.
   d) $\text{Range}(R) = B$.
   e) The cardinality of the set of relations from A to B and that of the set
      of those from B to A are the same.

6. Give a counterexample to each of the false statements in problem 5,
   whenever possible.

7. Let R be the relation defined on

$$A = \{3,4,6\} \text{ by } R = \{(x,y) \in A \times A \mid x - y \text{ is odd}\}$$

   Which of the following are true?

   a) $(3,4) \in R$ b) $(6,4) \in R$
   c) $4R3$ d) $4\cancel{R}4$

8. Use the roster method to display the relation

$$R = \{(x,y) \in A \times A \mid x - y \text{ is even}\}$$

   defined on the set $A = \{2,3,5\}$. Graph the relation.

9. Rewrite the relation $t = \{(a,b) \in A \times B \mid a \text{ is the capital of } b\}$ in the roster
   method, where $A = $ set of capitals of New England states and $B = $ set of
   New England states.

★**10.** Graph the relation g defined on the set R of real numbers by
  a) $g = \{(x,y) \in R \times R \mid y = x + 1\}$
  b) $g = \{(x,y) \in R \times R \mid 2y + x = 4\}$

★**11.** Graph the relation u defined on the set R of real numbers by
  a) $u = \{(x,y) \in R \times R \mid x^2 + y^2 = 9\}$
  b) $u = \{(x,y) \in R \times R \mid x \geq 0 \text{ and } x^2 + y^2 = 9\}$
  c) $u = \{(x,y) \in R \times R \mid x \geq 0, \ y \geq 0 \text{ and } x^2 + y^2 = 9\}$
  Find the domain and range of each of these relations.

★**12.** Find the domain and range of t by graphing the relation, where

$$t = \{(x,y) \in R \times R \mid 4x^2 + 9y^2 = 36\}.$$

10.2 INVERSE RELATIONS

Let R be a relation from A to B. By interchanging the first element with the second element in each ordered pair in R, we obtain a relation from B to A. This relation is called the *inverse* of the relation R, denoted by R^{-1}. Thus,

$$R^{-1} = \{(y,x) \mid (x,y) \in R\}$$

Example 10.7 Consider the sets $A = \{a,b,c\}$ and $B = \{1,2,3,4\}$. Then

$$R = \{(a,1),(b,2),(c,1)\}$$
$$S = \{(a,2),(b,2)\}$$
$$T = \{(a,1),(b,1),(a,3),(c,3)\}$$

are relations from A to B. Their inverse relations from B to A are, respectively,

$$R^{-1} = \{(1,a),(2,b),(1,c)\}$$
$$S^{-1} = \{(2,a),(2,b)\}$$
$$T^{-1} = \{(1,a),(1,b),(3,a),(3,c)\}$$

The graphs of the relations R and R^{-1} are displayed in Fig. 10.6.

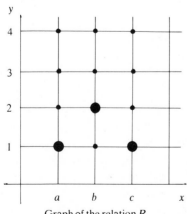
Graph of the relation R

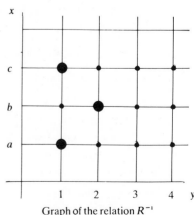
Graph of the relation R^{-1}

Fig. 10.6

The domain of a relation is the range of its inverse relation, and the range of the relation is the domain of its inverse relation.

Example 10.8   Let r be the relation from $A = \{1,2,3\}$ to $B = \{2,3,4\}$, defined by $r = \{(x,y) \in A \times B \mid x$ is a divisor of $y\}$. In the roster method,

$$r = \{(1,2),(1,3),(1,4),(2,2),(2,4)\}$$

and $$r^{-1} = \{(2,1),(3,1),(4,1),(2,2),(4,2)\}$$

Notice that $r^{-1} = \{(y,x) \in B \times A \mid y$ is a multiple of $x\}$. Graphs of these relations are given in Fig. 10.7.

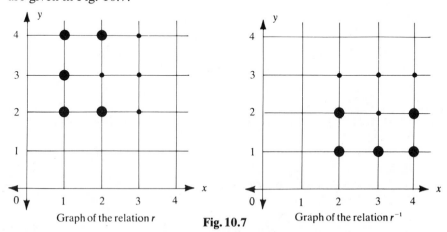

Graph of the relation r **Fig. 10.7**   Graph of the relation r^{-1}

Example 10.9   Let t be the relation defined on the set R of real numbers by $t = \{(x,y) \in R \times R \mid y \geq 0$ and $x^2 + y^2 = 4\}$. Then $(\pm 2,0),(0,2),(\pm\sqrt{2},\sqrt{2})$ are some members belonging to t, while $(0,\pm 2),(2,0),(\sqrt{2},\pm\sqrt{2})$ are some elements in t^{-1}. Notice that t^{-1} can also be given by $t^{-1} = \{(x,y) \in R \times R \mid x \geq 0$ and $x^2 + y^2 = 4\}$. The graphs of these relations are given in Fig. 10.8. Notice that

$$\text{dom}(t) = [-2,2] = \text{range}\,(t^{-1}) \text{ and range}(t) = [0,2] = \text{dom}(t^{-1})$$

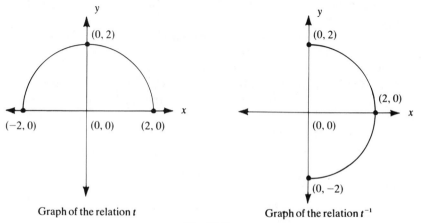

Graph of the relation t Graph of the relation t^{-1}

Fig. 10.8

Observe that the inverse of the relation "is a parent of" on the set of people is "is a child of," and that of the relation "is a divisor of" on the set of natural numbers is "is a multiple of."

Exercise 10.2

1. Find the inverse of the following relations from $A = \{0,1,2,3\}$ to $B = \{3,6,7\}$.
   a) $R_1 = \{(0,3),(1,3),(2,6),(3,6)\}$
   b) $R_2 = \{(0,6),(1,6),(2,6)\}$
   c) $R_3 = \{(2,3),(3,3)\}$
   d) $R_4 = \{(3,3)\}$
   e) $R_5 = \{(1,7),(2,7),(2,6)\}$
   f) $R_6 = \{(0,3),(1,3)\}$

   Find the domain, range, and graph of each of the inverse relations.

2. Let $R = \{(1,2),(2,1),(1,1),(2,2)\}$ and $S = \{(1,1),(2,2),(3,3),(4,4)\}$ be two relations on the set $A = \{1,2,3,4\}$. Find
   a) R^{-1}
   b) S^{-1}
   c) $R \cup R^{-1}$
   d) $R \cap R^{-1}$
   e) $S \cup S^{-1}$
   f) $S \cap S^{-1}$
   g) $R \cap S$
   h) $R \cap S^{-1}$

3. Find the domain and range of the inverse of each of the following relations from $A = \{1,3,5\}$ to $B = \{2,4,6\}$.
   a) $S_1 = \{(x,y) \in A \times B \mid y = x+1\}$
   b) $S_2 = \{(x,y) \in A \times B \mid x < y\}$
   c) $S_3 = \{(x,y) \in A \times B \mid x \text{ divides } y\}$
   d) $S_4 = \{(x,y) \in A \times B \mid x^2 < y\}$
   e) $S_5 = \{(x,y) \in A \times B \mid x - y \text{ is even}\}$
   f) $S_6 = \{(x,y) \in A \times B \mid x - y \text{ is divisible by } 3\}$

4. Mark *true* or *false* (R is any relation from an arbitrary set A to any set B).
   a) R is a relation from A to B implies so is R^{-1}.
   b) R is a relation on A implies so is R^{-1}.
   c) $\text{Dom}(R) = \text{range}(R^{-1})$
   d) $\text{Range}(R) = \text{domain}(R^{-1})$
   e) If R is a relation on A, then $R^{-1} = R$.
   f) $(R^{-1})^{-1} = R$

5. Give a counterexample to each of the false statements in problem 4, whenever possible.

★6. Find the domain and range of the inverse of the relation g defined on the set R of real numbers by $g = \{(x,y) \in R \times R \mid x \geq 0 \text{ and } x^2 + y^2 = 9\}$. Graph g and g^{-1}.

★7. Find the domain and range of the inverse of the relation $h = \{(x,y) \in R \times R \mid 4x^2 + 9y^2 = 36\}$ defined on the set R of real numbers.

10.3 SPECIAL RELATIONS

So far we have been discussing relations in general. Let's now dicuss relations with some special properties.   Consider the relation R defined on the set B of all people by

$$R = \{(x,y) \in B \times B \mid x \text{ has the same color hair as } y\}$$

Two people x and y are related by this relation R if and only if x has the same color hair as *y*. Since everyone has the same color hair as himself, we observe that every member of the set *B* is related to itself by the relation *R*. In other words, $(x,x) \in R$ for every *x* in *B*.

Definition 10.2   A relation *R* on a set *A* is said to be *reflexive* if (x,x) belongs to *R* for every *x* in *A*, that is, if every member of *A* is related to itself by the relation *R*. The relation *R* is said to be *irreflexive* if it is not reflexive.

Thus, the relation *R* defined on the set of all people by the phrase "has the same color hair as" is reflexive.

Example 10.10   Consider the relations

$$r_1 = \{(1,1),(2,1),(2,2),(2,3),(3,3),(4,4)\}$$
$$r_2 = \{(1,1),(2,2),(3,3),(4,4)\}$$
$$r_3 = \{(1,1),(2,2),(4,4)\}$$

defined on the set $A = \{1,2,3,4\}$. If *x* is in *A*, then *x* has four choices: $x = 1$, $x = 2$, $x = 3$, or $x = 4$. Since (x,x) belongs to both r_1 and r_2 for each of these choices of *x*, the relations r_1 and r_2 are reflexive. But r_3 is not reflexive, since for $x = 3$ in *A*, $(x,x) = (3,3)$ does not belong to r_3.

Example 10.11   Consider the relations *R* and *S* on the set of integers defined by the phrases "less than" and "less than or equal to," respectively. Then *R* is irreflexive (why?), whereas *S* is reflexive (why?).

The "less than" relation on any set of numbers is irreflexive whereas the equality relation is always reflexive.

Consider the relation *t* defined on the set *A* of all students on campus by the phrase "takes a course with." Thus (x,y) belongs to *t* if and only if *x* takes a course with *y*. Observe that this relation has the property that whenever *x* is related to *y*, then *y* is also related to *x* (by the same relation). This is true since, if *x* takes a course with *y*, then naturally *y* takes a course with *x*. In other words, if (x,y) belongs to *t*, then (y,x) also belongs to *t*.

Definition 10.3   A relation *R* on a set *A* is said to be *symmetric* if whenever (x,y) belongs to *R*, then (y,x) also belongs to *R*; otherwise it is *asymmetric*. In other words, *R* is symmetric if and only if, given an ordered pair in *R*, the ordered pair obtained by interchanging the first element with the second element is also in *R*.

Example 10.12   Consider the relations

$$S_1 = \{(x,y),(x,z),(y,x),(z,x)\}$$
$$S_2 = \{(x,y),(y,x)\}$$
$$S_3 = \{(x,y),(y,z),(z,y)\}$$
$$S_4 = \{(w,w),(x,x),(y,y),(z,z)\}$$

defined on the set $B = \{w,x,y,z\}$. The relations S_1, S_2, and S_4 are symmetric. But the relation S_3 is not symmetric, since if we interchange the first entry x with the second entry y in the ordered pair (x,y) in S_3, the resulting ordered pair (y,x) is not in S_3.

Example 10.13   Let r be the relation defined on the set of all people by the phrase "is a brother of." This relationship is asymmetric since x being a brother of y does not imply that y is a brother of x (why?).

The relations r_2 and r_3 defined in Example 10.10 are symmetric, whereas r_1 is not (why?). The "less than" and "less than or equal to" relations on the set of real numbers are asymmetric (why?).

Consider now the relation R defined on the set of all people, by the phrase "has the same color car as." A person x is related to a person y if and only if x has the same color car as y. If x has the same color car as y and y has the same color car as z, then is it not true that x has the same color car as z? Yes. Consequently, if (x,y) is in R and (y,z) is in R, then (x,z) is in R. That is, if x is related to y and y is related to z, then x is related to z, by the same relation.

Definition 10.4   A relation R on a set A is said to be *transitive* if $(x,y) \in R$ and $(y,z) \in R$ implies $(x,z) \in R$; otherwise it is *intransitive*. In other words, R is transitive if and only if whenever x is related to y and y is related to z, then x is related to z.

Example 10.14   Consider the relations

$$t_1 = \{(1,3),(3,5),(1,5)\}$$
$$t_2 = \{(1,5),(3,5),(5,5),(5,3),(1,3),(3,3)\}$$
$$t_3 = \{(3,5),(1,3),(1,5)\}$$
$$t_4 = \{(1,3),(3,5),(5,1)\}$$

on the set $A = \{1,3,5,7\}$. We must look at the possible ordered pairs in which the second entry y of an ordered pair (x,y) is the same as the first entry y of another ordered pair (y,z), to check if the new ordered pair (x,z) is in the relation or not. Notice that $(1,3)$ is in t_1 and $(3,5)$ is in t_1. Therefore, $(1,5)$ must be in t_1 for the relation to be transitive. Notice that $(1,5)$ *is* in t_1. Also, $(1,3)$ and $(3,5)$ are the only ordered pairs in t_1 where the second element of one is the same as the first element of the other. Thus, the relation t_1 is transitive. The relations t_2 and t_3 are also transitive. Now, $(1,3)$ belongs to t_4 and $(3,5)$ belongs to t_4, but $(1,5)$ does not belong to t_4. Therefore, t_4 is not transitive.

Example 10.15   The relation t defined on the set N of natural numbers by the phrase "divides" is transitive, since if x divides y and y divides z, then x divides z.

Example 10.16   The relation u defined on the set of students on campus by the phrase "takes a course with" is intransitive since if x takes a course with y and y takes a course with z, then x need not be taking a course with z (why?).

The "less than" relation and the equality relation on the set of real numbers are transitive. So is the inclusion relation "is a subset of" on sets. The relations, "is the father of," "is the mother of," "is a friend of," etc., on the set of people are intransitive.

Definition 10.5   A relation on a set A is called an *equivalence relation* if it is reflexive, symmetric, and transitive.

Example 10.17   Consider the relations defined on the set $A = \{1,3,5\}$:

$$r_1 = \{(1,1),(1,3),(3,3),(3,1),(5,5)\}$$
$$r_2 = \{(1,1),(1,5),(5,5),(5,1)\}$$
$$r_3 = \{(1,1),(3,3),(5,5)\}$$
$$r_4 = \{(1,1),(3,1),(3,3),(5,5)\}$$
$$r_5 = \{(1,1),(1,3),(3,1),(3,3),(3,5),(5,5),(5,3)\}$$

a) r_1 is reflexive, symmetric, transitive, and hence an equivalence relation.
b) r_2 is irreflexive, symmetric, transitive, not an equivalence relation.
c) r_3 is reflexive, symmetric, transitive (why?), and hence an equivalence relation.
d) r_4 is reflexive, asymmetric, transitive, not an equivalence relation.
e) r_5 is reflexive, symmetric, intransitive, not an equivalence relation.

Example 10.18   The relation t defined on the set of all people by the phrase "has the same color hair as" is an equivalence relation.

Example 10.19   Consider the relation R defined on the set I of integers by $R = \{(x,y) \in I \times I \mid x - y \text{ is divisible by } 3\}$. For instance, $(3,0)$, $(2,2)$, $(4,1)$, $(-7,2)$ belong to R. Since $x - x = 0$ is divisible by 3, every integer is related to itself by the relation R. Therefore, R is reflexive.

Now let (x,y) be in R; that is, $x - y$ is divisible by 3. Then $y - x = -(x - y)$ is divisible by 3 and hence (y,x) belongs to R. Thus, (x,y) is in R implies (y,x) is in R, showing that R is symmetric.

Finally, let (x,y) and (y,z) be in R; that is, both $x - y$ and $y - z$ are divisible by 3. Then, $x - z = (x - y) + (y - z)$ is divisible by 3. Therefore, (x, z) belongs to R and hence R is transitive.

Thus, R is an equivalence relation.

Observe that if we write the set R in the above example in the roster method,

$$R = \{..., (-2,1),(-1,2),(0,0),(1,1),(2,2),(3,0),(4,1),(5,2),(6,0),...\}$$

If we look at the exhibited ordered pairs carefully, we notice that the first element in each ordered pair is related to either 0, 1, or 2. Recall that by the division algorithm, these are the possible candidates for the remainder when an integer is divided by 3. Let $\bar{0}$, $\bar{1}$, and $\bar{2}$ denote the set of integers related to 0, 1, and 2, respectively. For example, $0 \in \bar{0}$, $3 \in \bar{0}$, $-3 \in \bar{0}$, $1 \in \bar{1}$, $4 \in \bar{1}$, $-5 \in \bar{1}$, $2 \in \bar{2}$,

$5 \in \overline{2}, -4 \in \overline{2}$, etc. Notice that every element of the set $\overline{0}$ is of the form $3m$; every element of the set $\overline{1}$ is of the form $3m+1$, and that of the set $\overline{2}$ is of the form $3m+2$, for some integer m. Each of the sets $\overline{0}$, $\overline{1}$, and $\overline{2}$ is nonempty. Every integer belongs to exactly one of the sets, that is, $\overline{0} \cup \overline{1} \cup \overline{2} = I$, and these sets are mutually disjoint. These sets are called *equivalence classes* and are said to form a *partitioning* of the set of integers, as shown in Fig. 10.9. Thus, the equivalence relation R on I divides (partitions) I into a family of nonempty disjoint subsets whose union is the same as I.

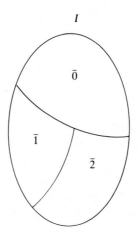

Fig. 10.9

Exercise 10.3

1. Which of the following relations on $A = \{a,b,c\}$ are reflexive, symmetric, or transitive? Which of them are equivalence relations?
   a) $R_1 = \{(a,b),(a,a),(b,a),(b,b)\}$
   b) $R_2 = \{(a,a),(a,b),(b,a),(b,b),(c,c)\}$
   c) $R_3 = \{(a,a),(b,b),(c,c),(b,c),(a,b)\}$
   d) $R_4 = \{(a,a),(a,b),(b,b)\}$
   e) $R_5 = \{(a,a),(a,b),(b,b),(b,c)\}$
   f) $R_6 = \{(a,a),(a,c),(b,c),(c,c),(c,a),(c,b)\}$
   g) $R_7 = \{(a,a),(b,b),(c,c)\}$
   h) $R_8 = A \times A$

2. List the properties of the following relations on the set $B = \{1,2,3,4,5,6\}$.
   a) $t_1 = \{(x,y) \mid x \le y\}$
   b) $t_2 = \{(x,y) \mid x+y \le 5\}$
   c) $t_3 = \{(x,y) \mid y = 2x\}$
   d) $t_4 = \{(x,y) \mid x \text{ and } y \text{ are relatively prime}\}$

3. Let h be the relation defined on the family of all subsets of a nonempty set by the phrase "is a subset of." Is h reflexive? Symmetric? Transitive? An equivalence relation?

4. a) Use the roster method to describe the relation r defined on the set $A = \{1,3,5,6\}$ by the phrase "is a factor of."

   b) Is r an equivalence relation on A? Why?

5. Is the relation $t = \{(x,y) \mid x$ is perpendicular to $y\}$ defined on the set A of all straight lines on a plane reflexive? Symmetric? Transitive? An equivalence relation?

6. Mark *true* or *false* (A is any nonempty set and R an arbitrary relation on A).

   a) The relation $\varnothing$ on A is irreflexive.

   b) The relation $\varnothing$ on A is asymmetric.

   c) The relation $\varnothing$ on A is intransitive.

   d) R is reflexive if and only if R^{-1} is reflexive.

   e) R is symmetric if and only if R^{-1} is symmetric.

   ★f) R is transitive if and only if R^{-1} is transitive.

   ★g) R is an equivalence relation if and only if R^{-1} is an equivalence relation.

   ★h) $A \times A$ is an equivalence relation.

7. a) Show that the relation h on the set of integers defined by $h = \{(a,b) \mid a - b$ is even$\}$ is an equivalence relation.

   b) Same as (*a*) with $h = \{(a,b) \mid a - b$ is divisible by 5$\}$.

8. When is a relation r on a set A irreflexive? Asymmetric? Intransitive?

9. Define a relation on the set $A = \{1,2,3\}$ that is

   a) reflexive, symmetric, and transitive

   b) reflexive, symmetric, and intransitive

   c) reflexive, asymmetric, and transitive

   d) irreflexive, symmetric, and transitive

   e) reflexive, asymmetric, and intransitive

   f) irreflexive, symmetric, and intransitive

   g) irreflexive, asymmetric, and transitive

   h) irreflexive, asymmetric, and intransitive

★10. Are the union and intersection of two

   a) reflexive relations on a set reflexive?

   b) symmetric relations on a set symmetric?

   c) transitive relations on a set transitive?

   Give a counterexample when the answer is no.

10.4 FUNCTIONS

One of the most fundamental and powerful concepts that is used practically in all branches and at all levels of mathematics is the concept of a function. A function is only a special type of relation. Before we make a formal definition of a function, let's try to introduce the concept by means of what is usually called a *function machine*, as shown in Fig. 10.10.

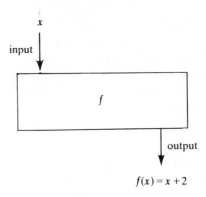

$$f(x) = x + 2$$

Function machine

Fig. 10.10

Assume that we have a machine f that is capable of adding the number 2 to any number x that is fed into it and produces $f(x) = x + 2$ as the (unique) output. For example, if the number 1 is fed as an input, the machine f performs the operation of adding 2 to it and gives out $f(1) = 1 + 2 = 3$ as the output. If the inputs are allowed to come from the set $A = \{0,1,2,3,4\}$, then we obtain the following outputs from our machine:

$$f(0) = 0 + 2 = 2$$
$$f(1) = 1 + 2 = 3$$
$$f(2) = 2 + 2 = 4$$
$$f(3) = 3 + 2 = 5$$

and
$$f(4) = 4 + 2 = 6$$

Thus $B = \{2,3,4,5,6\}$ is the set of all outputs. The set of all ordered pairs (x,y), where x is an input and y the corresponding output, is given by $f = \{(0,2),(1,3),(2,4),(3,5),(4,6)\}$, which is clearly a relation from A to B. Notice that in this relation, *every* element of A is paired with a *unique* element of B; that is, every element of A occurs as a first element and there are no two ordered pairs with the *same* first element. In other words, the function machine f associates each element (input) of the set A with a unique element (output) of the set B.

Consider the relation $r = \{(x,2),(y,4),(z,0)\}$ from $A = \{x,y,z\}$ to $B = \{0,2,4,6\}$. Observe that in the relation r, every element of A occurs and occurs exactly once. Consequently, each element in A is associated with a unique element in B by the relation r and there are no two distinct ordered pairs with the same first element in the relation r.

Definition 10.6 A relation f from a set A to a set B is called a *function* from A to B, denoted by $f : A \to B$, if *every* element of A is paired with a *unique* element of B, that is, if

1. dom(f) $= A$ and

2. no element of A is paired with more than one element of B.

The set B is called the *codomain* of the function. A function from A to A is called a function *on A*.

Example 10.20   Consider the following relations from $A = \{a,b,c\}$ to $B = \{1,2,3,4,5\}$:

$$f = \{(a,3),(b,2),(c,1)\}$$
$$g = \{(a,5),(b,1),(c,1)\}$$
$$r = \{(a,1),(b,2),(a,3),(c,1)\}$$
$$s = \{(a,3),(b,1)\}$$
$$h = \{(a,3),(b,3),(c,3)\}$$

Notice that dom$(f) =$ dom$(g) =$ dom$(h) = A$. Also, each element of A is paired with exactly one element of B by each of the relations f, g, and h. Thus f, g, and h are functions from A to B. The set B is the codomain of these functions. Notice that range$(f) = \{1,2,3\}$, range$(g) = \{1,5\}$, and range$(h) = \{3\}$. These functions are represented geometrically in Fig. 10.11.

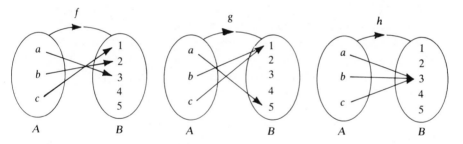

Fig. 10.11

Now, dom$(r) = \{a,b,c\} = A$; however, r is not a function, since the same element a in A is paired with two distinct elements of B. Geometrically, this means that there are two distinct arrows originating from the same element a in A. The relation s also is not a function from A to B since dom$(s) = \{a,b\} \neq A$. Geometrically, this means that no arrow originates from at least one element of A.

Example 10.21   Consider the following relations on $A = \{1,2,3,4\}$:

$$f = \{(x,y) \mid y = x\}$$
$$t = \{(x,y) \mid y < x\}$$
$$g = \{(x,y) \mid y = x+2\}$$

Observe that

$$f = \{(1,1),(2,2),(3,3),(4,4)\}$$
$$t = \{(1,2),(1,3),(1,4),(2,3),(2,4),(3,4)\}$$
$$g = \{(1,3),(2,4)\}$$

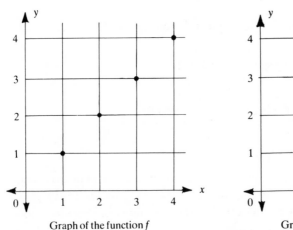

Graph of the function f

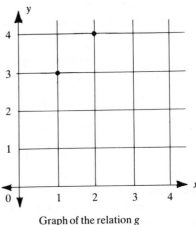

Graph of the relation g

Fig. 10.12

Notice that f is a function on A, whereas g and t are not. (Why?) The graphs of the function f and the relation g are given in Fig. 10.12.

Let $f: A \to B$. If $(x,y) \in f$, we usually write $y = f(x)$. The quantity y is called the *image* of x under the function f or the *value* of the function f at x. The quantity x is called a *preimage* of y under the function f. In example 10.20,

$$
\begin{array}{lll}
f(a) = 3 & \text{since} & (a,3) \in f \\
g(b) = 1 & \text{since} & (b,1) \in g \\
h(c) = 3 & \text{since} & (c,3) \in h
\end{array}
$$

Here 3 is the image of a under f and b is a preimage of 1 under g. Notice that the element 1 has two preimages, b and c, under g. Using the above notation, the functions f and g in example 10.20 could very well be defined by

$$f(a) = 3, \quad f(b) = 2, \quad f(c) = 1$$

and

$$g(a) = 5, \quad g(b) = 1 = g(c)$$

This is not a convenient way of defining a function, especially when its domain is fairly large. For example, consider the function g defined on the set R of real numbers by

$$g = \{(x,y) \in R \times R \mid y = 2x + 1\}$$

This function can also be written as $g(x) = 2x + 1$ for every $x \in R$, where we have defined the function g by stating its general behavior; it associates every real number x with 1 more than twice the value of x. The image of the real number 0 under g is given by $g(0) = 2 \cdot 0 + 1 = 1$ and hence $(0,1) \in g$. Similarly, the images of -1, $\frac{2}{3}$, and $-\frac{3}{5}$ are given by

$$g(-1) = 2(-1) + 1 = -1$$
$$g(\tfrac{2}{3}) = 2(\tfrac{2}{3}) + 1 = \tfrac{7}{3}$$
$$g(-\tfrac{3}{5}) = 2(-\tfrac{3}{5}) + 1 = -\tfrac{1}{5}$$

Example 10.22   Let $f : R \to R$ defined by $f(x) = \frac{2}{3}x - \frac{3}{5}$. Find $f(0)$, $f(-\frac{1}{2})$, $f(\frac{2}{3})$, and $f(-\frac{3}{4})$.

Solution:

$$f(0) = \tfrac{2}{3}(0) - \tfrac{3}{5} = -\tfrac{3}{5}$$
$$f(-\tfrac{1}{2}) = \tfrac{2}{3}(-\tfrac{1}{2}) - \tfrac{3}{5} = -\tfrac{1}{3} - \tfrac{3}{5} = -\tfrac{14}{15}$$
$$f(\tfrac{2}{3}) = \tfrac{2}{3}(\tfrac{2}{3}) - \tfrac{3}{5} = \tfrac{4}{9} - \tfrac{3}{5} = -\tfrac{7}{45}$$
$$f(-\tfrac{3}{4}) = \tfrac{2}{3}(-\tfrac{3}{4}) - \tfrac{3}{5} = -\tfrac{1}{2} - \tfrac{3}{5} = -\tfrac{11}{10}$$

Example 10.23   The concept of a function is used in our everyday life, whenever we go to the post office to find the postage of a letter addressed to, say, some place in the United States. Let x denote the weight of a letter in ounces. If $0 < x \leq 1$, then those letters require 13¢ postage. Letters with $x \in (1,2]$ require 26¢ postage; letters with $x \in (2.3]$ require 39¢ postage, etc. Let g be the function that associates every letter addressed to any place in the United States with exactly one element of the set $\{13,26,39,...\}$. g is called the *post office function*, the graph of which is given in Fig. 10.13.

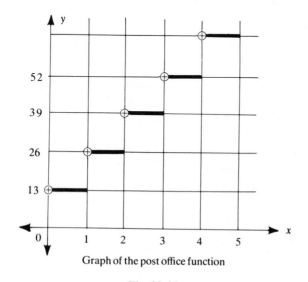

Graph of the post office function

Fig. 10.13

Example 10.24 Consider the function $g : R \rightarrow R$ defined by

$$g(x) = |x|$$

where $|x|$ denotes the absolute value of x (Section 7.10). For example, $g(0) = 0$, $g(-1) = 1 = g(1)$, $g(-\frac{1}{2}) = \frac{1}{2} = g(\frac{1}{2})$, etc. The function g is called the *absolute value function*, the graph of which is exhibited in Fig. 10.14.

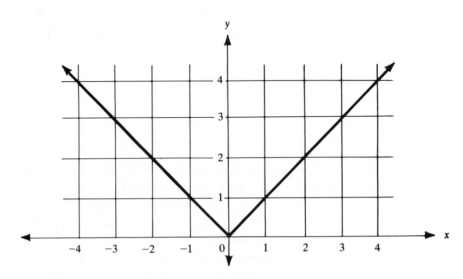

Graph of the absolute value function

Fig. 10.14

Exercise 10.4

1. Which of the following relations from $A = \{a,b,c,d\}$ to $B = \{1,2,3,4\}$ are functions from A to B? Find the range of each of them.
   a) $t_1 = \{(a,2),(b,3),(c,2),(d,1)\}$ b) $t_2 = \{(a,4),(b,3),(c,1),(d,3),(a,1)\}$
   c) $t_3 = \{(a,3),(b,3),(c,1)\}$ d) $t_4 = \{(a,1),(b,1),(c,1),(d,1)\}$
   e) $t_5 = \{(a,2),(b,1),(c,2),(a,3)\}$ f) $t_6 = \{(a,4),(b,1),(c,4),(d,1)\}$

2. Is the relation $r = \{(4,y) \mid y \in R\}$ on R a function on R? Why?

3. Find the inverses of the relations in problem 1. Which of them are functions from B to A?

4. Mark *true* or *false* (A and B are any nonempty sets).
   a) Every function from A to B is a relation from A to B.
   b) Every relation from A to B is a function from A to B.
   c) $f : A \rightarrow B$ implies range$(f) = B$.
   d) $f : A \rightarrow B$ implies range$(f) \subset B$.
   e) $\varnothing$ is a function from A to B.
   f) $f : R \rightarrow R$ implies $f(0) = 0$.

g) Every element in A under a function from A to B has a unique image.

h) Every element in B under a function from A to B has a unique preimage.

5. Give a counterexample to each of the false statements in problem 4, whenever possible.

6. Let $g : R \to R$ be defined by $g(x) = x^2 - 1$. Find:
   a) the images of $0, \frac{1}{2}, -\frac{1}{2},$ and $\frac{2}{3}$ under g.
   b) the set of all preimages of 0 and 3 under g.

7. Let h be the function defined by $h = \{(x,y) \in R \times R \mid y = x^2 - x + 1\}$. Find $h(0), h(-1), h(-\frac{2}{3})$ and $h(\frac{3}{5})$.

8. With the functions g and h as in problems 6 and 7, evaluate each of the following:
   a) $g(0) + h(0)$ b) $g(2) + h(-3)$ c) $g(-1) - h(-1)$
   d) $g(\frac{1}{2}) + h(-\frac{1}{3})$ e) $g(-1) \cdot h(-1)$ f) $g(\frac{1}{3}) \cdot g(-\frac{1}{2})$

9. Let f be the function defined on the set of nonzero real numbers by $f(x) = \dfrac{x}{|x|}$. Evaluate each of the following:
   a) $f(-1)$ b) $f(-\frac{1}{2})$ c) $f(1)$ d) $f(\frac{1}{2})$

10. Let g be the function defined on the set of positive real numbers by $g(x) = \dfrac{x}{x + |x|}$. Evaluate each of the following:
   a) $g(1)$ b) $g(\frac{1}{2})$ c) $g(\frac{2}{3})$ d) $g(\frac{3}{2})$

11. Let $h : R - \{2\} \to R$ be defined by $h(x) = \dfrac{2}{2-x}$. Evaluate:
   a) $h(-1)$ b) $h(-\frac{1}{2})$ c) $h(0)$ d) $h(1)$

12. Find the range of the function $f : W \to W$ defined by $f(x) = x + 1$.

13. Let $f : R \to R$ be defined by $f(x) = \begin{cases} 1 \text{ if } x \in Q \\ 0 \text{ if } x \notin Q \end{cases}$. Find $f(0), f(1), f(-\frac{1}{2}), f(\sqrt{2})$ and $f(\sqrt{4})$. What is the range of f?

14. Find the domain and range of the post office function.

15. Let $t : R \to R$ be defined by $t(x) = 5$. Find $t(0), t(1), t(-\frac{1}{2})$ and $t(\frac{2}{3})$. Also find range(t).

16. Consider the function f on R defined by $f(x) = 4x$. Verify that
   a) $f(x + y) = f(x) + f(y)$
   b) $f(x - y) = f(x) - f(y)$

★17. Let $f : R \to R$ such that $f(x) + f(y) = f(x + y)$ for every x, y in R. Show that
   a) $f(0) = 0$ b) $f(3) = 3f(1)$
   c) $f(-4) = -4f(1)$ d) $f(\frac{2}{3}) = \frac{2}{3}f(1)$

★18. Which of the following relations are functions on the set R of real numbers?
   a) $r_1 = \{(x,y) \mid x^2 + y^2 = 4\}$

b) $r_2 = \{(x,y) \mid y \geq 0 \text{ and } x^2 + y^2 \doteq 4\}$

c) $r_3 = \{(x,y) \mid x \geq 0 \text{ and } x^2 + y^2 = 4\}$

d) $r_4 = \{(x,y) \mid x \geq 0, \ y \geq 0, \text{ and } x^2 + y^2 = 4\}$

Draw the graph of each. Find the range of each.

★**19.** **a)** How many functions can be defined from $A = \{a,b,c\}$ to $B = \{0\}$? Find all of them.

  b) Same as part (a) with $A = \{a,b,c\}$ and $B = \{0,1\}$.

★**20.** Let's define a function T on the set S of all statements (Section 2.1) by

$$T(x) = \begin{cases} 1 \text{ if } x \text{ is a true statement} \\ 0 \text{ if } x \text{ is a false statement} \end{cases}$$

The function T is called the *truth function*.

  a) Evaluate $T(x)$ if

   1. x: Paris is in France.

   2. x: $2 + 3 = 5$

   3. x: California is the largest state in the United States.

   4. x: Montreal is the capital of Canada.

  b) Find the domain and range of the truth function.

  c) Show that

   1. $T(x \wedge y) = T(x)T(y)$

   2. $T(\sim x) = 1 - T(x)$

   3. $T(x \vee y) = 1 - [1 - T(x)][1 - T(y)]$

★**21.** Let $\varphi(n)$ denote the number of positive integers less than or equal to n and relatively prime to n, where n is a positive integer. Since 1 is the only positive integer less than or equal to 1 and relatively prime to 1, $\varphi(1) = 1$. Similarly, $\varphi(2) = 1$, $\varphi(3) = \varphi(4) = \varphi(6) = 2$. The function φ, called *Euler's phi function*, has numerous applications in number theory. Evaluate:

  a) $\varphi(5)$ **b)** $\varphi(7)$ **c)** $\varphi(13)$

  d) $\varphi(10)$ **e)** $\varphi(12)$ **f)** $\varphi(p)$ where p is prime

10.5   SPECIAL FUNCTIONS

In the previous section we discussed functions in general. This section is devoted to the discussion of functions with some restrictions.

Definition 10.7   A function $f : A \to B$ is called a *constant function* if its range consists of a single element.

Example 10.25   Consider the following functions defined from $A = \{a,b,c\}$ to $B = \{1,2,3,4,5\}$:

$$f_1 = \{(a,3),(b,3),(c,3)\}$$

$$f_2 = \{(a,2),(b,3),(c,2)\}$$

$$f_3 = \{(a,1),(b,1),(c,1)\}$$

The functions f_1 and f_3 are constant functions, whereas f_2 is not since range(f_2) = {2,3} consists of more than one element.

Example 10.26   The function $t : R \to R$ defined by $t(x) = 4$ is a constant function. Range(t) = {4}. The graph of the function t, given in Fig. 10.15, is a straight line parallel to the x-axis and 4 units above it.

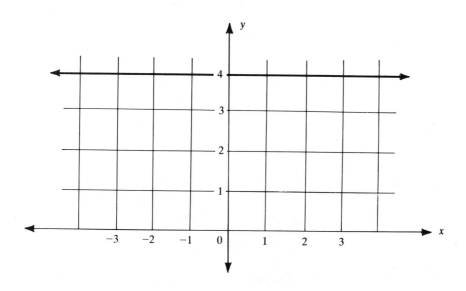

Graph of the constant function $t(x) = 4$

Fig. 10.15

Definition 10.8   Every function $f : R \to R$ of the form $f(x) = ax + b$ is called a *linear function*.

The graph of a linear function f is always a straight line. Since a straight line is determined by two points, we need only find two ordered pairs belonging to f to determine its graph.

Example 10.27   The function g on R defined by $g(x) = 2x + 1$ is linear. Notice that $g(0) = 1$, $g(1) = 3$, $g(\frac{1}{2}) = 2$, etc. Thus $(0,1) \in g$, $(1,3) \in g$, etc. Consequently, $(0,1)$ and $(1,3)$ are two points on the graph of g, which is the straight line obtained by joining these two points, as shown in Fig. 10.16.

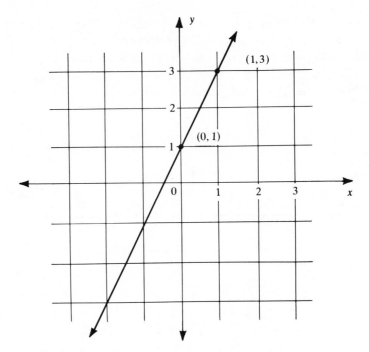

Graph of the linear function $g(x) = 2x + 1$

Fig. 10.16

Observe that a constant function on R is also a linear function on R. (Why?)

Definition 10.9 A function $f : A \to A$ is called the *identity function* if $f(x) = x$ for every x in A.

Recall that addition of zero leaves every real number unchanged. Similarly, the identity function leaves every element of its domain unchanged.

Example 10.28 Let t be the identity function on R, that is, $t(x) = x$ for every real number x. Then $t(0) = 0$, $t(3) = 3$, $t(-\frac{1}{2}) = -\frac{1}{2}$, etc. The graph of this identity function is given in Fig. 10.17.

Let's now consider the functions f and g from $A = \{a,b,c\}$ to $B = \{0,1,2\}$, and the function h from $A = \{a,b,c\}$ to $C = \{0,1\}$:

$$f = \{(a,2),(b,1),(c,0)\}$$
$$g = \{(a,1),(b,2),(c,1)\}$$
$$h = \{(a,0),(b,0),(c,1)\}$$

Can you find a property that f has and g doesn't have? Well, notice that range$(f) = B$, the codomain, whereas range$(g) \neq B$. Consequently, every element of B has a preimage under the function f but not under g. Observe that the function *h* also possesses the same property: range(*h*) = codomain. However, a closer look at the functions *f* and *h* shows that *f* has an additional property: every element of *B* has a *unique* preimage in *A*.

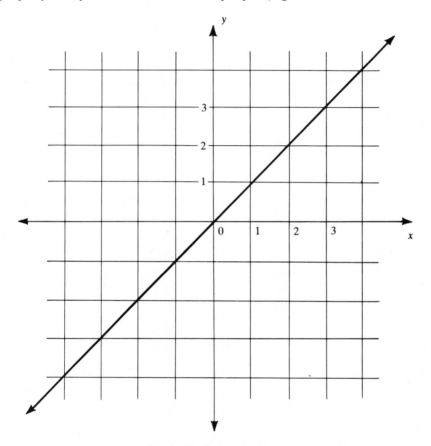

Graph of the identity function

Fig. 10.17

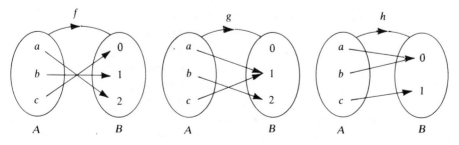

Fig. 10.18

Consider the following functions from $A = \{a,b,c\}$ to $B = \{0,1,2\}$:

$$h_1 = \{(a,1),(b,2),(c,0)\}$$
$$h_2 = \{(a,2),(b,0),(c,0)\}$$
$$h_3 = \{(a,2),(b,0),(c,1)\}$$

Notice that distinct elements in A are paired with distinct elements in B by the functions h_1 and h_3. Also, range$(h_1) = $ range$(h_3) = B$. Consequently, every element in B has exactly one preimage in A. Such functions are called one-to-one (1-1) correspondences. Observe that h_2 is not a 1-1 correspondence. (Why?)

Definition 10.10 A function $f : A \to B$ is called a *one-to-one correspondence* if *every* element in B has *exactly one* preimage in A.

How is this definition related to Definition 1.12? If $f : A \to B$ is a 1-1 correspondence, then every element of A is paired with a unique element of B and every element of B is paired with a unique element of A, so there exists a 1-1 correspondence between A and B. Conversely, if there is a 1-1 correspondence between A and B, then there exists a suitable function $f : A \to B$ that is that 1-1 correspondence.

Definition 10.11 A *permutation* is a 1-1 correspondence from a set to itself.

Example 10.29 The functions

$$f = \{(1,2),(2,3),(3,1)\}$$

and
$$g = \{(1,3),(2,2),(3,1)\}$$

are 1-1 correspondences on the set $A = \{1,2,3\}$. Thus, they are permutations of the set A.

Exercise 10.5

1. Mark *true* or *false* (A is any nonempty set).
   a) The graph of every linear function is a straight line.
   b) Every straight line is the graph of a linear function.
   c) Every constant function is a linear function.
   d) The range of a constant function contains exactly one element.
   e) The range of a constant function is empty.
   f) The identity function on A matches every element in A with a unique element of A.
   g) The range of the identity function on A contains exactly one element.
   h) The range of the absolute value function is R.
   i) The range of the post office function is R.
   j) The range of the post office function is N.
2. Give a counterexample to each of the false statements in problem 1, whenever possible.

3. How many constant functions can be defined from the set $A = \{a,b,c\}$ to the set $B = \{1,2\}$? What are they?

4. Same as problem 3 with $A = \{x,y,z\}$ and $B = \{0,1,2,3\}$.

5. Which of the following functions from $A = \{a,b,c,d\}$ to $B = \{1,2,3,4,5\}$ have the property that distinct elements in A are paired with distinct elements in B?

a) $t_1 = \{(a,2),(b,3),(c,4),(d,1)\}$

b) $t_2 = \{(a,3),(b,1),(c,2),(d,1)\}$

c) $t_3 = \{(a,4),(b,2),(c,1),(d,4)\}$

d) $t_4 = \{(a,3),(b,4),(c,5),(d,2)\}$

6. Which of the following functions from $A = \{a,b,c,d\}$ to $B = \{0,1,2\}$ have the property that every element in B has at least one preimage in A?

a) $f_1 = \{(a,0),(b,0),(c,1),(d,0)\}$

b) $f_2 = \{(a,0),(b,1),(c,0),(d,2)\}$

c) $f_3 = \{(a,2),(b,1),(c,2),(d,0)\}$

d) $f_4 = \{(a,1),(b,2),(c,1),(d,2)\}$

7. Which of the following functions from $A = \{x,y,z\}$ to $B = \{1,2,3\}$ are 1-1 correspondences?

a) $g_1 = \{(x,3),(y,2),(z,1)\}$ b) $g_2 = \{(x,3),(y,2),(z,3)\}$

c) $g_3 = \{(x,1),(y,2),(z,3)\}$ d) $g_4 = \{(x,2),(y,2)(z,2)\}$

8. At what point does the graph of the function $f(x) = 2x - 4$ intersect the x-axis? The y-axis?

9. When can a constant function $f : A \to B$ have the property that every element in B has exactly one preimage in A?

10. Does the function $g : R \to R$ such that $g(x) = x^2$ have the property that every element in R has exactly one preimage in R? Why?

11. Same as problem 10 with the function $v = \{(x,y) \mid y = x^2\}$ from $A = [-2,2]$ to $B = [0,4]$.

★12. Same as problem 10 with the function $u = \{(x,y) \in R \times R \mid y \geq 0$ and $x^2 + y^2 = 4\}$.

★13. a) How many 1-1 correspondences can be defined from $A = \{a,b\}$ to $B = \{1,2\}$? Find all of them.

b) Same as (a) with $A = \{a,b,c\}$ and $B = \{1,2,3\}$.

★14. How many 1-1 correspondences can be defined from A to B, where $c(A) = c(B) = m(\neq 0)$.

10.6   LINEAR INEQUALITIES

In the preceding section, we observed that the graph of a function of the form $f(x) = ax + b$ is a straight line. In other words, the equation $y = ax + b$ or, more generally, the equation $lx + my + n = 0$ with l and m not both zero, defines a straight line. Such an equation is called a *linear equation in two variables*. For example, $x + y - 7 = 0$ is a linear equation in two variables. Observe that if $x = 7$

then $y = 0$, and if $x = 0$ then $y = 7$. Thus, $(7,0)$ and $(0,7)$ are two solutions of the equation $x + y - 7 = 0$. Consequently, $A(7,0)$ and $B(0,7)$ are two points on the line. The graph of the equation $x + y - 7 = 0$ is exhibited in Fig. 10.19. Notice that $\{(x,y) \mid x + y - 7 = 0\}$ is the solution set of the equation $x + y - 7 = 0$.

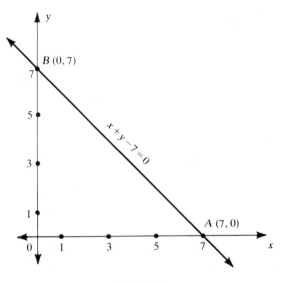

Fig. 10.19

An inequality obtained from the linear equation $lx + my + n = 0$ by replacing the equality sign $(=)$ by an inequality sign $(<, \leq, >, \text{ or } \geq)$ is called a *linear inequality in two variables*. Thus, $lx + my + n < 0$, $lx + my + n \leq 0$, $lx + my + n > 0$, and $lx + my + n \geq 0$ are all linear inequalities. A *solution* of a linear inequality is an ordered pair (x,y) of real numbers for which the inequality is true. For example, $(0,0)$ is a solution of the inequality $2x + 3y - 6 \leq 0$. (Why?)
Example 10.30   Graph the solution set of the inequality $x + y - 7 \leq 0$.
Solution: Observe that the line $x + y - 7 = 0$ meets the x-axis at $(7,0)$ and the y-axis at $(0,7)$. Since these two points determine the line $x + y - 7 = 0$, we first graph the line. This line divides the plane into three mutually disjoint sets: the set of points above the line, the set of points on the line, and the set of points below the line. Since the coordinates of every point on the line satisfy the equation $x + y - 7 = 0$, the line will be part of the graph of the solution set. Notice that $(0,0)$ lies below the line and satisfies the inequality $x + y - 7 \leq 0$, hence the coordinates of *any* point below the line also satisfy the given inequality. Thus, the shaded region in Fig. 10.20 (including the line) represents

the solution set of the inequality $x+y-7\leq0$. It may be noted that coordinates of points above (and not including) the line, (5,7) for example, satisfy the inequality $x+y-7>0$.

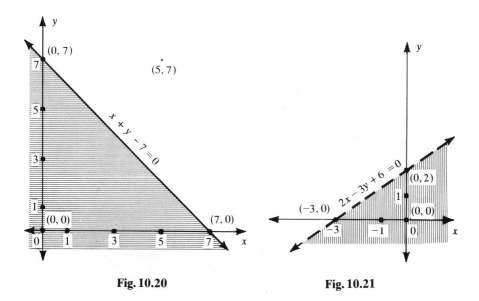

Fig. 10.20   Fig. 10.21

Example 10.31 Sketch the graph of the inequality $2x-3y+6>0$.
Solution: First, we choose two convenient points on the line $2x-3y+6=0$, for example, the points where it intersects the axes: $(x=0,y=2)$ and $(x=-3,y=0)$. Thus, the points (0,2) and (-3,0) fix the line $2x-3y+6=0$. In Fig. 10.21 it is shown as a broken line, since points on the line do not belong to the graph of the given inequality. (Why?) Do the coordinates $x=0$, $y=0$ of the origin satisfy the inequality? Yes. Therefore, every point below the line belongs to the graph, as shown in Fig. 10.21.
Example 10.32 Graph the solution set of the pair of inequalities

$$x+y-7\leq0$$
$$2x-3y+6>0$$

Solution: Let A be the solution set of the inequality $x+y-7\leq0$ and B that of the inequality $2x-3y+6>0$. Then, $A\cap B$ is the solution set of the given pair of inequalities (why?). The solution set A is represented by the region with horizontal lines and the solution set B by the region with vertical lines in Fig. 10.22. Therefore, the crosshatched region represents the solution set of the given pair of inequalities.

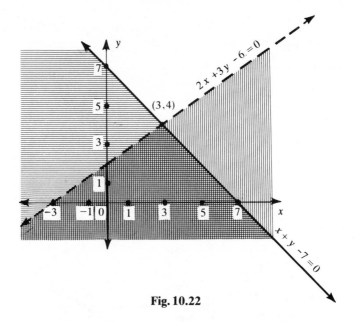

Fig. 10.22

How do we now find the coordinates of the point of intersection of the lines $x+y-7=0$ and $2x-3y+6=0$? Notice that these equations can be written $y = 7-x$ and $y = \dfrac{2x+6}{3}$, respectively. Equating these y-values, we have

$$7-x = \frac{2x+6}{3}$$

$$3(7-x) = 2x+6$$

$$15 = 5x$$

$$x = 3$$

Substituting this value of x in $y = 7-x$, we get $y = 7-3 = 4$. Thus, the two lines intersect at the point (3,4). In other words, (3,4) is the solution of the pair of equations $x+y-7=0$ and $2x-3y+6=0$.

Exercise 10.6

1. Graph the solution set of the following inequalities.
   a) $x \geq 0$ b) $y \geq 0$ c) $x-y \leq 2$
   d) $2x+y-3 \geq 0$ e) $3x-5y > 15$ f) $x-2y-4 \geq 0$
2. Sketch the graph of the solution set of the following pairs of inequalities.
   a) $x > 0, y > 0$ b) $x < y, x \geq 0$
   c) $x+y-3 > 0$ d) $x \geq 2, y \leq 3$
      $2x-3y+4 \geq 0$
   e) $y < x+1$ f) $y \leq x+1$
      $3x+2y > 12$ $x+y+1 \geq 0$

3. Exhibit the graph of the solution set of the following inequalities.

    a) $x \geq 0,\ y \geq 0,$ **b)** $x \geq 0,\ y \geq 0,$

        $x \leq 2,\ y \leq 3$ $x + y \leq 1,\ y + 1 \geq 2x$

4. Find the solution set of the following pairs of equations.

    a) $x + y - 3 = 0$ **b)** $x = 2$

        $2x - 3y + 4 = 0$ $y = 3$

    c) $x - y + 1 = 0$ **d)** $x - y + 1 = 0$

        $3x + 2y = 12$ $x + y + 1 = 0$

10.7   LINEAR PROGRAMMING

Linear programming is a relatively new branch of mathematics. Its development was influenced by the scientific need to solve some logistic military problems in World War II, for example, deploying aircraft and submarines at strategic positions, airlifting supplies and personnel, etc. Besides military applications, this fascinating field has numerous applications in economics, home economics, airline scheduling, theory of communications, inventory, etc.

Generally speaking, linear programming problems consist of finding the *maximum* or *minimum value* of a linear function, called the *objective function*, subject to some linear conditions, called *constraints*. For example, we may want to maximize the production or profit of a company, or to minimize the cost of production or of transportation, all subject to certain restrictions.

Example 10.33   Maximize the function $f(x,y) = 5x + 7y$ subject to the constraints

$$x \geq 0, \qquad y \geq 0$$

$$x + y - 7 \leq 0$$

$$2x - 3y + 6 \geq 0$$

Solution: First, we find the set of all possible pairs (x,y) that satisfy all four inequalities. Such a solution is called a *feasible solution* of the problem. For example, $(0,0)$ is a feasible solution, since $x = 0,\ y = 0$ satisfy the given conditions.

Second, we want to pick that feasible solution for which the given function $f(x,y)$ is a maximum (or minimum). Such a feasible solution is called an *optimal solution*.

Notice that the region defined by the inequality $x \geq 0$ consists of all points on and to the right of the y-axis. Similarly, the region given by $y \geq 0$ contains points lying on or above the x-axis. It now follows, by example 10.32, that the given constraints define the *polygonal region* bounded by the lines $x = 0,\ y = 0$, $x + y - 7 = 0$, and $2x - 3y + 6 = 0$, as shown in Fig. 10.23. This shaded region is a *convex set*: that is, a set S of points having the property that the line segment joining any two points of S lies within the set. More about convex sets is discussed in Section 13.3. A fundamental result in linear programming states that the objective function $f(x,y)$ attains a maximum or minimum value at a corner point O, A, B, *or* C of the polygon $OABC$. Therefore, let's evaluate the function $f(x,y)$ at these corner points:

$$f(O) = f(0,0) = 5 \cdot 0 + 7 \cdot 0 = 0$$
$$f(A) = f(7,0) = 5 \cdot 7 + 7 \cdot 0 = 35$$
$$f(B) = f(3,4) = 5 \cdot 3 + 7 \cdot 4 = 43$$
$$f(C) = f(0,2) = 5 \cdot 0 + 7 \cdot 2 = 14$$

Since 43 is the largest of these four functional values, the given function takes on the maximum value, 43, for $x = 3$ and $y = 4$, subject to the given constraints. Notice that (3,4) is the optimal solution of the problem.

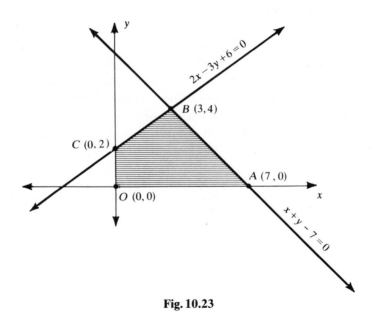

Fig. 10.23

Example 10.34 Minimize the function $g(x,y) = 13x - 15y$ subject to the constraints in the previous example.

Solution: Since the set of feasible solutions for this problem remains the same as that in the previous example (why?), let's now evaluate $g(x,y)$ at the corner points O, A, B, and C:

$$g(O) = g(0,0) = 13 \cdot 0 - 15 \cdot 0 = 0$$
$$g(A) = g(7,0) = 13 \cdot 7 - 15 \cdot 0 = 91$$
$$g(B) = g(3,4) = 13 \cdot 3 - 15 \cdot 4 = -21$$
$$g(C) = g(0,2) = 13 \cdot 0 - 15 \cdot 2 = -30$$

Thus, the function $g(x,y)$ has the minimum value -30 for $x = 0$ and $y = 2$.

Let's now summarize the procedure for solving a linear programming problem in the following steps.

1. Graph the polygonal region determined by the constraints.
2. Find the coordinates of the corner points of the polygon.
3. Evaluate the objective function at the corner points.
4. Identify the corner point at which the function has an optimal value.

We now give a typical linear programming problem that is of practical importance. At first glance, it may appear confusing and complicated. In a problem like this, we must first translate the given data into mathematical equations and inequalities; then, we follow the golden steps given above.

Example 10.35   A manufacturing company makes two types of television sets: one is black and white, and the other is color. The company has resources to make at most 300 sets a week. It costs \$180 to make a black and white set and \$270 to make a color set, but the company does not want to spend more than \$64,800 a week to make television sets. To make a profit of \$170 per black and white set and \$225 per color set, how many sets of each type should the company make to have maximum profit?

Solution: Let's first rewrite the problem in terms of mathematical symbols. Let x denote the number of black and white sets and y the number of color sets made each week. Then the total number of sets made per week is $x + y$, where

$$x + y \le 300$$

(Why?) The total cost per week is

$$180x + 270y \le 64800$$

that is,
$$2x + 3y \le 720$$

Also, since the company cannot make a negative number of sets, we have $x \ge 0$ and $y \ge 0$. Thus, the problem is to maximize the *profit function* $p(x,y) = 170x + 225y$ subject to the constraints

$$x \ge 0, \qquad y \ge 0$$
$$x + y \le 300$$
$$2x + 3y \le 720$$

Notice that these four linear constraints define the shaded region in Fig. 10.24 with corner points O, A, B, and C. Evaluating the profit function $p(x,y)$ at these points, we have

$$p(O) = p(0,0) = 170 \cdot 0 + 225 \cdot 0 = 0$$
$$p(A) = p(300,0) = 170 \cdot 300 + 225 \cdot 0 = 51,000$$
$$p(B) = p(180,120) = 170 \cdot 180 + 225 \cdot 120 = 57,600$$
$$p(C) = p(0,240) = 170 \cdot 0 + 225 \cdot 240 = 54,000$$

Consequently, the company should make $x = 180$ black and white sets, and $y = 120$ color sets to make a maximum profit of \$57,600.

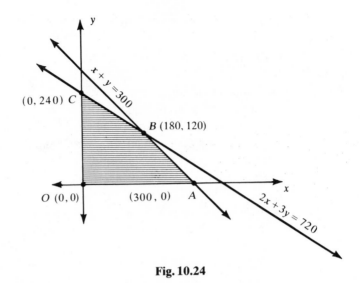

Fig. 10.24

Incidentally, it is interesting to discuss the problem if the company wants to make a profit of $150 on every black and white set, and $225 on every color set (try!).

Exercise 10.7
1. Find the value of $f(x,y)$ at the given points.
   a) $f(x,y) = 3x + 5y$ at $(x,y) = (0,-3)$
   b) $f(x,y) = \frac{3}{4}x - \frac{5}{6}y$ at $(x,y) = (8,12)$
   c) $f(x,y) = \frac{5}{3}(x+y) - \frac{3}{5}(x-y)$ at $(x,y) = (-2,3)$
2. Find the corner points of the polygons determined by the following inequalities.
   a) $x \geq 0, y \geq 0$ b) $x \geq 0, y \geq 0$
      $x \leq 2$ $x + y \leq 5$
      $2x + y \leq 6$ $x - y \leq 3$
                                   $2x - y \leq 4$
   c) $x \geq 0, y \geq 0$ d) $x \geq 0, y \geq 0$
      $x \leq 3, y \leq 2$ $x \leq 6, y \leq 6$
                                   $2x + 3y \leq 24$
3. Maximize the function $f(x,y) = 11x + 7y$ subject to the constraints in problem 2.
4. Minimize the function $g(x,y) = 13x - 5y$ subject to the conditions in problem 2.
5. A certain company makes two types of sweaters: type A and type B. It costs $9 to make a type A sweater and $3 to make a type B sweater. The company can make at most 300 sweaters and spend at most $1800 a day.

The number of sweaters of type B cannot exceed by more than 100 the sweaters of type A. It makes a profit of $5 for every sweater of type A and $3 for every sweater of type B.

a) Find the linear constraints.

b) Find the corner points.

c) What is the objective function?

d) How many sweaters of each type should the company make a day to yield a maximum profit?

e) What is the maximum profit?

6. Paul wants to invest part of his $4000 in business and deposit the rest at a savings bank that pays 6% interest annually. Since investing in business is riskier, he does not want to invest more than three times what he would deposit in the bank. He expects to make a profit of 15% from investment in business. How should he divide his money to make maximum total income? What will be the total return?

7. The Alpha and Omega Company makes two types of batteries: type A and type B. It wants to make at least 600 batteries, but not more than 800, a week. But the company does not have enough resources to make more than 500 type A batteries or 700 type B batteries a week. About 8% of the type A batteries and 5% of the type B batteries are found to be damaged after production. How many batteries of each type should be manufactured a week if the company wants to minimize the number of damaged batteries and still wants to make at least 100 type A batteries a week?

8. An oil company has 30,000 gallons of gasoline at storage place A and 60,000 gallons of gasoline at storage place B. It wants to transport 20,000 gallons to Boston and 50,000 to New York. It costs 4¢ and 5¢ a gallon to transport from location A to Boston and New York, respectively, while it costs 3¢ and 6¢ a gallon to transport from location B to these two cities, respectively. If the company has to transport at least some gasoline from each of the locations, how should gasoline be distributed so as to minimize the cost of transportation? What is the minimum cost?

10.8 SUMMARY AND COMMENTS

A relation R from a set A to a set B is a subset of the product set $A \times B$, so there are as many relations from A to B as there are subsets of $A \times B$. The set of all first elements of the ordered pairs in R is the domain of the relation and the set of all second elements gives its range. Thus, $\text{dom}(R) \subseteq A$ and $\text{range}(R) \subseteq B$. A relation on A is simply a relation from A to A.

The inverse R^{-1} of a relation $R \subseteq A \times B$ is a relation from B to A obtained by interchanging the first element with the second element of each ordered pair in R. $\text{Dom}(R^{-1}) = \text{range}(R)$ and $\text{range}(R^{-1}) = \text{dom}(R)$.

The following important special relations on a set A were discussed and illustrated in detail:

a) A relation R on A is reflexive if $(x,x) \in R$ for every $x \in A$.

b) A relation R on A is symmetric if $(x,y) \in R$ implies $(y,x) \in R$ for every $x,y \in A$.

c) A relation R on A is transitive if $(x,y) \in R$ and $(y,z) \in R$ implies $(x,z) \in R$ for every $x,y,z \in A$.

d) A relation is an equivalence relation if it is reflexive, symmetric, and transitive.

These properties are of great importance in advanced mathematics. We observed that the relation "is divisible by 3," which is only a special case of the relation "is divisible by a positive integer m," is an equivalence relation and partitions the set of integers into three mutually disjoint nonempty sets. This relation, its important properties, and applications are the topics of discussion in the next chapter.

A function is only a special case of a relation. The concept of a function is so basic, useful, and powerful that it is used in all branches of mathematics on a large scale.

A relation $f \subseteq A \times B$ is a function from A to B if $\mathrm{dom}(f) = A$ and every element of A is associated with a unique element of B. The set B is the codomain of the function. If $(x,y) \in f$, we usually write $y = f(x)$. We say that y is the image of x and x a preimage of y under f.

The post office function, which is an example of a class of functions called *step functions*, was discussed and graphed. The absolute value function was also discussed and graphed.

A few special functions were discussed:

a) $f : A \to B$ is a constant function if its range contains a single element.

b) $f : R \to R$ is a linear function if it is of the form $f(x) = ax + b$ with a,b in R.

c) $f : A \to A$ is the identity function if $f(x) = x$ for every x in A.

d) $f : A \to B$ is a 1-1 correspondence if every element in B is paired with a single element in A.

e) $f : A \to A$ is a permutation if it is a 1-1 correspondence.

More about permutations will be discussed in Chapter 12. They will be used to form new sets with several interesting properties.

In Section 10.6, we introduced linear inequalities in two variables:

$$lx + my + n < 0,\ lx + my + n \leq 0,\ lx + my + n > 0,\ \text{or}\ lx + my + n \geq 0$$

A solution of such an inequality is an ordered pair (x,y) of real numbers. We observed that a set of linear inequalities defines a polygonal region.

Linear inequalities play a significant role in solving linear programming problems, as observed in Section 10.7. A linear programming problem consists of finding the optimal value of an objective function, subject to a set of linear constraints.

SUGGESTED READING

Einhorn, E., "Graphing and the Centigrade-Fahrenheit Relationship," *School Science and Mathematics*, vol. 69 (Feb. 1969), pp. 89–91.

Miller, C. D., and V. E. Heeren, *Mathematical Ideas, an Introduction* (2nd ed.), Scott, Foresman and Co., Glenview, Ill. (1973), pp. 126–159.

Pierson, Robert C., "Elementary Graphing Experiences," *The Arithmetic Teacher*, vol. 16 (March 1969), pp. 199–201.

Read, C. B., "The Treatment of the Concept of Functions," *School Science and Mathematics*, vol. 69 (Nov. 1969), pp. 695–696.

Shurlow, Harold, J., "The Greatest Integer Function," *The Mathematics Teacher*, vol. 57 (April 1964), pp. 226–227.

Sorgenfrey, Robert H., "Relations," *The Arithmetic Teacher*, vol. 14 (Oct. 1967), pp. 468–469.

Willerding, M. F., and R. A. Hayward, *Mathematics, the Alphabet of Science* (2nd ed.), John Wiley and Sons, Inc., New York (1972), pp. 141–163.

Wiscamb, Margaret, "Graphing True-False Statements," *The Mathematics Teacher*, vol. 62 (Nov. 1969), pp. 553–556.

"The invention of the symbol ≡ by Gauss affords a striking example of the advantage which may be derived from an appropriate notation, and marks an epoch in the development of the science of arithmetic."

G. B. MATHEWS

11/Modular
Systems

After studying this chapter, you should be able to:
- *check if an integer is congruent to another integer modulo a positive integer*
- *find the sum and the product of any two clock numbers*
- *name and use the basic properties of the clock number system*
- *find the sum and the product of any two calendar numbers*
- *name and use the basic properties of the calendar number system*
- *solve equations of the form $ax + b = c$ in I_{12} and I_7*
- *find the remainder when an integer is divided by another using the congruence relation*
- *check if an integer is divisible by* 2, 3, 4, 6, 8, 9, 10, *or* 11

11.0 INTRODUCTION

Recall that the number systems we discussed in Chapters 3–8 share several interesting properties. All are examples of infinite mathematical systems. It is the purpose of this chapter to introduce and discuss mathematical systems that, though they have several fascinating properties in common with the number systems, differ from them in at least one way—they are finite mathematical systems. They are introduced with the help of an extremely useful and important relation in number theory. Some simple and useful applications of this relation will also be discussed. Everyone is familiar with these mathematical systems and uses them daily without even realizing it.

11.1 THE CONGRUENCE RELATION

The concept of congruence relation in number theory has been introduced in Section 10.3. We now present it a little more carefully and discuss some of its important properties.

Recall that an integer x is said to be divisible by a nonzero integer y if there exists an integer k such that $x = k \cdot y$. Notice that $23 - 1$ is divisible by 2, since $23 - 1 = 22 = 2 \cdot 11$. We then say that 23 *is congruent to* 1 *modulo* 2. Observe that 1 is the remainder obtained on dividing 23 by 2 using the long division method. Similarly, 38 is congruent to 2 modulo 6, since $38 - 2 = 36$ is divisible by 6. Also, 45 is congruent to 0 modulo 5, since $45 - 0 = 45$ is divisible by 5. Let's now make the following definition:

Definition 11.1 Let x and y be any integers and m a fixed positive integer. Then x *is said to be congruent to* y *modulo m* if and only if the difference $x - y$ is divisible by m. We then write $x \equiv y(\bmod m)$. If x is not congruent to y modulo m, we write $x \not\equiv y(\bmod m)$. The relation "$\equiv$" defined by the phrase "is congruent to" on the set I of integers is called the *congruence relation*. The positive integer m is called the *modulus* of the congruence relation.

Gauss has been given the credit of introducing the symbol "$\equiv$" to number theory. This notation is so useful and convenient that proofs of results concerning the congruence relation can be presented easily and elegantly, as will be seen in our later discussions.

Example 11.1
$11 \equiv 1(\bmod 5)$ since 5 divides $11 - 1 = 10$
$11 \equiv 4(\bmod 7)$ since 7 divides $11 - 4 = 7$
$23 \equiv 2(\bmod 7)$ since 7 dviides $23 - 2 = 21$
$17 \equiv 2(\bmod 3)$ since 3 divides $17 - 2 = 15$
$3 \equiv 42(\bmod 13)$ since 13 divides $3 - 42 = -39$
$15 \not\equiv 4(\bmod 6)$ since 6 does not divide $15 - 4 = 11$

Example 11.2 Solve the equation $x \equiv 2(\bmod 7)$.

Solution: We wish to find all those integers x having the property that $x - 2$ is divisible by 7. Now, $x - 2$ is divisible by 7 if and only if $x - 2 = 7k$ for some integer k, i.e., if and only if $x = 2 + 7k$. Thus $x \equiv 2(\bmod 7)$ if and only if x is of the form $2 + 7k$ for some integer k. For example,

$$2 \equiv 2(\bmod 7) \quad (\text{where } k = 0)$$
$$9 \equiv 2(\bmod 7) \quad (\text{where } k = 1)$$
$$-5 \equiv 2(\bmod 7) \quad (\text{where } k = -1)$$
$$-12 \equiv 2(\bmod 7) \quad (\text{where } k = -2)$$

Thus, every integer of the form $2 + 7k$ is a solution of the given equation and $\{2 + 7k \mid k \in I\}$ is the solution set of the given equation.

Example 11.3   Solve the equation $x \equiv -1 (\text{mod } 5)$.

Solution: x is a solution of $x \equiv -1 (\text{mod } 5)$ if and only if $x + 1$ is a multiple of 5. Now, $x + 1$ is a multiple of 5 if and only if $x + 1 = 5n$ for some integer n, i.e., if and only if $x = -1 + 5n = 4 + 5(n - 1)$, where n and hence $(n - 1)$ are arbitrary integers. Thus, every integer of the form $4 + 5k$ is a solution and $\{4 + 5k \mid k \in I\}$ is the solution set of the given equation.

Notice that $29 \equiv 14 (\text{mod } 5)$. Also, 29 and 14 have the same remainder, 4, when divided by 5. More generally, we have the following result:

Theorem 11.1   *Let a and b be any two integers and m a fixed positive integer. Then $a \equiv b \ (mod \ m)$ if and only if they have the same remainder when divided by m.*

Proof: We will prove only one-half of the theorem, leaving the other half for the reader. Let $a \equiv b (\text{mod } m)$. Let q_1 and q_2 be the quotients, r_1 and r_2 the remainders when a and b are divided by m. Recall, by the division algorithm (Theorem 5.1), that $a = q_1 m + r_1$ and $b = q_2 m + r_2$, where $0 \leq r_1 < m$ and $0 \leq r_2 < m$. We want to show that $r_1 = r_2$. We have

$$a - b = (q_1 m + r_1) - (q_2 m + r_2)$$
$$= (q_1 - q_2)m + (r_1 - r_2) \qquad \text{where } 0 \leq (r_1 - r_2) < m$$

Since $a - b$ is given to be a multiple of m and $r_1 - r_2 < m$, it follows that $r_1 - r_2 = 0$. Thus $r_1 = r_2$.

Example 11.4   Observe that $38 \equiv 17 (\text{mod } 7)$ since 7 divides $38 - 17 = 21$. The remainder when 38 is divided by 7 is 3 and that when 17 is divided by 7 is also 3.

Example 11.5   The integers 43 and 28 have the same remainder, 3, when divided by 5. Thus $43 \equiv 28 (\text{mod } 5)$.

Theorem 11.1 shows that remainders play an important role in the study of congruences. It can be used as an alternate definition of the congruence relation; however, Definition 11.1 is more easy to work with.

The relations of equality and congruence have several properties in common. It is well known that the equality relation is an equivalence relation. That the congruence relation is also an equivalence relation is given by the following theorem:

Theorem 11.2   *The congruence relation is an equivalence relation.*

Before we close this section, we discuss two more properties of the congruence relation that are important in our development of finite mathematical systems.

Theorem 11.3   *If $a \equiv b (mod \ m)$ and $c \equiv d (mod \ m)$, then*
1. $a + c \equiv b + d (mod \ m)$ *(addition property),*
2. $ac \equiv bd (mod \ m)$ *(multiplication property).*

Proof: We shall prove only the first half of the theorem, leaving the other half for the reader.

1. $a \equiv b(\bmod\ m)$ implies $a - b = lm$ for some integer l.

$c \equiv d(\bmod\ m)$ implies $c - d = km$ for some integer k.

$$(a + c) - (b + d) = (a - b) + (c - d)$$
$$= lm + km$$
$$= (l + k)m$$

Therefore, $a + c \equiv b + d(\bmod\ m)$.

This theorem means roughly that two congruences can be added as well as multiplied.

Exercise 11.1

1. Mark *true* or *false* (x, y, k are any integers).
   a) $15 \equiv 0(\bmod\ 5)$
   b) $10 \equiv 1(\bmod\ 3)$
   c) $29 \not\equiv 3(\bmod\ 5)$
   d) $10^2 \equiv 1(\bmod\ 3)$
   e) The congruence relation is symmetric.
   f) The congruence relation is an equivalence relation.
   g) If $x \equiv 0(\bmod\ 2)$ then x is even.
   h) If $x \equiv 1(\bmod\ 3)$ then x is odd.
   i) If $x \equiv y(\bmod\ m)$ then $2 + x \equiv 2 + y(\bmod\ m)$.
   j) If $x \equiv y(\bmod\ m)$ then $2x \equiv 2y(\bmod\ m)$.
2. Give a counterexample to each of the false statements in problem 1.
3. Find 3 values for x satisfying each of the following equations:
   a) $x \equiv 1(\bmod\ 4)$ b) $x \equiv -3(\bmod\ 5)$
   c) $x \equiv 7(\bmod\ 8)$ d) $x \equiv -12(\bmod\ 7)$
   e) $x \equiv 12(\bmod\ 7)$ f) $x \equiv 0\ (\bmod\ 5)$
   g) $x + 3 \equiv 5(\bmod\ 6)$ h) $x - 7 \equiv 3(\bmod\ 10)$
   i) $x + 15 \equiv 11(\bmod\ 25)$
4. Find the solution set of
   a) $x \equiv -1(\bmod\ 3)$ b) $x \equiv 3(\bmod\ 7)$
   c) $x \equiv 0\ (\bmod\ 6)$ d) $x + 2 \equiv 5(\bmod\ 7)$
   e) $x + 5 \equiv -2(\bmod\ 10)$ f) $x - 8 \equiv -10(\bmod\ 11)$
5. *True* or *false*:   Let b be the remainder obtained on dividing a by m. Then $a \equiv b(\bmod\ m)$.
6. *True* or *false*: $a^2 \equiv b^2(\bmod\ m)$ implies $a \equiv b(\bmod\ m)$.
7. Let $a \equiv b(\bmod\ m)$ and $-a \equiv b(\bmod\ m)$. Prove that $2b$ is divisible by m.
★8. Prove that d is a factor of a if and only if $a \equiv 0\ (\bmod\ d)$.

★**9.** Let a and b be integers having the same remainder when divided by the positive integer m. Prove that $a \equiv b \pmod{m}$.

★**10.** If $a \equiv b \pmod{m}$, prove that
   a) $a + c \equiv b + c \pmod{m}$
   b) $a - c \equiv b - c \pmod{m}$

★**11.** If $a \equiv b \pmod{m}$ and $c \equiv d \pmod{m}$, prove that
   a) $ac \equiv bd \pmod{m}$
   b) $ax \equiv bx \pmod{m}$
   c) $a^2 \equiv b^2 \pmod{m}$

★**12.** Prove that every odd prime is congruent to 1 or 3 (mod 4).

11.2 THE CLOCK NUMBER SYSTEM

Let's assume that it is 5 o'clock now, as shown in Fig. 11.1. What will be the time on the clock in 321 hours? The problem, at first glance, looks somewhat difficult. In fact it is not so. To find the time in 321 hours, we move the hour hand of the clock through 321 spaces. Observe that every time we move the hour hand through 12 spaces, we come back where we started, namely at 5 o'clock. After the hour hand makes 26 complete revolutions, that is, in 312 hours, it still comes to 5 o'clock. After 9 more hours, that is, $312 + 9 = 321$ hours, we find that the time is 2 o'clock, as in Fig. 11.2.

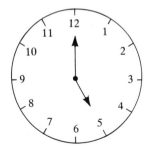

Fig. 11.1

Fig. 11.2

Obviously, this is not the best way to find the answer to the above problem. The wise reader will let the congruence relation do the job for him. By the division algorithm, we have $321 = 12 \cdot 26 + 9$. Therefore $321 \equiv 9 \pmod{12}$ and $321 + 5 \equiv 9 + 5 \pmod{12}$ by Theorem 11.3. Since $9 + 5 \equiv 2 \pmod{12}$, it follows (why?) that $321 + 5 \equiv 2 \pmod{12}$. Hence in 321 hours, the time will be 2 o'clock.

If it is 10 o'clock now, what will be the time in 1331 hours? Observe that $1331 = 12 \cdot 110 + 11$ and hence $1331 \equiv 11 \pmod{12}$. Therefore, $1331 + 10 \equiv 11 + 10 \equiv 21 \equiv 9 \pmod{12}$. Thus, the time will be 9 o'clock in 1331 hours.

We use the numbers $1, 2, \ldots, 12$ to tell the time of day. Based on our experience with counting on a clock, let's define the operation of addition, denoted by + for convenience, on the set {1,2,...,12} and construct the addition table as shown in Table 11.1. First we write the numbers 1 through 12 across the top and down the left side of the table. We enter the sum $a + b$ of a and b in the cell belonging to the row headed a and column headed b. For example, the sum $7 + 9 = 4$ is entered in the cell belonging to the row headed 7 and column headed 9. From the table, it is clear that $x + 12 = x$ for every x in {1,2,...,12}, so that 12 acts like an identity for addition. Just for convenience, let's replace 12 by 0 and then the table can be rewritten as shown in Table 11.2. The numbers 0, 1, 2, . . . , 11 can now be used to tell the time of any day. Accordingly, they are called *clock numbers* or *integers mod* 12. Recall that these are the possible candidates for the remainder when an integer is divided by 12, by the division algorithm. If we look at Table 11.2 carefully, it is clear that the operation of addition on the set I_{12} of clock numbers is defined as follows: the sum $a + b$ of two clock numbers a and b is defined as the remainder obtained on dividing their ordinary sum by 12. The operation + we defined on I_{12} is called *addition mod* 12.

Table 11.1

+	1	2	3	4	5	6	7	8	9	10	11	12
1	2	3	4	5	6	7	8	9	10	11	12	1
2	3	4	5	6	7	8	9	10	11	12	1	2
3	4	5	6	7	8	9	10	11	12	1	2	3
4	5	6	7	8	9	10	11	12	1	2	3	4
5	6	7	8	9	10	11	12	1	2	3	4	5
6	7	8	9	10	11	12	1	2	3	4	5	6
7	8	9	10	11	12	1	2	3	4	5	6	7
8	9	10	11	12	1	2	3	4	5	6	7	8
9	10	11	12	1	2	3	4	5	6	7	8	9
10	11	12	1	2	3	4	5	6	7	8	9	10
11	12	1	2	3	4	5	6	7	8	9	10	11
12	1	2	3	4	5	6	7	8	9	10	11	12

Table 11.2

+	0	1	2	3	4	5	6	7	8	9	10	11
0	0	1	2	3	4	5	6	7	8	9	10	11
1	1	2	3	4	5	6	7	8	9	10	11	0
2	2	3	4	5	6	7	8	9	10	11	0	1
3	3	4	5	6	7	8	9	10	11	0	1	2
4	4	5	6	7	8	9	10	11	0	1	2	3
5	5	6	7	8	9	10	11	0	1	2	3	4
6	6	7	8	9	10	11	0	1	2	3	4	5
7	7	8	9	10	11	0	1	2	3	4	5	6
8	8	9	10	11	0	1	2	3	4	5	6	7
9	9	10	11	0	1	2	3	4	5	6	7	8
10	10	11	0	1	2	3	4	5	6	7	8	9
11	11	0	1	2	3	4	5	6	7	8	9	10

It is clear from Table 11.2 that I_{12} is closed under addition mod 12. That the operation is commutative and associative can be easily verified; for instance, $5+9=2=9+5$; $2+(5+11)=6=(2+5)+11$. Notice that the table is symmetric about the main diagonal going from the upper left-hand corner to the lower right-hand corner. The set of clock numbers has an identity element, 0, for addition mod 12. Every element a of I_{12} has an additive inverse, $-a$, in I_{12}; for example, $-3=9$; $-7=5$. We now ask the reader to find the additive inverses of the remaining clock numbers. To find the additive inverse of an element a from the addition table, locate the row headed by a and the column that contains the additive identity 0 in that row; the number that heads this column gives the additive inverse of a.

Recall that during the development of the system of integers, we observed that subtracting y from x is equivalent to adding the additive inverse $-y$ of y to x; that is, $x-y=x+(-y)$. Let's make use of this fact to define the operation of subtraction on I_{12} as follows: $x-y=x+(-y)$, where $-y$ denotes the additive inverse of y. For example, $3-2=3+(-2)=3+10=1$; $2-7=2+(-7)=2+5=7$. Notice that $3-0=3+(-0)=3+0=3$, whereas $0-3=0+(-3)=$

$0 + 9 = 9$, showing that 0 is not an identity for subtraction and the operation of subtraction is not commutative. Also, it is not associative.

Thus, we observe that the set I_{12} of integers mod 12 possesses the same properties as the set I of integers under the usual addition.

Recall that every equation of the form $a = b + x$ is solvable in the set of integers. We have a similar result in I_{12}, as illustrated by the following examples.

Example 11.6   Solve the equation $x + 3 = 8$ in the clock number system.

Solution: To solve the equation $x + 3 = 8$, we want to isolate the x. This can be done by adding the additive inverse of 3 to both sides of the equation.

$x + 3 = 8$	*given*
$(x + 3) + 9 = 8 + 9$	*addition property*
$x + (3 + 9) = 8 + 9$	*associative property of addition mod* 12
$x + 0 = 5$	*definition of addition mod* 12
$x = 5$	*definition of additive identity*

It can easily be checked that 5 is actually the solution of the given equation.

Example 11.7   Solve the equation $x + 7 = 2$ in the system I_{12}.

Solution: To solve the equation $x + 7 = 2$, let's add the additive inverse of 7 to both sides of the equation.

$x + 7 = 2$	*given*
$(x + 7) + 5 = 2 + 5$	*addition property*
$x + (7 + 5) = 2 + 5$	*associative property of addition mod* 12
$x + 0 = 7$	*definition of addition mod* 12
$x = 7$	*definition of additive identity*

We ask the reader to check that 7 is in fact the solution of the given equation.

We now define a second operation, called *multiplication mod* 12, denoted by $\cdot$, on the set I_{12} of clock numbers. We define $a \cdot b$ of any two clock numbers a and b as the remainder obtained on dividing the ordinary product ab by 12. Table 11.3 illustrates the operation of multiplication mod 12 on I_{12}. It can easily be verified that multiplication mod 12 is repeated addition mod 12, as in the case of the system of whole numbers; for example, $3 \cdot 7 = 7 + 7 + 7 = 2 + 7 = 9$; $4 = 5 + 5 + 5 + 5 = 10 + 10 = 8$. It follows from Table 11.3 that I_{12} is closed under multiplication mod 12. Multiplication mod 12 is both commutative and associative. Multiplication mod 12 is distributive over addition mod 12. The set I_{12} contains an identity element, 1, with respect to multiplication mod 12. Does every element of I_{12} have a multiplicative inverse in I_{12}? It is clear from the multiplication table that 5 is its own multiplicative inverse since

$5 \cdot 5 = 1$; 7 is the multiplicative inverse of itself since $7 \cdot 7 = 1$; however, 2 has no multiplicative inverse since there exists no clock number x such that $2 \cdot x = 1$. Notice that the row (column) headed by 2 does not contain 1 in Table 11.3. A careful study of the multiplication table tells us that those and only those numbers that are relatively prime to 12 have multiplicative inverses.

Table 11.3

·	0	1	2	3	4	5	6	7	8	9	10	11
0	0	0	0	0	0	0	0	0	0	0	0	0
1	0	1	2	3	4	5	6	7	8	9	10	11
2	0	2	4	6	8	10	0	2	4	6	8	10
3	0	3	6	9	0	3	6	9	0	3	6	9
4	0	4	8	0	4	8	0	4	8	0	4	8
5	0	5	10	3	8	1	6	11	4	9	2	7
6	0	6	0	6	0	6	0	6	0	6	0	6
7	0	7	2	9	4	11	6	1	8	3	10	5
8	0	8	4	0	8	4	0	8	4	0	8	4
9	0	9	6	3	0	9	6	3	0	9	6	3
10	0	10	8	6	4	2	0	10	8	6	4	2
11	0	11	10	9	8	7	6	5	4	3	2	1

Recall from Chapter 4 that the product of two integers is zero if and only if one of them is zero (product law). This property, surprisingly enough, does not hold in I_{12}; for example, $2 \neq 0$, $6 \neq 0$, however $2 \cdot 6 = 0$; $3 \neq 0$, $4 \neq 0$, but $3 \cdot 4 = 0$. Also, the cancellation property with respect to multiplication mod 12 does not hold in I_{12}; for example, $2 \cdot 3 = 6 \cdot 3$, both being equal to 6, however, $2 \neq 6$.

The operation of division on I_{12} can be defined as in the case of subtraction. We define $x \div y = x \cdot y^{-1}$, where y^{-1} denotes the multiplicative inverse of y. For example, $4 \div 5 = 4 \cdot 5^{-1} = 4 \cdot 5 = 8$.

Is every equation of the form $ax = b$ with a and b in I_{12} solvable in I_{12}? Clearly $ax = b$ has a solution if a^{-1} exists, in which case the solution is unique and given by $x = a^{-1} \cdot b$. Does this mean that if a^{-1} does not exist, the equation has no solution? Well, let's not jump to any conclusions so soon. Even if a^{-1} fails to exist, it is possible to have no solutions or more than one solution, as demonstrated by the following examples.

Example 11.8   Solve the equation $7x = 3$ in the clock number system.

Solution: To solve the equation $7x = 3$ in I_{12}, we want to isolate the x. Notice that since 7 and 12 are relatively prime, 7^{-1} exists and $7^{-1} = 7$.

$7x = 3$	*given*
$7 \cdot (7x) = 7 \cdot 3$	*multiplication property*
$(7 \cdot 7)x = 7 \cdot 3$	*associative property of multiplication mod* 12
$1 \cdot x = 9$	*definition of multiplication mod* 12
$x = 9$	*definition of multiplicative identity*

We advise the reader to verify that 9 is indeed the actual solution of the given equation.

Example 11.9   Solve the equation $3x = 6$ in I_{12}.

Solution: Since 3 and 12 are not relatively prime, 3^{-1} does not exist and the method illustrated in the above example clearly fails. A solution, as remarked earlier, may or may not exist. From the multiplication table, we notice that $3 \cdot 2 = 6$, $3 \cdot 6 = 6$, and $3 \cdot 10 = 6$. Consequently, the given equation has three solutions: 2, 6, and 10.

Example 11.10   Solve the equation $4x = 3$ in I_{12}.

Solution: Since 4 and 12 are not relatively prime, 4^{-1} does not exist. Consequently, the method demonstrated in Example 11.8 does not work, even if a solution exists. From Table 11.3, it follows that there exists no x such that $4x = 3$. Accordingly, there exists no solution of the gien equation in I_{12}.

Exercise 11.2

1. What will be the time on a 12-hour clock in 245 hours if it is 2 o'clock now? In 520 hours? In 735 hours?
2. What will be the time on a 24-hour clock in 320 hours if it is 2 p.m. now? In 420 hours? In 720 hours?
3. If it is 4 o'clock now on a 6-hour clock, what will be the time on the clock in 110 hours? In 205 hours? In 340 hours?
4. Use Table 11.2 to verify the following.
   a) $2 + 3 = 3 + 2$ b) $5 + 11 = 11 + 5$
   c) $8 + 10 = 10 + 8$ d) $3 + 7 = 7 + 3$
5. Use Table 11.2 to verify the following.
   a) $2 + (3 + 8) = (2 + 3) + 8$ b) $(3 + 11) + 7 = 3 + (11 + 7)$
   c) $5 + (7 + 9) = (5 + 7) + 9$ d) $(2 + 4) + 6 = 2 + (4 + 6)$

6. Use Table 11.3 to verify the following.

   a) $2 \cdot 3 = 3 \cdot 2$

   b) $5 \cdot 11 = 11 \cdot 5$

   c) $8 \cdot 10 = 10 \cdot 8$

   d) $3 \cdot 7 = 7 \cdot 3$

7. Use Table 11.3 to verify each of the following.

   a) $2 \cdot (3 \cdot 8) = (2 \cdot 3) \cdot 8$

   b) $(3 \cdot 11) \cdot 7 = 3 \cdot (11 \cdot 7)$

   c) $5 \cdot (7 \cdot 9) = (5 \cdot 7) \cdot 9$

   d) $2 \cdot (4 \cdot 6) = (2 \cdot 4) \cdot 6$

8. Use Tables 11.2 and 11.3 to verify:

   a) $2 \cdot (3+5) = 2 \cdot 3 + 2 \cdot 5$

   b) $3 \cdot (5+11) = 3 \cdot 5 + 3 \cdot 11$

   c) $4 \cdot (6+8) = 4 \cdot 6 + 4 \cdot 8$

   d) $8 \cdot (7+4) = 8 \cdot 7 + 8 \cdot 4$

9. Find the additive inverse of 1, 4, 5, 6, 8, 9, and 11 in I_{12}.

10. Evaluate each of the following in I_{12}:

   a) $8-3$

   b) $3-8$

   c) $5-11$

   d) $7-3$

11. Find the multiplicative inverse of 1, 3, 5, 6, 7, 8, and 11 in I_{12}, if one exists.

12. Evaluate each of the following in I_{12}, if defined:

   a) $4 \div 5$

   b) $3 \div 3$

   c) $1 \div 7$

   d) $8 \div 4$

13. Solve each of the following equations in I_{12}.

   a) $x+3 = 8$

   b) $x+7 = 3$

   c) $5x = 3$

   d) $4x = 4$

   e) $7x+2 = 8$

   f) $5x+7 = 11$

   g) $x^2 = 0$

   h) $x^2 = 1$

   i) $x^2 = x$

14. Mark *true* or *false*.

   a) The set of clock numbers is closed under addition mod 12.

   b) Addition mod 12 is commutative.

   c) I_{12} possesses an identity under addition mod 12.

   d) Every equation of the form $a = b+x$ with a, b in I_{12} is solvable in I_{12}.

   e) Multiplication mod 12 on I_{12} is commutative.

   f) Multiplication mod 12 on I_{12} is associative.

   g) Multiplication mod 12 is distributive over addition mod 12.

   h) Every element of I_{12} has a multiplicative inverse.

   i) Every equation of the form $a = bx$ with a, b in I_{12} is solvable in I_{12}.

   j) I_{12} is closed under subtraction mod 12.

   k) There is a 1-1 correspondence between the set of clock numbers and the set of months of the year.

15. Give a counterexample to each of the false statements in problem 14.

16. a) Write down addition and multiplication tables for the set I_6 of integers mod 6.

   b) Is the set I_6 of integers mod 6 closed under addition mod 6?

   c) Is the set I_6 of integers mod 6 closed under multiplication mod 6?

   d) Does every element of I_6 have an additive inverse?

   e) Find the additive inverse of 2 and 4.

   f) Which elements of I_6 have multiplicative inverses?

   g) Find the multiplicative inverse of 5.

h) Does I_6 satisfy the cancellation property with respect to multiplication mod 6?

i) Does I_6 satisfy the product law with respect to multiplication mod 6?

11.3 THE CALENDAR NUMBER SYSTEM

Assume that we have a 7-hour clock and it is 2 o'clock now, as shown in Fig. 11.3. We wish to find the time in 120 hours on this clock. Since $120 + 2 = 122 = 7 \cdot 17 + 3$, we have $122 \equiv 3 \pmod 7$. Accordingly, it will be 3 o'clock on the 7-hour clock, as shown in Fig. 11.4.

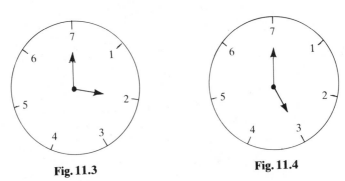

Fig. 11.3   **Fig. 11.4**

Let's now consider the following problem of the same nature. Assume today is Tuesday. What day will it be in 129 days? For convenience, let's label the days Sunday through Saturday by the digits 0 through 6, respectively. To find the day 129 days from Tuesday (labeled 2), we need only find the whole number less than 7 and congruent to $129 + 2 = 131 \pmod 7$. Since $131 = 7 \cdot 18 + 5$, we have $131 \equiv 5 \pmod 7$. Consequently, it will be Friday in 129 days.

The numbers $0, 1, \ldots, 6$ used to label the days of the week are called *calendar numbers* or *integers mod* 7. Recall that these are the possible remainders when an integer is divided by 7. As in the case of the operations of addition and multiplication mod 12 on the set of clock numbers, we define the operations of addition and multiplication mod 7 on the set $I_7 = \{0,1,2,3,4,5,6\}$ of calendar numbers as follows: the sum $a + b$ of two calendar numbers a and b is defined as the remainder obtained on dividing their ordinary sum by 7; for example, $5 + 6 = 4$ and $4 + 5 = 2$; the product $a \cdot b$ of two calendar numbers a and b is defined as the remainder when their ordinary product is divided by 7; for example, $3 \cdot 4 = 5$ and $5 \cdot 6 = 2$. Tables 11.4 and 11.5 show all possible ways of adding and multiplying any two calendar numbers mod 7.

Let's now discuss the properties of the set I_7 of integers mod 7 under these operations. It follows clearly from Tables 11.4 and 11.5 that I_7 is closed under addition and multiplication mod 7. This is so since when an integer is divided by 7, the possible candidates for the remainder are 0, 1, 2, 3, 4, 5, and 6. That the operations of addition and multiplication mod 7 are both commutative and associative can be verified. The set I_7 possesses an identity element with respect to addition and multiplication mod 7, namely, 0 and 1, respectively. Every element a of I_7 has an additive inverse, $-a$, in I_7; for example, $-3 = 4$ and $-2 = 5$. Every nonzero element a of I_7 has a multiplicative inverse a^{-1} in I_7; for example, $3^{-1} = 5$ and $4^{-1} = 2$. We now advise the reader to find the inverses of the other nonzero calendar numbers. Also, multiplication mod 7 distributes over addition mod 7.

Table 11.4

+	0	1	2	3	4	5	6
0	0	1	2	3	4	5	6
1	1	2	3	4	5	6	0
2	2	3	4	5	6	0	1
3	3	4	5	6	0	1	2
4	4	5	6	0	1	2	3
5	5	6	0	1	2	3	4
6	6	0	1	2	3	4	5

Table 11.5

·	0	1	2	3	4	5	6
0	0	0	0	0	0	0	0
1	0	1	2	3	4	5	6
2	0	2	4	6	1	3	5
3	0	3	6	2	5	1	4
4	0	4	1	5	2	6	3
5	0	5	3	1	6	4	2
6	0	6	5	4	3	2	1

We now define the operations of subtraction and division on I_7 as $x - y = x + (-y)$ and $x \div y = x \cdot y^{-1}$ whenever y^{-1} exists. For example, $5 - 3 = 5 + (-3) = 5 + 4 = 2$; $3 - 5 = 3 + (-5) = 3 + 2 = 5$; $5 \div 3 = 5 \cdot 3^{-1} = 5 \cdot 5 = 4$; and $3 \div 4 = 3 \cdot 4^{-1} = 3 \cdot 2 = 6$. Since every element of I_7 has an additive inverse, it follows (why?) that I_7 is closed under subtraction. Since every nonzero element of I_7 has a multiplicative inverse, the set of nonzero calendar numbers is closed under division (why?).

Unlike the clock number system, the cancellation property with respect to multiplication mod 7 holds in the calendar number system (why?). Also, the product law holds; that is, the product of two nonzero calendar numbers can never be zero.

Example 11.11   Solve the equation $x + 2 = 5$ in the calendar number system.
Solution:

$$x + 2 = 5 \qquad \textit{given}$$

$$(x + 2) + 5 = 5 + 5 \qquad \textit{addition property}$$

$x+(2+5)=5+5$ *associative property of addition mod* 7

$x+0=3$ *definition of addition mod* 7

$x=3$ *definition of additive identity*

It can be verified that 3 is in fact the solution of the given equation.

Example 11.12   Solve the equation $3x=5$ in I_7.

Solution: To solve the equation $3x=5$ in I_7, we have to isolate the x. Notice that 3 has a multiplicative inverse, namely 5.

$3x=5$ *given*

$5\cdot(3x)=5\cdot5$ *multiplication property*

$(5\cdot3)x=5\cdot5$ *associative property of multiplication mod* 7

$1\cdot x=4$ *definition of multiplication mod* 7

$x=4$ *definition of multiplicative identity*

We ask the reader to verify that 4 is indeed the solution of the given equation.

Exercise 11.3

1. Mr. Smith left on a Friday for a world tour. If he took 46 days for the tour, on what day did he come back?
2. If today is Tuesday, what day will it be in 120 days? In 210 days? In 350 days?
3. Use Table 11.4 to verify the following.
   a) $5+4=4+5$ b) $3+6=6+3$
   c) $3+1=1+3$ d) $5+6=6+5$
   e) $4+(2+3)=(4+2)+3$ f) $5+(4+2)=(5+4)+2$
   g) $2+(4+6)=(2+4)+6$ h) $5+(3+6)=(5+3)+6$
4. Use Table 11.5 to verify each of the following.
   a) $5\cdot4=4\cdot5$ b) $3\cdot6=6\cdot3$
   c) $3\cdot1=1\cdot3$ d) $5\cdot6=6\cdot5$
   e) $4\cdot(2\cdot3)=(4\cdot2)\cdot3$ f) $5\cdot(4\cdot2)=(5\cdot4)\cdot2$
   g) $2\cdot(4\cdot6)=(2\cdot4)\cdot6$ h) $5\cdot(3\cdot6)=(5\cdot3)\cdot6$
5. Use Tables 11.4 and 11.5 to verify each of the following.
   a) $2\cdot(3+5)=2\cdot3+2\cdot5$ b) $3\cdot(4+5)=3\cdot4+3\cdot5$
6. Find the additive inverse of 3, 4, and 6 in the calendar number system.
7. Find the multiplicative inverse of 2, 5, and 6 in the calendar number system.
8. Evaluate each of the following in the system of integers mod 7.
   a) $1-3$ b) $3-1$ c) $5-3$ d) $3-5$
   e) $1\div3$ f) $3\div1$ g) $5\div3$ h) $3\div5$
9. Solve each of the following equations in I_7.
   a) $x+5=2$ b) $x+6=3$ c) $4x=5$

d) $3x = 3$ **e)** $3x + 4 = 5$ **f)** $4x + 2 = 6$

g) $x^2 = 0$ **h)** $x^2 = 1$ **i)** $x^2 = x$

10. Mark *true* or *false*.

   a) I_7 is a finite set.

   b) I_7 is closed under the usual addition.

   c) I_7 is closed under addition mod 7.

   d) The set of nonzero integers mod 7 is closed under multiplication mod 7.

   e) Multiplication mod 7 on I_7 is commutative.

   f) Every element of I_7 has an inverse under addition mod 7.

   g) Every element of I_7 has an inverse under multiplication mod 7.

   h) I_7 is closed under subtraction mod 7.

   i) I_7 is closed under division mod 7.

   j) The set of nonzero calendar numbers is closed under division mod 7.

   k) Product law holds in I_7.

   l) The cancellation property with respect to multiplication mod 7 holds in I_7.

   m) There exists a 1-1 correspondence between the set of calendar numbers and the set of days of a week.

11. Give a counterexample to each of the false statements in problem 10.

12. a) Construct addition and multiplication tables for the set I_5 of integers mod 5.

   b) Is I_5 closed under addition mod 5?

   c) Is I_5 closed under multiplication mod 5?

   d) Does every element of I_5 have an additive inverse?

   e) Find the additive inverse of 2 and 4 in I_5.

   f) Which elements of I_5 have multiplicative inverses?

   g) Find the multiplicative inverse of 2 and 4 in I_5.

   h) Does I_5 satisfy the cancellation property with respect to multiplication mod 5?

   i) Does I_5 satisfy the product law with respect to multiplication mod 5?

11.4 CASTING OUT NINES

The remaining sections of this chapter are devoted to the discussion of some of the simple and useful applications of the theory of congruences. One of the simplest applications is to check the accuracy of arithmetical operations in computational work. The procedure of *casting out nines*, which originated among the Arabs of the eighth century, is based on Theorem 11.4, given below. It did not become popular in this country until the eighteenth century. The procedure might seem silly with the advent of modern, high-speed calculating machines, but the principle is fascinating.

Consider the integer $x = 2{,}387$. The sum of its digits is $y = 2+3+8+7 = 20$. Observe that $x - y = 2387 - 20 = 2367 = 9 \cdot 263$. Thus $x - y$ is a multiple of 9 or $x \equiv y \pmod 9$. Is this true in general? Is every integer congruent to the sum of its digits modulo 9? This is answered in the affirmative by the following theorem:

Theorem 11.4 *Every integer N is congruent to the sum of its digits mod* 9.

Example 11.13 Consider the integer $N = 53,872$. By the above theorem,

$$53872 \equiv (5+3+8+7+2)(\text{mod } 9)$$

$$\equiv 25(\text{mod } 9)$$

which is true since $53872 - 25 = 9 \cdot 5983$, a multiple of 9.

Theorem 11.4 tells us that under congruence mod 9, we can replace an integer by the sum of its digits. This procedure is usually referred to as *casting out nines*, since in the sum of the digits, digits that add up to 9 can be cancelled or "cast out." For example,

$$53872 \equiv (5+3+8+\cancel{7}+\cancel{2})(\text{mod } 9)$$

$$\equiv (5+2+\cancel{1}+\cancel{8})(\text{mod } 9)$$

$$\equiv 7(\text{mod } 9)$$

Thus 7 is the remainder r (recall $0 \leq r < 9$) when 53,872 is divided by 9. This illustrates a simple procedure to find the remainder when an integer is divided by a positive integer.

Example 11.14 Find the remainder when 1,234,321 is divided by 9.

Solution: $\quad 1234321 \equiv (1+2+3+\cancel{4}+\cancel{3}+\cancel{2}+1)(\text{mod } 9)$

$$\equiv (1+2+3+1)(\text{mod } 9)$$

$$\equiv 7(\text{mod } 9)$$

Thus the remainder is 7 when 1,234,321 is divided by 9.

Example 11.15 Find the remainder when 16^{53} is divided by 7.

Solution: Notice that $16 \equiv 2(\text{mod } 7)$. Thus, by Theorem 11.3,

$$16^{53} \equiv 2^{53}(\text{mod } 7)$$

Since $2^3 \equiv 1(\text{mod } 7)$, we have

$$2^{53} = 2^{51+2} = 2^{3 \cdot 17 + 2} = (2^3)^{17} \cdot 2^2$$

$$\equiv (1)^{17} \cdot 4(\text{mod } 7)$$

$$\equiv (1) \cdot 4(\text{mod } 7)$$

$$\equiv 4(\text{mod } 7)$$

Thus the remainder when 16^{53} is divided by 7 is 4.

Notice the tremendous advantage of the congruence relation in computing the remainder even when a large integer is divided by a positive integer.

The procedure of casting out nines to check the accuracy of computational work is illustrated in the following examples.

Example 11.16 The sum of the integers 351, 3569, and 24387 is given as 28307. Check this computation by casting out nines.

Solution:

$$351 \equiv 3+5+1 \qquad\qquad \equiv 0 \pmod 9$$
$$3569 \equiv 3+5+6+9 \qquad \equiv 5 \pmod 9$$
$$24387 \equiv 2+4+3+8+7 \equiv 6 \pmod 9$$
$$\text{sum} \equiv \qquad\qquad\qquad\qquad \equiv 11 \equiv 2 \pmod 9$$

We have expressed each integer congruent to a nonnegative integer less than 9, using Theorem 11.4. Adding the integers on the right of the congruences, we obtain $0+5+6 = 11 \equiv 2 \pmod 9$. Thus, if the given sum 28307 of the given integers is correct, then it must also be congruent to $2 \pmod 9$. We now observe that

$$28307 \equiv 2+8+3+0+7 \equiv 2 \pmod 9$$

Since both are congruent to the same number, 2, we have some guarantee that the given answer *could probably be correct.* This procedure is only a partial check; it is not, by any means, foolproof. However, if the two numbers are not the same, we could have immediately concluded that some error had occurred somewhere in the computation.

Example 11.17 Use casting out nines to check if the sum of the numbers 3521, 6783, 29352, and 83721 is 132,377.

Solution: We have

$$3521 \equiv 3+5+2+1 \qquad \equiv 2 \pmod 9$$
$$6783 \equiv 6+7+8+3 \qquad \equiv 6 \pmod 9$$
$$29352 \equiv 2+9+3+5+2 \equiv 3 \pmod 9$$
$$83721 \equiv 8+3+7+2+1 \equiv 3 \pmod 9$$
$$\text{sum} \equiv \qquad\qquad\qquad\qquad \equiv 14 \equiv 5 \pmod 9$$

Consequently, if the given sum is to be the correct answer, it must be congruent to $5 \pmod 9$. Observe now that 132377 is indeed congruent to $5 \pmod 9$, since

$$132377 \equiv 1+3+2+3+7+7 \pmod 9$$
$$\equiv 5 \pmod 9$$

This gives us some confidence that 132,377 *could probably be the correct sum;* indeed, it is not; the actual sum is 123,377.

From this example, we observe that any rearrangement of the digits of an integer does not affect the sum of its digits under congruence.

Example 11.18 Use casting out nines to determine if the sum of the integers 2316, 3893, and 21837 is 27,046.

Solution: Using Theorem 11.4, we have

$$2316 \equiv 2+3+1+6 \qquad \equiv 3(\text{mod } 9)$$
$$3893 \equiv 3+8+9+3 \qquad \equiv 5(\text{mod } 9)$$
$$\underline{21837 \equiv 2+1+8+3+7 \equiv 3(\text{mod } 9)}$$
$$\text{sum} \equiv \qquad\qquad\qquad \equiv 11 \equiv 2(\text{mod } 9)$$

Thus, if 27,046 is to be the correct sum, then it must be congruent to 2(mod 9); however, $27046 \equiv 2+7+0+4+6 \equiv 1(\text{mod } 9)$. Consequently, we conclude that 27,046 is definitely not the actual sum; the correct sum is 28,046.

Example 11.19 Use casting out nines to determine if the difference $7823 - 2732$ is 5091.

Solution: We have

$$7823 \equiv 7+8+2+3 \equiv 2(\text{mod } 9)$$
$$\underline{2732 \equiv 2+7+3+2 \equiv 5(\text{mod } 9)}$$
$$\text{difference} \equiv \qquad\qquad \equiv 2-5 \equiv -3 \equiv 6(\text{mod } 9)$$

Since $5091 \equiv 5+0+9+1 \equiv 6(\text{mod } 9)$, the given difference could probably be correct; in fact, it is correct.

Example 11.20 Use casting out nines to determine if the product of 328 and 35 is 11,380.

Solution:

$$328 \equiv 3+2+8 \equiv 4(\text{mod } 9)$$
$$\underline{35 \equiv 3+5 \qquad \equiv 8(\text{mod } 9)}$$
$$\text{product} \equiv \qquad\qquad \equiv 32 \equiv 5(\text{mod } 9)$$

Thus the product of 328 and 35 must be congruent to 5(mod 9). Since

$$11380 \equiv 1+1+3+8+0 \equiv 4(\text{mod } 9)$$

11,380 cannot be the product of 328 and 35. It can be seen that their product is 11,480.

Before we close this section, we discuss a simple and, at the same time, interesting puzzle that is known even to some high school students. Consider an integer $d_n d_{n-1} \ldots d_1 d_0$; rearrange its digits in any way; subtract the smaller number from the larger; if except for one nonzero digit, all the digits, including zeros if any, of this new number are stated, then the remaining nonzero digit can be given easily. Let's now illustrate this puzzle. Assume Don wants to try this puzzle with his girl friend, Linda. She thinks of the number 5862 and rearranges its digits to get 2658. Subtracting 2658 from 5862, she gets 3204. She then tells Don the digits 3, 0, and 4 of this new number 3204. What does Don have to do to find the remaining nonzero digit? Well, observe that

$$5862 \equiv 5+8+6+2 \equiv 3(\text{mod } 9)$$
$$\underline{2658 \equiv 2+6+5+8 \equiv 3(\text{mod } 9)}$$
$$\text{difference} \equiv \qquad\qquad \equiv 0(\text{mod } 9)$$

Therefore, his problem now is simply to find the digit x such that $3+0+4+x \equiv 0 \pmod 9$. Obviously $x = 2$. Thus the remaining digit is 2!

Exercise 11.4

1. Find the remainder when
   a) 36982 is divided by 9
   b) 463891 is divided by 9
   c) 2^{35} is divided by 7
   d) 5^{31} is divided by 12
   e) 3^{21} is divided by 11
   f) 5^{13} is divided by 11
   g) $3^{21}+5^{13}$ is divided by 11
   h) 5^{13} is divided by 13

2. Which of the following numbers are divisible by 9?
   a) 2356
   b) 813267
   c) 123321
   d) 5315678

3. Use casting out nines to classify each of the following computations as probably correct or definitely wrong.

   a)
   $$
   \begin{array}{r}
   357 \\
   2648 \\
   +13879 \\
   \hline
   16884
   \end{array}
   $$

   b)
   $$
   \begin{array}{r}
   8325 \\
   3773 \\
   1523 \\
   +4972 \\
   \hline
   18564
   \end{array}
   $$

   c)
   $$
   \begin{array}{r}
   1776 \\
   1948 \\
   1962 \\
   +1976 \\
   \hline
   7762
   \end{array}
   $$

   d)
   $$
   \begin{array}{r}
   8356 \\
   -7267 \\
   \hline
   1089
   \end{array}
   $$

   e)
   $$
   \begin{array}{r}
   3918 \\
   -1976 \\
   \hline
   1042
   \end{array}
   $$

   f)
   $$
   \begin{array}{r}
   123 \\
   \times 45 \\
   \hline
   3535
   \end{array}
   $$

   g)
   $$
   \begin{array}{r}
   235 \\
   \times 26 \\
   \hline
   6110
   \end{array}
   $$

   ★h)
   $$
   \begin{array}{r}
   3250 \\
   \div 26 \\
   \hline
   125
   \end{array}
   $$

4. Use casting out nines to find the missing nonzero digit d in each of the following:
   a) $7961 - 1976 = 59d5$
   b) $7167 - 1776 = 53d1$
   c) $253 \cdot 86 = 2d758$
   d) $123 \cdot 98 = 120d4$

★5. Let m be the integer obtained by rearranging the digits of an integer n. Prove that $m - n$ is divisible by 9.

11.5 DIVISIBILITY TESTS

We wish to know if there are simple tests that can be applied to check if an integer is divisible by another integer. It is well known among high school students that an integer is divisible by 2 if and only if its last digit is divisible by 2, divisible by 5 if and only if its last digit is either 0 or 5, and divisible by 10 if and only if its last digit is 0. But have you ever asked yourself, why do they work? The theory of congruences provides us with the answer.

Recall that any integer in base 10 can be written

$$N = d_n 10^n + d_{n-1} 10^{n-1} + \cdots + d_2 10^2 + d_1 10^1 + d_0, \text{ where}$$

$$0 \le d_k \le 9 \text{ and } d_n \ne 0$$

$$\equiv d_0 \pmod 2$$

(Why?) Consequently, N is divisible by 2 if and only if its units digit is divisible by 2. Since

$$N \equiv 10d_1 + d_0 (\text{mod } 4)$$

N is divisible by 4 if and only if the two-digit number $d_1 d_0$ is divisible by 4. More generally, N is divisible by 2^k if and only if the k-digit number $d_{k-1} d_{k-2} \ldots d_1 d_0$ is divisible by 2^k.

Example 11.21   Consider the integer 735,428. Since its last digit is even, it is divisible by 2. Since 28 is divisible by 4, we know 735,428 is divisible by 4. Since 8 does not divide 428, we know 735,428 is not divisible by 8. Notice that $735,428 = 8 \cdot 91928 + 4$.

Since $N \equiv d_0 (\text{mod } 5)$, it follows that N is divisible by 5 if and only if $d_0 \in \{0,5\}$.

Example 11.22   The integers 1960 and 32895 are divisible by 5, whereas 9601 is not. (Why?)

Theorem 11.4 provides us with a divisibility test for 9. Recall that $N \equiv (d_n + d_{n-1} + \cdots + d_1 + d_0)(\text{mod } 9)$. Therefore N is divisible by 9 if and only if the sum of its digits is divisible by 9. Since $10 \equiv 1(\text{mod } 3)$, it follows that N is divisible by 3 if and only if the sum of its digits is divisible by 3.

Example 11.27   The integer 735,891 is divisible by 3 since the sum of its digits $= 7+3+5+8+9+1 = 33 \equiv 0(\text{mod } 3)$. It is not divisible by 9 since $7+3+5+8+9+1 = 33 \not\equiv 0(\text{mod } 9)$.

In order to check if an integer is divisible by 6, we need only check if it is divisible by 2 and 3.

Example 11.27   Determine if 725,718 is divisible by 6.

Solution: 725,718 is divisible by 2 since its last digit is even. Since the sum of its digits, 30, is divisible by 3, it is divisible by 3 also. Therefore 725,718 is divisible by 6.

We now give a simple test to check if an integer is divisible by 11.

Theorem 11.5   *An integer is divisible by 11 if and only if the sum of its digits in the even places minus the sum of its digits in the odd places is divisible by 11; that is, if $N = d_n d_{n-1} \ldots d_2 d_1 d_0$, then N is divisible by 11 if and only if*

$$(\cdots + d_4 + d_2 + d_0) - (\cdots + d_3 + d_1) \text{ is divisible by 11.}$$

Example 11.25   Determine if 648,362 is divisible by 11.

Solution: The sum of the digits in the even places minus that in the odd places is

$$(4+3+2) - (6+8+6) = -11$$

$$\equiv 0(\text{mod } 11)$$

Therefore, 648,362 is divisible by 11, by Theorem 11.5.

Example 11.26   Determine if 28,372 is divisible by 11.
Solution: Since

$$(2+3+2)-(8+7)=7-15$$
$$=-8$$
$$\equiv 3(\text{mod } 11)$$
$$\not\equiv 0(\text{mod } 11)$$

28,372 is not divisible by 11.

Recall from exercise 5.6, problem 10 that a *palindrome* is a natural number that reads the same backward as well as forward. For example, 121 and 2332 are palindromes. If follows from Theorem 11.5 that every palindrome with an even number of digits is divisible by 11 (Why?). For example, 2,332 is divisible by 11, since

$$(2+3)-(3+2)=0$$
$$\equiv 0(\text{mod } 11)$$

Also, 123,321 is divisible by 11, since

$$(2+3+1)-(1+3+2)=0$$
$$\equiv 0(\text{mod } 11).$$

Finally, we remark that it is theoretically possible to find a divisibility test for every positive integer n; however, in most cases it is much easier simply to divide by n than to apply the test.

Exercise 11.5

1. Mark *true* or *false*.
  a) Any integer divisible by 9 is certainly divisible by 3.
  b) Any integer divisible by 3 is certainly divisible by 9.
  c) Any integer divisible by 6 is divisible by 3.
  d) Every integer divisible by 5 is odd.
  e) Every integer divisible by 11 is odd.
  f) 11 is a palindrome.
  g) Every palindrome is divisible by 11.
  ★**h)** $10^{100}-1$ is divisible by 9.
  ★**i)** $10^{100}-1$ is divisible by 11.

2. Give a counterexample to each of the false statements in problem 1.

3. Which of the following numbers are divisible by 2? By 4? By 8?

a) 248	**b)** 425	**c)** 1948	**d)** 1776
e) 1976	**f)** 327	**g)** 2000	**h)** 31264
i) 23426	**j)** 427364	**k)** 2380	**l)** 3240

4. Which of the following numbers are divisible by 3? By 9?
   a) 108 **b)** 121 **c)** 327 **d)** 1776
   e) 1948 **f)** 21921 **g)** 33264 **h)** 425781
5. Which of the following numbers are divisible by 6?
   a) 3282 **b)** 1776 **c)** 1976 **d)** 87654 **e)** 327723
6. Which of the following numbers are divisible by 11?
   a) 121 **b)** 2948 **c)** 5678 **d)** 43978 **e)** 548152
7. Which of the following palindromes are divisible by 11?
   a) 232 **b)** 2332 **c)** 57075 **d)** 66066 **e)** 123321
8. Find a four-digit number divisible by 4, 5, and 6.
9. Find a six-digit number divisible by 4, 5, and 6.
10. Find a four-digit number divisible by 5, 6, and 8.
11. Find a six-digit number divisible by 5, 6, and 8.
12. Find a four-digit number divisible by 4, 5, and 9.
13. Find a six-digit number divisible by 4, 5, and 9.

★11.6 BINARY CODES

Observe that the set $I_2 = \{0,1\}$ is closed under the operation of addition and multiplication mod 2, as illustrated by Tables 11.6 and 11.7. The numbers 0 and 1 are called *binary numbers.* It can be verified that the system I_2, usually called the *binary number system,* has all the basic properties of the calendar number system discussed in Section 11.3.

Table 11.6

+	0	1
0	0	1
1	1	0

Table 11.7

·	0	1
0	0	0
1	0	1

It is clear from Section 2.7 that there is close relationship between switching networks and binary numbers. This gives us a clue to that fact that the binary number system has tremendous application value in the theory of communications.

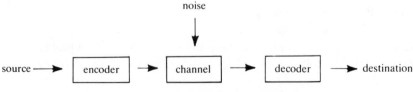

A typical communication system

Fig. 11.5

Most of us know that messages or data are transmitted from one place (*source*) to another (*destination* or *receiver*) in codes. The source may be a person or a machine, like a digital computer. The medium through which messages are sent is called a *channel*. A typical example of a transmission channel is a telephone line. Messages are converted (*encoded*) into signals of binary digits before transmission. These signals are then sent through the channel. At the decoder, every attempt is made to recover (*decode*) the original message from the received information, and then it is delivered to the destination. While in the channel, the signals can be disturbed at random by "noise." For example, a noise could be caused by lightning, operation of switches, thermal noise, cross-talk, etc. Tape defects are considered disturbances in magnetic-tape units. If the signals in the channel are not subjected to any possible noise, the message will be received at the destination without any error. However, it is more likely that this will not be the case, in general. Consequently, these noises cause *errors* in the original message and the receiver may not get all the information that was sent, which can cause serious troubles, confusion, and even embarrassment. It is said that the following passage from the Bible was sent to someone on her wedding. "There is no room for fear in love; perfect love banishes fear. . . ." (1 John 4:18). To our surprise, the message received by her read: ". . .that you have no husband, for although you have had five husbands, the man with whom you are now living is not your husband." (John 4:18).* If a 0 is sent, a 0 will probably be received; but occasionally a 1 can be received instead of 0, because of channel noise, as illustrated in Fig. 11.6. This concludes that messages should be encoded and decoded suitably and efficiently so as to minimize the chances for errors and to maximize the chances for receiving the original message intact. Errors, if any, in the received message must be detected and then corrected.

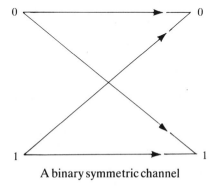

A binary symmetric channel

Fig. 11.6

* From *The New English Bible*, The Delegates of the Oxford University Press and the Syndics of the Cambridge University Press, 1961, 1970. Reprinted by permission.

As the English language uses the alphabet $\{a,b,c,...,x,y,z\}$ to convey mes-sages between people, a code makes use of the *binary alphabet* $\{0,1\}$ to send messages from source to destination. A *code* is thus a set of sequences of elements chosen from the binary alphabet. Every sequence of the code is called a *code word*. Each code word represents a message to be transmitted. For example, 10110 is a typical code word.

The process of *casting out twos* (see Section 11.4) plays a very significant role in detecting and correcting errors in codes. Before we transmit a code word, we annex a 0 or 1 to its right end according as the number of ones in the code word is even or odd. Recall that this is casting out twos. (Why?) Since there are seven ones in 1101011101, we annex a 1, which is called the *check digit*, to get 11010111011. This procedure, which makes the total number of ones even, is called *even parity check*. Let's assume we receive 11010111001 as the output. Since there is an odd number of ones in the received message, it follows clearly that an error has occurred during transmission. Consequently, if a single error occurs in the code word during transmission, then it changes the number of ones from even to odd and hence an error can be detected.

Assume Sam flips 4 coins 3 times and wants to transmit the outcomes of his experiment to his girlfriend. For each coin that falls heads up, he transmits a 1; for each one that falls tails up, he transmits a 0. One possible way the four coins can fall is as HTTH, which can be encoded 1001. Let's assume that the other three outcomes are HTTT, THTH, and HHHT, which can be translated in terms of 0's and 1's as 1000, 0101, and 1110, respectively. Let's arrange the outcomes of his experiment in a rectangular array, as follows:

$$
\begin{array}{cccc}
1 & 0 & 0 & 1 \\
1 & 0 & 0 & 0 \\
0 & 1 & 0 & 1 \\
1 & 1 & 1 & 0
\end{array}
$$

1	0	0	1	0		1	0	0	1	0
1	0	0	0	1		1	0	0	0	1
0	1	0	1	0		0	1	0	1	0
1	1	1	0	1		1	1	1	0	1
1	0	1	0			1	0	1	0	0

Fig. 11.7   **Fig. 11.8**

Using even parity check, a check digit is annexed at the end of each row and column (casting out twos again!). The resulting array is given in Fig. 11.7. We then insert a 0 at the lower right-hand corner to have an even number of 1's in the last row (column), as shown in Fig. 11.8. Notice that the number of 1's in each row and column is even. Assume the entire array in Fig. 11.8 is transmitted row by row (column by column) and his girlfriend receives the array in Fig. 11.9. When we check for even parity, we observe that all the row sums and columns check except the fourth row and first column, showing that there is an error at the intersection of the fourth row and first column. Thus this type of rectangular-array codes is not only single-error-detecting, but single-error-correcting. Such codes are common in magnetic-tape storage systems of today's large computers.

$$
\begin{array}{cccc|c}
1 & 0 & 0 & 1 & 0 \\
1 & 0 & 0 & 0 & 1 \\
0 & 1 & 0 & 1 & 0 \\
0 & 1 & 1 & 0 & 1 \\
\hline
1 & 0 & 1 & 0 & 0
\end{array}
$$

Fig. 11.9

11.7 SUMMARY AND COMMENTS

This chapter introduces and discusses in some detail one of the most important relations in mathematics: the congruence relation. An integer x is congruent to an integer y modulo a positive integer m if and only if $x - y$ is divisible by m, written as $x \equiv y \pmod{m}$. If $x \equiv y \pmod{m}$, then both x and y leave the same remainder on division by m. The congruence relation is an equivalence relation. Also, if $a \equiv b \pmod{m}$ and $c \equiv d \pmod{m}$, then

$$a + c \equiv b + d \pmod{m} \qquad \text{(addition property)}$$

and $\qquad\qquad ac \equiv bd \pmod{m} \qquad \text{(multiplication property)}$

This chapter, for the first time, provides us with examples of finite mathematical systems via the congruence relation.

The set I_{12} of clock numbers under addition and multiplication mod 12 possesses the following important properties:

1. I_{12} is closed under both addition and multiplication mod 12.
2. Both operations are commutative and associative.
3. I_{12} contains a unique identity, 0, under addition mod 12.
4. I_{12} possesses a unique identity, 1, under multiplication mod 12.
5. Every clock number has a unique additive inverse.

6. Clock numbers that are relatiely prime to 12 have multiplicative inverses.

7. The cancellation property with respect to multiplication mod 12 does not hold.

8. The product law does not hold.

9. Not every equation of the form $a = bx$ is solvable in I_{12}.

10. Multiplication mod 12 is distributive over addition mod 12.

The following are the important properties satisfied by the set I_7 of calendar numbers under addition and multiplication mod 7.

1. I_7 is closed under both addition and multiplication mod 7.

2. Addition and multiplication mod 7 are both associative and commutative.

3. I_7 possesses a unique identity, 0, under addition mod 7.

4. It contains a unique identity, 1, under multiplication mod 7.

5. Every calendar number has a unique additive inverse.

6. Every nonzero calendar number has a unique multiplicative inverse.

7. Cancellation properties hold in I_7.

8. The product law holds in I_7.

9. Every equation of the form $a = bx$ in I_7 *has a unique solution in I_7.*

10. Multiplication mod 7 is distributive over addition mod 7.

The congruence relation helps us to find the remainder when one integer is divided by another, a method that is more efficient than the familiar long division method.

The procedure of casting out nines, known to the early Arabs, gives a necessary condition to check the accuracy of the four fundamental arithmetical operations in computations. It makes use of the fact that every integer is congruent to the sum of its digits mod 9.

Table 11.8

An integer is divisible by	If and only if
2	its last digit is even
3	the sum of its digits is divisible by 3
4	the two-digit number obtained from its last two digits is divisible by 4
5	it ends in 0 or 5
6	it is divisible by 2 and 3
8	the three-digit number obtained from its last three digits is divisible by 8
9	the sum of its digits is divisible by 9
10	it ends in 0
11	the sum of its digits in the even places minus that in the odd places is divisible by 11

As a further application of congruences, we discussed some well known divisibility tests, which are summarized in Table 11.8.

The binary number system plays a very important role in theory of communications. Messages are translated in terms of 0's and 1's before transmission over a (noisy) channel. Channel noises cause errors in the received information. A good code is one that minimizes the probability of errors and maximizes the probability for getting the original message back. Abstract algebra, the topic of discussion of Chapter 12, plays a very significant part in the theory of error-correcting codes.

SUGGESTED READING

Berenson, Lewis, "A Divisibility Test for Amateur Discoverers," *The Arithmetic Teacher*, vol. 17 (Jan. 1970), pp. 39–41.

Heath, F. G., "Origin of the Binary Code," *Scientific American*, vol. 227 (Aug. 1972), pp. 76–83.

Hohn, Franz E., "Automatic Addition," *The Arithmetic Teacher*, vol. 10 (March 1963), pp. 127–132.

Kennedy, Robert E., "Divisibility by Integers Ending in 1, 3, 7, or 9," *The Mathematics Teacher*, vol. 54 (Feb. 1971), pp. 137–138.

Matthews, E. Rebecca, "A Simple 7 Divisibility Rule," *The Mathematics Teacher*, vol. 62 (Oct. 1969), pp. 461–464.

Singer, R., "Modular Arithmetic and Divisibility Criteria," *The Mathematics Teacher*, vol. 63 (Dec. 1970), pp. 653–656.

Smith, F., "Divisibility Rules for the First 15 Primes," *The Arithmetic Teacher*, vol. 18 (Feb. 1971), 85–87.

"Modern mathematics, that most astounding of intellectual creations, has projected the mind's eye through infinite time and the mind's hand into boundless space."

N. M. BUTLER

12/Algebraic Systems

After studying this chapter, you should be able to:
- *check if an operation on a set is a binary operation*
- *check the basic properties of a binary operation*
- *decide if a set together with a binary operation is a group*
- *identify a few special groups*
- *check if a set together with two binary operations is a ring or a field*

12.0 INTRODUCTION

We have discussed a few algebraic systems whose elements are all numbers and hence are named number systems: system of whole numbers, system of integers, system of rational numbers, system of real numbers, and system of complex numbers. We observed that each system, except the first, is richer than the previous system. All of these are examples of infinite algebraic systems; the underlying set in each case is infinite. In the previous chapter we discussed examples of finite algebraic systems. Both kinds of systems have several fascinating common properties. In general, the elements of an algebraic system need not be numbers; we will discuss some examples in later sections. Also, there may exist one or more operations in the system.

In this chapter, we categorize the different systems according to their fundamental properties and discuss a few basic properties of each category. The three basic algebraic systems, *groups*, *rings*, and *fields*, are the topics of discussion in this chapter. They form part of what is currently called abstract algebra or modern algebra. To some people, it is abstract, while to some others it is concrete. To some people, it is still modern, while to others it is no longer modern.

12.1 BINARY OPERATIONS

The concept of binary operations has been introduced intuitively in Chapters 1 and 2. Since we now have more mathematical tools, we shall give a definition of binary operation. But first let's consider the following examples.

Consider the operations of addition and multiplication on the set of integers. We have,

$$2+3=5 \qquad\qquad 2\cdot 3=6$$
$$(-3)+5=2 \qquad\qquad (-3)\cdot 5=-15$$
$$(-1)+(-3)=-4 \qquad (-1)\cdot(-3)=3$$

In each of these cases, the operation of addition (multiplication) associates with each pair of integers on the left-hand side of the equation a unique integer written on the right-hand side, as shown below:

$$(2,3)\to 5 \qquad\qquad (2,3)\to 6$$
$$(-3,5)\to 2 \qquad\qquad (-3,5)\to -15$$
$$(-1,-3)\to -4 \qquad (-1,-3)\to 3$$

Thus, in terms of the terminologies introduced in Chapter 10, the operation of addition (multiplication) on the set I of integers can be considered as a function from $I\times I$ to I. It associates with every ordered pair (a,b) in $I\times I$ a unique element $a+b$ (or $a\cdot b$) in I. In either case, the operation combines two elements of I to obtain a unique element of I. Now, we are in a position to make the following definition:

Definition 12.1 A *binary operation* $*$ on a set A is a function from $A\times A$ to A.

Example 12.1 Consider the set $A=\{1,2,3\}$. Then $A\times A=\{(1,1),(1,2),(1,3),(2,1),(2,2),(2,3),(3,1),(3,2),(3,3)\}$. The function $f:A\times A\to A$ defined by

$$f((1,1))=2 \qquad f((2,1))=3 \qquad f((3,1))=1$$
$$f((1,2))=3 \qquad f((2,2))=1 \qquad f((3,2))=2$$
$$f((1,3))=1 \qquad f((2,3))=2 \qquad f((3,3))=3$$

is a binary operation on the set A. If we denote $f((a,b))$ by $a*b$, then the above values can be rewritten as follows:

$$1*1=2 \qquad 2*1=3 \qquad 3*1=1$$
$$1*2=3 \qquad 2*2=1 \qquad 3*2=2$$
$$1*3=1 \qquad 2*3=2 \qquad 3*3=3$$

When the set A is small enough, tables can be used to define binary operations on the set A. The following table describes the operation $*$ we defined on the set $A = \{1,2,3\}$.

Table 12.1

$*$	1	2	3
1	2	3	1
2	3	1	2
3	1	2	3

Example 12.2   The operation $*$ defined on the set $B = \{0,1\}$ by Table 12.2 is not a binary operation on B. (Why?)

Table 12.2

$*$	0	1
0	0	1
1	1	2

The statement that $*$ is a binary operation on a set A is not the same as saying that the set A is closed under the operation $*$.

Recall that the operations of addition and multiplication on the set of real numbers are commutative, whereas subtraction and division are not: $a + b = b + a$ and $a \cdot b = b \cdot a$ for all real numbers a and b; but $a - b \neq b - a$ and $a \div b \neq b \div a$, in general. When do we say that a binary operation on a set is commutative?

Definition 12.2   A binary operation $*$ on a set A is said to be *commutative* if $a * b = b * a$ for *every* a,b in A.

Example 12.3   Consider the binary operation $\circ$ defined on the set of real numbers by $a \circ b = a + b - ab$. Since

$$a \circ b = a + b - ab$$
$$= b + a - ba$$
$$= b \circ a$$

the operation $\circ$ is commutative.

Example 12.4   The operation θ defined on the set of integers by $a\,\theta\,b = a - b + ab$ is not commutative, since $a\,\theta\,b = a - b + ab \neq b - a + ba = b\,\theta\,a$, in general.

Recall that addition and multiplication on the set of real numbers are associative, whereas subtraction and division are not: $a + (b + c) = (a + b) + c$ and $a(bc) = (ab)c$; but $a - (b - c) \neq (a - b) - c$ and $a \div (b \div c) \neq (a \div b) \div c$, in general. Also, the union and the intersection operations on sets are associative. When do we say that an operation on a set is associative?

Definition 12.3   A binary operation $*$ on a set A is said to be *associative* if $a * (b * c) = (a * b) * c$ for *every* a,b,c in A.

Example 12.5   Consider the binary operation $*$ defined on the set of real numbers by $a * b = a + b + ab$. Let's check whether this operation is associative.

$$b * c = b + c + bc$$

$$a * (b * c) = a * (b + c + bc)$$

$$= a + b + c + bc + a(b + c + bc)$$

$$= a + b + c + ab + bc + ac + abc$$

$$a * b = a + b + ab$$

$$(a * b) * c = (a + b + ab) * c$$

$$= a + b + ab + c + (a + b + ab)c$$

$$= a + b + c + ab + bc + ac + abc$$

Thus, $a * (b * c) = (a * b) * c$, showing that the operation $*$ is associative.

Recall that the set of real numbers possesses an identity element, 0, with respect to addition: $a + 0 = 0 + a$; the number 1 is an identity for multiplication on the set of real numbers: $a \cdot 1 = a = 1 \cdot a$. Therefore, when do we say that an element is an identity with respect to an operation on a set?

Definition 12.4   Let $*$ be a binary operation on a set A. An element e in A is called an *identity* with respect to the operation $*$ if $a * e = a = e * a$ for every element a in A.

Example 12.6   Consider the binary operation Δ defined on the set of integers by $a \Delta b = a + b + 1$. An integer e is an identity under the operation Δ if and only if $a \Delta e = a = e \Delta a$. Since $a \Delta e = e \Delta a = a + e + 1$, e will be an identity if and only if $a + e + 1 = a$, that is, if and only if $e = -1$. Thus, -1 is an identity under the operation Δ on the set of integers.

Recall that 0 is not an identity for subtraction of real numbers, since $a - 0 = a \neq 0 - a$ for every real number a.

We observed in Chapter 7 that every real number x has an inverse, $-x$, with respect to addition: $x + (-x) = 0 = (-x) + x$; every nonzero real number x has an inverse $x^{-1} = 1/x$ with respect to multiplication: $x \cdot x^{-1} = 1 = x^{-1} \cdot x$. When can we say that a certain element is an inverse of another element with respect to an operation on a set?

Definition 12.5   Let e be the identity element for a binary operation $*$ on a set A. An element a^{-1} in A is called an *inverse* of a with respect to $*$ if $a^{-1} * a = e = a * a^{-1}$.

Example 12.7   Consider the set $\{1, -1\}$. Under multiplication, the inverse of 1 is itself; the inverse of -1 is itself.

Example 12.8   Consider the set $A = \{0,1,2,3,4,5\}$ under $\oplus$, addition mod 6. Zero is the identity under $\oplus$ (see Table 12.3). Every element a in A has an inverse with respect to $\oplus$, which can be read from the table as follows: locate the column containing 0 that lies in the row headed by a. The element that heads this column (above the top horizontal line) gives the inverse of a. Thus, the inverse of 0 is 0; the inverse of 1 is 5 and vice versa; the inverse of 2 is 4 and vice versa, and the inverse of 3 is itself.

Table 12.3

$\oplus$	0	1	2	3	4	5
0	0	1	2	3	4	5
1	1	2	3	4	5	0
2	2	3	4	5	0	1
3	3	4	5	0	1	2
4	4	5	0	1	2	3
5	5	0	1	2	3	4

Recall that the only integers having inverses with respect to ordinary multiplication are ± 1. (Why?)

It is well known that multiplication of real numbers is distributive over addition, but not vice versa: $a(b + c) = ab + ac$, but $a + (bc) \neq (a + b)(a + c)$, in general. The union operation on sets is distributive over intersection and vice versa:

$$A \cup (B \cap C) = (A \cup B) \cap (A \cup C) \quad \text{and} \quad A \cap (B \cup C) = (A \cap B) \cup (A \cap C)$$

Definition 12.6   Let $*$ and Δ be two binary operations on a set A. Then $*$ is *left distributive* over Δ if $a * (b \Delta c) = (a * b)\Delta(a * c)$ for every a, b, c in A, and *right distributive* over Δ if $(a \Delta b) * c = (a * c) \Delta (b * c)$ for every a, b, c in A. The operation $*$ is *distributive* over Δ if it is both left and right distributive over Δ.

Example 12.9 Consider the binary operations $*$ and Δ defined on the set R of real numbers by $a * b = a + b + ab$ and $a \Delta b = a$. Since

$$a * (b \Delta c) = a * b = a + b + ab$$

and

$$(a * b) \Delta (a * c) = a * b = a + b + ab$$

we have

$$a * (b \Delta c) = (a * b) \Delta (a * c)$$

showing that the operation $*$ is left distributive over Δ. Also,

$$(a \Delta b) * c = a * c = (a * c) \Delta (b * c)$$

Consequently, $*$ is right distributive over Δ. Thus, $*$ distributes over Δ. Also, since

$$(a * b) \Delta c = a * b = (a \Delta c) * (b \Delta c)$$

Δ is right distributive over $*$. However, it is not left distributive over $*$. (Why?)

Exercise 12.1

1. Consider the operation $\circ$ defined on the set I of integers by $a \circ b = a + b - 1$.
   a) Is I closed under $\circ$?
   b) Is the operation $\circ$ commutative?
   c) Is the operation $\circ$ associative?
   d) Does there exist an identity? If yes, what is it?
   e) Does every element have an inverse with respect to $\circ$? If not, which elements do have inverses?
2. Same as problem 1 with the operation $*$ on R defined by $a * b = a - b + 1$.
3. Same as problem 1 with the operation Δ on R defined by $a \Delta b = a + b - ab$.
4. Same as problem 1 with the operation ω on R defined by $a \, \omega \, b = |a + b|$.
5. Same as problem 1 with the operation $\bullet$ on R defined by $a \bullet b = a$.
6. Same as problem 1 with the operation $\otimes$ on Q defined by $a \otimes b = a^2 + b^2$.
7. Same as problem 1 with the operation $\oplus$ on N defined by

$$a \oplus b = \begin{cases} a \text{ if } a \leq b \\ b \text{ if } b < a \end{cases}$$

8. Mark *true* or *false* (A is an arbitrary nonempty set).
   a) Every function on A is a binary operation on A.
   b) Every binary operation on A is a function from $A \times A$ to A.
   c) Subtraction is a binary operation on the set of natural numbers.
   d) The operation $*$ defined by $a * b = a + b + \frac{2}{3}$ is a binary operation on I.
   e) Addition is a binary operation on the set of real numbers.
   f) Division on the set of real numbers is a binary operation.

9. Give a counterexample to each of the false statements in problem 8.

10. Is the operation Δ on R defined by $a \,\Delta\, b = \dfrac{d(a,b)}{1 + d(a,b)}$ commutative, $d(a,b)$ being the distance between a and b? Justify your answer.

11. Consider the binary operations Δ and $*$ on R defined by $a \,\Delta\, b = 0$ and $a * b = a + b\sqrt{2}$.
  a) Is Δ left distributive over $*$?
  b) Is Δ right distributive over $*$?
  c) Is $*$ left distributive over Δ ?
  d) Is $*$ right distributive over Δ ?

12. Same as problem 11 with Δ and $*$ defined on R by $a \,\Delta\, b = 0$ and $a * b = a$.

13. Same as problem 11 with Δ and $*$ defined on R by $a \,\Delta\, b = a$ and $a * b = b$.

★14. Same as problem 1 with the operation θ on N defined by $a \,\theta\, b = \cdot \gcd\{a,b\}$.

★15. Same as problem 1 with the operation θ on N defined by $a \,\theta\, b = \operatorname{lcm}\{a,b\}$.

★16. How many binary operations can be defined on the set A if
  a) $A = \{1\}$?   **b)** $A = \{1,2\}$?   **c)** $A = \{1,2,3\}$?

12.2 THE CONCEPT OF A GROUP

The concept of a *group* forms a cornerstone of the study of all algebraic systems. It acts like a thread that interconnects several branches of mathematics.

Historically, the theory of groups came into existence from an attempt to solve equations in terms of their coefficients. Evariste Galois (1811–1832), considered a radical by his fellow citizens, is widely respected as one of the pioneers in the development of the theory of groups. Unfortunately, Galois was killed in a duel on May 30, 1832, at the age of 20, over an affair related to a former girlfriend. On the eve of his death, he wrote a long letter of 31 pages containing very significant contributions to mathematics, which have kept mathematicians around the world busy ever since. Galois has been given the credit for coining the word "group" and for introducing abstraction in mathematics.

Consider the sets of integers, rational numbers, and real numbers we obtained by enlarging the set of whole numbers. All of these sets have four fundamental properties in common with respect to addition:
a) Each set is closed under addition.
b) Addition is associative.
c) Each set possesses an identity with respect to addition.
d) Every element in each set has an inverse under addition.

Consider now the sets of nonzero rational numbers and nonzero real numbers. Recall that

a) Each set is closed under multiplication.
b) Multiplication is associative.
c) Each set possesses an identity under multiplication.
d) Every element in each set has an inverse with respect to multiplication.

What observations can we make from these concrete cases? In each case, we have a nonempty set together with a binary operation $*$ satisfying four different properties: the closure property, the associative property, existence of an identity under $*$, and existence of an inverse with respect to $*$ for each element. These observations lead us now to make the following definition:

Definition 12.7 A nonempty set G together with a binary operation $*$ on G is called a *group* if:

a) G is closed under $*$.
b) The operation $*$ is associative.
c) G possesses an identity e with respect to $*$.
d) Every element in G has an inverse in G with respect to $*$.

We now emphasize that a group consists of two components: a nonempty set G and a binary operation $*$ on G satisfying the four above axioms. A set by itself does not form a group. Particular attention must be given to mention the operation with respect to which a given set is a group.

Observe that the set of integers, the set of rational numbers, and the set of real numbers, all with addition as an operation, are examples of a group. So are the sets of nonzero rational numbers, nonzero real numbers, and nonzero complex numbers under multiplication. However, the set of integers is not a group under multiplication! (Why?) The same is the case with the set of rational numbers under multiplication. (Why?)

A group G can have more properties than the four specified axioms. For instance, the binary operation on G can very well be commutative. Properties of such groups were studied extensively by the Norwegian mathematician Niels Henrik Abel (1802–1829). Accordingly, we make the following definition:

Definition 12.8 A group G is said to be *abelian* (*commutative*) if the binary operation $*$ on G is commutative.

The set of integers is an abelian group under the usual addition, as are the sets of rational numbers and real numbers.

Examples of nonabelian groups will be discussed in the next section.

Definition 12.9 A group G is called a *finite group* if it contains only a finite number of elements; otherwise it is *infinite*.

The additive group of integers is an infinite group. So are the additive group of rational numbers and the multiplicative group of nonzero rational numbers. The set {0,1,2,3} of integers under addition mod 4 is a finite group whose addition table is given in Table 12.4.

Table 12.4

$\oplus$	0	1	2	3
0	0	1	2	3
1	1	2	3	0
2	2	3	0	1
3	3	0	1	2

Table 12.5

*	e	a	b	c
e	e	a	b	c
a	a	e	c	b
b	b	c	e	a
c	c	b	a	e

The set of integers {0,1,2,3,4} under addition mod 5 is also a finite group. We now give a less trivial example of a finite abelian group.

Example 12.10   Consider the set $G = \{e,a,b,c\}$. Define a binary operation $*$ on G as given by Table 12.5. It can be verified that G is a group under $*$. The element e is the identity of the group. It is a finite abelian group. This group is called the *four group*.

Let's now discuss a few simple properties of a group. In all the number systems, we observed that the additive identity 0 is unique. Is this true in every group? Can a group have more than one identity element? The answer is provided by the following theorem:

Theorem 12.1   *Every group has a unique identity.*

Theorem 12.2   (**Cancellation Property**)   *Let a, b, and c be arbitrary elements of a group G with binary operation $*$. Then*

1. $a * b = a * c$ *implies* $b = c$ *and*

2. $b * a = c * a$ *implies* $b = c$.

Proof: We will prove only (1) and leave (2) for the reader as an exercise.

$a * b = a * c$	given
$a^{-1} * (a * b) = a^{-1} * (a * c)$	$*$ is a binary operation
$(a^{-1} * a) * b = (a^{-1} * a) * c$	associative property
$e * b = e * c$	definition of inverse
$b = c$	definition of identity

Recall that the cancellation property holds in the sets of integers, rational numbers, and real numbers under addition. It also holds in the sets of nonzero integers, nonzero rational numbers, and nonzero real numbers under multiplication. We observed in Section 11.2 that the cancellation property does not hold in the set I_{12} of integers mod 12 under multiplication mod 12.

Theorem 12.2 can be used to prove that every element in a group has a unique inverse (see problem 20).

Exercise 12.2

1. Is the set of even integers a group under addition?
2. Is the set of odd integers a group under addition? Why?
3. Is the set of multiples of 3 a group under addition?
4. Show that $I_2 = \{0,1\}$ with addition mod 2 is an abelian group. What is the identity of the group? What is the inverse of 1?
5. Is the set $\{-1,0,1\}$ a group under addition? Justify your answer.
6. Show that the set $H = \{-1,+1\}$ is a group under multiplication. Is it abelian?
7. Show that the set $\{1,2,3,4\}$ of integers forms a group under multiplication mod 5. Is it abelian?
8. Show that the set $\{1,2,4\}$ of integers is a group under multiplication mod 7. Is it abelian?
9. Show that the set $\{1,5\}$ of integers is a group under multiplication mod 6. Is it abelian?
10. Show that the four group is abelian.
11. Does the set Q of rational numbers with binary operation $*$ defined by $a * b = a + b + ab$ form a group? Justify your answer.
12. Mark *true* or *false*.
    a) The empty set is trivially a group.
    b) The set of natural numbers is a group under addition.
    c) The set of integers is a group under addition.
    d) Every group has at least one element.
    e) $\{0\}$ is a group under addition.
    f) $\{1\}$ is a group under multiplication.
    g) $\{0,1\}$ is a group under addition.
    h) The set of real numbers is a group under multiplication.
    i) The cancellation property holds in every group.
    j) Every group is finite.
    k) Every finite group is abelian.
    l) Every abelian group is finite.
    m) The identity in a group has no inverse.
13. Give a counterexample (whenever possible) to each of the false statements in problem 12.

14. How should the following table be completed so that $G = \{e, a\}$ forms a group under the operation $*$?

$*$	e	a
e	e	a
a	a	

15. Same as problem 14 with $G = \{e, a, b\}$ and

$*$	e	a	b
e	e	a	b
a	a		
b	b		

16. The set $G = \{e, a, b, c\}$ is a group with binary operation $*$ defined by

$*$	e	a	b	c
e	e	a	b	c
a	a	b	c	e
b	b	c	e	a
c	c	e	a	b

Find the inverse of each element in G. Is G abelian?

17. Give an example of a group with 3 elements; 4 elements; 5 elements.

18. Let a, b, and c be arbitrary elements of a group G with binary operation $*$ such that $b * a = c * a$. Prove that $b = c$.

★**19.** Let a and b be arbitrary elements of a group G with binary operation $*$. Is the equation $a * x = b$ solvable in G? What is the solution, if it exists?

★**20.** Prove that every element in a group has a unique inverse.

★**21.** In any group with operation $*$, prove that (a^2 means $a * a$):

    a) If $a * b = b$, then $a = e$.

    b) If $a^2 = a$, then $a = e$.

12.3 PERMUTATION GROUPS

Recall that in Section 10.5, we defined a *permutation* of a set S as a 1-1 correspondence from S to itself. Consider now the set of all permutations of the set $S = \{1,2,3\}$. Recall that there are six permutations of the set S:

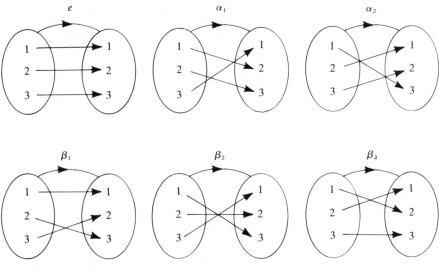

Fig. 12.1

Definition 12.10   A *permutation on n symbols* is a permutation of a set with n elements.

Let S_n denote the set of all permutations of the set $S = \{1,2,3,...,n\}$. Thus, $S_3 = \{e, \alpha_1, \alpha_2, \beta_1, \beta_2, \beta_3\}$. Observe from Fig. 12.1 that, for example,

$$e(1) = 1 \qquad \alpha_1(1) = 2 \qquad \alpha_2(1) = 3$$
$$e(2) = 2 \qquad \alpha_1(2) = 3 \qquad \alpha_2(2) = 1$$
$$e(3) = 3 \qquad \alpha_1(3) = 1 \qquad \alpha_2(3) = 2$$

For convenience, let's denote these six permutations as follows:

$$e = \begin{pmatrix} 1 & 2 & 3 \\ 1 & 2 & 3 \end{pmatrix} \qquad \alpha_1 = \begin{pmatrix} 1 & 2 & 3 \\ 2 & 3 & 1 \end{pmatrix} \qquad \alpha_2 = \begin{pmatrix} 1 & 2 & 3 \\ 3 & 1 & 2 \end{pmatrix}$$

$$\beta_1 = \begin{pmatrix} 1 & 2 & 3 \\ 1 & 3 & 2 \end{pmatrix} \qquad \beta_2 = \begin{pmatrix} 1 & 2 & 3 \\ 3 & 2 & 1 \end{pmatrix} \qquad \beta_3 = \begin{pmatrix} 1 & 2 & 3 \\ 2 & 1 & 3 \end{pmatrix}$$

where each element in the second row is the image of the element just above it.

We define a binary operation $\circ$, called *composition* on S_n, as follows: if f and g are elements of S_n, then $(f \circ g)(x) = f(g(x))$ for every x in $\{1,2,3,...,n\}$. The reader is warned that in $f \circ g$, first g is applied and then f. For example, in the set

$S_3 = \{e, \alpha_1, \alpha_2, \beta_1, \beta_2, \beta_3\}$ of all permutations of the set $S = \{1, 2, 3\}$,

$$(\beta_1 \circ \beta_3)(1) = \beta_1(\beta_3(1)) = \beta_1(2) = 3$$
$$(\beta_1 \circ \beta_3)(2) = \beta_1(\beta_3(2)) = \beta_1(1) = 1$$
$$(\beta_1 \circ \beta_3)(3) = \beta_1(\beta_3(3)) = \beta_1(3) = 2$$

where β_3 takes 1 to 2 and β_1 takes 2 to 3, hence $\beta_1 \circ \beta_3$ takes 1 to 3; and β_3 takes 2 to 1 and β_1 takes 1 to 1, hence $\beta_1 \circ \beta_3$ takes 2 to 1. Finally, 3 goes to 3 under β_3 and 3 goes to 2 under β_1; thus 3 goes to 2 under $\beta_1 \circ \beta_3$. Thus,

$$\beta_1 \circ \beta_3 = \begin{pmatrix} 1 & 2 & 3 \\ 1 & 3 & 2 \end{pmatrix} \circ \begin{pmatrix} 1 & 2 & 3 \\ 2 & 1 & 3 \end{pmatrix}$$

$$= \begin{pmatrix} 1 & 2 & 3 \\ 3 & 1 & 2 \end{pmatrix} = \alpha_2$$

Similarly,

$$\beta_3 \circ \beta_1 = \begin{pmatrix} 1 & 2 & 3 \\ 2 & 1 & 3 \end{pmatrix} \circ \begin{pmatrix} 1 & 2 & 3 \\ 1 & 3 & 2 \end{pmatrix}$$

$$= \begin{pmatrix} 1 & 2 & 3 \\ 2 & 3 & 1 \end{pmatrix} = \alpha_1$$

and

$$\beta_1 \circ \alpha_2 = \begin{pmatrix} 1 & 2 & 3 \\ 1 & 3 & 2 \end{pmatrix} \circ \begin{pmatrix} 1 & 2 & 3 \\ 3 & 1 & 2 \end{pmatrix}$$

$$= \begin{pmatrix} 1 & 2 & 3 \\ 2 & 1 & 3 \end{pmatrix} = \beta_3$$

That $\beta_1 \circ \beta_3 = \alpha_2 \neq \alpha_1 = \beta_3 \circ \beta_1$ shows that the operation of composition is not commutative. Table 12.6 exhibits all possible compositions of elements in S_3. At this point we advise the reader to check all the compositions himself. It follows from the table that S_3 is closed under composition. That the operation is associative can be verified using the table (there are $6 \cdot 6 \cdot 6 = 216$ cases to be

Table 12.6

$\circ$	e	α_1	α_2	β_1	β_2	β_3
e	e	α_1	α_2	β_1	β_2	β_3
α_1	α_1	α_2	e	β_3	β_1	β_2
α_2	α_2	e	α_1	β_2	β_3	β_1
β_1	β_1	β_2	β_3	e	α_1	α_2
β_2	β_2	β_3	β_1	α_2	e	α_1
β_3	β_3	β_1	β_2	α_1	α_2	e

checked). For example,

$$\alpha_1 \circ (\beta_1 \circ \alpha_2) = \alpha_1 \circ \beta_3 = \beta_2$$

$$(\alpha_1 \circ \beta_1) \circ \alpha_2 = \beta_3 \circ \alpha_2 = \beta_2$$

Thus,

$$\alpha_1 \circ (\beta_1 \circ \alpha_2) = (\alpha_1 \circ \beta_1) \circ \alpha_2$$

The permutation e is an identity for composition. Also, every element in S_3 has an inverse: $e^{-1} = e$, $\alpha_1^{-1} = \alpha_2$, $\alpha_2^{-1} = \alpha_1$, $\beta_1^{-1} = \beta_1$, $\beta_2^{-1} = \beta_2$, $\beta_3^{-1} = \beta_3$. Thus, S_3 is a group under the operation of composition; it is not abelian. The group S_3 is called the *permutation group (symmetric group)* on three elements.

A geometrical interpretation of this group will be given in the next section.

More generally, S_n, the set of all permutations of a set with n elements, is a group under composition. It is the *permutation group (symmetric group)* on n elements. The identity of this group is given by

$$e = \begin{pmatrix} 1 & 2 & 3 \dots n \\ 1 & 2 & 3 \dots n \end{pmatrix}$$

To find the inverse of an element in S_n, we need only interchange the first row with the second row in the permutation. For example, if

$$\alpha = \begin{pmatrix} 1 & 2 & 3 & 4 & 5 \\ 4 & 1 & 3 & 5 & 2 \end{pmatrix}$$

then

$$\alpha^{-1} = \begin{pmatrix} 4 & 1 & 3 & 5 & 2 \\ 1 & 2 & 3 & 4 & 5 \end{pmatrix}$$

which can be rewritten

$$\alpha^{-1} = \begin{pmatrix} 1 & 2 & 3 & 4 & 5 \\ 2 & 5 & 3 & 1 & 4 \end{pmatrix}$$

Exercise 12.3

1. Use Table 12.6 to determine if the set $\{e, \beta_1\}$ is a group with respect to composition. If it is a group, is it abelian?
2. Same as problem 1 with $\{e, \alpha_1, \alpha_2\}$.
3. Use Table 12.6 to compute each of the following (α^2 means $\alpha \circ \alpha$ and α^{-2} means $(\alpha^{-1})^2$).
   a) $\alpha_1 \circ \beta_1^2$
   b) $(\alpha_1 \circ \beta_1)^2$
   c) $\alpha_1^2 \circ \beta_1^2$
   d) $\alpha_1 \circ \beta_1^2 \circ \alpha_2$
   e) $\alpha_2^{-1} \circ \beta_2$
   f) $\alpha_1^{-1} \circ \alpha_2^{-1}$
   g) $\alpha_2^{-2} \circ \beta_2^{-2}$
   h) $(\alpha_1 \circ \alpha_2 \circ \beta_2)^2$
   i) all distinct powers of β_1
4. With the following permutations on four elements,

$$\alpha = \begin{pmatrix} 1 & 2 & 3 & 4 \\ 2 & 3 & 4 & 1 \end{pmatrix}, \qquad \beta = \begin{pmatrix} 1 & 2 & 3 & 4 \\ 3 & 1 & 4 & 2 \end{pmatrix}, \qquad \gamma = \begin{pmatrix} 1 & 2 & 3 & 4 \\ 4 & 3 & 2 & 1 \end{pmatrix}$$

compute:

a) $\alpha \circ \beta$ b) $\alpha \circ (\beta \circ \gamma)$ c) $\beta^{-1} \circ \alpha^{-1}$

d) $\alpha \circ \beta^{-1} \circ \alpha$ e) $(\alpha \circ \beta \circ \gamma)^{-1}$ f) $\alpha^{-1} \circ \beta^{-1} \circ \gamma^{-1}$

5. Mark *true* or *false*.

   a) The group of permutations on three letters has six elements.

   b) The group of permutations on three elements is commutative.

   c) The group of permutations on three elements is not infinite.

   d) The symmetric group S_5 has five elements.

   e) The cancellation property does not hold in the permutation group S_3.

6. Give a counterexample to each of the false statements in problem 5, if possible.

7. Compute each of the following, where f, g, and h are permutations on five elements, given by

$$f = \begin{pmatrix} 1 & 2 & 3 & 4 & 5 \\ 3 & 2 & 5 & 1 & 4 \end{pmatrix}, \quad g = \begin{pmatrix} 1 & 2 & 3 & 4 & 5 \\ 2 & 3 & 4 & 5 & 1 \end{pmatrix}, \quad h = \begin{pmatrix} 1 & 2 & 3 & 4 & 5 \\ 3 & 4 & 5 & 1 & 2 \end{pmatrix}$$

a) $f \circ g$ b) $g \circ f$ c) $g \circ f \circ h$

d) $g \circ f^2$ e) f^{-1} f) $f^{-1} \circ g^{-1}$

g) $f^{-2} \circ g^{-2}$ h) $f^{-1} \circ g^{-1} \circ h^{-1}$

i) all distinct powers of f

12.4 GROUPS OF ROTATIONS

In this section, we study groups obtained from rotations of some geometric figures.

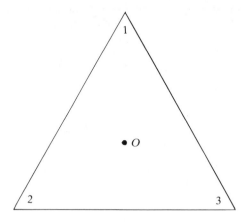

Fig. 12.2

Consider an equilateral triangle on a plane whose vertices are labeled 1, 2, and 3, as in Fig. 12.2. Let's now study the set of all rotations (rigid motions) of the triangle that will take the triangle to itself, perhaps with a possible change in the positions of the vertices. We now advise the reader to have a piece of

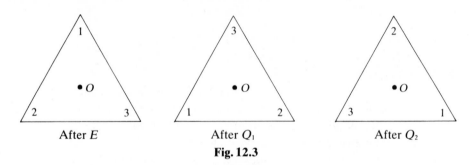

After E   After Q_1 After Q_2

Fig. 12.3

cardboard in the shape of an equilateral triangle at his side to study the various motions. Let E denote the rotation of the triangle through an angle of 0° (zero degrees) about a line perpendicular to its plane and passing through its center, O. Then the rotation E leaves the triangle in the same initial position, as shown in Fig. 12.3. Let Q_1 and Q_2 denote rotations of the triangle about 120° and 240° in the counterclockwise direction. The effects of these rotations are demonstrated in Fig. 12.3. Each of these rotations can be considered as a permutation of the elements 1, 2, and 3. For example, the rotation Q_1 sends 1 to 2, 2 to 3, and 3 to 1; the rotation Q_2 takes 1 to 3, 2 to 1, and 3 to 2. Accordingly, the rotations E, Q_1, and Q_2 can be represented as follows:

$$E = \begin{pmatrix} 1 & 2 & 3 \\ 1 & 2 & 3 \end{pmatrix}, \qquad Q_1 = \begin{pmatrix} 1 & 2 & 3 \\ 2 & 3 & 1 \end{pmatrix}, \qquad Q_2 = \begin{pmatrix} 1 & 2 & 3 \\ 3 & 1 & 2 \end{pmatrix}$$

Here E is the "identity motion" since it sends every vertex to itself.

We now define three more rigid motions of the equilateral triangle. Let R_1, R_2, and R_3 denote the rotations (reflections) of the triangle through 180° about the perpendicular bisectors b_1, b_2, and b_3, respectively.

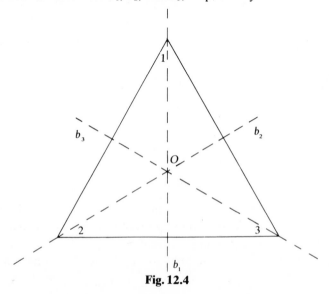

Fig. 12.4

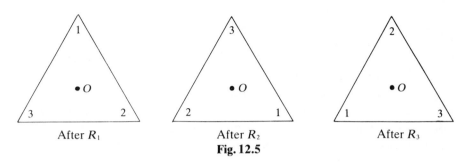

After R_1 After R_2 After R_3
Fig. 12.5

The effects of these motions are exhibited in Fig. 12.4. The reflection R_1 sends 1 to 1, 2 to 3, and 3 to 2; R_2 takes 1 to 3, 2 to 2, and 3 to 1; R_3 takes 1 to 2, 2 to 1, and 3 to 3. These three notations can now be represented as follows:

$$R_1 = \begin{pmatrix} 1 & 2 & 3 \\ 1 & 3 & 2 \end{pmatrix} \qquad R_3 = \begin{pmatrix} 1 & 2 & 3 \\ 3 & 2 & 1 \end{pmatrix} \qquad R_3 = \begin{pmatrix} 1 & 2 & 3 \\ 2 & 1 & 3 \end{pmatrix}$$

Consider now the set $T = \{E, Q_1, Q_2, R_1, R_2, R_3\}$ of all the rigid motions of an equilateral triangle. Define an operation $\circ$, called *composition* on T, as follows: $f \circ g$ is the rigid motion of the triangle obtained by performing first g and then f, for every f and g in T. For example, $Q_1 \circ Q_2$ is the rigid motion obtained by rotating the triangle first through $240°$ and then through $120°$ in the counterclockwise direction first. The effect of this is the rotation E; in other words, $Q_1 \circ Q_2 = E$. Similarly, $Q_2 \circ Q_1 = E$. The composition $Q_1 \circ R_1$ is the motion obtained by taking the reflection R_1 of the triangle and then rotating it through $120°$ in the counterclockwise sense. The result of this composition is R_3. Table 12.7 gives all the possible ways of taking the composition of two elements of T. It now follows that T is closed under composition. That the operation is associative can be verified (there are $6 \cdot 6 \cdot 6 = 216$ cases to be checked). The rotation E is an identity for the operation. Every element in T has an inverse: $E^{-1} = E$, $Q_1^{-1} = Q_2$, $Q_2^{-1} = Q_1$, $R_1^{-1} = R_1$, $R_2^{-1} = R_2$, and $R_3^{-1} = R_3$. Thus T is a group with respect to composition.

Table 12.7

$\circ$	E	Q_1	Q_2	R_1	R_2	R_3
E	E	Q_1	Q_2	R_1	R_2	R_3
Q_1	Q_1	Q_2	E	R_3	R_1	R_2
Q_2	Q_2	E	Q_1	R_2	R_3	R_1
R_1	R_1	R_2	R_3	E	Q_1	Q_2
R_2	R_2	R_3	R_1	Q_2	E	Q_1
R_3	R_3	R_1	R_2	Q_1	Q_2	E

Notice that if we replace the rigid motions E, Q_1, Q_2, R_1, R_2, and R_3 by e, α_1, α_2, β_1, β_2, and β_3, respectively, then Tables 12.6 and 12.7 are identical. Consequently, the group T of rotations of an equilateral triangle is the same as the permutation group S_3 on three symbols.

Let's now study the rigid motions of a square. Label its vertices with 1, 2, 3, and 4, as shown in Fig. 12.6. As before, we advise the reader to keep a piece of cardboard cut in the shape of a square and labeled like Fig. 12.6, for clarity and better understanding of the various motions.

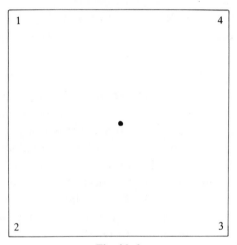

Fig. 12.6

Let e, α_1, α_2, and α_3 denote the rotations of the square about a line through its center and perpendicular to its plane, through an angle of 0°, 90°, 180°, and 270° in the counterclockwise sense, respectively. The effects of these rigid motions are illustrated in Fig. 12.7.

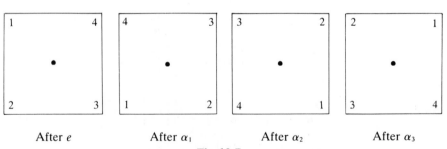

After e After α_1 After α_2 After α_3

Fig. 12.7

For example, α_1 sends 1 to 2, 2 to 3, 3 to 4, and 4 to 1. These rotations may now be represented as follows:

$$e = \begin{pmatrix} 1 & 2 & 3 & 4 \\ 1 & 2 & 3 & 4 \end{pmatrix} \qquad \alpha_1 = \begin{pmatrix} 1 & 2 & 3 & 4 \\ 2 & 3 & 4 & 1 \end{pmatrix}$$

$$\alpha_2 = \begin{pmatrix} 1 & 2 & 3 & 4 \\ 3 & 4 & 1 & 2 \end{pmatrix} \qquad \alpha_3 = \begin{pmatrix} 1 & 2 & 3 & 4 \\ 4 & 1 & 2 & 3 \end{pmatrix}$$

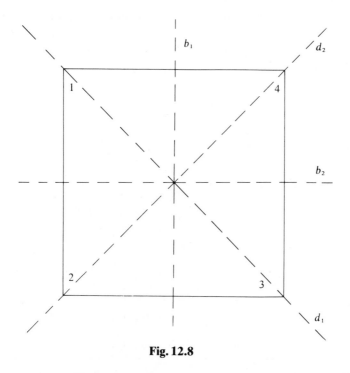

Fig. 12.8

Let β_1 and β_2 denote the reflections of the square about the perpendicular bisectors b_1 and b_2 of the sides, γ_1 and γ_2 the reflections about the diagonals d_1

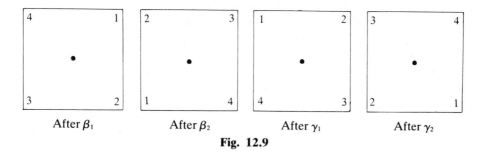

After β_1 After β_2 After γ_1 After γ_2

Fig. 12.9

and d_2. Then, these rotations can be represented as follows:

$$\beta_1 = \begin{pmatrix} 1 & 2 & 3 & 4 \\ 4 & 3 & 2 & 1 \end{pmatrix} \qquad \beta_2 = \begin{pmatrix} 1 & 2 & 3 & 4 \\ 2 & 1 & 4 & 3 \end{pmatrix}$$

$$\gamma_1 = \begin{pmatrix} 1 & 2 & 3 & 4 \\ 1 & 4 & 3 & 2 \end{pmatrix} \qquad \gamma_2 = \begin{pmatrix} 1 & 2 & 3 & 4 \\ 3 & 2 & 1 & 4 \end{pmatrix}$$

Now, $D_4 = \{e, \alpha_1, \alpha_2, \alpha_3, \beta_1, \beta_2, \gamma_1, \gamma_2\}$ is the set of all possible rigid motions of the square. It can be verified that D_4 is a group under composition, with e as the identity. We call D_4 the *group of rigid motions of a square*. The multiplication table of the group is given in Table 12.8. We advise the reader to check a few of the entries of the table.

Table 12.8

○	e	α_1	α_2	α_3	β_1	β_2	γ_1	γ_2
e	e	α_1	α_2	α_3	β_1	β_2	γ_1	γ_2
α_1	α_1	α_2	α_3	e	γ_1	γ_2	β_2	β_1
α_2	α_2	α_3	e	α_1	β_2	β_1	γ_2	γ_1
α_3	α_3	e	α_1	α_2	γ_2	γ_1	β_1	β_2
β_1	β_1	γ_2	β_2	γ_1	e	α_2	α_3	α_1
β_2	β_2	γ_1	β_1	γ_2	α_2	e	α_1	α_3
γ_1	γ_1	β_1	γ_2	β_2	α_1	α_3	e	α_2
γ_2	γ_2	β_2	γ_1	β_1	α_3	α_1	α_2	e

Exercise 12.4

1. Use Table 12.7 to determine if the set $A = \{E, Q_1, Q_2\}$ is a group under composition.
2. Same as problem 1 with $A = \{E, R_1\}$.
3. Same as problem 1 with $A = \{E, R_1, R_2, R_3\}$.
4. Use Table 12.8 to show that $A = \{e, \alpha_1, \alpha_2, \alpha_3\}$ is a group under composition. Is this group abelian?
5. Use Table 12.8 to verify:
   a) $\alpha_1 \circ (\alpha_2 \circ \alpha_3) = (\alpha_1 \circ \alpha_2) \circ \alpha_3$
   b) $\alpha_1 \circ (\beta_1 \circ \gamma_1) = (\alpha_1 \circ \beta_1) \circ \gamma_1$
   c) $\alpha_2 \circ (\beta_2 \circ \gamma_2) = (\alpha_2 \circ \beta_2) \circ \gamma_2$
   d) $\alpha_3 \circ (\beta_1 \circ \gamma_2) = (\alpha_3 \circ \beta_1) \circ \gamma_2$

6. Use Table 12.8 to compute:

a) $\alpha_1 \circ \alpha_2^{-1}$

b) $\alpha_1 \circ \beta_1 \circ \alpha_1^{-1}$

c) $\alpha_1 \circ \beta_1^{-1}$

d) $\alpha_1^{-1} \circ \beta_1$

e) $(\alpha_1 \circ \alpha_2 \circ \alpha_3)^{-1}$

f) $\alpha_3^{-1} \circ \alpha_2^{-1} \circ \alpha_1^{-1}$

g) $\beta_1^{-1} \circ \gamma_1^{-1}$

h) $\beta_1^{-2} \circ \gamma_1^{-2}$

i) all distinct powers of α_1

7. Mark *true* or *false*.

a) The group of rotations of an equilateral triangle is finite.

b) The group of rotations of an equilateral triangle is abelian.

c) The group of rotations of a square is abelian.

8. Give a counterexample to each of the false statements in problem 7.

9. Is the group D_4 of rotations of a square the same as the symmetric group S_4? Why?

★**10.** Consider the rigid motions of a regular pentagon given by e, α_1, α_2, α_3, α_4, the rotations of $0°$, $72°$, $144°$, $216°$, and $288°$, respectively, about an axis through its center; and β_1, β_2, β_3, β_4, β_5, the reflections about the lines joining the center to the vertices.

a) Express each of the rotations as a permutation by labeling the vertices with 1, 2, 3, 4, and 5.

b) Show that these permutations form a group, D_5, under composition.

c) Show that D_5 is not abelian.

12.5 SUBGROUPS OF A GROUP

Recall that the set I of integers is a group with addition as an operation. The set E of even integers, which is a subset of I, is also a group under addition. So is the set of multiples of any fixed integer n.

We observed in Section 12.2 that the set $I_4 = \{0,1,2,3\}$ of integers is a group under addition mod 4. Consider now the subset $H = \{0,2\}$ if I_4. It can easily be verified that H itself is a group under the same operation, addition mod 4.

Thus, we observe that it is possible for a group G to have subsets that are also groups under the very same operation as on G. Accordingly, we make the following definition:

Definition 12.11   A nonempty subset H of a group G is called a *subgroup* of G if H itself is a group under the same operation as on G.

The set of integers is a subgroup of the additive group of rational numbers. The set of multiples of a fixed integer n is a subgroup of the additive group of integers. The set $H = \{1\}$ is a subgroup of the multiplicative group $\{1, -1\}$. The sets $K = \{0,3\}$ and $L = \{0,2,4\}$ are subgroups of the group $I_6 = \{0,1,2,3,4,5\}$ of integers under addition mod 6. Tables 12.9 and 12.10 show that K and L are closed under addition mod 6.

Table 12.9

$\oplus$	0	3
0	0	3
3	3	0

Table 12.10

$\oplus$	0	2	4
0	0	2	4
2	2	4	0
4	4	0	2

The associative property of $\oplus$ in K and L follows automatically, as it holds in the larger set I_6. The identity in each case is 0. Also, each element in K and L has an inverse in K and L, respectively. Thus, K and L are both groups under $\oplus$ and hence subgroups of I_6.

Notice that every group is a subgroup of itself. Also, $\{e\}$ itself is a subgroup of every group. Consequently, every group G with more than one element has at least two subgroups, G and $\{e\}$. These are called the *trivial subgroups* (*improper subgroups*) of G. Any other subgroups of G are called *proper subgroups* of G.

There is no single method to find all the subgroups of a group. However, Lagrange's theorem, named after Joseph Louis Lagrange (1736–1813), is very helpful in this direction. Lagrange was one of the two greatest mathematicians of the eighteenth century. In 1776, Frederick the Great invited Lagrange to Berlin to take the post held by Euler, the other greatest mathematician of the eighteenth century. In a letter to Lagrange, the king wrote that "the greatest king in Europe" wished to have "the greatest mathematician of Europe" at his court. Lagrange's theorem states that the number of elements in a subgroup of a finite group divides the number of elements in the group.

Let's now find all the subgroups of the group $I_6 = \{0,1,2,3,4,5\}$ of integers under addition mod 6. Let H be a subgroup of I_6. Then, the number of elements of H must divide 6, by Lagrange's theorem. Thus, H can have 1, 2, 3, or 6 elements. Clearly $H_1 = \{0\}$ is the only subgroup with one element. The sets $H_2 = \{0,3\}$ and $H_3 = \{0,2,4\}$ are the only subgroups with two and three elements each. Thus, the subgroups of I_6 are H_1, H_2, H_3, and I_6. Figure 12.10 gives the *lattice diagram* of the subgroups of I_6, where a line running down from group A to group B implies B is a subgroup of A.

It may be verified that the four group, discussed in example 12.10, has five subgroups: $G_1 = \{e\}$, $G_2 = \{e,a\}$, $G_3 = \{e,b\}$, $G_4 = \{e,c\}$, and $G_5 = G$ (see Fig. 12.11).

At this point, we like to emphasize that the converse of Lagrange's theorem need not be true: even though a natural number n may divide the number of elements of a finite group G, there is not necessarily a subgroup of G containing n elements.

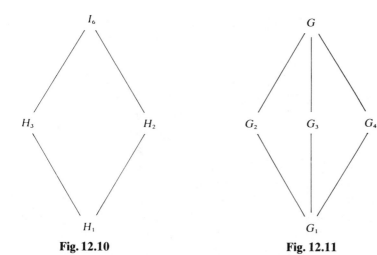

Fig. 12.10   Fig. 12.11

Exercise 12.5

1. Is the set $K = \{0,4,6,8\}$ a subgroup of the group I_{12} of integers under addition mod 12? Why?

2. Show that $H = \{1,2,4\}$ is a subgroup of the group of integers $\{1,2,3,4,5,6\}$ with multiplication mod 7 as the operation.

3. Show that $K = \{0,4\}$ and $L = \{0,2,4,6\}$ form subgroups of the group I_8 of integers with addition mod 8.

4. Show that the set of even integers is a subgroup of the additive group of integers.

5. Use Table 12.7 to show that $\{E,Q_1,Q_2\}$ is a subgroup of the group of rotations of an equilateral triangle.

6. Use Table 12.8 to show that $\{e,\alpha_1,\alpha_2,\alpha_3\}$ is a subgroup of D_4.

7. Give an example of an infinite group with finite subgroups.

8. Mark *true* or *false*.
   a) Every group has at least two subgroups.
   b) Every group with a prime number of elements has no subgroups.
   c) Every group with a prime number of elements has exactly two subgroups.
   d) The identity of a subgroup of a group G is the same as that of G.
   e) A group with two elements has exactly two subgroups.
   f) A group with three elements has exactly three subgroups.
   ★g) A nonabelian group cannot have abelian subgroups.
   ★h) Every subgroup of an infinite group is infinite.
   ★i) Every subgroup of a finite group is finite.

9. Give a counterexample to each of the false statements in problem 8.

10. If G is a group with identity e, show that $\{e\}$ is a subgroup of G.

11. If G is a group with five elements, how many subgroups does G have?

12. If G is a group with ten elements, what are the numbers of elements a subgroup of G can possibly have?

13. Same as problem 12 with 12 elements.

14. Find all the subgroups of the group $G = \{e,x,x^2\}$ with identity $e = x^3$.

15. Find all the subgroups of the multiplicative group $\{\pm 1\}$.

16. Same as problem 15 with the symmetric group S_3.

12.6 RINGS AND FIELDS

So far in our discussion of algebraic systems, we have been concerned with a system consisting of a set together with a single binary operation. It is certainly possible to have sets with two binary operations defined on them. For instance, we have the operations of addition and multiplication on the set of integers. It is the purpose of this section to introduce algebraic systems consisting of two binary operations.

The concept of a ring originates from the elementary properties of the set of integers under addition and multiplication. A ring consists of two binary operations defined on a nonempty set satisfying a set of axioms.

Definition 12.12   A nonempty set S with two binary operations, denoted by $+$ and $\cdot$, called addition and multiplication, defined on S is called a *ring* if

a) S is an abelian group with respect to addition.

b) S is closed under multiplication.

c) Multiplication is associative.

d) Multiplication is distributive over addition.

The reader is warned that the operations of addition and multiplication in this definition are just two operations we have defined on the set S. They should not be confused with the operations of addition and multiplication in number systems.

Clearly, the set of integers is a ring with the usual addition and multiplication. So are the sets of rational and real numbers. The set $I_5 = \{0,1,2,3,4\}$ of integers is a ring under addition and multiplication mod 5. More generally, the set $I_n = \{0,1,2,...,(n-1)\}$ of integers is a ring under addition and multiplication mod n, for any natural number n.

In a ring, the identity with respect to addition, usually denoted by 0, is called the *zero* of the ring. That a ring is an additive (abelian) group guarantees a unique zero for the ring and a unique additive inverse for every element of the ring.

Observe that Definition 12.12 guarantees the existence of neither a multiplicative identity, multiplicative inverses, nor the commutativity of multiplication.

Definition 12.13   A ring S is called a *ring with unit element* if there exists an element 1 in S such that $a \cdot 1 = a = 1 \cdot a$ for every a in S. The element 1 is called the *unit element* of the ring.

It is not hard to show that if a ring has a unit element, then that unit element is unique. The integers form a ring with unit element with the usual addition and multiplication, the unit element being the integer 1. The even integers also form a ring under the ordinary addition and multiplication; however, it contains no unit element! The set $I_6 = \{0,1,2,3,4,5\}$ under addition and multiplication mod 6 is a ring with unit element 1.

Before we proceed any further, let's discuss a few of the rudimentary properties of rings.

Theorem 12.3   *Let S be a ring. Then for all elements a, b in S,*
1. $a \cdot 0 = 0 = 0 \cdot a$
2. $a \cdot (-b) = -(a \cdot b) = (-a) \cdot b$
3. $(-a) \cdot (-b) = a \cdot b$

Proof:

1.
$$a \cdot 0 = a \cdot (0+0) \qquad \text{\textit{0 is the zero of the ring}}$$
$$= a \cdot 0 + a \cdot 0 \qquad \text{\textit{left distributive property}}$$
$$a \cdot 0 + 0 = a \cdot 0 + a \cdot 0 \qquad \text{\textit{0 is the zero of the ring}}$$
$$0 = a \cdot 0 \qquad \text{\textit{left cancellation property with respect to addition}}$$

That $0 \cdot a = 0$ follows similarly.

2.
$$0 = a \cdot 0 \qquad \text{\textit{part (1) above}}$$
$$= a \cdot [b + (-b)] \qquad \text{\textit{$-b$ is the additive inverse of b}}$$
$$= a \cdot b + a \cdot (-b) \qquad \text{\textit{left distributive property}}$$

Thus, $a \cdot (-b) = -(a \cdot b) \qquad \text{\textit{definition of (additive) inverse}}$

It now follows similarly that $(-a) \cdot b = -(a \cdot b)$.

3.
$$(-a) \cdot (-b) = -((-a) \cdot b) \qquad \text{\textit{part (2) above}}$$
$$= -(-(a \cdot b)) \qquad \text{\textit{part (2) above}}$$
$$= ab \qquad \text{\textit{uniqueness of inverses in a group}}$$

In addition, if we assume that the ring S possesses a unit element 1, it follows from the theorem that $(-1) \cdot a = -a$ for every a in S, and $(-1) \cdot (-1) = 1$.

The reader will recall that these results have already been discussed while studying the familiar number systems.

Definition 12.14   A ring S is said to be *commutative* if $a \cdot b = b \cdot a$ for every a, b in S.

The integers, rational numbers, and real numbers form commutative rings with the usual addition and multiplication as operations.

Recall from Chapter 4 that the product of two integers is zero if and only if at least one of them is zero. However, in the previous chapter, we observed that the set $I_6 = \{0,1,2,3,4,5\}$ of integers does not satisfy this property under multiplication mod 6; for instance, $2 \cdot 3 = 0$, but $2 \neq 0$ and $3 \neq 0$. Also, $2 \cdot 1 = 2 \cdot 4$, but $1 \neq 4$. Accordingly, we make the following definition:

Definition 12.15 A nonzero element a in a ring S is called a *zero divisor* if there exists a nonzero element b in S such that $a \cdot b = 0$.

Thus, 2, 3, and 4 are zero divisors in the ring I_6. The rings of integers and rational numbers have no zero divisors. (Why?)

Definition 12.16 A commutative ring S with unit element is called an *integral domain* if it has no zero divisors.

The rings of integers and rational numbers under addition and multiplication are integral domains. So are the rings I_3 and I_5 under addition and multiplication mod 3 and mod 5, respectively, but not the ring I_6 under addition and multiplication mod 6. (Why?)

Let's now discuss one of the extremely important algebraic systems: fields. A field is a special case of a ring. Fields play a vital role in modern algebra. Several important applications of theory of fields can be found in the theory of numbers, the theory of equations, geometry, information theory, etc.

Definition 12.17 A nonempty set F with two binary operations $+$ and $\cdot$, usually called addition and multiplication, defined on F is called a *field* if

1. F is an abelian group under addition.
2. $F - \{0\}$ is an abelian group under multiplication, where 0 is the additive identity of F.
3. Multiplication is distributive over addition.

In other words, a field is a commutative ring with unit element such that every nonzero element a in F has a multiplicative inverse a^{-1} in F.

The reader has, by now, come across several examples of fields: the set of rational numbers and the set of real numbers, with the usual addition and multiplication. All these are examples of *infinite fields*, fields containing infinitely many elements. We now give an example of a *finite field*, a field containing only a finite number of elements. Consider the set $I_5 = \{0,1,2,3,4\}$. Both I_5 and $I_5 - \{0\}$ are abelian groups under addition and multiplication mod 5. Also, both the distributive properties are satisfied. Consequently, I_5 is a field under addition and multiplication mod 5. However, I_6 is not a field under addition and multiplication mod 6. (Why?) In general, the set $I_n = \{0,1,2,...,(n-1)\}$ of integers is a field under addition and multiplication mod n if and only if n is a prime.

Cancellation properties with respect to both addition and multiplication hold in every field: $a + b = a + c$ implies $b = c$ and $a \cdot b = a \cdot c$ implies $b = c$ provided $a \neq 0$. (Why?) It follows from Theorem 12.3 that $a \cdot 0 = 0$ for every element a in a field. Recall that if the product of two real numbers is zero, then at least one of them must be zero. Is this true in every field? This is answered in the affirmative by the following theorem:

Theorem 12.4 *Let a and b be any element of a field F such that $a \cdot b = 0$. Then, either $a = 0$ or $b = 0$.*

Finally, if we consider multiplication by multiplicative inverses as division in a field F, then roughly speaking, we are allowed to divide any element of F by any nonzero element of F, as in the case of real numbers.

Exercise 12.6

1. Is the set of integers a ring under addition $\oplus$, defined by $a \oplus b = a + b - 1$, and multiplication as usual?
2. Is the set of integers a ring with addition $\oplus$ and multiplication $*$ defined by $a \oplus b = a + b - 1$ and $a * b = a + b - ab$? If yes,
   a) What is the zero of the ring?
   b) Does the ring have unit element?
   c) Is it commutative?
   d) Is it an integral domain?
3. Is $I_3 = \{0,1,2\}$ a ring under addition and multiplication mod 3? Is it an integral domain?
4. Is the set of even integers a ring with the ordinary addition and multiplication? Is it an integral domain?
5. Is the set of multiples of 3 a ring with the usual addition and multiplication? If yes, does it have a unit element? Is it an integral domain?
6. Mark *true* or *false*.
   a) Integers form a ring under addition and multiplication.
   b) Natural numbers form a ring under addition and multiplication.
   c) Even integers form a commutative ring with unit element under addition and multiplication.
   d) The set of odd integers is a ring under the operations of addition and multiplication.
   e) The set of even integers is an integral domain under addition and multiplication.
   f) $\{0,1\}$ is an integral domain under addition and multiplication mod 2.
   g) There exist zero divisors in the ring I_8.
   h) Integers form a field under the usual addition and multiplication.
   i) Addition in every ring is commutative.
   j) Multiplication in every ring is commutative.
   k) Multiplication in every field is commutative.

    l) Every field is also a ring.

    m) Every field is infinite.

   ★**n)** Every field contains at least two elements.

7. Give a reason why each of the false statements in problem 6 is false.

8. Is $I_4 = \{0,1,2,3\}$ a field under addition and multiplication mod 4? Why?

9. Same as problem 8 with $I_3 = \{0,1,2\}$.

10. Show that $K = \{a + b\sqrt{2} \mid a,b \in Q\}$ is a field under the usual addition and multiplication. What is $(2 + 3\sqrt{2})^{-1}$?

11. Is the set $L = \{a + b\sqrt{2} \mid a,b \in I\}$ a field under the usual addition and multiplication? Why?

12. Is the set $M = \{a + b\sqrt{-3} \mid a,b \in Q\}$ a field under the usual addition and multiplication?

13. Let a and b be arbitrary elements of a ring S. Show that $(a + b)^2 = a^2 + ab + ba + b^2$.

14. Let S be a ring such that $a^2 = a$ for every element a in S. Show that $2a = 0$. [*Hint*: use problem 13.]

15. Let S be a ring. Prove that $(-a) \cdot b = -(a \cdot b)$ for every a,b in S.

16. Let S be a ring with unit element 1. Prove that $(-1) \cdot a = -a$ for every a in S.

★**17.** Prove that the cancellation property with respect to multiplication holds in every integral domain.

★**18.** Let S be a commutative ring with unit element and satisfying the cancellation property with respect to multiplication. Prove that S is an integral domain.

★**19.** Prove that every field is an integral domain.

★**20.** Use the cancellation property to prove Theorem 12.4.

    [*Hint*: $a \cdot b = a \cdot 0$.]

12.7 SUMMARY AND COMMENTS

This chapter provides an introduction to the so-called abstract systems in mathematics. Galois has been given the credit of introducing abstraction to mathematics.

The concept of a binary operation on a set has been defined using the concept of function. A binary operation on a set A is a function from $A \times A$ to A. That $*$ is a binary operation on A simply means that A is closed under $*$. We defined commutative and associative operations on a set, the identity with respect to an operation $*$ on a set, the inverse a^{-1} of an element a with respect to $*$, and the distributive properties.

The algebraic system called a group was discussed and illustrated extensively. A group G is a nonempty set together with a binary operation $*$ on G such that (1) $*$ is associative, (2) there exists an identity e for $*$ in G (e is the first letter of the German word *einheit*, which means unity, unit, etc.), and (3) every element in G has an inverse with respect to $*$ in G. In addition, if the operation $*$ is commutative, the group is said to be (commutative) abelian. A group G is called a finite or infinite group according as G contains a finite number of or infinitely many elements. The identity element in a group is unique and every element has a unique inverse. Cancellation properties hold in every group. The set of all permutations of a nonempty set forms a group under composition, called a permutation group. In general, it is not commutative. The group of rotations of an equilateral triangle is the same as the permutation group on three elements, while that of a square is a subgroup of the permutation group on four elements.

A ring S is an algebraic system consisting of two binary operations $+$ and $\cdot$, such that (1) S is an abelian group with respect to $+$, (2) the operation $\cdot$ is associative, and (3) the operation $\cdot$ distributes over $+$. In every ring S, we have for every a, b in S,

1. $a \cdot 0 = 0 = 0 \cdot a$
2. $a \cdot (-b) = (-a) \cdot b = -(a \cdot b)$
3. $(-a) \cdot (-b) = a \cdot b$

A ring is said to be commutative if the operation $\cdot$ is commutative. A commutative ring with unit element is called an integral domain.

A field, one of the most important algebraic systems, is a special case of a ring. A field F is a commutative ring with unit element such that every nonzero element of F has a multiplicative inverse in F. There are two kinds of fields: infinite and finite fields. If a and b are any two elements in a field, then $a \cdot b = 0$ implies that $a = 0$ or $b = 0$, corresponding to the product law in number systems.

There was a time when it was believed that these algebraic systems were too abstract and had no practical use and application. But this is no longer true. These systems form a unifying thread that intertwines several branches of mathematics. They have great applications in computers, coding theory, theory of communications, number theory, theory of equations, the study of ruler and compass constructions in Euclidean geometry, physics, chemistry, etc. The names of Abel and Galois will be written in golden letters in the history of the so-called modern mathematics.

SUGGESTED READING

Hudson, F. M., and D. W. Adlong, *Introduction to Mathematics*, Addison-Wesley Publishing Co., Reading, Mass. (1970), pp. 235–265.

Larsen, M. D., *Introduction to Modern Algebraic Concepts*, Addison-Wesley Publishing Co., Reading, Mass. (1969), pp. 1–40, 57–67, 88–91.

Miller, C. D., and V. E. Heeren, *Mathematical Ideas, an Introduction* (2nd ed.), Scott, Foresman and Co., Glenview, Ill. (1973), pp. 71–87.

Oosse, W. J., "Properties of Operations: A Meaningful Study," *The Arithmetic Teacher*, vol. 16 (April 1969), pp. 271–275.

Triola, M. F., *Mathematics and the Modern World*, Cummings Publishing Co., Menlo Park, Calif. (1973), pp. 317–326.

Willerding, M. F., and R. A. Hayward, *Mathematics: The Alphabet of Science*, (2nd ed.), John Wiley and Sons, New York (1972), pp. 166–208.

"Geometry has been, throughout, of supreme importance in the history of knowledge."
                                                    B. RUSSELL

13/Introduction to Geometry

After studying this chapter, you should be able to:
- *state a few properties associated with lines and planes*
- *identify the interior and the exterior of an angle*
- *name a few special angles*
- *identify simple closed curves*
- *decide if a given set of points is a convex set*
- *name a few special polygons and polyhedrons*
- *state and apply Euler's formula*
- *find the lengths of some familiar curves*
- *compute the areas of regions bounded by a few well-known curves*
- *compute the volumes of solids enclosed by a few familiar surfaces*

13.0 INTRODUCTION

Geometry is one of the oldest branches of mathematics. It is believed that geometry originated at the hands of the early Egyptians during the process of land surveying. They used geometric ideas and concepts to build the monumental pyramids. It is said that the Babylonians of 2000–1600 B.C. knew how to find the areas of regions bounded by rectangles, special triangles, and trapezoids.

The Greeks, who contributed substantially to the development of geometry, learned it from the Egyptians. The well-known Greek mathematician Euclid (Section 5.0) was at the forefront of all Greek geometers. Most of his contributions appear in his classical work, *Elements*. The Greeks have been given credit for introducing logical reasoning and mathematical abstraction to the study of geometry.

The language of sets has been a significant factor for the revolutionary changes that have taken place in the whole approach to geometry in the past few years. At this point, we ask the reader to review Sections 1.5 and 1.6 before proceeding to the next section.

In this chapter, we outline and discuss a few rudimentary ideas and concepts in geometry on a rather informal basis. Throughout our discussion, we rely on the intuition of the reader.

13.1   POINTS, LINES, PLANES, AND SPACE

Most of us have some intuitive ideas about points, lines, planes, and space. As any attempts to define them can cause only more problems, we shall accept them as undefined terms. A point is represented by a small dot on the paper. We use the letters A, B, C, etc., to label points. As we have a universal set in any discussion of sets, space may be thought of as the set consisting of all points under consideration.

A line is a special collection of points in space and hence is a subset of the space. It has no thickness and extends in both directions infinitely. Figure 13.1 represents a line. The arrowheads on either end simply indicate that the line is considered to extend in either direction infinitely. The line in Fig. 13.1 *contains* the points P, Q, and R. Since the points P, Q, and R lie on the same line, we also say that these points are *collinear*. Observe that if three points are collinear, then exactly one of them lies *between* the other two. We all know that two points determine a unique line. Consequently, the above line may be denoted by $\overleftrightarrow{PQ}$, $\overleftrightarrow{QR}$, $\overleftrightarrow{PR}$, etc.

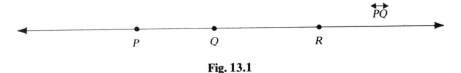

Fig. 13.1

How many lines pass through a given point? The answer can be arrived at intuitively by noting the various positions of the second hand of a wristwatch, namely, infinitely many.

Observe that every point P on a line partitions it into three disjoint sets: the set A of points lying to the right of P, the set B of points lying to the left of P, and the set $\{P\}$ consisting of the point P alone. The sets A and B are called *half-lines*. The sets $A \cup \{P\}$ and $B \cup \{P\}$ are called *rays*, denoted in Fig. 13.2 by $\overrightarrow{PQ}$ and $\overrightarrow{PR}$, respectively. The point P is called the *endpoint* of each of these

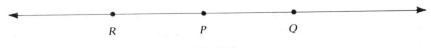

Fig. 13.2

rays. The rays $\overrightarrow{PQ}$ and $\overrightarrow{PR}$ are called *opposite rays*. Observe that $\overrightarrow{PQ} \cup \overrightarrow{PR} = \overleftrightarrow{QR}$ and $\overrightarrow{PQ} \cap \overrightarrow{PR} = \{P\}$. Unlike the case of a line, a ray extends indefinitely only in one direction. It is uniquely determined by the endpoint P and some other point Q on it, as shown in Fig. 13.3.

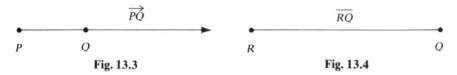

$$\overrightarrow{PQ} \qquad\qquad\qquad\qquad \overline{RQ}$$

$$P \qquad\qquad Q \qquad\qquad\qquad\qquad R \qquad\qquad\qquad\qquad Q$$

Fig. 13.3   **Fig. 13.4**

What can we say about $\overleftrightarrow{QR} \cap \overrightarrow{RQ}$ in Fig. 13.2? It consists of all points of the line lying between Q and R, and the points Q and R. It is called a *line segment*, denoted by $\overline{RQ}$, as pictured in Fig. 13.4. The points R and Q are the endpoints of the line segment $\overline{RQ}$. Observe now that

$$\overline{RQ} = \overline{QR} = \overrightarrow{QR} \cap \overrightarrow{RQ} \qquad \overrightarrow{QP} \cup \overrightarrow{PR} = \overleftrightarrow{RQ}$$

$$\overrightarrow{QP} \cap \overrightarrow{PR} = \{P\} \qquad\qquad \overline{PQ} \qquad \subset \overrightarrow{PQ}$$

$$\overrightarrow{PQ} \cup \overline{PQ} = \overrightarrow{PQ} \qquad\qquad \overrightarrow{PQ} \cap \overrightarrow{PQ} = \overrightarrow{PQ}$$

Also, the line segment $\overline{RQ}$ is completely determined by the endpoints R and Q.

When can the line segments $\overline{AB}$ and $\overline{CD}$ be the same? Recall that two sets are equal if and only if they contain exactly the same elements. Thus, $\overline{AB} = \overline{CD}$ if and only if they contain the same points. Consequently, if $\overline{AB} = \overline{CD}$, then the endpoints A and B of $\overline{AB}$ will be the same as the endpoints C and D in some order; that is, $\{A,B\} = \{C,D\}$.

Two line segments $\overline{AB}$ and $\overline{CD}$ are said to be *congruent* if when one segment is placed on the other with one endpoint of $\overline{AB}$ coinciding with an endpoint of $\overline{CD}$, then the other endpoints of $\overline{AB}$ and $\overline{CD}$ also coincide. If $\overline{AB}$ is congruent to $\overline{CD}$, we write $\overline{AB} \cong \overline{CD}$.

A plane is also a subset of space. It may be thought of as a fully flat surface extending indefinitely in all directions. However, for practical purposes, the surface of a table top, a blackboard, the floor of a room, etc., may be considered visual examples of planes.

How many points determine a plane? How many planes can pass through the same line? Notice that the different pieces of paper in this book have a common edge that may be considered part of a line. Intuitively, this makes us feel that there are infinitely many planes containing the same line. Thus, two points are not enough to fix a plane. How about three points? Since a child can ride a tricycle on any flat surface, we are tempted to say that three distinct noncollinear points determine a unique plane. Indeed, that is the case. A set of points is said to be *coplanar* if they lie in the same plane.

Recall that there is a unique line containing two distinct points. Consequently, a line and a point not on the line determine a unique plane (why?).

As a point in a line divides it into three disjoint sets, a line ℓ in a plane E partitions it into three disjoint sets: the line ℓ and the *half-planes* E_1 and E_2. Also, if P and Q are points of the half-planes E_1 and E_2, respectively, then the line segment $\overline{PQ}$ intersects the line ℓ, as in Fig. 13.5. On the other hand, if P and Q belong to the same half-plane, the $\overline{PQ}$ lies completely in that half-plane.

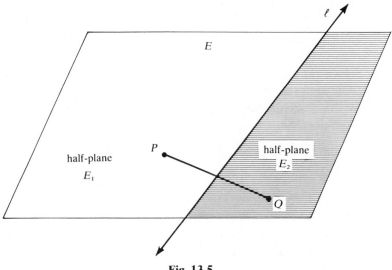

Fig. 13.5

What can we say about the intersection of two lines $\overleftrightarrow{AB}$ and $\overleftrightarrow{PQ}$ lying in the same plane? Clearly, there are two possible cases to be considered: $\overleftrightarrow{AB} \cap \overleftrightarrow{PQ} = \varnothing$ and $\overleftrightarrow{AB} \cap \overleftrightarrow{PQ} \neq \varnothing$. If $\overleftrightarrow{AB} \cap \overleftrightarrow{PQ} = \varnothing$, then the lines are said to be *parallel* and we write $\overleftrightarrow{AB} \parallel \overleftrightarrow{PQ}$. If $\overleftrightarrow{AB} \cap \overleftrightarrow{PQ} \neq \varnothing$ but $\overleftrightarrow{AB} = \overleftrightarrow{PQ}$, then the lines are said to be *coincident*. On the other hand, if $\overleftrightarrow{AB} \cap \overleftrightarrow{PQ} \neq \varnothing$ and $\overleftrightarrow{AB} \neq \overleftrightarrow{PQ}$, then they intersect at a unique point. All three of these situations are illustrated in Fig. 13.6.

If $\overleftrightarrow{AB}$ and $\overleftrightarrow{PQ}$ are two lines in different planes such that $\overleftrightarrow{AB} \cap \overleftrightarrow{PQ} = \varnothing$, they are called *skew lines*. For example, the lines ℓ_1 and ℓ_2 in Fig. 13.7 are skew lines.

What is the intersection of a line ℓ and a plane E? If any two points of ℓ already belong to E then ℓ is clearly contained in E (why?) and $\ell \cap E = \ell$. For example, the line $\overleftrightarrow{PQ}$ in Fig. 13.8 is part of the plane of this paper. If ℓ and E have no points in common, then ℓ and E are *parallel*. If ℓ and E are not parallel and ℓ is not contained in E, then the line intersects the plane in exactly one point, as shown in Fig. 13.9. For example, if we let a sharpened pencil touch a sheet of paper vertically, then the pencil meets the plane of the paper at a unique point.

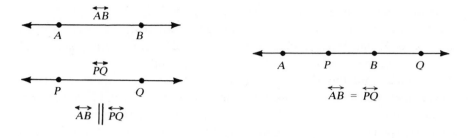

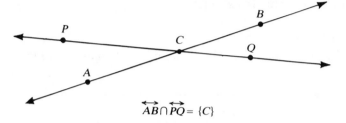

$$\overleftrightarrow{AB} \cap \overleftrightarrow{PQ} = \{C\}$$

Fig. 13.6

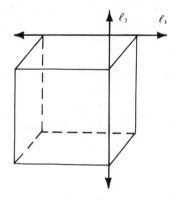

Fig. 13.7

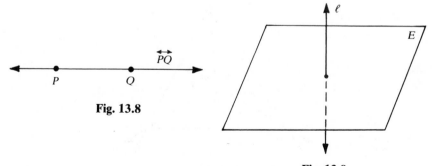

Fig. 13.8

Fig. 13.9

How about the intersection of two planes E_1 and E_2? If E_1 and E_2 have at least three noncollinear points in common, then clearly, $E_1 = E_2$ (why?); otherwise, E_1 and E_2 are distinct planes. If E_1 and E_2 are different planes and $E_1 \cap E_2 = \emptyset$, then they are *parallel planes*. For example, considering the floor of a room and the top of a table in it as parts of two planes, these two planes never meet and hence are parallel. On the other hand, if E_1 and E_2 are distinct planes and $E_1 \cap E_2 \neq \emptyset$, then $E_1 \cap E_2$ is a straight line. For example, the floor and a wall of a room intersect along a line; likewise, the intersection of any two adjacent walls in a room is also a line.

Exercise 13.1

1. With the points P, Q, R, and S as in Fig. 13.10, find each of the following sets.

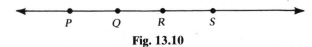

$$P \qquad Q \qquad R \qquad S$$

Fig. 13.10

a) $\overline{PQ} \cup \overline{QR}$ b) $\overline{PR} \cap \overline{QR}$ c) $\overline{PQ} \cap \overline{QR}$

d) $\overline{PQ} \cap \overline{RS}$ e) $\overline{PQ} \cup \{Q\}$ f) $\overline{PQ} \cap \{Q\}$

g) $\overrightarrow{PS} \cap \overrightarrow{RP}$ h) $\overleftrightarrow{QS} \cap \overrightarrow{RQ}$ i) $\overleftrightarrow{QS} \cap \overline{QP}$

j) $\overline{PR} \cap \overline{QS}$ k) $\overline{PQ} \cap \overrightarrow{RS}$ l) $\overline{PQ} \cup \overleftrightarrow{RS}$

2. Find the following sets, where A, B, and C are three noncollinear points.

a) $\overline{AB} \cap \overline{AC}$ b) $\overleftrightarrow{AB} \cap \overleftrightarrow{AC}$ c) $\overline{AB} \cap \overline{BC} \cap \overline{CA}$

d) $\overline{AB} \cap \{C\}$ e) $(\overline{AB} \cup \overline{BC}) \cap \overline{AC}$ f) $(\overleftrightarrow{AB} \cup \overleftrightarrow{BC}) \cap \overleftrightarrow{AC}$

3. With A, B, C, and D as in Fig. 13.11, find the following sets.

a) $\overrightarrow{AD} \cap \overrightarrow{CD}$

b) $\overline{AB} \cap \overline{BD}$

c) $\overline{AD} \cap \overleftrightarrow{BC}$

d) $(\overline{AD} \cup \overline{CD}) \cap \overline{AC}$

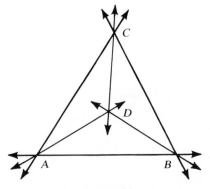

Fig. 13.11

4. Using Fig. 13.12, identify the following as *true* or *false*.

   a) $\overline{AB} \cap \overline{BC} = \{B\}$
   b) $\overline{AB} \cup \overline{BC} = \overline{AC}$
   c) $\overleftrightarrow{DE} = \overleftrightarrow{BE}$
   d) $\overrightarrow{DE} = \overrightarrow{BE}$
   e) $\overline{AB} \cap \overline{BC} \cap \overline{AC} = \varnothing$
   f) $\overline{AB} \cap \overline{BC} \cap \overrightarrow{AC} = \overrightarrow{AD} \cap \overrightarrow{DC} \cap \overrightarrow{AC}$

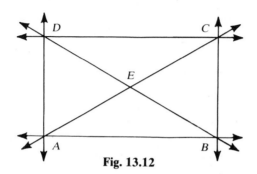

Fig. 13.12

5. Let $\overleftrightarrow{AB}$ and $\overleftrightarrow{CD}$ be two lines such that $\overleftrightarrow{AB} \parallel \overleftrightarrow{CD}$. If $\overrightarrow{PQ}$ intersects $\overleftrightarrow{AB}$, does it intersect $\overleftrightarrow{CD}$ also? If not, when?

6. Recall that a line ℓ in a plane E divides it into three disjoint sets: the line ℓ and the half-planes E_1 and E_2. Let A and B be points of E_1 and C a point of E_2.

   a) Does $\overline{AB}$ intersect ℓ?   **b)** Does $\overline{BC}$ meet ℓ?
   c) Is $E_1 \cup E_2 = E$?   **d)** Is $E_1 \cup E_2 = E - \ell$?
   e) Is $\overline{AB} \subset E_1$?   **f)** Is $E_1 \cap \ell = \varnothing$?

7. Mark *true* or *false*.

   a) Two distinct points determine a line.
   b) If A and B are distinct points on a line ℓ, then $\overleftrightarrow{AB} = \ell$.
   c) $\overleftrightarrow{AB} = \overleftrightarrow{BA}$
   d) If $C \in \overleftrightarrow{AB}$, then $\overleftrightarrow{AB} = \overleftrightarrow{AC}$.
   e) If $\overleftrightarrow{AB} = \overleftrightarrow{CD}$, then $\{A,B\} = \{C,D\}$.
   f) A and B are the endpoints of $\overleftrightarrow{AB}$.
   g) Every point on a line divides it into three disjoint sets.
   h) The endpoint of a ray belongs to the ray.
   i) The ray $\overrightarrow{AB}$ has exactly one endpoint.
   j) If $C \in \overrightarrow{AB}$, then $\overrightarrow{AB} = \overrightarrow{AC}$.
   k) $\overrightarrow{AB} = \overrightarrow{BA}$
   l) $\overrightarrow{AB} \subset \overleftrightarrow{AB}$
   m) $\overrightarrow{AB} \cap \overrightarrow{BA} = \overleftrightarrow{AB}$
   n) $\overrightarrow{AB} \cup \overrightarrow{BA} = \overleftrightarrow{AB}$

o) $\overline{AB} = \overline{BA}$

p) If $\overline{AB} = \overline{CD}$, then $\{A,B\} = \{C,D\}$.

q) If $\overline{AB} = \overline{CD}$, then $A = C$ and $B = D$.

r) $\overline{AB}$ is a finite set.

s) A and B are the endpoints of $\overline{AB}$.

t) $\overline{AB} \subset \overleftrightarrow{AB}$

u) If C lies between A and B, then $\overline{AC} \cup \overline{CB} = \overline{AB}$.

v) If C lies between A and B, then $\overline{AC} \cap \overline{CB} = \varnothing$.

w) Any three points determine a plane.

x) Every line in a plane divides it into three disjoint sets.

y) Any two intersecting lines determine a plane.

8. Do any three points determine a plane? If not, when?

9. Do any two lines determine a plane?

10. Do two intersecting lines fix a plane?

★11. Is the relation "is parallel to" on the set of all straight lines in a plane symmetric? transitive?

★12. Is the relation "is congruent to" on the set of all line segments in a plane
- **a)** reflexive?
- **b)** symmetric?
- **c)** transitive?
- **d)** an equivalence relation?

★13. How many distinct lines can be drawn through
- **a)** two points
- **b)** three collinear points
- **c)** three noncollinear points
- **d)** four points, no three being collinear

★14. How many distinct planes can pass through
- **a)** two points
- **b)** three collinear points
- **c)** three noncollinear points
- **d)** four points, not all being coplanar

13.2   ANGLES

Most of us have some concept of angles. However, very often the *measure* of an angle is misunderstood for the angle. An angle, as will be seen shortly, is a set of points, whereas its measure is a number.

Consider two rays $\overrightarrow{AB}$ and $\overrightarrow{AC}$ with the same endpoint, A, as in Fig. 13.13a or b. Since $\overrightarrow{AB}$ and $\overrightarrow{AC}$ are two sets, we can very well talk about their union. An *angle* is the union of two rays with the same endpoint. The rays are called the *sides* of the angle and the common endpoint is called the *vertex* of the angle. The rays $\overrightarrow{AB}$ and $\overrightarrow{AC}$ are the sides and A the vertex of the angle in Fig. 13.13a. Such an angle is denoted by either $\angle BAC$ or $\angle A$. Thus,

$$\angle BAC = \overrightarrow{AB} \cup \overrightarrow{AC} = \overrightarrow{AC} \cup \overrightarrow{AB} = \angle CAB$$

Since the angle in Fig. 13.13b is formed by two opposite rays, it is called a *straight angle*.

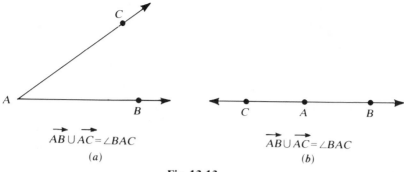

$\overrightarrow{AB} \cup \overrightarrow{AC} = \angle BAC$

(a)

$\overrightarrow{AB} \cup \overrightarrow{AC} = \angle BAC$

(b)

Fig. 13.13

As a line divides a plane into three disjoint sets, an angle also separates a plane into three disjoint sets, as shown in Fig. 13.14. The shaded region in Fig. 13.14 is called the *interior* of $\angle BAC$, denoted by $\text{Int}\angle BAC$. The set of points of the plane that do not belong to either $\angle BAC$ or $\text{Int}\angle BAC$ is called the *exterior* of $\angle BAC$, denoted by $\text{Ext}\angle BAC$.

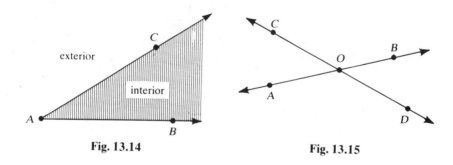

Fig. 13.14

Fig. 13.15

Consider now two lines $\overleftrightarrow{AB}$ and $\overleftrightarrow{CD}$ intersecting at the point O, as in Fig. 13.15. Observe that $\angle BOC$ and $\angle BOD$ have the property that they have a common side, $\overrightarrow{OB}$, and a common vertex, O. Two such angles are called *adjacent angles*. Notice that $\angle BOC$ and $\angle AOC$, $\angle AOC$ and $\angle AOD$, and $\angle AOD$ and $\angle BOD$ are also adjacent angles. Now, $\angle AOD$ and $\angle BOC$ are not adjacent angles (why?). Still, they have a common vertex; also, the rays forming $\angle AOD$ and those forming $\angle BOC$ are opposite rays. Two such angles are called *vertical angles*. Notice that $\angle AOC$ and $\angle BOD$ are also vertical angles.

The two angles $\angle BOC$ and $\angle BOD$ satisfy the property that their union is the line $\overleftrightarrow{CD}$ and the ray $\overrightarrow{OB}$ with endpoint O on the line $\overleftrightarrow{CD}$. Two such angles are called *supplementary angles*.

Two angles $\angle ABC$ and $\angle PQR$ are said to be *congruent*, denoted by $\angle ABC \cong \angle PQR$, if when one is superimposed on the other, then they coincide (see Fig. 13.16). Consider now the lines $\overleftrightarrow{AB}$ and $\overleftrightarrow{CD}$ meeting at O, as in Fig. 13.17. If $\angle BOC \cong \angle BOD$, then each is called a *right angle* and the lines $\overleftrightarrow{AB}$

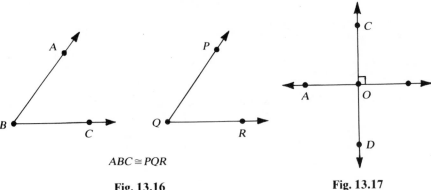

$ABC \cong PQR$

Fig. 13.16   **Fig. 13.17**

and $\overleftrightarrow{CD}$ are said to be *perpendicular*, denoted by $\overleftrightarrow{AB} \perp \overleftrightarrow{CD}$. Since vertical angles formed by two intersecting lines are always congruent, it now follows that if $\overleftrightarrow{AB} \perp \overleftrightarrow{CD}$, then $\angle AOC$ and $\angle AOD$ are also right angles.

A line $\overleftrightarrow{PS}$ intersecting two lines $\overleftrightarrow{AB}$ and $\overleftrightarrow{CD}$ in exactly two points Q and R, as in Fig. 13.18, is called a *transversal*. The angles $\angle BRS$ and $\angle DQR$ are called *corresponding angles*. The other pairs of corresponding angles are $\angle SRA$ and $\angle RQC$, $\angle ARQ$ and $\angle CQP$, and $\angle BRQ$ and $\angle DQP$. The angles $\angle ARQ$, $\angle BRQ$, $\angle CQR$, and $\angle DQR$ are called *interior angles*, while the angles $\angle SRA$, $\angle SRB$, $\angle PQC$, and $\angle PQD$ are called *exterior angles*. Angles $\angle ARQ$ and $\angle DQR$ are *alternate interior angles*, as are angles $\angle CQR$ and $\angle BRQ$.

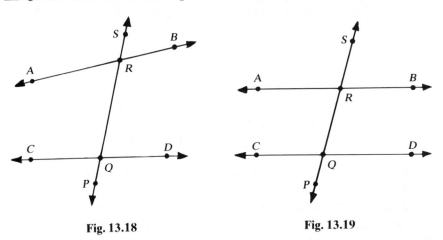

Fig. 13.18   **Fig. 13.19**

In particular, if $\overleftrightarrow{AB} \parallel \overleftrightarrow{CD}$, as in Fig. 13.19, then every pair of corresponding angles and alternate interior angles are congruent. Conversely, if a pair of corresponding angles or alternate interior angles are congruent, then the lines are parallel. It now follows that if $\overleftrightarrow{AB} \parallel \overleftrightarrow{CD}$ and $\overleftrightarrow{PQ} \perp \overleftrightarrow{AB}$, then $\overleftrightarrow{PQ} \perp \overleftrightarrow{CD}$ also. (Why?)

Exercise 13.2

1. Identify each of the following as *true* or *false* with respect to Fig. 13.20.
   a) $\angle CAB \cup \angle BAD = \angle CAD$
   b) $\angle CAB \cup \angle BAD = \angle CAD \cup \overrightarrow{AB}$
   c) $\angle CAB \cap \angle BAD = \varnothing$
   d) $\angle CAB \cap \angle BAD = \{A\}$
   e) $\angle CAB \cap \angle BAD = \overrightarrow{AB}$
   f) $\angle BAC = \overrightarrow{AB} \cup \overrightarrow{AC}$
   g) $\angle BAC = \angle CAB$
   h) $\angle BAC$ and $\angle BAD$ are adjacent angles.
   i) $\angle BAC$ and $\angle DAC$ are supplementary angles.
   j) $\overrightarrow{AB} \cup \overrightarrow{AC} = \angle BAC$

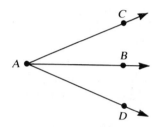

Fig. 13.20

2. a) How many angles are formed by two intersecting lines?
   b) How many pairs of them are vertical angles?
   c) Supplementary angles?
   d) Adjacent angles?
3. In Fig. 13.21, identify pairs of
   a) adjacent angles
   b) supplementary angles
   c) vertical angles

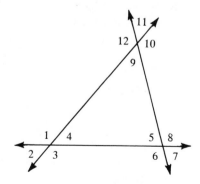

Fig. 13.21

4. Can the intersection of two angles (*a*) be empty? (*b*) Consist of a single point? (*c*) Be a ray?

5. How many angles are formed when two parallel lines are cut by a transversal?

6. *True* or *false*: A line perpendicular to one of two parallel lines is perpendicular to the other also.

7. Mark *true* or *false*.

   a) An angle is a set of points.
   b) The union of any two rays is an angle.
   c) The interior of an angle is disjoint from its exterior.
   d) Two angles with a common vertex are adjacent angles.
   e) The vertical angles formed by two intersecting lines are congruent.
   f) If $\angle A \cong \angle B$, then $\angle B \cong \angle A$.
   g) If $\angle A \cong \angle B$ and $\angle B \cong \angle C$, then $\angle A \cong \angle C$.
   h) If $\overleftrightarrow{AB} \perp \overleftrightarrow{CD}$, then $\overleftrightarrow{CD} \perp \overleftrightarrow{AB}$.
   i) If $\overleftrightarrow{AB} \perp \overleftrightarrow{CD}$ and $\overleftrightarrow{CD} \perp \overleftrightarrow{EF}$, then $\overleftrightarrow{AB} \perp \overleftrightarrow{EF}$.
   j) The union of two angles is also an angle.

★8. Is the relation "is perpendicular to" on the set of straight lines in a plane symmetric? Transitive?

★9. Consider $\angle BAC$ in a plane E with points P and Q in Int$\angle BAC$.

   a) Is $\overline{PQ} \subset$ Int$\angle BAC$?   **b)** Does $\overline{PQ}$ intersect $\angle BAC$?
   c) Does $\overleftrightarrow{PQ}$ intersect $\angle BAC$?   **d)** Does $\overline{PQ}$ intersect Ext$\angle BAC$?
   e) Does $\overleftrightarrow{PQ}$ intersect Ext$\angle BAC$?
   f) Are Int$\angle BAC$ and Ext$\angle BAC$ disjoint?
   g) Is Int$\angle BAC \cup$ Ext$\angle BAC = E$?

13.3 CURVES

We have some intuitive idea about how a curve looks. A *curve* is a set of points that can be marked by a pencil without lifting it. Lines, line segments, circles, triangles, and rectangles are all examples of curves. The figures in Fig. 13.22 are curves, whereas those in Fig. 13.23 are not (why?). In this section, we shall confine ourselves to the discussion of curves that can be represented on a plane. Such curves are called *plane curves*. From now on, the word curve will stand for a plane curve.

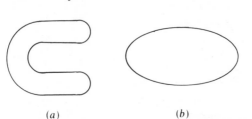

(*a*)   (*b*)   (*c*)   (*d*)

Fig. 13.22

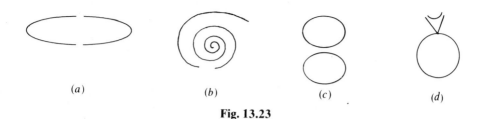

(*a*)   (*b*)   (*c*)   (*d*)

Fig. 13.23

A curve that does not pass through a point more than once is called a *simple curve*. The curves (*a*) and (*b*) in Fig. 13.22 are simple, whereas (*c*) and (*d*) are not. Also, there can be curves whose initial and terminal points coincide. Such curves are called *closed curves*. For example, the curves (*a*), (*b*), and (*c*) in Fig. 13.22 are closed, while (*d*) is not. Curves that are simple and closed are called *simple closed curves*. The curves (*a*) and (*b*) in Fig. 13.22 are simple closed curves. Notice that circles, triangles, and rectangles are also simple closed curves.

One of the remarkable properties associated with simple closed curves is the *Jordan curve theorem*, which intuitively looks trivial. It states that every simple closed curve partitions a plane into three disjoint sets: the set of points inside the curve, called its *interior*; the set of points outside the curve, called its *exterior*; and the curve itself, as exhibited in Fig. 13.24. The curve forms the *boundary* to both the interior and the exterior of the curve. Observe from Fig. 13.25 that if P is in the interior and Q in the exterior of a simple closed curve C, then $\overleftrightarrow{PQ}$ intersects the curve at some point R.

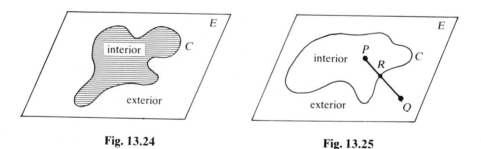

Fig. 13.24   **Fig. 13.25**

One of the most familiar simple closed curves is a circle. A *circle C* consists of a set of points such that if P and Q are any two points on C, then $\overline{OP} \cong \overline{OQ}$ for some fixed point O. The fixed point O is called the *center* of the circle. The line segment $\overline{OP}$ is called a *radius* of the circle. Any line segment $\overline{AB}$ with its endpoints A and B on the circle is a *chord* of the circle. A chord that passes through the center is a *diameter* of the circle. For example, $\overline{XY}$ is a diameter of the circle in Fig. 13.26.

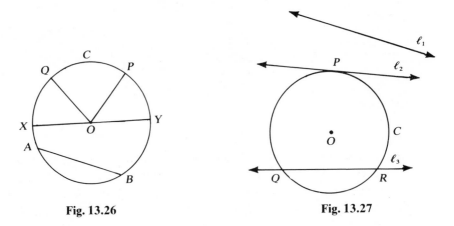

Fig. 13.26   Fig. 13.27

Any part of a circle joining two points on it is called an *arc* of the circle. For example, the points B and P on the circle in Fig. 13.26 determine two arcs: the arc BYP containing the point Y and the arc BXP containing the point X, denoted by $\overparen{BYP}$ and $\overparen{BXP}$, respectively. An arc whose endpoints are the endpoints of a diameter is called a *semicircle*. The arcs $\overparen{XQY}$ and $\overparen{XBY}$ in Fig. 13.26 are semicircles.

Notice that a line in the plane of a circle can intersect it at either no points, exactly one point, or exactly two points, as in Fig. 13.27. If the intersection of a line with a circle consists of a single point, then the line is called a *tangent* to the circle at that point; if the intersection contains two points, then the line is called a *secant*. In Fig. 13.27, the line ℓ_2 is a tangent to the circle at P and the line ℓ_3 is a secant.

Let's now consider the interior of a circle. It has the fascinating property that if P and Q are any two points in its interior, then the line segment $\overline{PQ}$ lies completely within the interior, as in Fig. 13.28a. Such a set of points is called a *convex set*. Thus, a set A of points is called a *convex set* if the line segment $\overline{PQ}$ is contained in A for any two points P and Q in A. For example, the interior of a circle, the interior of a triangle, a line segment, and a line are all convex sets. Notice that the interiors of the curves in Fig. 13.29 are not convex.

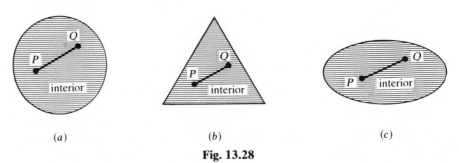

(a)   (b)   (c)

Fig. 13.28

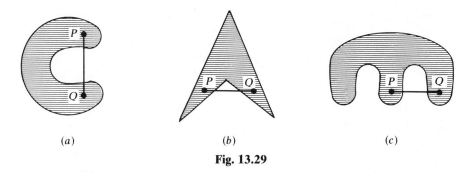

(a)   (b)   (c)

Fig. 13.29

Finally, we remark that two curves C_1 and C_2 are *congruent*, denoted by $C_1 \cong C_2$, if when one is placed on the other, then they coincide. Intuitively, two curves are congruent if and only if they have the same shape and size.

Exercise 13.3

1. Which of the following curves are simple?

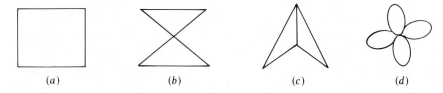

(a)   (b)   (c)   (d)

2. Which of the following curves are closed?

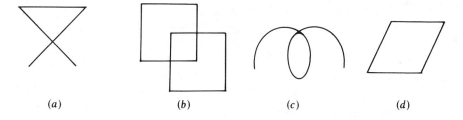

(a)   (b)   (c)   (d)

3. Is any of the curves in problem 1 closed? If yes, which ones?
4. Is any of the curves in problem 2 simple? If yes, which ones?
5. Which of the following are simple closed curves?

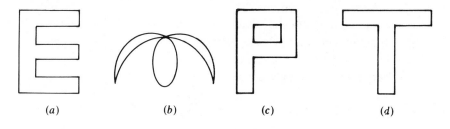

(a)   (b)   (c)   (d)

6. Draw a simple closed curve.

7. Draw a curve that is simple but not closed.

8. Draw a curve that is closed but not simple.

9. Draw a curve that is neither simple nor closed.

10. If $\overline{OA}$ and $\overline{OB}$ are two radii of a circle C with center O, find

    a) $\overline{OA} \cap \overline{OB}$ **b)** $\overline{OA} \cap C$ **c)** $\overline{OB} \cap C$

11. If $\overline{AB}$ is a chord of a circle C with center O, find

    a) $\overline{AB} \cap C$ **b)** $\overline{OA} \cap \overline{AB}$ **c)** $\overleftrightarrow{AB} \cap C$

12. Use Fig. 13.30 to identify the following as *true* or *false*.

    a) $\overline{OC}$ is a radius of the circle.

    b) $\overline{OB} \cong \overline{OC}$

    c) $\overline{OA} \cup \overline{OB} = \overline{AB}$

    d) The center O is part of the circle.

    e) The arc $\overset{\frown}{CBD}$ is a semicircle.

    f) $\angle AOC$ and $\angle BOC$ are adjacent angles.

    g) $\overline{AB}$ is a secant of the circle.

    h) $\overleftrightarrow{AB}$ is a chord of the circle.

    i) The point O belongs to the interior of the circle.

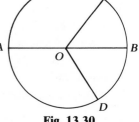

Fig. 13.30

13. If $\overline{AB}$ and $\overline{PQ}$ are two distinct diameters of a circle C with center O, find

    a) $\overline{AB} \cap \overline{PQ}$ **b)** $\overline{AB} \cap C$ **c)** $\overline{PQ} \cap C$

14. If $\overline{AB}$ and $\overline{PQ}$ are diameters of a circle with center O such that $\overline{OA} = \overline{OP}$, what is your conclusion?

15. If $\overleftrightarrow{PQ}$ is a secant of a circle C intersecting at P and Q, find $C \cap \overleftrightarrow{PQ}$.

16. In how many points can two distinct circles intersect?

17. If P and Q are points in the interior of a circle C, find

    a) $\overline{PQ} \cap C$ **b)** $\overline{PQ} \cap \text{Int } C$

    c) $\overline{PQ} \cap \text{Ext } C$ **d)** $\text{Int } C \cap \text{Ext } C$

18. Mark *true* or *false*.

    a) Every curve is simple.

    b) Every curve is closed.

    c) Every simple curve is closed.

    d) Every closed curve is simple.

    e) A circle is a simple closed curve.

    f) The center of a circle belongs to the circle.

    g) Every semicircle is part of a circle.

    h) Every circle has a unique diameter.

    i) All radii of a circle are congruent.

    j) All diameters of a circle are congruent.

    k) The interior of every circle is a convex set.

    l) Every line segment is a convex set.

    m) Any ray is convex.

    n) Every circle is a convex set.

    o) Every triangle is a convex set.

19. What is the union of all radii of a circle *C*?

20. Which of the following curves have the property that their interiors are convex sets?

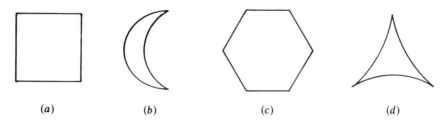

(*a*)   (*b*)   (*c*)   (*d*)

13.4 POLYGONS

Triangles, rectangles, and squares are examples of simple closed curves formed by line segments. Such curves constitute a class of curves called polygons. A *polygon* is a simple closed curve formed by the union of (a finite number of) line segments. These line segments are called the *sides* of the polygon; the endpoints of the sides are called the *vertices* of the polygon; the line segments formed by joining any two nonadjacent vertices are called the *diagonals* of the polygon. The curve in Fig. 13.31 is a polygon with five sides: $\overline{AB}$, $\overline{BC}$, $\overline{CD}$, $\overline{DE}$, and $\overline{EA}$; it has five vertices, namely, *A*, *B*, *C*, *D*, and *E*. Consequently, it is called the polygon *ABCDE*. Likewise, the curve in Fig. 13.32 is the polygon *PQRS*. The line segments $\overline{AC}$, $\overline{AD}$, $\overline{BD}$, $\overline{BE}$, and $\overline{CE}$ are the diagonals of the polygon *ABCDE*. Even though $\angle ABC = \overrightarrow{BA} \cup \overrightarrow{BC}$, it has become customary to call $\angle ABC$ an *angle* of the polygon *ABCDE*. Similarly, the other angles of the polygon are $\angle BCD$, $\angle CDE$, $\angle DEA$, and $\angle EAB$.

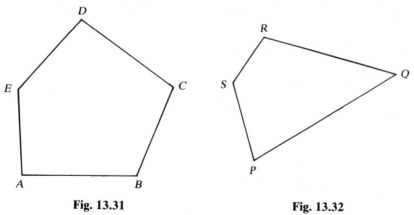

Fig. 13.31   **Fig. 13.32**

A polygon together with its interior constitutes a *polygonal region*. Most of us use the word polygon instead of "polygonal region" for convenience, even though it is not correct. Notice that a polygon is only the boundary of a polygonal region.

Specific names are given to polygons for convenience, depending on their number of sides. For example, a *triangle* is a three-sided polygon; a *quadrilateral* is a four-sided polygon; a *pentagon* is a polygon with five sides; a *hexagon* is a six-sided polygon; a polygon with seven sides is a *heptagon*; and an *octagon* is a polygon with eight sides.

A triangle with vertices A, B, and C is denoted by $\triangle ABC$. Thus, $\triangle ABC = \overline{AB} \cup \overline{BC} \cup \overline{CA}$. We now list some special triangles and quadrilaterals.

1. A triangle with two of its sides congruent is an *isosceles triangle*.
2. *An equilateral triangle* is a triangle with all of its sides congruent.
3. A triangle is *equiangular* if all its angles are congruent.
4. A triangle is called a *right triangle* if one of its angles is a right angle.
5. A quadrilateral with a pair of opposite sides on parallel lines is a *trapezoid*.
6. A quadrilateral with opposite sides on parallel lines is called a *parallelogram*.
7. A parallelogram is called a *rectangle* provided one of its angles is a right angle.
8. A rectangle with any two adjacent sides congruent is a *square*.
9. A *rhombus* is a parallelogram with any two adjacent sides congruent.

It is well known that in an isosceles triangle, the angles opposite the congruent sides are congruent. Conversely, if two angles of a triangle are congruent, then the sides opposite them are congruent. It now follows that every equilateral triangle is equiangular and, conversely, every equiangular triangle is equilateral.

Exercise 13.4

1. Identify each of the following as true or false with respect to the $\triangle ABC$ in Fig. 13.33.

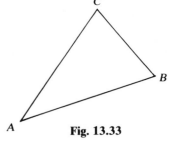

Fig. 13.33

a) $\triangle ABC = \overline{AB} \cup \overline{BC} \cup \overline{AC}$
b) $\overline{AB} \cap \overline{AC} = \{A\}$
c) $\overline{AB} \cup \overline{AC} = \angle BAC$
d) $\{A\} \cap \overline{BC} = \varnothing$
e) $(\triangle ABC) \cap \overline{BC} = \{A\}$
f) $\angle A \cap \angle B = \varnothing$

2. Use Fig. 13.34 to find the following sets.

a) $\overline{AB} \cap \overline{CD}$
b) $\overline{AB} \cap \overline{AF}$
c) $\overline{AD} \cap \overline{BE}$
d) $\angle ABC \cap \angle CDE$
e) $\angle BOC \cap \angle EOF$
f) $\angle BOC \cup \angle COD$

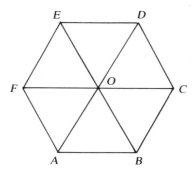

Fig. 13.34

3. Use Fig. 13.34 to answer the following.
   a) Are $\overline{AB}$ and $\overline{BC}$ two adjacent sides?
   b) Are A and C two adjacent vertices?
   c) Is $\overline{AD}$ a diagonal of the hexagon?
   d) Does the point O belong to the hexagon?
   e) Does the point O belong to the interior of the hexagon?

13.5   SURFACES

Thus far in our discussion, we have been primarily concerned with simple closed plane curves. As in the case of curves, we have some idea as to what we mean by a surface in space. Words like, "surface of the earth," "surface of water," "surface of the moon," "surface of a ball," "surface of a donut," etc., are not new to us. The type of surfaces we shall be concentrating on in this section is the one similar to simple closed curves. Recall that a simple closed curve in a plane partitions it into three disjoint sets. Likewise, a surface in space that divides space into three disjoint sets—the interior, the exterior, and the surface itself—is a *simple closed surface*. A simple closed surface together with its interior forms a *solid*.

Recall that a polygon is a simple closed curve formed by line segments. Analogously, a simple closed surface formed by polygons and polygonal regions is called a *polyhedron*. The vertices and sides of the polygons are called *vertices* and *edges* of the polyhedron, respectively, while the polygonal regions are called *faces* of the polyhedron. A few familiar polyhedrons are exhibited in Fig. 13.35. The line segments $\overline{AB}$, $\overline{BC}$, $\overline{CE}$, etc., are the edges and the polygonal regions ABF, $ABCD$, $BCEF$, etc., are the faces of the polyhedron in Fig. 13.35*a*.

One of the simplest and most familiar type of polyhedrons is a prism. A *prism* is a simple closed surface formed by two congruent polygonal regions lying in parallel planes, together with quadrilateral regions obtained by joining the vertices of the two polygonal regions. The two polygonal regions form the *bases*

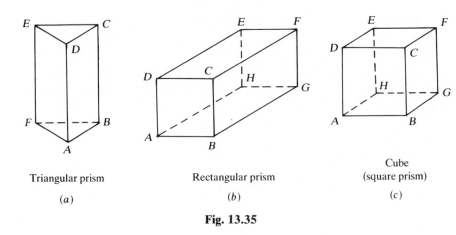

Triangular prism

(a)

Rectangular prism

(b)

Cube
(square prism)

(c)

Fig. 13.35

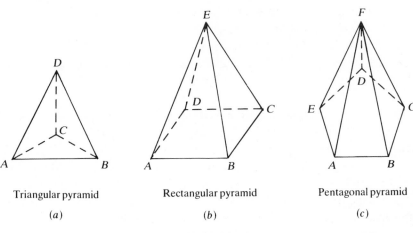

Triangular pyramid

(a)

Rectangular pyramid

(b)

Pentagonal pyramid

(c)

Fig. 13.36

and the quadrilateral regions the *lateral faces* of the prism. The bases and the lateral faces together form the *faces* of the prism.

Prisms are classified as *triangular prisms, rectangular prisms,* etc., depending on the shape of the base. For the triangular prism in Fig. 13.35*a*, the triangular regions *ABF* and *DEC* are the bases and the rectangular regions *ABCD, BCEF,* and *ADEF* are the lateral faces.

If V, F, and E denote the number of vertices, faces, and edges of a polyhedron, then for a triangular prism we have $V = 6$, $F = 5$, $E = 9$, and $V + F - E = 6 + 5 - 9 = 2$; for the rectangular prism in Fig. 13.35*b*, $V = 8$, $F = 6$, $E = 12$, and $V + F - E = 8 + 6 - 12 = 2$. More generally, we have the following fascinating result:

Euler's formula   If V, F, and E denote the number of vertices, the number of faces, and the number of edges of a polyhedron, respectively, then $V+F-E=2$.

The next class of polyhedrons we shall discuss is the pyramid. A *pyramid* is a simple closed surface consisting of one polygonal region and the triangular regions obtained by joining the vertices of the polygonal region to an external point. The polygonal region is the *base* and the triangular regions are the *lateral faces* of the pyramid.

Pyramids are classified *triangular pyramids, rectangular pyramids,* etc., depending on the shape of the base. Observe that for the rectangular pyramid in Fig. 13.36, $V=5$, $F=5$, $E=8$, and $V+F-E=5+5-8=2$, satisfying Euler's formula.

We now devote the remainder of this section to the discussion of three types of simple closed surfaces with which we are all familiar: spheres, right circular cylinders, and right circular cones. A *sphere* is a simple closed surface with the additional property that if A and B are any two points on the surface, then $\overline{OA} \cong \overline{OB}$ for some fixed point O in space. The fixed point O is the *center* of the sphere. If P is any point on the sphere, then $\overline{OP}$ is a *radius* of the sphere. Chords and diameters of a sphere can be defined as in the case of a circle.

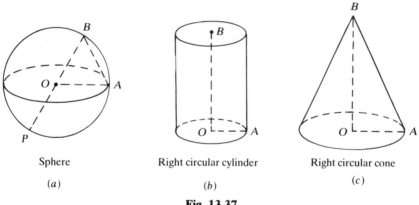

Sphere

(a)

Right circular cylinder

(b)

Right circular cone

(c)

Fig. 13.37

The simple closed surface exhibited in Fig. 13.37b is a *right circular cylinder.* It consists of two circular bases lying in two parallel planes and the surface obtained by joining every point of one base to the corresponding point of the other by line segments. These line segments have the property that they are perpendicular to the planes containing the bases.

A *right circular cone* is a simple closed surface consisting of a circular base, an external point B such that $\overleftrightarrow{OB}$ is perpendicular to the plane of the base, and the surface obtained by joining the point B to every point on the circle by line segments, as shown in Fig. 13.37c.

Exercise 13.5

1. Mark *true* or *false.*

    a) A pyramid is a simple closed surface.

    b) Every pyramid is a polyhedron.

    c) Every prism is a polyhedron.

    d) The bases of a cube lie in parallel planes.

    e) A triangular prism has two bases.

    f) A triangular pyramid has only one base.

    g) A triangular pyramid has five vertices.

    h) A pentagonal prism has 15 edges.

    i) A cube has only 6 vertices.

    j) A hexagonal prism has 8 faces.

    k) A square pyramid has 8 edges.

    l) A plane intersects a sphere in exactly two points.

2. If V, F, and E denote the number of vertices, the number of faces, and the number of edges of a prism, respectively, complete the following table.

Prism	V	F	E	$V+F-E$
Cube				
Pentagonal				
Hexagonal				

3. If V denotes the number of vertices, F the number of faces, and E the number of edges of a pyramid, complete the following table.

Pyramid	V	F	E	$V+F-E$
Triangular				
Square				
Pentagonal				
Hexagonal				

4. If a polyhedron has 6 vertices and 12 edges, how many faces does it have?

5. If a polyhedron has 12 faces and 30 edges, how many vertices does it have?

6. If a prism has 16 vertices and 10 faces, how many edges does it have?

7. If a pyramid has 5 edges in its base, how many edges does it have altogether?

8. If a pyramid has 10 edges in its base, how many
  a) faces does it have?
  b) vertices does it have?
  c) edges does it have?

9. If a prism has 6 edges in one base, how many edges does it have altogether?

10. If a prism has 10 edges in one base, how many
  a) faces does it have?
  b) vertices does it have?
  c) edges does it have?

★**11.** If the intersection of a plane and a sphere is nonempty, what can you say about the intersection?

★**12.** If a plane intersects a sphere, then the intersection is a circle. For what kind of a plane will this circle have the largest radius measure?

★**13.** If a plane parallel to the base of a right circular cone intersects it, what can you say about the intersection?

★**14.** If a plane parallel to a base of a right circular cylinder intersects it, what can be said about the intersection?

13.6   LENGTH

In Section 3.1, we discussed how to find the cardinal number of a finite set. Since a line segment consists of infinitely many points, we cannot count "how many" points there are in the segment. Consequently, we need a method to determine "how long" a segment is. Recall that in Section 7.6 we observed that there is a 1-1 correspondence between the set of points on a straight line and the set of real numbers. We used this fact to define the distance between any two points on the real line. But this was made possible only by choosing a suitable line segment as a *unit segment.* Thus, to find the *linear measure* of any line segment, we need an accepted unit line segment to begin with. It may be one centimetre, one inch, one foot, one kilometre, one mile, etc. Notice that "centimetre," "inch," "foot," "kilometre," "mile," etc., are a few units of linear measure. If the measure of a segment $\overline{AB}$ is n and the unit of measurement is inch, then the *length* of $\overline{AB}$ is n inches.

If $\overline{AB} \cong \overline{CD}$, we intuitively feel that the segments $\overline{AB}$ and $\overline{CD}$ have the same linear measure. Let's make use of this idea to find the linear measure of the segment $\overline{AB}$, using $\overline{PQ}$ as a unit segment, as shown in Fig. 13.38. Starting at A, we divide $\overline{AB}$ into segments congruent to $\overline{PQ}$ by the points C, D, E, and F. Since this procedure divides $\overline{AB}$ into five segments, $\overline{AC}$, $\overline{CD}$, $\overline{DE}$, $\overline{EF}$, and $\overline{FB}$, each being congruent to $\overline{PQ}$, we say that the measure of $\overline{AB}$ is 5; that is, $m(\overline{AB}) = 5$ or simply $AB = 5$.

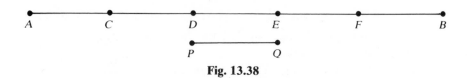

Fig. 13.38

Since we now know how to find the length of a line segment, let's discuss what we mean by and how to find the length of a polygon. The length of a polygon, usually called its *perimeter*, is the sum of the lengths of its sides. For example, the perimeter of $\triangle ABC = \overline{AB} \cup \overline{BC} \cup \overline{CA}$ is given by $AB + BC + CA$. In particular, if the lengths of a triangle are 2 inches, 3 inches, and 4 inches, then its perimeter is $2 + 3 + 4 = 9$ inches.

Let's now consider a right triangle $\triangle ABC$ right angled at B, as in Fig. 13.39. The side $\overline{AC}$ opposite the right angle $\angle B$ is called the *hypotenuse* and the other two sides are called the *legs* of the right triangle. One of the remarkable properties associated with right triangles is the well-known Pythagorean theorem:

Pythagorean theorem   *If $\triangle ABC$ is a right triangle, right angled at B, then $AC^2 = AB^2 + BC^2$.*

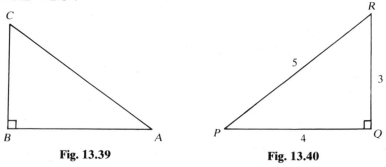

Fig. 13.39   **Fig. 13.40**

Conversely, if $AC^2 = AB^2 + BC^2$ in $\triangle ABC$, then $\triangle ABC$ is a right triangle, right angled at B. For example, consider the right triangle PQR in Fig. 13.40. We have $PQ = 4$, $QR = 3$, and $PR = 5$; also, $PR^2 = 5^2 = 25 = 16 + 9 = 4^2 + 3^2 = PQ^2 + QR^2$, satisfying the Pythagorean theorem.

How do we now find the length of a circle, called its *circumference*? In Section 6.7, we remarked that the ratio of the circumference c of a circle to the measure of its diameter d is always a constant, denoted by π; that is, $\pi = c/d$ and hence $c = \pi d = 2\pi r$, where r is the measure of its radius. Recall that π is an irrational number usually approximated by $\frac{22}{7}$ or 3.142.

Example 13.1   Find the circumference of a circle with a radius of measure 5.

Solution:

$$\text{circumference} = 2\pi r = 2\pi(5)$$

$$= 10\pi = 10(3.142) \qquad \text{(approximately)}$$

$$= 31.42$$

Example 13.2   Find the length r of a radius of a circle with circumference 55 inches.

Solution:

$$r = \frac{c}{2\pi} = \frac{55}{2(\frac{22}{7})} \quad \text{(approximately)}$$

$$= \frac{55 \times 7}{2 \times 22} = \frac{35}{4}$$

$$= 8.75 \text{ inches}$$

Exercise 13.6

1. Use Fig. 13.41 to compute the perimeter of the following polygons.
   a) *AED*
   b) *DEF*
   c) *ADEF*
   d) *BCEF*

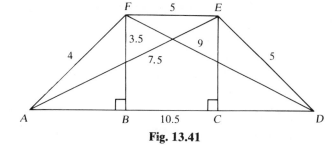

Fig. 13.41

2. Use Fig. 13.42 to find the measure of each of the following, where $AD = 8$, $BD = 5$, and $\overline{AB} \cong \overline{CD}$.
   a) $\overline{AB}$
   b) $\overline{BC}$
   c) $\overline{AC}$

Fig. 13.42

3. Mark *true* or *false*.
   a) If $\overline{AB} \cong \overline{CD}$, then $AB = CD$.
   b) If $\overline{AB} = \overline{CD}$, then $AB = CD$.
   c) If $AB = CD$, then $\overline{AB} = \overline{CD}$.
   d) If $AB = CD$, then $\overline{AB} \cong \overline{CD}$.
   e) If $AB = 5$ and $BC = 3$, then $AC = 8$.
   f) If $AB = 5$, $BC = 3$ and $\overline{AB} \cap \overline{BC} = \{B\}$, then $AC = 8$.
   g) If $AB = a$, $CD = b$ and $\overline{AB} \cap \overline{CD} = \varnothing$, then $AB + CD = a + b$.
4. Find the perimeter of a triangle with sides 4 inches, 8 inches, and 11 inches.
5. Find the length of a side of an equilateral triangle with perimeter 36 feet.
6. Compute the circumference of a circle with radius measure given by 14 inches.
7. Find the length of a radius of a circle with circumference 66 centimetres.
8. Find the length of a side of a square with perimeter 52 metres.
9. If the radii of two circles have measures 5 and 7, find the ratio of the measures of their circumferences.

10. If the ratio of the measures of circumferences of two circles is $\frac{3}{5}$, find the ratio of the measures of their radii.

13.7 AREA

As in the case of line segments, we need some criterion to measure the set of points bounded by a simple closed curve. The familiar measure of a region is its *area*. To measure the area of a region, we need to agree on a *unit area*. This is accepted as the interior of a *unit square*, a square of side one unit long. If the side of the square is one unit, then the area of the square region is one square unit; for example, if the length is one inch, then the area is one square inch.

To determine the area of a set of points bounded by a simple closed curve, we find how many unit squares are needed to cover the region. For example, consider the square region in Fig. 13.43 formed by a square with a side 4 inches long. Since 16 unit squares are needed to cover the region, the area of the region inside the square is $16 = 4 \cdot 4$ square inches. Likewise, the area of the rectangular region in Fig. 13.44 is $12 = 4 \cdot 3$ square inches. It is now almost intuitively obvious that the area of any rectangular region is given by the product of the measures of any two adjacent sides of the rectangle.

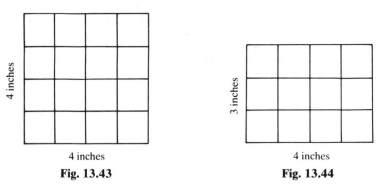

4 inches

4 inches
Fig. 13.43

3 inches

4 inches
Fig. 13.44

How do we now compute the area of the region bounded by a parallelogram? In Fig. 13.45, the line segments $\overline{CF}$ and $\overline{DE}$, drawn perpendicular to the base $\overline{AB}$ of the parallelogram $ABCD$, are called *altitudes* of the parallelogram. It now follows that $\triangle ADE \cong \triangle BCF$. Consequently, $\overline{AB} \cong \overline{EF}$ and the area of the region $ABCD$ is the same as that of the rectangular region $CDEF$, *namely* $EF \cdot DE = AB \cdot DE$. Thus, the area of the region bounded by a parallelogram is given by bh, where b and h are the linear measures of a base and an altitude to the base, respectively. For example, the area of the region enclosed by a parallelogram with base 5 inches long and altitude 3 inches long is $5 \cdot 3 = 15$ square inches.

Let's now discuss how to find the area of a triangular region ABC. Consider $\triangle ABC$, in Fig. 13.46, with base $\overline{AB}$ and altitude $\overline{CD}$. Now complete the

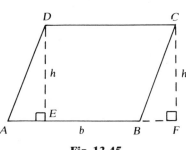

Fig. 13.45

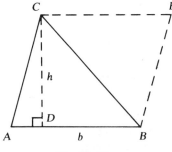

Fig. 13.46

parallelogram *ABEC*. Since $\triangle ABC \cong \triangle EBC$, the area of the triangular region is $\frac{1}{2}$ of the area bounded by the parallelogram $ABEC = \frac{1}{2}bh$. For example, the area of the region bounded by a triangle with one side 8 centimetres long and the corresponding altitude 5 centimetres long is given by

$$\text{area} = \frac{1}{2} \cdot 8 \cdot 5 = 20 \text{ square centimetres}$$

How do we compute the area of the region bounded by a trapezoid? Consider the trapezoid *ABCD* in Fig. 13.47 with $AB = a$ and $CD = b$. Let h be the measure of the altitude $\overline{DE}$. Since the area enclosed by the trapezoid

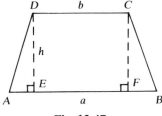

Fig. 13.47

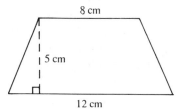

Fig. 13.48

ABCD is equal to the sum of the areas enclosed by the triangles *AED* and *BCF* and the rectangle *CDEF*, it can be shown that the required area $= \frac{1}{2}(a + b)h = \frac{1}{2}$(sum of the measures of bases) (measure of altitude).

For instance, the area of the region inside the trapezoid in Fig. 13.48 is given by

$$\text{area} = \frac{1}{2}(12 + 8)5 = 50 \text{ square centimetres}$$

Finally, we remark that the area of the circular region within a circle with radius measure r is πr^2. For example, the area of the region inside a circle of radius 3.5 inches long is given by

$$\text{area} = \pi r^2 = \frac{22}{7} \cdot \frac{7}{2} \cdot \frac{7}{2} \qquad \text{(approximately)}$$

$$= 38.5 \text{ square inches}$$

Exercise 13.7

1. Mark *true* or *false.*

  a) The area of a circle with a radius 2 inches long is 4π square inches.

  b) The area of a circle with a radius 2 inches long is zero square inches.

  c) The area of the circular region enclosed by a circle with a radius 2 inches long is 4π square inches.

  d) If A and B denote two polygonal regions, then area $(A \cup B) =$ area $A +$ area B.

  e) A square and a rectangle can have the same perimeter and enclose unequal areas.

  f) A square and a rectangle can enclose the same area and have different perimeters.

  g) If $\triangle ABC = \triangle DEF$, then they enclose the same area.

2. Find the area of the region bounded by a square with a side 3 inches long.

3. Find the length of a side of a square that encloses an area of 6.25 square feet.

4. What are the dimensions of a rectangle if one of its sides is 7 inches long and it encloses an area of 91 square inches.

5. What is the area measure of the region bounded by $\triangle ABC$ if $AC = 12$ and the altitude from B to $\overline{AC}$ has measure 5?

6. Find the length of the side $\overline{AB}$ of $\triangle ABC$ if it encloses an area of 60 square metres and the length of the altitude to $\overline{AB}$ is 8 metres.

7. Compute the area of the region inside a parallelogram with a base 8 inches long if the length of the corresponding altitude is 5 inches.

8. Compute the area of the region bounded by a trapezoid with bases 7 inches and 5 inches long and altitude 4 inches long.

9. What is the length of the altitude of a trapezoid if its bases are 10 centimetres and 13 centimetres long and it encloses an area of 92 square centimetres?

13.8 VOLUME

The concept of *volume* is associated with solids in exactly the same way as area is associated with regions. A cube with edges equal to one unit is called a *unit cube*. The volume of the solid enclosed by a unit cube is one cubic unit. As in the case of areas, the volume of a solid is obtained by finding how many unit cubes are required to fill in the whole solid.

Consider the rectangular prism in Fig. 13.49 with edges l (length), w (width), and h (height). Then, the volume of the solid enclosed by the rectangular prism is given by

$$V = lwh = (lw)h$$

$$= \text{base area} \times \text{height}$$

For example, if the edges are 5 inches, 4 inches, and 3 inches long, then

$$V = 5 \cdot 4 \cdot 3 = 60 \text{ cubic inches}$$

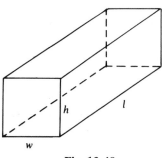

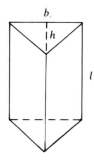

Fig. 13.49   Fig. 13.50

The formula volume = base area × height works in the case of every right prism. For example, the volume of the solid bounded by a triangular prism, as in Fig. 13.50, is given by

$$\text{volume} = \text{base area} \times \text{height} = (\tfrac{1}{2}bh)l$$

Consider a right circular cylinder with radius measure r and height h, as in Fig. 13.51. It can be shown that the volume bounded by this cylinder is given by

$$\text{volume} = \pi r^2 h = (\pi r^2)h$$
$$= (\text{base area}) \cdot (\text{height})$$

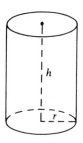

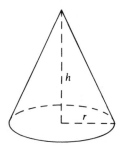

Fig. 13.51   Fig. 13.52

Also, the volume enclosed by the right circular cone in Fig. 13.52 is $\tfrac{1}{3}\pi r^2 h = \tfrac{1}{3}(\pi r^2)h = \tfrac{1}{3}(\text{base area})(\text{height})$. For instance, if $r = 3$ and $h = 5$, then the volume enclosed by the right circular cylinder is

$$\pi r^2 h = \tfrac{22}{7} \cdot 9 \cdot 5$$
$$= 141.43 \text{ cubic units} \qquad (\text{approximately})$$

and that bounded by the right circular cone is

$$\tfrac{1}{3}\pi r^2 h = \tfrac{1}{3} \cdot \tfrac{22}{7} \cdot 9 \cdot 5$$
$$= 47.143 \text{ cubic units} \qquad (\text{approximately})$$

Lastly, the volume of the solid bounded by·a sphere of radius measure r is $\frac{4}{3}\pi r^3$. For example, a sphere with a radius 3 inches long encloses

$$\text{volume} = \tfrac{4}{3}\pi r^3 = \tfrac{4}{3} \cdot \tfrac{22}{7} \cdot 27$$

$$= 113.143 \text{ cubic inches} \qquad \text{(approximately)}$$

Exercise 13.8

1. Find the volume of the solid bounded by the rectangular prism with
   a) $l = 9$ cm, $w = 6$ cm, and $h = 3$ cm.
   b) $l = 15$ in., $w = 6$ in., and $h = 6$ in.
2. Compute the volume bounded by the triangular prism if the base of the base triangle is 8 inches long, the altitude is 3 inches long and the height of the prism is 5 inches.
3. Find the volume enclosed by the right circular cylinder with $r = 4$ and $h = 9$.
4. What is the volume of the solid enclosed by the right circular cone with $r = 4$ and $h = 9$?
5. Compute the volume of the solid bounded by a sphere of radius 3 cm long.

13.9 SUMMARY AND COMMENTS

In this chapter, we introduced the basic concepts of geometry on an informal basis. We left the terms "point," "line," "plane," and "space" undefined. A line extends infinitely in either direction, whereas a ray extends infinitely only in one direction. A point on a line partitions the line into three disjoint sets. A set of points is collinear or coplanar if all the points lie in the same line or in the same plane.

A plane is determined by three noncollinear points or by two intersecting lines. A line in a plane divides it into three disjoint sets.

An angle is the union of two rays with a common endpoint, called the vertex of the angle. We discussed a few special angles.

In our discussion about curves, we were primarily concerned with simple closed plane curves. Every simple closed curve in a plane separates it into three disjoint sets: the interior, the exterior, and the curve itself. We observed the existence of convex sets.

Polygons are simple closed curves formed by the union of line segments.

Polyhedrons are simple closed surfaces formed by polygons and polygonal regions. The vertices, faces, and edges of any polyhedron satisfy Euler's formula: $V + F - E = 2$.

The concepts of length, area, and volume were introduced informally. The well-known Pythagorean theorem gives a relationship connecting the measures of the sides of a right triangle. Its converse can be used as a criterion to check if a given triangle is a right triangle or not.

The area of the region bounded by a unit square was chosen as the yardstick to compute the areas of regions enclosed by polygons. Likewise, the volume of the solid enclosed by a unit cube was the basis for the discussion of volumes of solids bounded by familiar surfaces.

SUGGESTED READING

Buchman, A., "An Experimental Approach to the Pythagorean Theorem," *The Arithmetic Teacher*, vol. 17 (Feb. 1970), pp. 129–132.

Froelich, E., "An Investigation Leading to the Pythagorean Property", *The Arithmetic Teacher*, vol. 14 (Oct. 1967), pp. 500–504.

Prielipp, R. W., "The Area of Pythagorean Triangle and the Number Six," *The Mathematics Teacher*, vol. 62 (Nov. 1969), pp. 547–548.

Richards, P. L., "Tinkertoy Geometry," *The Arithmetic Teacher*, vol. 14 (Oct. 1967), pp. 468–469.

Teegarden, D. O., "Geometry via *T*-Board," *The Arithmetic Teacher*, vol. 16 (Oct. 1969), pp. 485–487.

Wheeler, R. E., *Modern Mathematics: An Elementary Approach* (3rd ed.), Brooks/Cole Publishing Co., Monterey, Calif. (1973), pp. 333–428.

"The theory of probabilities and the theory of errors now constitute a formidable body of great mathematical interest and of great practical importance."

R. S. WOODWARD

14/Introduction to Probability

After studying this chapter, you should be able to:
- *apply the fundamental counting principle in counting problems*
- *find the number of permutations of n things taken r at a time*
- *find the number of combinations of n things taken r at a time*
- *evaluate the probability of an event*

14.0 INTRODUCTION

Statements like, "There is a 50-50 chance for snow today," "Chances are good that the price of gasoline will still go up," "There is an eighty percent probability that Ron will go to college," "What is the probability that Dick will pass the mathematics course?" are no longer new to us. What do they really mean? How is it possible to make predictions like these? Are they merely guesswork?

Well, it is the purpose of this chapter to introduce the reader to the basic ideas and concepts of one of the most fascinating areas in mathematics, one that deals with problems and statements of the kind above. This area is called the theory of probability.

The groundwork toward the development of probability theory was laid accidentally when the following problem was proposed to Blaise Pascal (1623–1662) by the gambler Chevalier de Mere: if two players of equal skill are forced to quit a game before it is over, how should the stakes be divided between them? The problem sounds simple, doesn't it? The stakes should be divided so that the person who has the greater chance to win the game when they stop playing gets more stakes than his opponent. Pascal communicated this problem to Fermat in 1654. Both Pascal and Fermat solved the problem independently. This laid the foundation for probability theory.

Pascal was born at Chermont, France, on June 19, 1623, the son of a mathematician and judge. Even when Pascal was a boy, his mathematical talents and interests were amazing. He discovered many results in geometry by the age of 12. At 14, he started attending the weekly meetings of a group of French mathematicians, which became the French Academy in 1666. When he was 16, he wrote *Essay on Conic Sections*, containing over 400 theorems with proofs. Surprisingly enough, the history of modern calculating machines (Section 16.1) begins with Pascal, who invented the first such machine at the age of 19.

Pascal's unfailing interest and pioneering work were constantly interrupted by his ill health. On November 23, 1654, he miraculously escaped from an accident caused by his runaway horses. This forced Pascal to reorder his priorities in life. He devoted most of the rest of his life in religious pursuits.

Besides Pascal and Fermat, outstanding results in probability theory were contributed by the following brilliant mathematicians: C. Huygens (1629–1695), a Dutchman; J. Bernoulli (1654–1705), a Swiss; A. de Movire (1667–1754), a Frenchman who spent most of his life in England; P. S. Laplace (1749–1827), a Frenchman who is called the "Newton of France"; P. L. Chebychev (1821–1894), a Russian; A. A. Markov (1856–1922), a Russian; and A. N. Kolmogorov (1903–   ), a Russian.

14.1   THE COUNTING PRINCIPLE

We now discuss how to find answers to some problems using the technique of counting, applied in a somewhat sophisticated fashion. Before formally presenting the principle behind the technique, let's consider a few examples. We now advise the reader to review Section 1.12 on cartesian products.

Consider the sets $A = \{1,2\}$ and $B = \{a,b,c\}$. Recall that the cartesian product $A \times B$ of sets A and B is given by $A \times B = \{(1,a),(1,b),(1,c),(2,a),(2,b),(2,c)\}$. The cardinal number of each of these sets is given by $c(A) = 2$, $c(B) = 3$ and $c(A \times B) = 6$. Thus,

$$c(A \times B) = 6 = 2 \cdot 3 = c(A) \cdot c(B)$$

More generally, if $c(A) = m$ and $c(B) = n$, then

$$c(A \times B) = mn = m \cdot n = c(A) \cdot c(B)$$

This fact plays a significant role in our discussions to follow.

Example 14.1   If John can travel by bus, plane, or ship from Boston to New York and then on to Paris by plane or ship, in how many ways can he make a trip from Boston to Paris via New York?

Solution: The problem is to find the number of different means of transportation John can choose for his trip. If he goes by bus from Boston to New York, then he has two choices for his trip from New York to Paris: choose plane or ship. Even if he goes by plane or ship to New York, still he can choose plane or ship to go from New York to Paris. Thus, there are $2+2+2=6$ different means of transportation he can employ for his entire trip.

How can the concept of cartesian product of two sets be used in a problem like this? Let A denote the set of means of transportation John can choose from Boston to New York and B that he can choose from New York to Paris. Thus, $A = \{bus,plane,ship\}$, $B = \{plane,ship\}$, and

$$A \times B = \{(bus,plane),(bus,ship),(plane,plane),(plane,ship),$$
$$(ship,plane),(ship,ship)\}$$

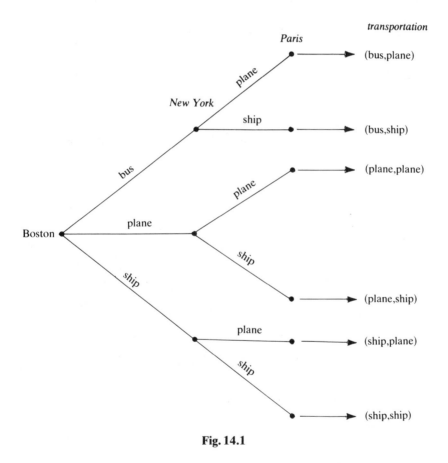

Fig. 14.1

The set $A \times B$ now gives the set of all possible different types of transportation John can select for his trip. Observe that our answer, 6, is also given by

$$c(A \times B) = 6 = 3 \cdot 2 = c(A) \cdot c(B)$$

These ordered pairs and hence the different ways of making the whole trip can be represented in a tree diagram, as in Fig. 14.1.

Example 14.2 In how many ways can a committee consisting of a boy and a girl be formed from the set {Fred,Ned,Ted,Ed,Mary,Sally}?

Solution: With Fred as a member of the committee, any one of the two girls can serve as a second member of the committee. Thus, two committees can be formed with Fred as one member, namely, {Fred,Mary} and {Fred,Sally}. With Ned as a member, two more committees can be formed: {Ned,Mary} and {Ned,Sally}. Two committees can be formed with Ted as a member: {Ted,Mary} and {Ted,Sally}. Finally, we can form two more committees with Ed as a member: {Ed,Mary} and {Ed,Sally}. Now, no more boys are left and hence we can form $2 + 2 + 2 + 2 = 4 \cdot 2 = 8$ committees. The tree diagram in Fig. 14.2 illustrates the various ways of forming these committees.

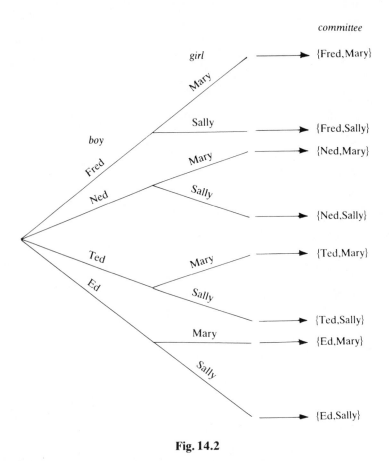

Fig. 14.2

Example 14.3 How many even two-digit numbers can be formed using the digits 0, 1, 2, 3, 4, 5, and 6?

Solution: Recall that for a two-digit number xy to be even, the units digit y must be even and the tens digit can be any of the given digits except 0 (why?). Thus, the possible candidates for x are the elements of the set $A = \{1,2,3,4,5,6\}$, while those for y are the elements of the set $B = \{0,2,4,6\}$. With each of the six digits in A as the tens digit, we can choose any of the four digits in B as the units digit. Since

$$c(A \times B) = c(A) \cdot c(B) = 6 \cdot 4 = 24$$

we can form 24 even two-digit numbers.

If we study these examples carefully, it is almost clear that the procedure to find the answer in each case follows the same pattern: in example 14.1, John has three modes of transportation from Boston to New York and for each of these choices, there are two ways of traveling from New York to Paris; we found that he can make the trip in $3 \cdot 2 = 6$ different ways. In example 14.2, a committee consisting of a boy and a girl was to be selected; a boy can be chosen in four different ways and for each of these choices, a girl can be selected in two different ways; we could form $4 \cdot 2 = 8$ different committees. In example 14.3, we wanted to find the number of even two-digit numbers that can be formed using the digits 0, 1, 2, 3, 4, 5, and 6. Since the tens digit can be chosen in six different ways and the units digit in four different ways, we can form $6 \cdot 4 = 24$ even two-digit numbers.

These observations now lead us to the following fundamental property, which forms the cornerstone of all our later discussions.

Fundamental counting principle   *If there are* m *different ways of doing something and, after doing this, a second thing can be done in* n *different ways, then these two things can together be done in* m $\cdot$ n *different ways. More generally, if an event* E_1 *can take place in* m_1 *different ways, a second event* E_2 *in* m_2 *different ways after the first has occurred, a third event* E_3 *in* m_3 *different ways after the first two have taken place,* ..., *a* k*th event* E_k *in* m_k *different ways after* $E_1, E_2, \ldots, E_{k-1}$ *have occurred, then these events can occur in* $m_1 \cdot m_2 \cdot m_3 \cdots m_k$ *different ways.*

Example 14.4   If a nickel, a dime, and a quarter (all assumed to be fair coins) are tossed, in how many different ways can they fall?

Solution: Each of the three coins can fall in two different ways: heads up (H) or tails up (T). Thus, they can fall in $2 \cdot 2 \cdot 2 = 8$ different ways: *HHH, HHT, HTH, HTT, THH, THT, TTH,* and *TTT*, as illustrated by the tree diagram in Fig. 14.3.

Example 14.5   A car dealer has four models of cars, all available in seven colors. How many choices does a customer have to buy a car from him?

Solution: Since there are four models of cars and each model is available in seven colors, a customer can buy a car in $4 \cdot 7 = 28$ different ways.

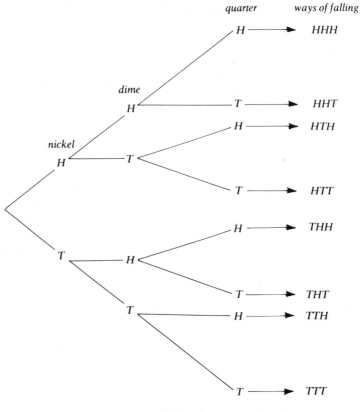

Fig. 14.3

Exercise 14.1

1. In how many ways can a committee consisting of a doctor and a lawyer be chosen from a group of 10 doctors and 12 lawyers?

2. How many different committees, consisting of a Democrat and a Republican, can be formed from a group of 21 Democrats and 17 Republicans?

3. In how many ways can four coins fall?

4. How many ordered pairs (a,b) can be formed using the digits 0 and 1?

5. How many ordered triplets (a,b,c) can be formed using the digits 0 and 1?

6. a) How many three-digit numbers can be formed using the digits 1, 2, and 3?
   b) How many of them are even?
   c) How many of them are odd?

7. a) How many three-digit numbers can be formed using the digits 0, 1, 2, and 4?

b) How many of them are even?

c) How many of them are odd?

d) How many of them are divisible by 10?

e) How many of them are divisible by 5?

8. How many three-digit numerals can be formed using the digits 0, 1, 2, 3, 4, and 5 if no digit is used more than once?

9. How many three-letter "words" can be formed using the vowels of the English alphabet?

10. In how many ways can four boys sit on (*a*) four chairs in a row? (*b*) five chairs in a row?

11. In how many ways can five boys sit on five chairs in a row?

12. Find the number of ways six mathematics books can be arranged on a shelf.

13. A telephone number usually contains seven digits. How many telephone numbers can be formed if 8, 7, 2 are the first three digits of each number and the fourth digit is nonzero?

14. How many (nonzero) telephone numbers of seven digits each can be formed (*a*) using all the digits? (*b*) if repetitions are not allowed?

15. A social security number contains three parts: the first part contains three digits, the second part consists of two digits, and the third part consists of four digits. How many social security numbers can be formed if the second part is 42 and neither the first part nor the third part is zero?

★16. In how many ways can a committee consisting of a doctor and two lawyers be chosen from a group of 10 doctors and 12 lawyers?

★17. How many code words of five symbols each can be formed (*a*) using the digits 0 and 1? (*b*) using the digits 0, 1, and 2?

14.2 PERMUTATIONS

Consider the set $A = \{a,b\}$. What are the different ways we can arrange the elements of the set A? They are clearly ab and ba. Thus, the elements of the set A, which contains two elements, can be arranged in $2 = 2 \cdot 1$ different ways.

In how many ways can we arrange the elements of the set $B = \{a,b,c\}$ containing three elements? Observe that the first place can be occupied by any one of the three elements a, b, and c; once the first place is occupied by a letter, only two elements are left and hence the second position can be filled only in two different ways; consequently, the first two places can be filled in $3 \cdot 2$ different ways. Now, since only one element is left, the third (last) element has only one choice; thus the positions can be filled in $3 \cdot 2 \cdot 1 = 6$ different ways. The six different arrangements of the three letters are abc, acb, bac, bca, cab, and cba. The tree diagram in Fig. 14.4 can be used as an aid to arrive at these different arrangements.

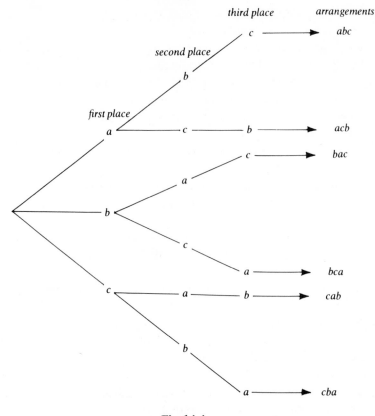

Fig. 14.4

We now introduce a useful and convenient notation to denote products of a special kind. We define $n!$ (*n factorial*) as the product $n(n-1)\ldots 3\cdot 2\cdot 1$ of the natural numbers n, $(n-1)$, etc., down to 1. That is,

$$n! = n(n-1)\ldots 3\cdot 2\cdot 1$$

For example,

$$1! = 1$$
$$2! = 2\cdot 1$$
$$3! = 3\cdot 2\cdot 1$$
$$4! = 4\cdot 3\cdot 2\cdot 1$$
$$5! = 5\cdot 4\cdot 3\cdot 2\cdot 1$$

Since we will be coming across the symbol $0!$, it is only appropriate that we assign a meaning to it. The natural way to define $0!$ is equal to 1.

Thus, we observe that the three letters a, b, and c can be arranged in $3! = 6$ different ways; two letters can be arranged in $2! = 2$ different ways.

To sum up, the letters a, b, and c, taken all at the same time, can be arranged in six different ways: abc, acb, bac, bca, cab, and cba; taking only two at a time, they can be arranged in six different ways: ab, ac, ba, bc, ca, and cb; there are only three different arrangements if we use only one letter at a time: a, b, and c. Each of these arrangements of the three letters is called a permutation, taking all letters at a time, taking two at a time, and taking only one letter at a time, respectively.

Definition 14.1   A *permutation* of a set of n elements taken r at a time is an arrangement of r elements of the set. The number of permutations of n items taken r at a time is denoted by $P(n,r)$.

The number of permutations of three elements using all at the same time is $P(3,3) = \dfrac{3!}{(3-3)!} = 6$ (why?); there are $P(3,2) = \dfrac{3!}{(3-2)!} = 6$ (why?) permutations of three letters taken two at a time, and $P(3,1) = \dfrac{3!}{(3-1)!} = 3$ permutations of three letters using only one letter at a time. More generally, we have the following fascinating result:

Theorem 14.1   *The number of permutations of a set of n elements taken r at a time is given by* $P(n,r) = \dfrac{n!}{(n-r)!}$.

It now follows that $P(n,n) = \dfrac{n!}{(n-n)!} = \dfrac{n!}{0!} = \dfrac{n!}{1} = n!$. In other words, n elements can be rearranged in $n!$ different ways.

Example 14.6   Evaluate $P(8,3)$ and $P(11,5)$.

Solution:
$$P(8,3) = \frac{8!}{(8-3)!} = \frac{8!}{5!}$$
$$= \frac{8 \cdot 7 \cdot 6 \cdot \cancel{5!}}{\cancel{5!}} = 8 \cdot 7 \cdot 6$$
$$= 336$$
$$P(11,5) = \frac{11!}{(11-5)!} = \frac{11!}{6!}$$
$$= \frac{11 \cdot 10 \cdot 9 \cdot 8 \cdot 7 \cdot \cancel{6!}}{\cancel{6!}} = 11 \cdot 10 \cdot 9 \cdot 8 \cdot 7$$
$$= 55{,}440$$

Example 14.7   Find the number of permutations of a set of ten elements taken three at a time.

Solution: Observe that the answer is given by $P(10,3)$:

$$P(10,3) = \frac{10!}{(10-3)!} = \frac{10!}{7!}$$

$$= \frac{10 \cdot 9 \cdot 8 \cdot 7!}{7!} = 10 \cdot 9 \cdot 8$$

$$= 720$$

Example 14.8   In how many different ways can you scramble the letters of the word "scramble"?

Solution: Since the word "scramble" contains eight letters, they can be rearranged in $P(8,8) = 8! = 40,320$ different ways.

Example 14.9   How many three-digit numerals can be formed using the digits 1, 2, 3, 4, 5, 6, and 7, provided no repetitions are allowed?

Solution: Observe that the number of numerals that can be formed is the same as the number of permutations of seven items taken three at a time, namely, $P(7,3)$:

$$P(7,3) = \frac{7!}{(7-3)!} = \frac{7!}{4!}$$

$$= \frac{7 \cdot 6 \cdot 5 \cdot 4!}{4!} = 7 \cdot 6 \cdot 5$$

$$= 210$$

Example 14.10   How many "words" can be formed by rearranging the letters of the word "cheese"?

Solution: Since the letter e appears more than once (three times) in the word "cheese," it is obvious that the number of words that can be formed will not be $6!$, but definitely less than $6!$. The letter c has six choices since it can occupy any of the six positions; the letter h now has five positions to choose; s has only four choices; the remaining three places have only one choice each, since each has to be filled in by the letter e. Thus, the total number of words (permutations) that can be formed is

$$6 \cdot 5 \cdot 4 \cdot 1 \cdot 1 \cdot 1 = 120$$

Notice that 120 can also be written as $\dfrac{6!}{3!}$ (why?).

More generally, we have the following remarkable result:

Theorem 14.2   *The number of permutations of n elements, where n_1 of them are alike, n_2 of them are alike, ..., n_k of them are alike, is given by*

$$\frac{n!}{n_1! n_2! n_3! \cdots n_k!}$$

Example 14.11   Find the number of ways the letters of the word "Cincinnati" can be scrambled.

Solution: "Cincinnati" contains ten letters, of which two are alike (two c's), three are alike (three i's), three others are alike (three n's), and the other two (a and t) are unlike. Hence, by the preceding theorem, we can rearrange the letters in

$$\frac{10!}{2!3!3!} = \frac{10 \cdot 9 \cdot 8 \cdot 7 \cdot 6 \cdot 5 \cdot 4 \cdot 3!}{2 \cdot 6 \cdot 3!} = \frac{10 \cdot 9 \cdot 8 \cdot 7 \cdot 6 \cdot 5 \cdot 4}{2 \cdot 6}$$

$$= 50{,}400 \text{ different ways}$$

Example 14.12   Find the number of permutations of the letters of the word "Honolulu."

Solution: The word "Honolulu" contains eight letters: two o's, two l's, two u's, one h, and one n. Consequently, by Theorem 14.2, the number of permutations is given by

$$\frac{8!}{2!2!2!} = \frac{8 \cdot 7 \cdot 6 \cdot 5 \cdot 4 \cdot 3 \cdot 2 \cdot 1}{2 \cdot 2 \cdot 2}$$

$$= 5{,}040$$

Exercise 14.2

1. Evaluate each of the following.
   a) $P(7,3)$ b) $P(13,4)$ c) $P(8,6)$ d) $P(11,1)$
2. Mark *true* or *false*.
   a) 0! is undefined.
   b) $0! = 0$
   c) $1! = 1$
   d) $P(5,2) = P(5,3)$
   e) $P(3,0) = 0$
   f) $P(n,n) = n$
   g) $P(n,1) = 1$
   h) $n!$ is even.
   i) $(m+n)! = m! + n!$
   j) $(2+3)! = 2! + 3!$
   k) $(m \cdot n)! = m!n!$
   l) $(2 \cdot 3)! = 2!3!$
   m) $5 \cdot 4! = 5!$
3. In how many ways can you rearrange the letters of the following words?
   a) on   b) word   c) logic   d) lamb
   e) orange   f) relation   g) computer   h) France
4. How many two- and three-digit numerals can be formed using the digits 1, 2, 3, 4, and 5 if repetitions are not allowed?

5. How many five-digit numerals can be formed using the digits 1, 2, 3, 4, and 5 if no digit is used more than once in the same numeral?

6. In how many ways can
   a) seven girls be seated on seven chairs?
   b) five girls be seated on seven chairs?

7. How many committees consisting of a president, vice-president, secretary, and treasurer can be formed from a club containing 13 people, assuming none can serve two positions at the same time?

8. There are 12 players in a basketball team. If every player can play every position, in how many ways can the coach of the team choose a group of five players at a time?

9. In how many different ways can five people be seated around a round table?

10. How many words can be formed by scrambling the letters of the following words?
    a) Alabama   b) Canada   c) Calcutta   d) calculus
    e) success   f) criticism   g) Tallahassee   h) Tennessee

11. How many different numerals can be formed by rearranging the digits of the following numerals?
    a) 27   b) 256   c) 1231   d) 12321

12. A license plate consists of the letters T and K, and the digits 1, 2, 3, and 4. How many license plates can have the same letters and digits if each letter appears at each end and each digit is used only once in the same plate?

13. Find the number of ways in which
    a) five algebra books and three calculus books can be arranged on a shelf;
    b) five algebra books and three calculus books can be arranged on a shelf such that all books on the same subject sit together.

14. Find the number of ways seven boys and three girls can be seated in a row if
    a) a boy sits on either end of the row;
    b) a girl sits on either end of the row;
    c) the girls sit together at an end of the row.

15. In how many ways can three mathematics books, four physics books, and five chemistry books be arranged on a shelf such that all books on the same subject sit together and the set of physics books appears between the other two sets?

14.3 COMBINATIONS

In the preceding section, we were concerned with different ways of arranging elements of a set. In this section, our interest is in finding how many subsets of a given size can be formed from a given set. For example, assume we wish to form a committee of three people from a group of five people,

{Peter,Paul,Tom,Dick,Harry}. One possible committee is {Tom,Dick,Harry}. Observe that this committee is the same as {Tom,Harry,Dick} or {Dick,Harry,Tom}. In other words, in a problem like this, we are not concerned about arrangements any more, but merely about the membership of the committee. Consequently, the order of the people is unimportant.

How many subsets of two elements each can be formed from the set {a,b,c}? Clearly, they are {a,b}, {a,c}, and {b,c}. Hence, three subsets of two elements each can be formed from a set with three elements. Each of these is called a combination of the three elements a, b, and c taken two at a time. Since we can form three subsets of one element each, namely, {a}, {b}, and {c}, the number of combinations of three elements using only one element at a time is three.

Definition 14.2   A subset containing r elements of a set with n elements is called a *combination* of n elements taken r at a time. We denote the number of combinations of n things taken r at a time by $C(n,r)$.

We now warn the reader that in a permutation, the order of the elements is important, whereas in a combination, the order is insignificant. For example, ab and ba are two permutations of the elements a and b, whereas {a,b} is the only combination of the same elements.

It now follows from our above discussion that $C(3,1) = 3$ since three subsets of one element each can be formed from a set with three elements; $C(3,2) = 3$; and $C(3,3) = 1$, since only one subset of three elements can be formed from a set of three elements.

Observe that $C(3,1)$ can also be written

$$C(3,1) = 3 = \frac{3!}{1!(3-1)!}; \qquad C(3,2) = 3 = \frac{3!}{2!(3-2)!}$$

and
$$C(3,3) = 1 = \frac{3!}{3!(3-3)!}$$

More generally, we have the following result:

Theorem 14.3   *The number of combinations of n things taken r at a time is given by* $C(n,r) = \dfrac{n!}{r!(n-r)!}.$

Example 14.13   Evaluate $C(5,3)$ and $C(10,7)$.

Solution:
$$C(5,3) = \frac{5!}{3!(5-3)!} = \frac{5!}{3!2!}$$

$$= \frac{5 \cdot 4 \cdot 3!}{3!2!} = \frac{5 \cdot 4}{2}$$

$$= 10$$

$$C(10,7) = \frac{10!}{7!(10-7)!} = \frac{10!}{7!3!}$$

$$= \frac{10 \cdot 9 \cdot 8 \cdot 7!}{7!3!} = \frac{10 \cdot 9 \cdot 8}{6}$$

$$= 120$$

Example 14.14 In how many ways can a committee of five be selected from a group of 11 doctors?

Solution: Observe that the number of committees that can be formed is the same as the number of combinations of 11 things taken five at a time; namely, $C(11,5)$:

$$C(11,5) = \frac{11!}{5!(11-5)!} = \frac{11!}{5!6!}$$

$$= \frac{11 \cdot 10 \cdot 9 \cdot 8 \cdot 7 \cdot 6!}{5!6!} = \frac{11 \cdot 10 \cdot 9 \cdot 8 \cdot 7}{5 \cdot 4 \cdot 3 \cdot 2 \cdot 1}$$

$$= 462$$

Example 14.15 How many committees of five students and three teachers can be selected from a group of 11 students and eight teachers?

Solution: Five students can be selected from 11 students in $C(11,5)$ different ways; three teachers can be chosen from eight teachers in $C(8,3)$ different ways. Hence, by the fundamental principle of counting, the total number of committees that can be formed is

$$C(11,5) \cdot C(8,3) = \frac{11!}{5!6!} \cdot \frac{8!}{3!5!}$$

$$= \frac{11 \cdot 10 \cdot 9 \cdot 8 \cdot 7 \cdot 6!}{5!6!} \cdot \frac{8 \cdot 7 \cdot 6 \cdot 5!}{3!5!}$$

$$= \frac{11 \cdot 10 \cdot 9 \cdot 8 \cdot 7}{5 \cdot 4 \cdot 3 \cdot 2 \cdot 1} \cdot \frac{8 \cdot 7 \cdot 6}{3 \cdot 2 \cdot 1}$$

$$= 25872$$

Example 14.16 How many groups of marbles can be selected from a set of seven marbles if each group is to contain at least three marbles?

Solution: Since every group contains three or more marbles, it can contain 3, 4, 5, 6, or 7 marbles. A group of three marbles can be chosen in $C(7,3) = 35$ different ways, four marbles in $C(7,4) = 35$ different ways, five marbles in $C(7,5) = 21$ different ways, six marbles in $C(7,6) = 7$, and seven marbles in $C(7,7) = 1$ way. Hence the total number of groups we can form is

$$C(7,3) + C(7,4) + C(7,5) + C(7,6) + C(7,7) = 35 + 35 + 21 + 7 + 1 = 99$$

Exercise 14.3

1. Evaluate each of the following.
    a) $C(11,3)$ **b)** $C(5,5)$ **c)** $C(5,0)$ **d)** $C(11,8)$
2. Verify each of the following.
    a) $C(10,3) = C(10,7)$ **b)** $C(8,3) = C(8,5)$
3. Verify that
    a) $C(2,0) + C(2,1) + C(2,2) = 2^2$
    b) $C(3,0) + C(3,1) + C(3,2) + C(3,3) = 2^3$
    c) $C(4,0) + C(4,1) + C(4,2) + C(4,3) + C(4,4) = 2^4$
4. How many committees of five girls can be selected from a group of nine girls?
5. In how many ways can four cards be selected from a deck of 52 cards?
6. In how many ways can an ace be chosen from four aces?
7. Find the number of ways a committee of five can be chosen from a group of seven boys and four girls.
8. In how many ways can a committee of three boys and two girls be formed from seven boys and four girls?
9. In how many ways can two red and three black cards be selected from a deck of 52 cards?
10. Find the number of subsets of three elements of the set of all letters of the following words.
    a) calculus   **b)** Tennessee   **c)** tintinnabulation
11. In how many ways can three red balls and four green balls be selected from a bag of five red and six green balls?
12. How many committees consisting of one doctor, two teachers, and three lawyers can be selected from a group of five doctors, six teachers, and seven lawyers?
13. How many straight lines can be drawn using seven points?
14. How many triangles can be formed using seven points, assuming no three of them lie on the same straight line?
15. In how many ways can you choose three coins from a collection of four nickels, five dimes, and seven quarters?
16. Find the number of ways a sum of 40 cents can be formed from a collection of four nickels, five dimes, and seven quarters?
17. In how many ways can you form a sum of 30 cents from a collection of ten nickels and five quarters?
18. In how many ways can a committee of five be formed from a group of five boys and four girls if each committee is to consist of (*a*) two boys? (*b*) at least two boys? (*c*) at least two girls? (*d*) at least one boy and at least one girl? (*e*) at most one boy? (*f*) at most one girl?
★19. Prove each of the following.
    a) $C(n,0) = 1$ **b)** $C(n,1) = n$
    c) $C(n,n-1) = n$ **d)** $C(n,r) = C(n,n-r)$

14.4   THE PROBABILITY OF AN EVENT

We often use words like "probability," "chance," etc., in our daily life. What do they really mean? It is the purpose of this section to introduce the concept of probability in a somewhat informal fashion.

Suppose we toss a (fair) coin. It clearly can land in two different ways: heads up (H) or tails up (T). Each of these is called an *outcome* of the experiment of tossing a coin. In other words, $\{H,T\}$ is the set of all possible outcomes of the experiment of tossing a coin; thus, there are two possible outcomes. Then the event E that it shows heads up can occur exactly one way. The ratio

$$\frac{\text{number of ways the event } E \text{ can happen}}{\text{total number of possible outcomes}} = \frac{1}{2}$$

can be used as a norm to measure the chance of showing heads up. This ratio is called the probability of showing heads up in tossing a coin.

Let E be the event of drawing a jack from a deck of 52 cards. Since there are 52 cards in the deck, there are 52 possible outcomes of the experiment of drawing a card. Since there are four jacks in the deck, the event E can take place in four different ways. Consequently, the ratio

$$\frac{\text{number of ways the event } E \text{ can happen}}{\text{total number of possible outcomes}} = \frac{4}{52}$$

gives the probability of drawing a jack from a deck of 52 cards.

Definition 14.3   Assume there are n possible outcomes of an experiment and m of them correspond to a certain event E. Then the *probability* that the event E will happen, denoted by $P(E)$, is defined by

$$P(E) = \frac{\text{number of ways the event } E \text{ can happen}}{\text{total number of possible outcomes}} = \frac{m}{n}$$

Since $0 \le m \le n$, it follows that $0 \le P(E) \le 1$ for any event E. Consequently, an event E will certainly happen if and only if $P(E) = 1$ and will certainly not occur if and only if $P(E) = 0$.

Example 14.17   If two coins are tossed, find the probability that (*a*) both will show heads up; (*b*) one coin will show heads up and the other tails up.

Solution: Since each coin can fall in two different ways, the two coins can fall in $2 \cdot 2 = 4$ different ways; that is, the total number of outcomes is 4.

a) Since both coins can fall heads up exactly one way (see Fig. 14.5), the probability that they will show heads up is given by $P(E) = \frac{1}{4}$.

b) Since there are two ways one coin can fall heads up and the other tails up (HT or TH), the probability that one coin will show heads up and the other tails up is given by $P(E) = \frac{2}{4} = \frac{1}{2}$.

Example 14.18   Suppose a card is drawn from a deck of 52 cards. What is the probability that it will be (*a*) a spade? (*b*) a red card?

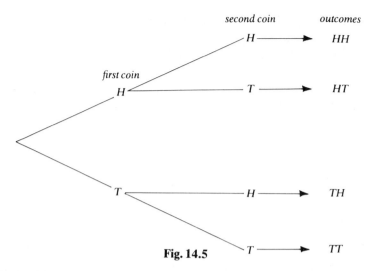

second coin   outcomes

first coin

H   H   →   HH

     T   →   HT

T   H   →   TH

     T   →   TT

Fig. 14.5

Solution: Since there are 52 cards in the deck, the number of possible outcomes of drawing a card is $C(52,1) = 52$.

a) There are 13 spades in the deck. With the result, a spade can be chosen in $C(13,1) = 13$ different ways. Hence, the probability that the card drawn is a spade is given by

$$P(E_1) = \frac{C(13,1)}{C(52,1)} = \frac{13}{52} = \frac{1}{4}$$

b) Since there are 26 red cards (13 hearts and 13 diamonds) in the deck, the event E_2 that the card drawn is a red card can happen in $C(26,1) = 26$ different ways. Consequently, the probability that the event E_2 will occur is given by

$$P(E_2) = \frac{C(26,1)}{C(52,1)} = \frac{26}{52} = \frac{1}{2}$$

Example 14.9 If five marbles are drawn from a bag containing seven green marbles and four red marbles, what is the probability that (*a*) all will be green? (*b*) three will be green and two red?

Solution: The bag contains $7 + 4 = 11$ marbles, of which five are drawn. Consequently, the total number of outcomes is $C(11,5)$.

a) The event E_1 that all five marbles are green can happen in $C(7,5)$ different ways. Therefore,

$$P(E_1) = \frac{C(7,5)}{C(11,5)} = \frac{7!}{5!2!} \div \frac{11!}{5!6!}$$

$$= \frac{7!}{5!2!} \cdot \frac{5!6!}{11!} = \frac{7!6!}{2 \cdot 11 \cdot 10 \cdot 9 \cdot 8 \cdot 7!}$$

$$= \frac{6 \cdot 5 \cdot 4 \cdot 3 \cdot 2 \cdot 1}{2 \cdot 11 \cdot 10 \cdot 9 \cdot 8} = \frac{1}{22}$$

b) Since three green marbles can be selected in $C(7,3)$ ways and two red marbles in $C(4,2)$ ways, they can together be chosen in $C(7,3) \cdot C(4,2)$ different ways, by the counting principle. Therefore, the required probability is given by

$$P(E_2) = \frac{C(7,3) \cdot C(4,2)}{C(11,5)}$$

$$= \left(\frac{7!}{3!4!} \cdot \frac{4!}{2!2!} \right) \div \frac{11!}{5!6!}$$

$$= \frac{4!5!6!7!}{2!2!3!4!11!} = \frac{5}{11}$$

Exercise 14.4

1. What are the possible outcomes when two coins are tossed? How many possible outcomes are there?
2. Same as problem 1 with three coins tossed. [*Hint*: a tree diagram showing the outcomes is helpful.]
3. What are the possible outcomes when a die is rolled? How many outcomes are there?
4. How many possible outcomes are there when two dice are rolled?
5. Find the number of possible outcomes when three dice are rolled?
6. Which of the following can be the probability of an event?
   a) $\frac{2}{3}$ **b**) 0 **c**) $-\frac{2}{3}$ **d**) $\frac{3}{2}$
7. When a coin is tossed, what is the probability that it will fall heads?
8. When two coins are tossed, what is the probability that both will show heads?
9. If $\frac{2}{11}$ is the probability that an event will happen, what is the probability that it will not occur?
10. When three coins are tossed, find the probability of getting
    a) all heads   **b**) all tails
    c) one head and two tails   **b**) one tail and two heads
11. If a card is drawn from a deck of 52 cards, find the probability of getting
    a) a king   **b**) a club
    c) a king or a queen   **d**) a club or a diamond
12. Two cards are drawn from a deck of 52 cards. Find the probability of getting
    a) two kings   **b**) two clubs
    c) a king and a queen   **d**) a club and a diamond
13. If a ball is drawn from a bag that contains four green and three white balls, what is the probability that it will be white?

14. A subcommittee of five people is to be formed from a committee of 11 Democrats and seven Republicans. What is the probability that three of them will be

  a) Democrats?   **b)** Republicans?

15. A committee of three is to be formed from a club consisting of five doctors and seven engineers. What is the probability that

  a) all will be doctors?   **b)** all will be engineers?

  c) two of them will be doctors and the other an engineer?

16. If two marbles are drawn at random from a box containing three blue, four green, and five red marbles, find the probability that

  a) they are both green   **b)** one is green and the other red

  c) they are neither green nor red

17. If five balls are selected from a bag containing seven white and six red balls, find the probability of getting

  a) all white balls   **b)** all red balls

  c) 3 white and 2 red balls   **d)** 2 white and 3 red balls

18. If a die is rolled, find the probability of getting

  a) a three   **b)** a four

  c) an even number   **d)** an odd number

19. If a die is rolled, what is the probability of getting a number less than 6?

20. Two dice are rolled. Find the probability of getting

  a) 2 fives   **b)** a five and a six

  c) a sum less than five   **d)** a sum of four

21. A box contains 100 pieces of paper, containing numbers 1 through 100. If a piece of paper is drawn at random from the box, find the probability of getting

  a) 23   **b)** a one-digit number   **c)** 123

  d) a two-digit number   **e)** a three-digit number.

14.5 MUTUALLY EXCLUSIVE EVENTS

Suppose, we draw one marble from a box containing blue, green, and red marbles. Let E_1 be the event of drawing a blue marble, E_2 the event of drawing a green marble, and E_3 that of drawing a red marble. If the marble drawn is blue, then it certainly cannot be a green or a red marble. That is, if the event E_1 takes place, then neither event E_2 nor event E_3 can happen. In other words, the event E_1 excludes the possible occurrences of the events E_2 and E_3. Similarly, the event E_2 precludes the chances for the occurrences of the events E_1 and E_3, and E_3 precludes E_1 and E_2. Such events are said to be mutually exclusive.

Definition 14.4   Two or more events are said to be *mutually exclusive* if the occurrence of each of them excludes the possible occurrence of the other, that is, if they cannot occur at the same time.

Assume that we form a committee consisting of a boy and a girl out of a group of three boys and four girls. Let E_1 be the event of selecting a boy and E_2 that of choosing a girl. Then E_1 and E_2 are not mutually exclusive events. (Why?)

The following result is useful in our discussion of the probability of mutually exclusive events.

Theorem 14.4 *Let $E_1, E_2, \ldots, E_k$ be a set of mutually exclusive events in a certain experiment. Then the probability that at least one of them will occur is given by $P(E_1) + P(E_2) + \cdots + P(E_k)$.*

Example 14.20 A marble is drawn at random from a box containing four blue marbles, five green marbles, and six white marbles. What is the probability that it will be either (*a*) blue or green? (*b*) green or white? (*c*) blue or green or white?

Solution: Since we are drawing only one marble from the box, the number of possible outcomes is $4 + 5 + 6 = 15$. Let E_1, E_2, and E_3 be the event of drawing a blue marble, green marble, and white marble, respectively. Then,

$$P(E_1) = \frac{C(4,1)}{C(15,1)} = \frac{4}{15}$$

$$P(E_2) = \frac{C(5,1)}{C(15,1)} = \frac{5}{15}$$

$$P(E_3) = \frac{C(6,1)}{C(15,1)} = \frac{6}{15}$$

Let $E_1 \cup E_2$ denote the event that at least one of the events E_1 and E_2 will occur, and similarly for $E_2 \cup E_3$, etc. Then, by Theorem 14.4,

a) $P(E_1 \cup E_2) = P(E_1) + P(E_2) = \frac{4}{15} + \frac{5}{15}$

  $= \frac{9}{15} = \frac{3}{5}$

b) $P(E_2 \cup E_3) = P(E_2) + P(E_3) = \frac{5}{15} + \frac{6}{15}$

  $= \frac{11}{15}$

c) $P(E_1 \cup E_2 \cup E_3) = P(E_1) + P(E_2) + P(E_3) = \frac{4}{15} + \frac{5}{15} + \frac{6}{15}$

  $= \frac{15}{15} = 1$

Observe that this agrees with the property that the probability of an event that is certain to happen is 1. (Why?)

Exercise 14.5

1. If the probability that Peter will go to a film this afternoon is $\frac{1}{3}$ and the probability that he will mow his lawn this afternoon is $\frac{1}{2}$, find the probability that he will either go to a film or mow his lawn this afternoon.

2. A ball is drawn at random from a bag containing black, white, and red balls. The probability of drawing a black ball is $\frac{2}{11}$ and that of drawing a white ball is $\frac{5}{13}$. Find the probability that the ball drawn is
   a) black or white   b) red.

3. If a ball is drawn from a bag containing three black balls and four white balls, what is the probability that it will be black or white?

4. A fruit is drawn from a box containing three apples, four oranges, and five peaches. What is the probability that it will be

a) an orange?   **b)** an apple or an orange?

c) an apple or a peach?   **d)** at least one of the fruits?

5. There is an opening in a committee of a club consisting of four engineers, five doctors, and six lawyers. Find the probability that the opening will be filled by

a) an engineer   **b)** a doctor

c) an engineer or a doctor   **d)** a doctor or a lawyer

e) a lawyer or an engineer

f) an engineer or a doctor or a lawyer

6. A card is drawn at random from a deck of 52 cards. Find the probability that it will be

a) a king   **b)** a queen

c) a king or a queen   **d)** a king or an ace

e) a club or a heart

f) a club or a heart or a diamond

7. If a die is rolled, what is the probability that it will show

a) a two?   **b)** a two or a five?

c) a three or a five or a seven?

d) a two or a three or a four?

8. Two dice are rolled. Find the probability that they will show

a) a sum of four   **b)** a sum of five

c) a sum of four or five   **d)** a sum of four or six

e) neither a sum of four nor a sum of five

14.6 DEPENDENT AND INDEPENDENT EVENTS

Suppose we draw two marbles from a box containing black and white marbles. Observe that if the first marble is not replaced before drawing the second marble, then the event E_1 of drawing the the first marble affects the probability of occurrence of the event E_2 of drawing the second marble. Two such events are called dependent events. On the other hand, if the first marble is replaced before the second marble is drawn, then the event E_1 has no effect on the probability of occurrence of event E_2. Two such events are called independent events.

Definition 14.5   Two events are said to be *dependent* if one event affects the probability of occurrence of the other event. They are said to be *independent* if one event has no bearing on the probability of occurrence of the other.

The following results are helpful in our discussion.

Theorem 14.5   *Let p_1 be the probability of an event E_1, p_2 the probability of event E_2 after E_1 has occurred, p_3 the probability of event E_3 after both E_1 and E_2 have occurred, . . . , p_k the probability of event E_k after all the events $E_1, E_2, \ldots, E_{k-1}$ have occurred. Then the probability that the events $E_1, E_2, \ldots, E_{k-1}$, and E_k will happen in that order is given by $p_1 p_2 \cdots p_{k-1}, p_k$.*

Theorem 14.6   *Let $E_1, E_2, \ldots, E_k$ be a set of independent events. Then the probability that all these events will take place is given by $P(E_1)P(E_2) \ldots P(E_k)$.*

Example 14.21   Two marbles are drawn from a box containing three black and four white marbles. Find the probability that both are black if the first marble is (*a*) replaced before the second drawing, (*b*) not replaced before the second drawing.

Solution: Let E_1 be the event of drawing the first black marble. Then $P(E_1) = C(3,1) \div C(7,1) = \frac{3}{7}$.

a) Let E_2 be the event of drawing the second black marble after replacement. Then $P(E_2) = C(3,1) \div C(7,1) = \frac{3}{7}$. Since E_1 and E_2 are independent events, the probability that both marbles are black is given by

$$P(E_1) \cdot P(E_2) = \tfrac{3}{7} \cdot \tfrac{3}{7} = \tfrac{9}{49}$$

b) Let E_3 be the event of drawing the second black marble without the first black marble being replaced. Notice that E_1 and E_3 are dependent events. Since there are only two black marbles left in the box, the total number of possible outcomes is $2 + 4 = 6$ and the event E_3 can happen only in two different ways. Therefore, $P(E_3) = \frac{2}{6}$. Consequently, the probability of getting two black marbles without replacement is given by

$$P(E_1) \cdot P(E_3) = \tfrac{3}{7} \cdot \tfrac{2}{6} = \tfrac{1}{7}$$

Example 14.22   Two cards are drawn at random successively from a deck of 52 cards. Find the probability of drawing

a) two kings if the first card is replaced before the second drawing;
b) two kings if the first card is not replaced before the second drawing;
c) a black king and then a red queen provided the first card is replaced before the second drawing;
d) a black king and then a red queen provided the first card is not replaced before the second drawing.

Solution:

a) Recall that there are four kings in the deck. The first king can be drawn in $C(4,1)$ different ways. Consequently, the probability that the first card drawn is a king is

$$\frac{C(4,1)}{C(52,1)} = \frac{4}{52} = \frac{1}{13}$$

Since the first card is replaced before drawing the second card, the probability that the second card is also a king is

$$\frac{C(4,1)}{C(52,1)} = \frac{1}{13}$$

Therefore, the probability that both cards are kings is $\frac{1}{13} \cdot \frac{1}{13} = \frac{1}{169}$.

b) If the first card is not replaced before drawing the second card, there are only 51 cards left, of which only three are kings. Consequently, the probability of drawing a second king is

$$\frac{C(3,1)}{C(51,1)} = \frac{3}{51}$$

Therefore, the probability that both cards are kings is $\frac{4}{52} \cdot \frac{3}{51} = \frac{1}{221}$.

c) The probability of drawing a black king from a deck of 52 cards is

$$\frac{C(2,1)}{C(52,1)} = \frac{2}{52} = \frac{1}{26}$$

(Why?) Since the first card is replaced before drawing a second card, the probability that the second card drawn is a red queen is also

$$\frac{C(2,1)}{C(52,1)} = \frac{2}{52} = \frac{1}{26}$$

Consequently, the required probability is $\frac{1}{26} \cdot \frac{1}{26} = \frac{1}{676}$.

d) The probability of drawing a black king is $\frac{2}{52} = \frac{1}{26}$. Since this card is not replaced before drawing the second card, there are only 51 cards left in the deck. Therefore, the probability of drawing a red queen is

$$\frac{C(2,1)}{C(51,1)} = \frac{2}{51}$$

Thus, the probability of drawing a black king and then a red queen without replacement is $\frac{2}{52} \cdot \frac{2}{51} = \frac{1}{663}$.

Exercise 14.6

1. If the probability that Paul will rake his lawn is $\frac{2}{5}$ and the probability that Jean will play tennis is $\frac{3}{7}$, find the probability that Paul will rake his lawn and Jean will play tennis.

2. If the probability that Ted will take a course in philosophy is $\frac{3}{11}$ and the probability that Ned will take a course in psychology is $\frac{11}{15}$, what is the probability that both these events will occur?

3. If a black king is drawn from a deck of 52 cards and not replaced, what is the probability of drawing
   a) a second king?   b) a queen?   c) a heart?

4. If a dime is tossed three times, what is the probability that it will show
   a) heads each time?   b) heads, heads, and then tails?

5. If a die is rolled twice, what is the probability of getting

    a) two 3's?   **b)** two 5's?

6. If two dice are rolled successively, find the probability of getting

    a) two 6's   **b)** a 2 and then a 3

7. If a coin is tossed, a die is rolled, and a card is drawn from an ordinary deck of cards, what is the probability of getting a head, a five, and an ace?

8. The probability that Mr. B. B. Bourbaki will get the Democratic presidential nomination is $\frac{3}{7}$. If he is nominated, the probability that he will choose Mr. W. W. Wronski as the vice-presidential candidate is $\frac{4}{5}$. What is the probability that Mr. Wronski will become a vice-presidential candidate?

9. A subcommittee of two people is to be formed from the members of a committee of five men and seven women. Find the probability that

    a) both are men   **b)** both are women

    c) one is a man and the other a woman

10. Two balls are drawn successively at random from a bag containing two black, three blue and four white balls. If the first ball is replaced before the second is drawn, what is the probability that

    a) both are black?   **b)** both are white?

    c) the first ball is black and the other is blue?

    d) the first ball is blue and the other is black?

    e) the first ball is blue and the other is white?

11. Same as problem 10 without any replacement.

12. If three marbles are drawn successively, with replacement, from a box containing three green, four red, and five white marbles, find the probability that

    a) all are green   **b)** all are white

    c) the first two marbles are green and the third is red

    d) the first two marbles are red and the other is green

    e) the first is green, the second is red, and the third is white

13. Same as problem 12 without any replacement.

14. Two cards are drawn at random from a deck of 52 cards. If the first card is replaced before the second drawing, find the probability that

    a) both are queens   **b)** both are clubs

    c) the first is a club and the second a spade

    d) the first is a red ace and the other a spade king

    e) the first is a heart and the other a green ace

15. Same as problem 14 without any replacement.

14.7 SUMMARY AND COMMENTS

In this chapter, we introduced one of the most fascinating areas in mathematics, the theory of probability. The outstanding mathematicians Pascal and Fermat are considered cofounders of this branch of learning.

We stated and illustrated the fundamental counting principle, which formed the basis of all our discussions in this chapter.

Products of the form $n(n-1) \ldots 3 \cdot 2 \cdot 1$ were expressed in a compact form, using the factorial notation:

$$n! = n(n-1) \ldots 3 \cdot 2 \cdot 1$$

For practical purposes, we defined $0! = 1$.

The terms permutation and combination were introduced and illustrated extensively. A permutation is an arrangement of things; a combination is merely a set, so that order plays no role there. There are $P(n,r) = \dfrac{n!}{(n-r)!}$ permutations of n things taken r at a time, while there are only $C(n,r) = \dfrac{n!}{r!(n-r)!}$ combinations of n things taken r at a time. The fact that $C(n,r) = C(n,n-r)$ is helpful in computations.

We introduced the following terms informally: event, outcome, and experiment. The probability of an event is the ratio of the number of ways the event can occur to the total number of possible outcomes. It is at least zero and never exceeds one.

Three kinds of events were discussed: mutually exclusive events, dependent events, and independent events. Two events are mutually exclusive if the occurrence of one event precludes the possible occurrence of the other; two events are independent if the occurrence of one event has no effect on the probability of occurrence of the other; otherwise they are dependent.

SUGGESTED READING

Flory, D. W., "What Are the Chances?" *The Arithmetic Teacher*, vol. 16 (Nov. 1969), pp. 581–582.

Gardner, M., "Amazing Mathematical Card Tricks That Do Not Require Prestidigitation," *Scientific American*, vol. 227 (July 1972), pp. 102–105.

Gardner, M., "On the Ancient Lore of Dice and the Odds Against Making a Point," *Scientific American*, vol 219 (Nov. 1968), pp. 140–146.

Gardner, M., "The Multiple Charms of Pascal's Triangle," *Scientific American*, vol. 215 (Dec. 1966), pp. 128–132.

Miller, C. D., and V. E. Heeren, *Mathematical Ideas, an Introduction* (2nd ed.), Scott, Foresman and Co., Glenview, Ill. (1973), pp. 161–197.

Triola, M. F., *Mathematics and the Modern World*, Cummings Publishing Co., Menlo Park, Calif. (1973), pp. 161–196.

*"The social sciences mathematically developed are
to be the controlling factors in civilization."*
                                        W. F. WHITE

15/Introduction to Statistics

After studying this chapter, you should be able to:
- organize data in a frequency distribution
- compute the mean, median, and mode of a set of values
- find the range, mean deviation, and standard deviation of a set of values
- draw a few conclusions about a normal distribution

15.0 INTRODUCTION

Statistics, the subject matter of this chapter, is very closely related to the topic of the preceding chapter. What is statistics? To a layman it might mean a display of charts and tables of large numbers. To a mathematician, it definitely means more than that. This branch of learning is concerned with collecting and organizing numerical data and then drawing logical inferences from them.

John Graunt (1620–1674), a London shopkeeper, did the groundwork for this branch of mathematics when he studied and analyzed the causes of death in London from 1604 to 1661. It turned out that his work was a significant factor in the formation of life insurance companies, the first one being established in 1669. Credit goes to de Moivre, Laplace, Gauss, and Karl Pearson (1857–1936) for their outstanding work in the field of statistics.

It is noteworthy that Charles Darwin (1809–1882), the English naturalist, and Gregor Mendel (1822–1884), the Austrian monk, botanist, and founder of genetics, had developed statistical tools in their pioneering work on evolution and genetics, respectively.

15.1 FREQUENCY DISTRIBUTION

When gathering data, it may be difficult or even impossible to use all the items. The totality of all items under consideration constitutes the *population*. For example, the set of all voters, the set of all governors, the set of all elephants, etc., can be a population. In taking an opinion poll about the choice of presidential candidates at the forthcoming election, it is not an easy task to ask every voter. Consequently, we depend on subsets of the population, called *samples*, to make inferences about the population. For instance, one might be interested in knowing what percent of adults prefer a certain brand of tooth-paste. Instead of asking the question to every adult, the pollster takes a sample of 100 or 500 or 1000 adults and asks each of them. If the sample selected is representative of the total population, then the conclusions made about the sample could be true about the population, too. If every element of the population has an equal probability of being selected as an element of the sample, it is called a *random sample*; otherwise it is a *selective sample*. That is, a sample is selective if its members are selected according to a certain proportion. For example, to find out the percentage of the population preferring a certain brand of toothpaste, a sample of people chosen by selecting their social security numbers at random will be a random sample, whereas a sample based on sex, color, creed, etc., will be a selective sample.

How do we organize numerical data in a systematic fashion? Numerical data can be given in different ways. Most of the time, it needs to be organized properly for the data to make any sense and to draw sensible conclusions from the data.

For example, consider the following list of weights (in pounds) of 30 students in a mathematics course:

$$160, 137, 143, 120, 137, 152, 126, 126, 143, 146,$$
$$163, 120, 120, 178, 126, 131, 137, 146, 143, 137,$$
$$163, 152, 137, 126, 168, 137, 163, 137, 131, 175$$

This list looks complicated. It seems difficult to draw any kind of conclusions from the list. Observe that 121 appears five times, 126 appears four times, etc., in the list. The number of occurrences of an entry in a list is called the *frequency*. Accordingly, the frequency of 120 is 3, that of 126 is 4, etc.

In order to make an intelligent study of the weights of the students under consideration, we construct a table (Table 15.1) whose first column contains these weights in ascending order of magnitude, whose second column contains tally marks, and whose third column contains the frequencies. From the table, which looks much simpler than the above list, we can easily read how many students there are with a certain weight. For example, there are three students weighing 163 pounds, nine students weighing at least 150 pounds, etc. Thus, $\frac{9}{30} = 30\%$ of the class weigh 150 pounds or more. This way of organizing data, as in Table 15.1, is called a *frequency distribution* or *frequency table*.

Table 15.1

Weight	Tally	Frequency
120					3
126						4
131				2
137	ЖII	7
143					3
146				2
152				2
160			1
163					3
168			1
175			1
178			1
		sum = 30

Observe that the table could get longer and longer as the number of entries in the data increases. This problem can be minimized at the risk of losing some information about the original data. We shorten the *frequency table* by grouping the weights into six classes: 120–130, 130–140, 140–150, 150–160, 160–170, and 170–180; making a tally; and counting the frequency for each class. The resulting frequency distribution is exhibited in Table 15.2. The disadvantage with a distribution of this kind is that we do not get any idea about the number of students with a certain weight. Nonetheless, it follows from the table that there are seven students whose weights lie between 120 and 130, inclusively. There are $9 + 7 = 16$ students weighing not more than 140 pounds, which is $\frac{16}{30} = 53\%$ (approximately) of all students.

Table 15.2

Class	Tally	Frequency
120–130	ЖII	7
130–140	ЖIIII	9
140–150	Ж	5
150–160	|||	3
160–170	||||	4
170–180	||	2
		sum = 30

The smallest and largest values of a class are called *class boundaries*; the difference between the class boundaries is called the *class width*; the *class mark* is obtained by taking one-half of the sum of the class boundaries. As an example, 140 and 150 are the class boundaries of the class 140–150; its class width is $150 - 140 = 10$ and its class mark is $\frac{1}{2}(140 + 150) = 145$.

It is not easy to give a general discussion of how data should be divided into a certain number of classes. It could depend on the nature of the problem and vary from individual to individual. In general, the data is divided into between five and 20 classes. Even though it is convenient, all classes need not be of the same size.

The following steps may be found useful in constructing a frequency distribution:

1. Compute the difference between the largest and smallest values in the data. This is called the *range* of the data. In our above example, the range is $178 - 120 = 58$.

2. Divide the range into a convenient number (between five and 20) of classes of equal size. We divided the data of our example into six classes, all of the same width, 10.

3. List the various classes in the first column of the table.

4. Put tally marks corresponding to every entry in each class, in the second column.

5. Keep a count of the tally marks and enter it in the third column.

We now discuss two ways of displaying a frequency distribution: the histogram and the frequency polygon. A *histogram* consists of a set of rectangles with lower bases on a horizontal axis; the midpoints of the bases coincide with class marks; the lengths of the bases are equal to the class widths and the heights of the rectangles are equal to the corresponding class frequencies, measured along a verticle axis. Figure 15.1 exhibits the histogram of the frequency distribution of Table 15.2.

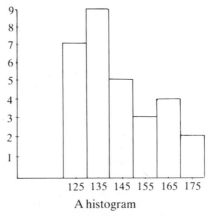

A histogram

Fig. 15.1

The *frequency polygon* of a frequency distribution is obtained by joining the midpoints of the upper bases of the rectangles in the histogram. The frequency polygon of the distribution in Table 15.2 is given in Fig. 15.2.

We now remark that histograms and frequency polygons are a convenient way of pictorially representing a frequency distribution. They provide us a good visual aid to analyze the data.

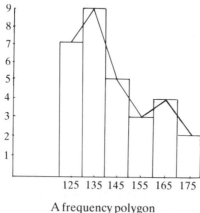

A frequency polygon

Fig. 15.2

Exercise 15.1

1. Construct a frequency distribution as in Table 15.1 for the heights (in inches) of 30 students in a mathematics course:

$$63, 65, 69, 78, 70, 64, 65, 78, 69, 72$$
$$71, 70, 63, 64, 66, 60, 63, 69, 69, 63$$
$$60, 68, 72, 82, 80, 71, 69, 80, 70, 72$$

a) What is the range of the distribution?
b) What is the most common height among the students?
c) How many students are at least 5 feet 9 inches tall?
d) What percent of the students in the class are at least 6 feet tall?
e) How many students are 5 feet 10 inches tall?

2. Construct a frequency distribution with five classes beginning with 58 and class width 5 for the data in problem 1.

3. Construct a histogram for the frequency distribution in problem 2.

4. Draw a frequency polygon for the frequency distribution in problem 2.

5. Here is a list of the scores obtained by 40 students in a mathematics test at Ding-Ling College:

$$63, 79, 70, 78, 95, 80, 90, 85, 65, 73,$$
$$60, 75, 98, 73, 86, 56, 78, 72, 69, 84,$$
$$89, 85, 73, 90, 84, 73, 85, 73, 75, 58,$$
$$95, 69, 73, 70, 60, 93, 65, 50, 89, 98$$

a) Construct a frequency distribution with nine classes starting with 46 and class width 6 for this data.

b) What is the range of the scores in the test?

c) Find the most common score in the test.

d) What is the frequency of score 69?

e) Find the frequency of score 93.

f) How many students scored at least 90 in the test?

g) How many students scored at least 59?

h) What percent of students scored at least 80 points?

6. Construct (a) a histogram and (b) a frequency polygon for the frequency distribution in problem 5.

7. Here are the ages of 30 science teachers at King-Kong College:

$$35, 43, 65, 41, 38, 38, 55, 40, 38, 69,$$
$$38, 41, 25, 60, 50, 63, 52, 43, 48, 28,$$
$$40, 30, 36, 58, 35, 45, 36, 43, 38, 68$$

a) Find the range of the age distribution.

b) How many of the science teachers are 55 years or older?

c) What percent of the teachers are 40 years or younger?

d) Construct a frequency distribution with nine classes starting with 25 and class width 5 for this data.

e) Find the class with the highest frequency.

8. Construct (a) a histogram and (b) a frequency polygon for the frequency distribution in problem 7.

9. The approximate ages of the past 35 presidents (Washington to Johnson) of the United States at the time of their death are given below:

$$67, 91, 83, 85, 73, 81, 78, 80, 68, 72, 54, 66,$$
$$74, 65, 77, 56, 67, 63, 71, 50, 56, 71, 65, 58,$$
$$61, 73, 68, 58, 61, 90, 63, 89, 79, 45, 65$$

a) Use this information to construct a frequency distribution with six classes of size 10 beginning with 40.

b) How many presidents were in the age group 60–70 when they died?

c) Find the frequency of class 90–100.

d) What percent of presidents died before the age of 71?

15.2   MEASURES OF CENTRAL TENDENCY

In the preceding section, we discussed how to organize data in a frequency distribution and how to represent them graphically. But numerical measures are often necessary to make inferences about a population. The measures we will be discussing are called *averages* or *measures of central tendency*.

The term average is no longer new to us. We are all familiar with terms like average salary, average raise, average school, average teacher, etc. But it might mean different things to different people, depending on the nature of the situation.

There are five averages used in computation: arithmetic mean, median, mode, geometric mean, and harmonic mean. However, we shall discuss only the first three averages in this book. The one average most of us are familiar with is the arithmetic mean. The *arithmetic mean* or simply the *mean* of a set of n values $x_1, x_2, \ldots, x_n$, denoted by $\bar{x}$, is obtained by dividing their sum by n; that is,

$$\text{arithmetic mean} = \bar{x} = \frac{x_1 + x_2 + \cdots + x_n}{n}$$

Example 15.1   The weights (in pounds) of five players in a basketball team are 180, 195, 212, 178, and 195. Find the arithmetic mean of their weights.
Solution:

$$\text{arithmetic mean} = \frac{180 + 195 + 212 + 178 + 195}{5}$$

$$= \frac{960}{5} = 192$$

Example 15.2   The heights (in inches) of 10 students in a math course are 80, 76, 74, 63, 68, 74, 65, 80, 63, and 74. Find the mean height.
Solution:

$$\text{mean} = \frac{80 + 76 + 74 + 63 + 68 + 74 + 65 + 80 + 63 + 74}{10}$$

$$= \frac{717}{10} = 72 \quad \text{(approximately)}$$

Observe that in the data of example 15.2, the frequencies of 80, 76, 74, 68, 65, and 63 are 2, 1, 3, 1, 1, and 2, respectively. Consequently, the mean is also given by

$$\text{mean} = \frac{2 \cdot 80 + 1 \cdot 76 + 3 \cdot 74 + 1 \cdot 68 + 1 \cdot 65 + 2 \cdot 63}{2 + 1 + 3 + 1 + 1 + 2}$$

$$= \frac{717}{10} = 72 \quad \text{(approximately)}$$

More generally, if we have a set of n values $x_1, x_2, \ldots, x_n$ with frequencies $f_1, f_2, \ldots, f_n$, respectively, then their arithmetic mean is given by

$$\bar{x} = \frac{f_1 x_1 + f_2 x_2 + \cdots + f_n x_n}{f_1 + f_2 + \cdots + f_{n.}}$$

We now illustrate a method that will prove to be useful in computing the arithmetic mean, especially when we handle fairly large numbers. Consider the weights 180, 195, 212, 178, and 195 of example 15.1. First, we choose a possible candidate for the mean, say $a = 195$; compute now the *deviations* of the various weights from 195:

$$180 - 195 = -15, \qquad 195 - 195 = 0$$
$$212 - 195 = 17, \qquad 178 - 195 = -17$$
$$195 - 195 = 0$$

The sum of the deviations $= (-15) + 0 + 17 + (-17) + 0 = -15$, and if $b =$ the mean of the deviations, then $b = \frac{-15}{5} = -3$. Therefore, the required mean $= a + b = 195 - 3 = 192$, which agrees with the mean computed in example 15.1.

How do we now compute the mean for data given in a frequency distribution? The class mark of each class is considered a representative of all the entries in the class and is counted as many times as the class frequency, as illustrated by the following example.

Example 15.3 Compute the mean of the frequency distribution in Table 15.3, which contains the scores obtained by 30 students in a calculus exam.

Table 15.3

Class	Class mark (x)	Frequency (f)	$f \cdot x$
40–50	45	4	180
50–60	55	3	165
60–70	65	8	520
70–80	75	5	375
80–90	85	6	510
90–100	95	4	380
			sum $= 2130$

Solution: $\qquad$ arithmetic mean $= \frac{2130}{30} = 71$

We now remark that the mean of a set of values need not be the same as the one found from the frequency distribution. (Why?)

The next average we shall discuss is the median. The *median* of a set of values is the number that lies halfway between the smallest and largest values in the data when the values are arranged in increasing or decreasing order of magnitude. The median is the middle term if there is an odd number of terms in the data; it is the mean of the two middle values if there is an even number of values in the data.

Example 15.4 Find the median of the weights 180, 195, 212, 178, and 195 of example 15.1.

Solution: First of all, we arrange the weights in ascending order of magnitude:

$$178, 180, 195, 195, 212$$
$$\uparrow$$
median

Since there are five values in the data and five is odd, the median is obtained by picking the middle (third from either end) value; thus, median $= 195$.

Example 15.5 Find the median of the heights (in inches) of 10 students in a physics course: 80, 76, 70, 63, 68, 74, 65, 80, 63, and 74.

Solution: On arranging the heights in increasing order of magnitude, we have

$$63, 63, 65, 68, 70, 74, 74, 76, 80, 80$$
$$\uparrow$$
median

Since there are ten entries in the data and ten is even, the median is found by computing the mean of the two middle values, namely the fifth and sixth values. Thus,

$$\text{median} = \tfrac{1}{2}(70 + 74) = 72$$

In order to find the median of data tabulated in a frequency table, we use the class marks as representatives of values counted as many times as the class frequencies.

Example 15.6 Compute the median of the data in Table 15.3.

Solution: Since Table 15.3 gives the scores of 30 (even number) students, the mean of the fifteenth and sixteenth scores (class marks here) gives the median. The fifteenth score lies in class 60–70 and the sixteenth in class 70–80, with class marks 65 and 75, respectively. Hence the median is given by

$$\text{median} = \tfrac{1}{2}(65 + 75) = 70$$

The third and the last average we shall discuss is the mode. The *mode* of a set of values is the value that occurs more frequently than any other values. Compared with the mean and median, it is relatively easy to locate the mode. A set of values may not possess a mode. If no number appears more than once, then no mode exists for the data. Also, it is possible for data to have more than one mode. If there are two modes, then the data is said to be *bimodal.*

Example 15.7 Find the mode of weights (in pounds) of five basketball players: 180, 195, 212, 178 and 195.

Solution: Of the five numbers in the list, pick the number that has the maximum frequency. It is clearly 195 since its frequency, 2, is larger than that of the others. Consequently, the mode is 195.

Example 15.8 The heights of ten students in a chemistry class are 80, 76, 74, 63, 68, 74, 80, 63, 63, and 74. Find the mode of the heights.

Solution: For convenience, let's construct a frequency table, as in Table 15.4. We easily observe from the table that both 63 and 74 have frequency 3 each, while the others have frequency less than 3. As a result, the distribution contains two modes, 63 and 74, and hence is bimodal.

Table 15.4

Height	Tally	Frequency	
63	|||	3	←
68	|	1	
74	|||	3	←
76	|	1	
80	||	2	

In passing, we point out that if the data are given in terms of classes, then the class with greatest frequency gives the *modal class*.

Now that we have discussed and illustrated the three averages, it is tempting to ask the following question: which is the best average one should use? It is not easy to give a clear-cut and satisfactory answer to the question. The answer could vary from person to person and from problem to problem, as illustrated by the following examples.

Here is a list of scores obtained by 11 students in a calculus exam:

$$98, 79, 70, 74, 98, 60, 68, 75, 98, 70, 90$$

It can easily be verified that mean $= 80$, median $= 75$, and mode $= 98$. Of the 11 scores, 98 is the largest and hence the mode is not representative of the scores. Hence mode can be a poor choice of an average.

Consider a small company with ten employees whose annual salaries in dollars are 7200, 7500, 8300, 8600, 9000, 9300, 9300, 9700, 35000, and 50000; mean $= 15390$, median $= 9150$, and mode $= 9300$. In a situation like this, there is every chance that the manager of the company will claim that the average (the mean) salary at his company is 15390 dollars to corroborate his arguments, while the union leader will definitely quote the median, $9150, as the average salary to argue his case for the benefit of the employees at the bargaining table. Observe that in this case, even the mode is not a bad average for the union leader to use.

Suppose one of the commercial television networks takes a poll as to whether people prefer program *A*, program *B*, or program *C*. From a sample of 500 people, it is found that 145 people like program *A*, 120 people like program *B*, and 235 people like program *C*. In a situation like this, the network will certainly not be interested in any of the averages other than the mode, since logically they want programs that appeal to more people.

Exercise 15.2

1. Find the mean, median, and mode (if it exists) of each of the following sets of values.
   a) 2,5,3,2,5,4,8,5,1,8,1
   b) 2,3,13,23,3,11,13,3,7
   c) 1,2,3,4,4,4,3,2,1
   d) 1234,2341,1234,3412,1234,4123
   e) $a - 3d, a - d, a + d, a + 3d$
   f) $a - 2d, a - d, a, a + d, a + 2d$

2. Is it possible for a set of values to have
   a) no mean?   b) more than one mean?
   c) no median?   d) more than one median?
   e) no mode?   f) more than one mode?

3. Jean received the following scores in a calculus course: 96, 83, 89, 91, 87, and 97. Find (*a*) the mean score, (*b*) the median score.

4. Here is a list of annual salaries made by the employees at Tin Bin Company: 7500, 8600, 8300, 8600, 9200, 8900, 8600, 12000, 11500, 8600, 13000, 7900, 10500, 8600, and 7400.
   a) If Todd makes 2720 dollars more than the mean salary, what is his annual salary?
   b) If Peter makes 300 dollars less than the median salary, what is his annual salary?

5. The mean salary last year at Noody Moody Company of 100 employees was 9500 dollars. How much money did the company give as salary last year?

6. The instructor of a chemistry course announced at the beginning of the course that there would be five exams for the course and that a minimum average of 70 would be a C for the course. If Jane scored 65, 83, 68, and 56 points in the first four exams, what is the smallest score she can have in the last exam to receive a C or better for the course?

7. The salaries of teachers at Eveready University are summarized in Table 15.5. Find the modal class and median for the data.

Table 15.5

Class	Frequency
8000–11000	13
11000–14000	31
14000–17000	45
17000–20000	17
20000–23000	15
23000–26000	13
26000–29000	11
29000–32000	5

8. The scores obtained by Paul in five exams of a mathematics course are 89, 98, 79, 83, and 91. Find
   a) the mean score
   b) the deviations of the scores from the mean
   c) the sum of the deviations of the scores from the mean
   d) the mean of the deviations
9. The salaries of employees at a small company are 8600, 7900, 8700, 15000, 8600, 25000, and 31000. Find the mean, median, and mode. Which of these averages will better represent the interests of
   a) the management?   b) the employees?
10. A company that makes T-shirts sold 46000 T-shirts of small size, 93000 of medium size, 3100 of large size, and 9000 of extra large size last year. Since the company is definitely interested in making more profits, of what size T-shirts will they be inclined to make more?

15.3 MEASURES OF DISPERSION

Thus far, we have discussed three kinds of averages and observed that each has its own merits and demerits. None of these averages gives us any information as to how the various items in the data are scattered or dispersed. For example, it is possible to have two sets of data with the same averages, but at the same time, the way the values are scattered may be totally different. Consider the salaries of the employees at two small companies with eight employees each, given in Table 15.6. Observe that the averages of salaries at the two companies remain equal. However, a closer observation of the two sets of data reveals that the distribution of salaries at the two companies differs radically. The salaries at company A lie pretty close together around the mean, whereas those at company B are widely spread out or dispersed. Consequently, we need some measure of the dispersion of values in data.

Table 15.6

Company A	Company B
6800	4100
6900	4300
7200	7200
7200	7200
7300	7300
7900	9800
8300	10200
9200	10700

mean = 7600
median = 7250
mode = 7200

The simplest measure of dispersion is the range. Recall that the *range* of a set of values is obtained by subtracting the smallest value from the largest value. For example, the salary range at company A is $9200 - 6800 = 2400$, while that at company B is $10700 - 4100 = 6600$. Notice that the range is determined by the extreme values in the data. It is in no way related to any other item in the data and hence is not very useful as a measure of dispersion.

In the previous section, we discussed how to find deviations of values from some fixed value. Can we somehow use deviations from the mean to gauge the scattering nature of the data? From Table 15.6, we find that the deviations of salaries at company A from the mean salary are

$$6800 - 7600 = -800, \qquad 6900 - 7600 = -700$$
$$7200 - 7600 = -400, \qquad 7200 - 7600 = -400$$
$$7300 - 7600 = -300, \qquad 7900 - 7600 = 300$$
$$8300 - 7600 = 700, \qquad 9200 - 7600 = 1600$$

Observe that the sum of the deviations is zero. (Why?) It turns out that this is indeed always the case. Consequently, it cannot be taken as an indicator of dispersion. Nevertheless, let's not be disappointed. This problem can be overcome by taking the absolute values (Section 7.10) of the deviations.

The *mean deviation* of a set of values is the mean of the absolute values of the deviations from the arithmetic mean.

Example 15.9   Find the mean deviation of the salaries at company A in Table 15.6.

Solution: We already found that the deviations from the mean are -800, -700, -400, -400, -300, 300, 700 and 1600. Therefore,

$$\text{mean deviation} = \frac{800 + 700 + 400 + 400 + 300 + 300 + 700 + 1600}{8}$$
$$= \frac{5200}{8} = 650$$

Example 15.10   Find the mean deviation of the salaries of employees at company B given in Table 15.6.

Solution: Recall that the mean salary is 7600. The deviations are

$$4100 - 7600 = -3500, \qquad 4300 - 7600 = -3300$$
$$7200 - 7600 = -400, \qquad 7200 - 7600 = -400$$
$$7300 - 7600 = -300, \qquad 9800 - 7600 = 2200$$
$$10200 - 7600 = 2600, \qquad 10700 - 7600 = 3100$$

$$\text{mean deviation} = \frac{3500 + 3300 + 400 + 400 + 300 + 2200 + 2600 + 3100}{8}$$
$$= \frac{15800}{8} = 1975$$

The mean deviation gives us the average "distance" of each item from the mean. It varies with the dispersion of the data. If the mean deviation is large, then the values are widely spread out; if it is small, they are clustered around the mean. Observe that the above examples agree with this. The mean deviation of salaries at company *B* is much higher than that at company *A*, and the salaries at company *B* are radically spread out compared with those at company *A*.

The next and the most frequently used measure of dispersion is the standard deviation, which is closely related to mean deviation. The *standard deviation* of a set of values, denoted by the Greek letter σ (sigma), is the square root of the mean of squares of deviations from the mean. The following steps may be useful to compute the standard deviation:

1. Compute the arithmetic mean, $\bar{x}$.

2. Compute the deviations from $\bar{x}$.

3. Square each deviation obtained in (2).

4. Find the sum of the numbers in (3).

5. Divide the sum in (4) by the total number of values in the data. This result is called *variance.*

6. Extract the square root of the answer obtained in (5).

Example 15.11   Find the standard deviation of the weights 180, 195, 212, 178, and 195 of the five basketball players of example 15.1.

Solution: In example 15.1, we computed the mean weight to be $\bar{x} = 192$. It will now help to construct a table, as in Table 15.7.

Table 15.7

Weight (x)	Deviation $(x - \bar{x})$	(Deviation)^2 $(x - \bar{x})^2$
178	-14	196
180	-12	144
195	3	9
195	3	9
212	20	400
sum = 960	0	758

$$\text{variance} = \tfrac{758}{5} = 151.6$$

$$\text{standard deviation} = \sqrt{151.6} = 12.3 \quad \text{(approximately)}$$

Example 15.12   Find the standard deviation of the scores obtained by ten students in a mathematics exam: 71, 98, 76, 83, 90, 86, 80, 95, 81, 90.

Table 15.8

Score (x)	Deviation ($x - \bar{x}$)	(Deviation)² $(x - \bar{x})^2$
71	-14	196
76	-9	81
80	-5	25
81	-4	16
83	-2	4
86	1	1
90	5	25
90	5	25
95	10	100
98	13	169
sum $=$ 850	0	642

Solution: As before, let's construct a table, as in Table 15.8. From the table, the sum of the scores $= 850$. Hence

$$\text{mean} = \tfrac{850}{10} = 85$$

$$\text{variance} = \tfrac{642}{10} = 64.2$$

$$\text{standard deviation} = \sigma = \sqrt{64.2} = 8 \quad \text{(approximately)}$$

Exercise 15.3

1. Find the range, mean deviation, and standard deviation of each of the following sets of numbers.
   a) 1,2,3,4,5,6,7
   b) 3,3,13,23,5,11,13,3,7
   c) 1,2,3,4,4,5,4,4,3,2,1
   d) $a - 3d,\ a - d,\ a + d,\ a + 3d$
   e) $a - 2d,\ a - d,\ a,\ a + d,\ a + 2d$
2. Frank received the following scores on five tests in a mathematics course: 73, 89, 81, 95, and 87. Find the mean and standard deviations of the scores.
3. Is it possible for the standard deviation of a set of values to be zero? If yes, when?
4. Find the standard deviation of the values 5,5,5,5,5,5.
5. Is it possible for a set of values to have the standard deviation equal its mean deviation?
6. Find the mean and standard deviations of the values 3,3,3,3,3.
7. Tables 15.9 and 15.10 contain the scores received by 20 students in two physics exams.

a) Find the mean and standard deviations of scores of each exam.
b) Which has a higher standard deviation?
c) Which of the two exams has scores more scattered around the mean?

Table 15.9   Exam 1

Score	Frequency
63	2
69	2
72	1
78	2
79	3
83	2
85	3
91	1
93	2
95	1
98	1

Table 15.10   Exam 2

Score	Frequency
57	3
60	2
65	1
69	2
75	4
83	2
87	1
91	3
97	2

8. Compute the mean and standard deviations for the following frequency distribution.

Table 15.11

Class	Frequency
10–20	8
20–30	4
30–40	5
40–50	2
50–60	1

15.4 THE NORMAL DISTRIBUTION

Let's now discuss how the standard deviation of a set of values is used to derive more information about the data in a special case.

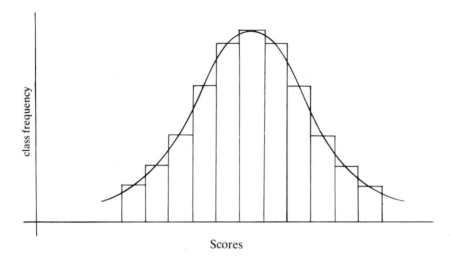

Scores

Fig. 15.3

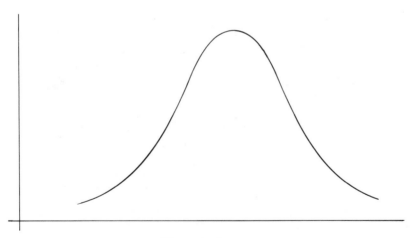

The normal curve

Fig. 15.4

Suppose that the histogram for the scores of a large group of students in a mathematics test looks as in Fig. 15.3, with scores on the horizontal axis and class frequency on the vertical axis, where the classes are assumed to be fairly small. Then clearly the histogram can be approximated by a nice smooth curve that is bell-shaped. This bell-shaped curve is usually called the *normal curve* (Fig. 15.4). Frequency distributions that have the property that their histograms can be approximated by bell-shaped curves are called *normal distributions*.

Let's now list some of the fascinating properties of the normal curve in addition to the fact that it is bell-shaped.

1. It is symmetric about a vertical line through the mean $\bar{x}$ of the distribution (Fig. 15.5). The highest frequency occurs for the value $x = \bar{x}$. Thus in a normal distribution, the three averages agree.

2. About 68.2% of all items in the data falls within one standard deviation of the mean, that is between $\bar{x} - \sigma$ and $\bar{x} + \sigma$; 95.4% of the total number of values lies within two standard deviations of the mean, that is, between $\bar{x} - 2\sigma$ and $\bar{x} + 2\sigma$; and 99.8% falls within three standard deviations of the mean, that is, between $\bar{x} - 3\sigma$ and $\bar{x} + 3\sigma$, as shown in Fig. 15.5.

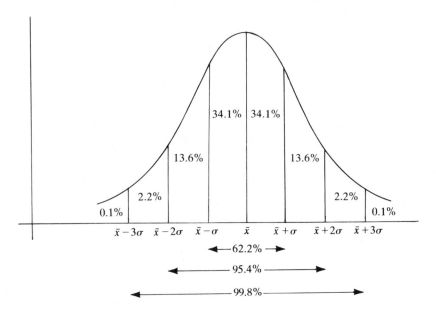

Fig. 15.5

3. The total area under the normal curve is one unit.

4. Since $13.6\% + 2.2\% + 0.1\% = 15.9\%$ of all values are greater than $\bar{x} + \sigma$, the probability that a value selected at random from data will be larger than $\bar{x} + \sigma$ is $\frac{15.9}{100} = 0.16$ approximately. Similarly, the probability that it will be less than $\bar{x} - 2\sigma$ is roughly $\frac{2.3}{100} = 0.023$.

5. These facts remain true with every normal distribution, even if distributions are associated with different means and standard deviations.

We now illustrate some of these remarks by the following examples.

Example 15.13   The scores in the final exam of a calculus class of 400 students at Zebra University are found to be normally distributed with mean 76 and standard deviation 8.

a) What percent of the students in the class scored between 68 and 84?
b) How many students received at least 84 points?
c) If 92 and above is an *A* for the exam, how many students made *A*'s?
d) If 59 or less is an *F*, how many students received *F*'s?
e) What is the probability that a student selected at random will have scored between 60 and 68?

Solution: Since the scores are normally distributed with mean 76 and standard deviation 8, the normal curve looks as in Fig. 15.6, from which our answers can easily be read.

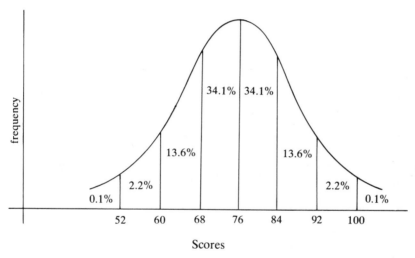

Fig. 15.6

a) 34.1% + 34.1% = 68.2% of the students received a score between 68 and 84.
b) Approximately 13.6% + 2.2% + 0.1% = 15.9% of the students scored 84 or more. Since 15.9% of 400 is $400 \times \frac{15.9}{100} = 63.6$, about 64 students received at least 84 points.
c) About 2.3% of all students scored 92 or more. Since 2.3% of 400 is $400 \times \frac{2.3}{100} = 9.2$, about nine students received *A*'s.
d) Roughly 2.3% of students got less than 60 points and 2.3% of 400 is about 9. Therefore, nine students got *F*'s.
e) 13.6% of students scored between 60 and 68. Hence the required probability is $\frac{13.6}{100} = 0.14$ approximately.

Example 15.14   The time in minutes for 300 students to take a sociology test was found to obey a normal distribution with mean 100 and standard deviation 10.

a) What percent of the students took 80 or more minutes to answer the test?

b) How many students took less than an hour and a half?

c) What is the probability that a student chosen at random will have finished the test in an hour and a half?

d) How many students needed more than two hours?

Solution: Since the distribution is normal with mean 100 and standard deviation 10, the normal curve looks as in Fig. 15.7.

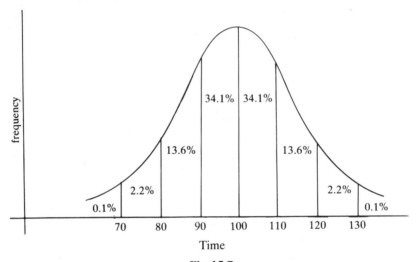

Fig. 15.7

a) Approximately $13.6\% + 34.1\% + 34.1\% + 13.6\% + 2.2\% + 0.1\% = 97.7\%$ of the students needed 80 or more minutes to answer the test.

b) About $0.1\% + 2.2\% + 13.6\% = 15.9\%$ (about 48) of the students took less than 90 minutes.

c) Since about 15.9% of the students completed the test in 90 minutes, the required probability is $\frac{15.9}{100} = 0.16$, approximately.

d) Roughly 2.3% of students took more than two hours, that is, about seven students.

Exercise 15.4

1. The scores in the final exam of a psychology course of 300 students at Rex University are found to be normally distributed with mean 70 and standard deviation 10.

   a) How many students scored between 70 and 80?

b) Find the frequency of the class 80–90.

c) What percent of the students got a score between 70 and 90?

d) If 90 and above is an *A*, how many students got *A*'s?

e) If 59 or less is *F*, how many got *F*'s?

f) If a student is selected at random, what is the probability that he scored between 60 and 80 points?

2. In the Ava Bowling Match among 100 people, the numbers of pins knocked down were found to be normally distributed with mean 180 and standard deviation 20.

  a) How many bowlers knocked down between 140 and 180 pins?

  b) What percent of all bowlers knocked down between 220 and 240 pins?

  c) How many bowlers are within one standard deviation of the mean?

  d) What is the probability that a bowler selected at random will have knocked down between 140 and 160 pins?

3. A study shows that the ages of the 150 science teachers at Rattle Snake University follow the normal distribution with normal curve as shown in Fig. 15.8.

  a) What is the average age of the science teachers?

  b) What is the standard deviation of the distribution?

  c) Find the median of the age distribution?

  d) What is the mode of the age distribution?

  e) What percent of teachers are in the age group 29–37?

  f) Find the probability that a randomly selected teacher will be in the age group 53–61.

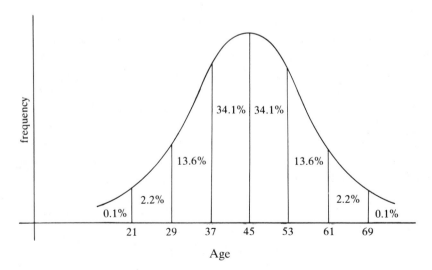

Fig. 15.8

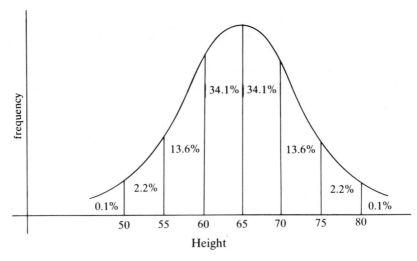

Fig. 15.9

4. The graph of heights in inches of the 200 students in a calculus class at Lex
University is given in Fig. 15.9.
   a) What is the shape of the curve?
   b) What is the curve called?
   c) Find the mean height of students in the class.
   d) How many students are at least five feet tall?
   e) What percent of the students are in the height group 75–80?
   f) How many students are in the height group 70–80?

15.5 SUMMARY AND COMMENTS

In this chapter, we introduced the following terms that are basic in the study
of statistics: population, sample, random sample, selective sample, frequency,
frequency distribution, class, class boundary, class width, range, histogram, and
frequency polygon.

We discussed and illustrated three averages: mean, median, and mode. The
arithmetic mean is a magnitude average. It is influenced by every item in
the data. The mean will reflect the data better if all the entries cluster around
the mean. The median is a positional average. Exactly half of the entries in the
data, when arranged in ascending order of magnitude, lies on either side of the
median. The mode is a popularity average. The mean is the most widely used
average.

Since the averages do not furnish any information about how the various
values in the data are scattered, we introduced three measures to gauge the
dispersion of the data: range, mean deviation, and standard deviation. Range is
poor measure since it depends only on the extreme values. The most commonly
used measure of dispersion is the standard deviation.

If the histogram of a frequency distribution can be approximated by a bell-shaped curve, then the distribution is a normal distribution and the curve is a normal curve. In a normal distribution, the three averages agree. About 68% of all values lie within one standard deviation of the mean, 95% within two standard deviations, and 99% within three standard deviations.

Finally, we point out that statistical tools are used on a large scale in biology, chemistry, economics, engineering, medicine, physics, psychology, and sociology.

SUGGESTED READING

Ball, J., "Finding Averages with Bar Graphs," *The Arithmetic Teacher*, vol. 16 (Oct. 1969), pp. 487–489.

Maxfield, J. E., and M. W. Maxfield, *Key to Mathematics*, W. B. Saunders Co., Philadelphia (1973), pp. 231–260.

Miller, C. D., and V. E. Heeren, *Mathematical Ideas, an Introduction* (2nd ed.), Scott, Foresman and Co., Glenview, Ill. (1973), pp. 201–237.

Triola, M. F., *Mathematics and the Modern World*, Cummings Publishing Co., Menlo Park, Calif. (1973), pp. 197–224.

Wheeler, R. E., *Modern Mathematics: An Elementary Approach* (3rd ed.), Brooks/Cole Publishing Co., Monterey, Calif. (1973), pp. 461–512.

"It is the man not the method that solves the problem."

H. MASCHKE

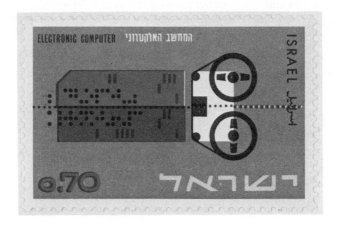

16/Introduction to Computers

After studying this chapter, you should be able to:
- *name a few early computers*
- *list the five basic components of a typical digital computer*
- *name a few computer languages*
- *draw flow charts of simple algorithms*
- *convert a fraction to base two*
- *express a binary numeral in base eight and base sixteen*
- *find the tens (twos) complement of a decimal (binary) number*
- *perform subtraction using complements*

16.0 INTRODUCTION

One of man's greatest achievements is the invention of computers. Almost every phase of our life is influenced by this monstrous device. Man's progress in industry, the running of government and every branch of learning have been rapidly and steadily increasing since the invention of this incredible machine. There is every reason to believe that we would not have been able to acquire so much knowledge about outer space and man could not have gone to the moon in 1969 if we had no access to these very fast and powerful "brains."

Some people think of a computer as just a complicated "number-crunching" machine used to solve highly complex numerical problems. Today, computers are used to prepare payrolls, to enter deposits and withdrawals in banks, to store and supply huge volumes of information, to collect data, to draw inferences about patients, to make weather predictions, to control flow of traffic at airports and in subway systems, to make seat reservations in planes, to prepare personalized books, to prepare indexes for books, to help students register for courses, to help students learn different subjects in schools, to simulate huge projects before they are built, to find the best possible route for garbage collection in big cities, to generate new designs, etc.

Because of the wide use of computers and the tremendous impact of computers on society, a new branch of learning called computer science has sprung up. Today, several major universities offer both undergraduate and graduate programs in this fast-growing discipline.

In this chapter, we briefly discuss the origin of these highly accurate and reliable machines, list a few languages understandable by them, discuss a pictorial representation of instructions given to computers to solve problems, and, finally, elucidate the arithmetic used by these remarkable devices.

16.1 HISTORICAL BACKGROUND

It is only natural that man wanted some device to relieve him of complicated calculations so that he could devote that time and energy to other constructive and interesting things. Man's thirst for fast and accurate machines that would do complicated computations resulted in the invention of computers.

The earliest known form of calculating machines is the well-known *abacus*, invented around 3000 B.C. Even now it is used in China, Japan, and Russia. This simple device can be used to perform the four fundamental arithmetic operations.

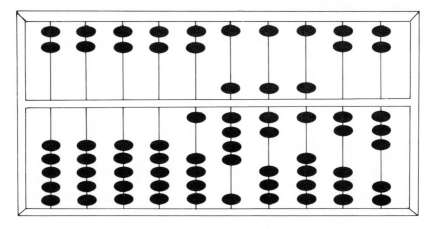

Fig. 16.1. Chinese Abacus representing 197,623

The abacus consists of a rectangular wooden frame strung with parallel strings. The Chinese abacus (Fig. 16.1) has seven beads on each string. These strings are divided into two parts by a wooden piece, so that on each string two beads lie always in the upper half and the other five in the lower half of the

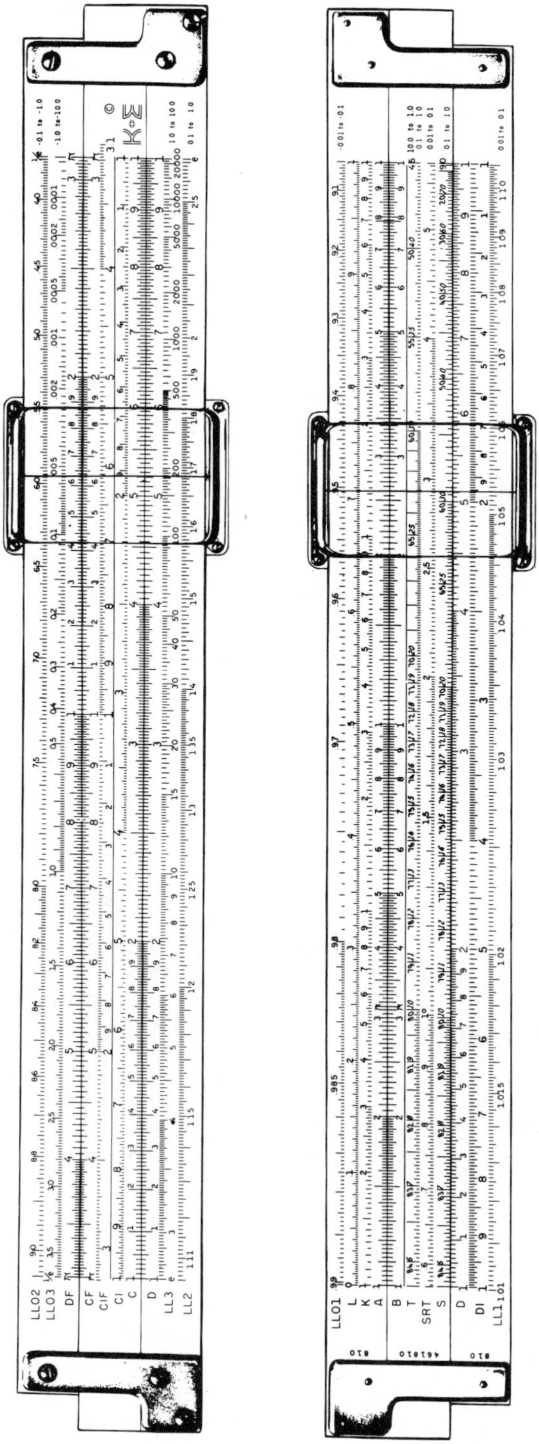

Fig. 16.2. A Typical Slide Rule

whole frame. Each bead above the middle bar represents five units, while each below the bar represents one unit. To represent a number on the abacus, which makes use of the concept of place-value, enough beads from above and below the middle bar are brought near the bar. Figure 16.1 displays a Chinese abacus representing the number 197,623. The Japanese abacus contains only five beads on each string with one above and the other four below the middle bar. The abacus is an example of a *digital computer* since it works on the concept of counting.

The *slide rule*, widely used even nowadays, especially by engineers and physicists, is an example of an *analogue computer*, a device that makes use of the concept of lengths (Fig. 16.2). The slide rule is based on the concept of *logarithms*, conceived by the Scottish mathematician John Napier (1550–1617). The slide rule is used to evaluate the product, quotient, powers, and roots of numbers.

The first mechanical calculating machine was built by Pascal (Section 14.0) at the age of 19. The laborious auditing of government accounts by his father motivated Pascal to invent this device (Fig. 16.3). The machine contained eight dials, each like a telephone dial, with digits 0 through 9 written around it. It could be used to add and subtract numbers up to six digits by turning these dials. The concept of carrying was accomplished automatically, since as one dial passed the digit 9, the dial on its left would advance one digit, as in the case of a gasoline-pump meter.

Fig. 16.3. Pascal's Calculating Machine

Leibniz (Section 2.0), realizing that intelligent men should not waste their time like slaves in tedious computations, invented a calculating machine in 1671. This machine, unlike Pascal's, could even multiply, divide, and extract square roots of numbers. But, nonetheless, it had its own limitations: it was slow and undependable. That he conceived the idea of using the binary system instead of the decimal system in computer arithmetic was a significant factor in the development of modern computing machines.

In 1812, the English mathematician Charles Babbage (1792–1871) conceived the idea of building a calculating machine to construct mathematical tables. This machine, built in 1822 and called a *difference engine*, could be used to generate mathematical tables correct to six decimal places. In 1833, he started building a steam-run engine, called an *analytical engine*, that would have worked without human intervention. In spite of the brilliant ideas he conceived in designing such a machine since they were well ahead of the technological tools available during his time, his machine did not see the light of day. The rest of his life ended in utter frustration. In any case, Babbage had the right idea and was on the right track of building a modern type of calculating machine.

Cards punched with small holes, which are used on a large scale to feed information into modern computers, were used in 1801 by Joseph Jacquard (1752–1834) in the weaving of fabrics with designs, thus bringing a major contribution to the development of computers. The concept of using punched cards drew the attention of Herman Hollerith of the U.S. Census Bureau. He invented machines that could process census data using punched cards. This was a revolutionary step in census taking. Even though there was about a 25% increase in population from 1880 to 1890, the 1890 census was completed in about one-third of the time taken by the 1880 census. In 1896, Hollerith founded the Tabulating Machine Company which later became part of IBM (International Business Machines Corporation).

In 1944, the first automatic and electromechanical digital computer was completed at Harvard University as the result of seven years of hard work by Howard Aiken with the help of IBM. It was called the IBM Automatic Sequence Controlled Calculator and, later, Mark I. It weighed five tons, was 51 feet long and eight feet high. It could add and subtract two 23-digit numbers in $\frac{3}{10}$ of a second, multiply in approximately six seconds, and divide in about 16 seconds. Most of Babbage's ideas could be seen in Aiken's work. However, it is said that he did not have any prior knowledge of Babbage's contributions until after the completion of Mark I.

The first all-electronic digital computer, called ENIAC (Electronic Numerical Integrator and Calculator), was built for the Ballistic Research Laboratories of the U.S. Army at Aberdeen, by J. P. Eckert and J. W. Mauchly of the University of Pennsylvania in 1946. It weighed 30 tons and occupied an area of

Fig. 16.4. Jacquard Loom Cards

1500 square feet. However, this monstrous machine could perform 5000 additions or 350 multiplications per second, making a considerable improvement over Mark I.

None of the above machines could store any information in them. John von Neumann (1903–1957), one of the greatest mathematicians of this century, suggested the possibility of storing both instructions and data in the machine. This prompted scientists to build stored-program computers. The first such digital computer was EDSAC (Electronic Delay Storage Arithmetic Calculator), built at Cambridge University, England, in 1949, even though EDVAC (Electronic Discrete Variable Computer) was started at the University of Pennsylvania, earlier than EDSAC.

UNIVAC I (Universal Automatic Computer), the first commercial computer, was developed in 1951 by the Eckert–Mauchly Corporation, which later became the Univac division of the Sperry-Rand Corporation.

Since then, more and more computers have been built that can perform operations at an incredible speed and that are much smaller than their predecessors. Many of today's computers can perform about three million operations per second. ILLIAC IV, designed at the University of Illinois and recently built by Burroughs Corporation for NASA (National Aeronautics and Space Administration) can perform about 100 million computations a second.

16.2 A TYPICAL DIGITAL COMPUTER

Roughly speaking, a computer is an electronic device that receives data together with instructions, processes the data, and then draws conclusions in accordance with the instructions fed into it.

A typical digital computer consists of five basic components: input unit, memory unit, arithmetic unit, control unit and output unit (Fig. 16.5).

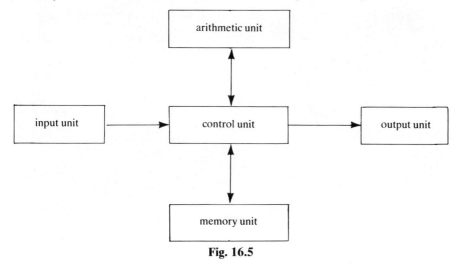

Fig. 16.5

(*a*) The *input unit* receives data and related instructions for processing the data. The set of instructions given to the computer is called a *program* and the person who writes programs is a *programmer*. The input can be fed into the computer using punched cards (Fig. 16.6), typewriter, punched paper tape (Fig. 16.7), magnetic tape, etc. Cards are punched on a *keypunch* that has a keyboard like that of a typewriter. Punched paper tapes do essentially the same job as punched cards. Magnetic tapes are definitely much faster and more efficient than paper tapes. They can be used to record over 600,000 characters per second. Also, the recording can be erased and hence magnetic tapes are reusable. Programs can also be entered into a computer by typing them on a typewriter connected to it.

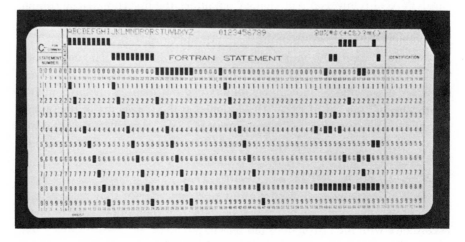

Fig. 16.6. A Punched Card

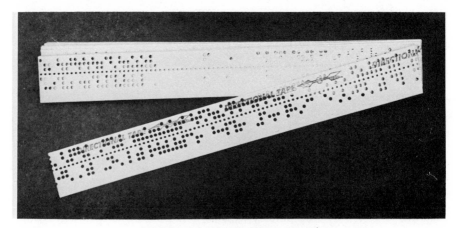

Fig. 16.7. A Punched Paper Tape*

(b) The *memory unit* is a storage device where data, instructions, and intermediate results, if any, are saved so that the information stored can be used as needed. Messages are stored in the memory unit in terms of the *binary digits* (called *bits*) 0 and 1.

(c) The *arithmetic unit* is where the actual arithmetic and logical operations take place according to the instructions received from the control unit.

(d) The *control unit* monitors the operations of the other four units. It dictates which unit does what and when. It has complete control over the following: When is information to be fed in? Where is it to be stored? When is it to be retrieved from storage? What kind of arithmetic and logical operations are to be performed and when? How is the final result to be given to the user? In short, the control unit enjoys a commanding position in the whole system.

The arithmetic, memory and control units constitute the *central processing unit* (CPU) of a computer.

(e) The *output unit* supplies the computer-user with the result obtained after processing the data. Output, as in the case of input, can be provided by using either punched cards, typewriter, paper tape, magnetic tape, etc. The input-output units (*I/O devices*) together act like a middleman between the computer and user.

Can we communicate with the computer in English or some other natural language? Since the same English word can have more than one meaning and the computer is not "intelligent" enough to identify the right meaning intended for the word, the English language is not precise enough and can create lots of problems. Consequently, artificial languages that are precise and concise have been developed as media of communications between the computer and programmer. The language used, in general, depends on the nature of the problem. Each computer language has its alphabet, grammar, sentence structure, etc. In any event, modern computers have a special device called a *compiler* that will automatically translate the instructions into a language consisting of 0's and 1's before they are processed.

FORTRAN (FORmula TRANslator), *ALGOL* (ALGOrithmic Language), *COBOL* (COmmon Business-Oriented Language), and *BASIC* (Beginners All-purpose Symbolic Instruction Code) are a few of the several computer languages developed so far, of which FORTRAN is the most widely used scientific language.

16.3   ALGORITHMS AND FLOW CHARTS

A computer cannot think independently. It does exactly what it has been instructed to do like a faithful servant. It executes the instructions in the program in a step-by-step pattern without going back and forth over other steps. Consequently, the instructions must be logical, precise, and concise, forming an algorithm for the problem. So programmers have to write efficient algorithms to instruct the computer to solve problems.

We are already familiar with step-by-step procedures even outside the mathematical world. If we want to make pie, bread, coffee, etc., we follow the recipe one step after another; if we buy a bike, generally it comes in parts in a carton and later we put the pieces together by following the step-by-step directions provided.

In the preceding chapters, we discussed several algorithms. Determining if a positive integer is prime, using the Euclidean algorithm, constructing the Fibonacci sequence, constructing magic squares, the square root algorithm, algorithms for the four arithmetic operations, and the ancient Egyptian algorithms for multiplication and division are a few of them. We now advise the reader to review these before proceeding any further.

Let's now write down an algorithm to determine if a positive integer n (>1) is prime. Recall from Section 5.3 that 2 is a prime and n (>2) is a prime if n is not divisible by any of the integers $2, 3, 4, \ldots, n-1$. This procedure is definitely lengthy, especially if n is fairly large. Indeed, we need only check if n is divisible by any positive integer k (>1) such that $k^2 \le n$. This procedure can be broken into the following steps (*prime-number algorithm*):

Step 1   If $n = 2$, then n is a prime and we are done; otherwise, proceed to step 2.

Step 2   Start with the value $k = 2$.

Step 3   Evaluate n/k. if n/k is a whole number, then k is a factor of n and hence n is not a prime. If n/k is not a whole number, go to step 4.

Step 4   Check if $k^2 \le n$. If $k^2 \le n$, then increase k by 1 and proceed as in step 3. If $k^2 > n$, then n is a prime and the procedure stops. (Observe that if $k^2 = n$, then n is not a prime!)

Example 16.1   Illustrate the prime-number algorithm for $n = 35$.

Solution:

1. Since $n = 35 \ne 2$, we check if $k = 2$ is a factor of $n = 35$.

2. $\dfrac{n}{k} = \dfrac{35}{2} = 17.5$, which is not a whole number. So 35 can still be a prime and we go to step 4.

3. Is $k^2 = 2^2 \le 35 = n$? Yes; therefore, we increase the value of k by 1.

4. We now let $k = 2 + 1 = 3$; we have $\dfrac{n}{k} = \dfrac{35}{3} = 11.6$ (approximately), again not a whole number. Hence, 35 is still a candidate for a prime.

5. Is $k^2 = 3^2 \le 35 = n$? Yes; therefore, we add 1 to the new value of k, yielding $k = 3 + 1 = 4$.

6. We have $\dfrac{n}{k} = \dfrac{35}{4} = 8.75$, not an integer. Therefore, 35 can still be a prime.

7. Is $k^2 = 4^2 \le 35 = n$? Yes; so we let $k = 4 + 1 = 5$.

8. Since $\dfrac{n}{k} = \dfrac{35}{5} = 7$, an integer, $k = 5$ is a factor of 35 and hence 35 is not a prime, as we already know.

Example 16.2   Determine if $n = 23$ is a prime using the prime-number algorithm.

Solution:

1. We start with $k = 2$, since $n = 23 > 2$.

2. Since $\dfrac{n}{k} = \dfrac{23}{2} = 11.5$ is not an integer, we try step 4 of the algorithm.

3. We have $k^2 = 4 \leq 23 = n$. Therefore, we increase the value of k by 1, giving $k = 2 + 1 = 3$.

4. Is $\dfrac{n}{k} = \dfrac{23}{3}$ an integer? No. So we go to step 4 again.

5. Since $k^2 = 9 \leq 23 = n$, let's try step 3 by letting $k = 3 + 1 = 4$.

6. Observe that $\dfrac{n}{k} = \dfrac{23}{4}$ is not an integer. As a result, we try step 4.

7. Since $k^2 = 16 \leq 23 = n$, we continue step 3 with $k = 4 + 1 = 5$.

8. We have $\dfrac{n}{k} = \dfrac{23}{5}$, still not an integer. So we proceed to step 4 again.

9. Is $k^2 = 25 \leq 23 = n$? No. Consequently, we stop here. Since we could not find a factor k (> 1) of 23 such that $k^2 \leq 23$, we conclude that 23 is a prime, a fact already known to us.

Algorithms can conveniently be represented by diagrams called flow charts. A *flow chart* is a pictorial representation of an algorithm consisting of circles, rectangles, and diamonds with arrows on connecting lines showing the order of operations. Circles show the beginning and end of a program; rectangles represent instructions, and diamonds denote stages where a decision is made regarding the rest of the execution of operations. Figure 16.8 represents a flow chart of the prime-number algorithm discussed above. The flow chart in Fig. 16.9 illustrates the algorithm for making a cup of black coffee!

Let's now write down the steps in the Euclidean algorithm (Theorem 5.12) to find the gcd of two positive integers, say, a and b:

Step 1   If $a = b$, then clearly $\gcd\{a,b\} = b$; otherwise, let's assume for convenience that $a > b$.

Step 2   Let q be the quotient and r the remainder obtained on dividing a by b:

$$a = bq + r \qquad \text{(division algorithm)}$$

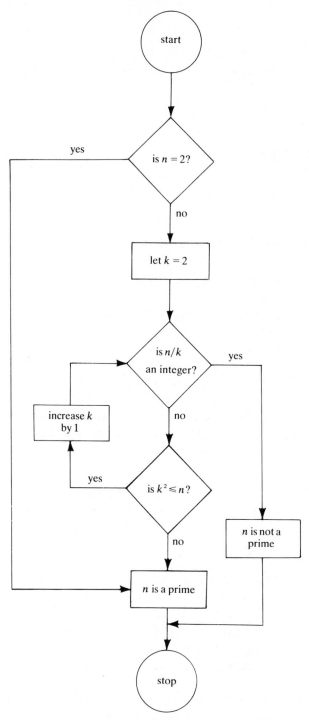

Fig. 16.8   Flow Chart for the Prime-Number Algorithm

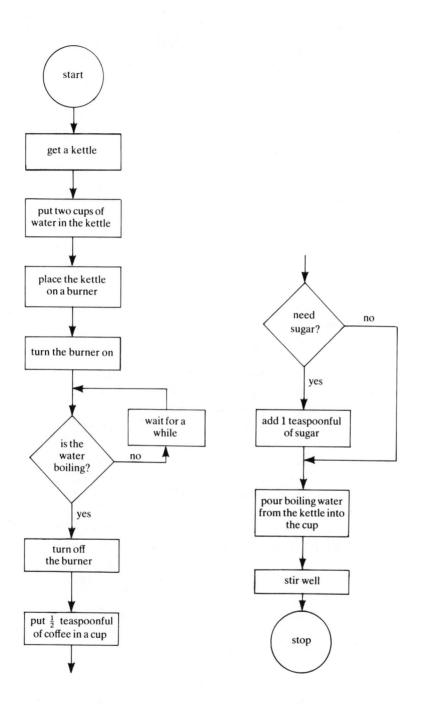

Fig. 16.9

504 An Elementary Approach to Mathematics

Step 3 If $r = 0$, gcd$\{a,b\} = b$ and we are done. If $r \neq 0$, proceed to step 4.

Step 4 Continue step 2 with b as the new dividend and r the new divisor.

The flow chart for the Euclidean algorithm is exhibited in Fig. 16.10, where we have assumed that $a \geq b$ for convenience.

Example 16.3 Find the gcd of 90 and 24.

Solution: Table 16.1 explains the various steps of the algorithm. Since $90 \neq 24$, we choose $a = 90$ and $b = 24$ in step 1. The gcd of 90 and 24 is 6.

Table 16.1

Step	a	b	$a = bq + r$
1	90	24	$\cdot\,\cdot$
2	90	24	$90 = 24 \cdot 3 + 18$
3	24	18	$24 = 18 \cdot 1 + \boxed{6}$
4	18	6	$18 = 6 \cdot 3 + 0$

Exercise 16.3

1. **a)** Write an algorithm to determine the truth value of the compound statement $p \wedge q$.

   b) Draw a flow chart to represent your algorithm.

2. Same as problem 1 with the statement $p \vee q$.

3. Same as problem 1 with the statement $p \rightarrow q$.

4. Write an algorithm to check if the sets $A = \{a,b,c\}$ and $B = \{w,x,y,z\}$ are disjoint.

5. **a)** Write an algorithm to check if an element x belongs to the set $A \cup B$.

   b) Represent your algorithm in a flow chart.

6. Same as problem 5 with the set $A \cap B$.

7. Same as problem 5 with the set $A - B$.

8. Draw a flow chart showing how to determine if an integer a is greater than an integer b. [*Hint:* recall that $a > b$ if and only if $a - b$ is a positive integer.]

9. Determine which of the following is a prime, exhibiting all your steps, as in example 16.1.

   a) 49   **b)** 123   **c)** 11

10. **a)** Find an algorithm to construct the first ten Fibonacci numbers.

    b) Draw a flow chart for your algorithm.

★11. The solutions of the equation $ax^2 + bx + c = 0$ are real and distinct, real and equal, or complex according as $b^2 > 4ac$, $b^2 = 4ac$, or $b^2 < 4ac$, respectively. Make a flow chart discussing the nature of the solutions of $ax^2 + bx + c = 0$.

★12. A nonempty set S has an operation $*$ defined on it. Draw a flow chart showing how to check if S is a group under this operation.

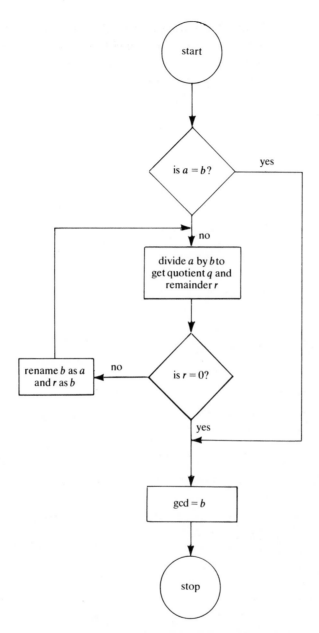

Fig. 16.10   Flow Chart for the Euclidean Algorithm

16.4   ELEMENTS OF COMPUTER ARITHMETIC

We are all familiar with a few nondecimal systems. One nondecimal system that deserves special attention is the binary system. We all have used systems that might sound nonmathematical in nature but at the same time are related to the binary system: "yes" and "no," "true" and "false," "on" and "off," "plus" and "minus," "positive" and "negative," etc. Since an electronic switch has exactly two states of existence—on and off—the binary system has received acceptance in the computer world. Most of today's computers make use of bits to store messages and perform arithmetic operations. We shall not discuss how information is encoded in terms of 0's and 1's.

We illustrated in Section 8.9 how to convert a decimal integer to a binary integer and vice versa, and in Sections 8.10 and 8.11 how to perform the four arithmetic operations in the binary system. For example, given $1776 = (11011110000)_2$ and $123 = (1111011)_2$

$$
\begin{array}{r}
(11011110000)_2 \\
+(\quad 11111011)_2 \\
\hline
(11111101011)_2
\end{array}
\qquad
\begin{array}{r}
(11011110000)_2 \\
-(\quad 11111011)_2 \\
\hline
(10111110101)_2
\end{array}
$$

However, we did not discuss how to represent fractions in the binary system. Since computers have to handle nonintegral numbers, too, we shall discuss that now.

Recall that to represent a decimal integer a in base b, we divided a and the successive quotients by b and picked the remainders in the reverse order. Representation of a proper fraction in base b is accomplished by *multiplying* it and successive fractional parts by b and then writing down the integral parts in that order:

Step 1   Multiply the fractional part by b to get an integral part and a new fractional part. This integral part becomes the first digit on the right of the reference point.

Step 2   Now multiply the new fractional part by b. The resulting integral part will be the second digit on the right of the reference point.

Step 3   Continue step 2 with the last fractional part until a zero fractional part is obtained or enough digits have been found.

Step 4   The required representation is obtained by writing the integral parts resulting from the above steps after the reference point.

Example 16.4   Convert 0.6875 to base two.

Solution: Following the above steps, we have

$$
\begin{array}{rl}
2 \times 0.6875 = \boxed{1} & .\ 3750 \qquad\qquad \text{read down} \\
2 \times 0.3750 = \boxed{0} & .\ 7500 \qquad\qquad\qquad \downarrow \\
2 \times 0.7500 = \boxed{1} & .\ 5000 \\
2 \times 0.5000 = \boxed{1} & .\ 0000
\end{array}
$$

Thus, by step 4, we have $0.6875 = (0.1011)_2$.

Example 16.5   Express 23.125 as a binary numeral.

Solution :   We express the integral part, 23, in base two by successive divisions and the fractional part, 0.125, by successive multiplications:

$$23 = 2 \cdot 11 + \boxed{1}$$
$$11 = 2 \cdot 5 \ + \boxed{1}$$
$$5 = 2 \cdot 2 \ + \boxed{1} \quad \uparrow$$
$$2 = 2 \cdot 1 \ + \boxed{0} \quad \text{read up}$$
$$1 = 2 \cdot 0 \ + \boxed{1}$$

$$2 \times 0.125 = \boxed{0} \ . \ 250 \quad \text{read down}$$
$$2 \times 0.250 = \boxed{0} \ . \ 500 \qquad \downarrow$$
$$2 \times 0.500 = \boxed{1} \ . \ 000$$

Thus, $23.125 = (10111.001)_2$.

Even though the binary system is ideal for computers to work with, one can easily misread a binary numeral if it contains several 0's and 1's. Also, it needs more space. For example,

$$3050845 = (1011101000110101011101)_2$$

These disadvantages are overcome by using the *octal* (base eight) and *hexadecimal* (base sixteen) systems, which are very closely related to the binary system. Again, conversion from binary to these systems or vice versa is fairly trivial. In the octal system, a number is represented using the digits 0 through 7, and in the hexadecimal system, we use the digits 0 through 9 and the letters A through F for the numbers 10 through 15.

Example 16.6   Rewrite 1976 in the octal and hexadecimal systems.

Solution: We can use our standard methods to express 1976 in base eight and base sixteen (try!). However, let's make use of its binary representation to answer the problem. Applying the familiar algorithm, we ask the reader to verify that

$$1976 = (11110111000)_2$$

Let's now group the bits in threes from right to left, appending one or two zeros on the left if necessary. Thus,

$$1976 = (011 \ \ 110 \ \ 111 \ \ 000)_2$$

Now, replace each group by its decimal equivalent:

$$011 = 3, \qquad 110 = 6, \qquad 111 = 7, \qquad 000 = 0$$

Thus, $1976 = (3670)_8$. This is indeed true, since

$$(3670)_8 = 3 \cdot 8^3 + 6 \cdot 8^2 + 7 \cdot 8^1 + 0 \cdot 8^0$$
$$= 1536 + 384 + 56 + 0$$
$$= 1976$$

Grouping the bits in fours, we have

$$1976 = (0111 \ 1011 \ 1000)_2$$

Substitute now the decimal equivalent of every group:

$$0111 = 7, \qquad 1011 = B, \qquad 1000 = 8$$

(Recall $(B)_{16} = 11$.) Thus, $1976 = (7B8)_{16}$. That this is indeed the case follows, since

$$(7B8)_{16} = 7 \cdot 16^2 + B \cdot 16^1 + 8 \cdot 16^0$$

$$= 1792 + 176 + 8$$

$$= 1976$$

Example 16.7 Express 250.15625 in base eight and base sixteen.
Solution: We now ask the reader to verify that

$$250.15625 = (11111010.00101)_2$$

Let's now group the bits in threes starting at the binary point:

$$250.15625 = (011 \ 111 \ 010 \ . \ 001 \ 010)_2$$
$$= (\ 3 \quad 7 \quad 2 \ . \ 1 \quad 2 \ )_8$$

replacing each group by the corresponding decimal equivalent. Collecting the bits in fours we have

$$250.15625 = (1111 \ 1010 \ . \ 0010 \ 1000)_2$$
$$= (\ F \quad A \ . \ 2 \quad 8 \ )_{16}$$

converting each set into its decimal equivalent.

We observe from the above examples that we can easily write down the octal and hexadecimal representations of a number from its binary representation.

The circuitry in most of today's computers is designed in such a way that subtraction is performed using the concept of complementation. The *tens complement* of a decimal number is obtained by subtracting each digit in the number from 9 and then adding 1 to the result. For example, the tens complement of 1776 is 8224, since

$$
\begin{array}{r}
9999 \\
-1776 \\
\hline
8223 \quad \leftarrow \text{ nines complement} \\
+ \quad 1 \quad \leftarrow \text{ add one} \\
\hline
8224 \quad \leftarrow \text{ tens complement}
\end{array}
$$

Similarly, the *twos complement* of a binary number is obtained by subtracting each bit from one and then adding a one to the resulting number. For example, the twos complement of 101011 is 010101, since

$$
\begin{array}{r}
111111 \\
-101011 \\
\hline
\end{array}
$$

$$
\begin{array}{rl}
010100 & \leftarrow \text{ ones complement} \\
+ \qquad 1 & \leftarrow \text{ add one} \\
\hline
010101 & \leftarrow \text{ twos complement}
\end{array}
$$

Observe that the twos complement of a binary number is also obtained by swapping 0's with 1's and 1's with 0's and then adding a 1 to the new number.

Since addition is much more easy to perform than subtraction, we will transform a subtraction problem to an addition problem using complements. For example, to evaluate $a - b$ in base ten (two), we need only add the tens (twos) complement of b to a and then delete the left-most digit in the resulting number.

Example 16.8 Evaluate $5871 - 3776$.

Solution: First, find the tens complement of 3776:

$$
\begin{array}{r}
9\ 9\ 9\ 9 \\
-3\ 7\ 7\ 6 \\
\hline
\end{array}
$$

$$
\begin{array}{rl}
6\ 2\ 2\ 3 & \leftarrow \quad \text{nines complement} \\
+ \qquad 1 & \leftarrow \quad \text{add one} \\
\hline
6\ 2\ 2\ 4 & \leftarrow \quad \text{tens complement}
\end{array}
$$

Now add the tens complement to 5871:

$$
\begin{array}{r}
5\ 8\ 7\ 1 \\
+6\ 2\ 2\ 4 \\
\hline
\end{array}
$$

$$
\text{delete} \quad \rightarrow \quad ①2\ 0\ 9\ 5
$$

Thus, $5871 - 3776 = 2095$. Observe that this is indeed true, since $3776 + 2095 = 5871$.

Example 16.9 Evaluate $(110111)_2 - (11011)_2$.

Solution: Changing 0's to 1's and 1's to 0's in $(11011)_2$, we get $(00100)_2$. Adding 1 to it, we have $(00101)_2$ as the twos complement of $(11011)_2$. Now add this to $(110111)_2$:

$$
\begin{array}{r}
1\ 1\ 0\ 1\ 1\ 1 \\
+\ \ 0\ 0\ 1\ 0\ 1 \qquad \leftarrow \quad \text{twos complement} \\
\hline
\end{array}
$$

$$
\text{delete} \quad \rightarrow \quad ①1\ 1\ 1\ 0\ 0
$$

Thus, $(110111)_2 - (11011)_2 = (11100)_2$.

Observe that these examples also show that an algorithm for an operation need not be unique.

Exercise 16.4

1. Find the decimal equivalent of each of the following.
   a) $(23)_8$ **b)** $(23)_{16}$ **c)** $(130.5)_8$
   d) $(0.01)_2$ **e)** $(10.101)_2$ **f)** $(BAD.E)_{16}$
2. Convert each of the following decimal numbers to base two.
   a) 23   **b)** 101   **c)** 343
   d) 13.25   **e)** 23.125   **f)** 41.625
3. Express the following binary numerals in the octal system.
   a) 1011   **b)** 10110   **c)** 11011110000
4. Rewrite the binary numerals of problem 3 in base sixteen.
5. Convert your answers in problem 2 to base eight.
6. Convert your answers in problem 2 to base sixteen.
7. Find the tens complement of the following.
   a) 31   **b)** 257   **c)** 1250   **d)** 1976
8. What are the twos complements of the following numbers?
   a) 101   **b)** 1011   **c)** 11001
   d) 110011.01
9. Evaluate each of the following using tens complements.
   a) $39-23$ **b)** $125-87$ **c)** $1976-1250$
10. Simplify each of the following using twos complements.
   a) $101-11$ **b)** $1011-101$ **c)** $10110-1101$
11. Evaluate each of the following.
   a) $(127)_8+(72)_8$ **b)** $(127)_8-(72)_8$
   c) $(879)_{16}+(76)_{16}$ **d)** $(CBA)_{16}-(ABC)_{16}$
   e) $(23)_8 \cdot (45)_8$ **f)** $(96)_{16} \cdot (35)_{16}$
12. Rewrite the binary numerals of problem 3 in base four.

16.5 SUMMARY AND COMMENTS

This chapter provides us with a brief introduction to one of man's greatest technological achievements: the computer. We are living in a society that has become very computer-dependent. Computers are used in many different ways, as listed in the introduction.

One of the fast-developing industries in the nation is the computer industry, which is overwhelmingly dominated by IBM. About 70% of all computers manufactured in this country are used here.

The concept of *time-sharing*, developed at Massachusetts Institute of Technology (MIT) in the early 1960s, is becoming increasingly popular. Roughly speaking, it means the sharing of the time of a computer by several people. This is achieved by using terminals that may be located far away from the computer

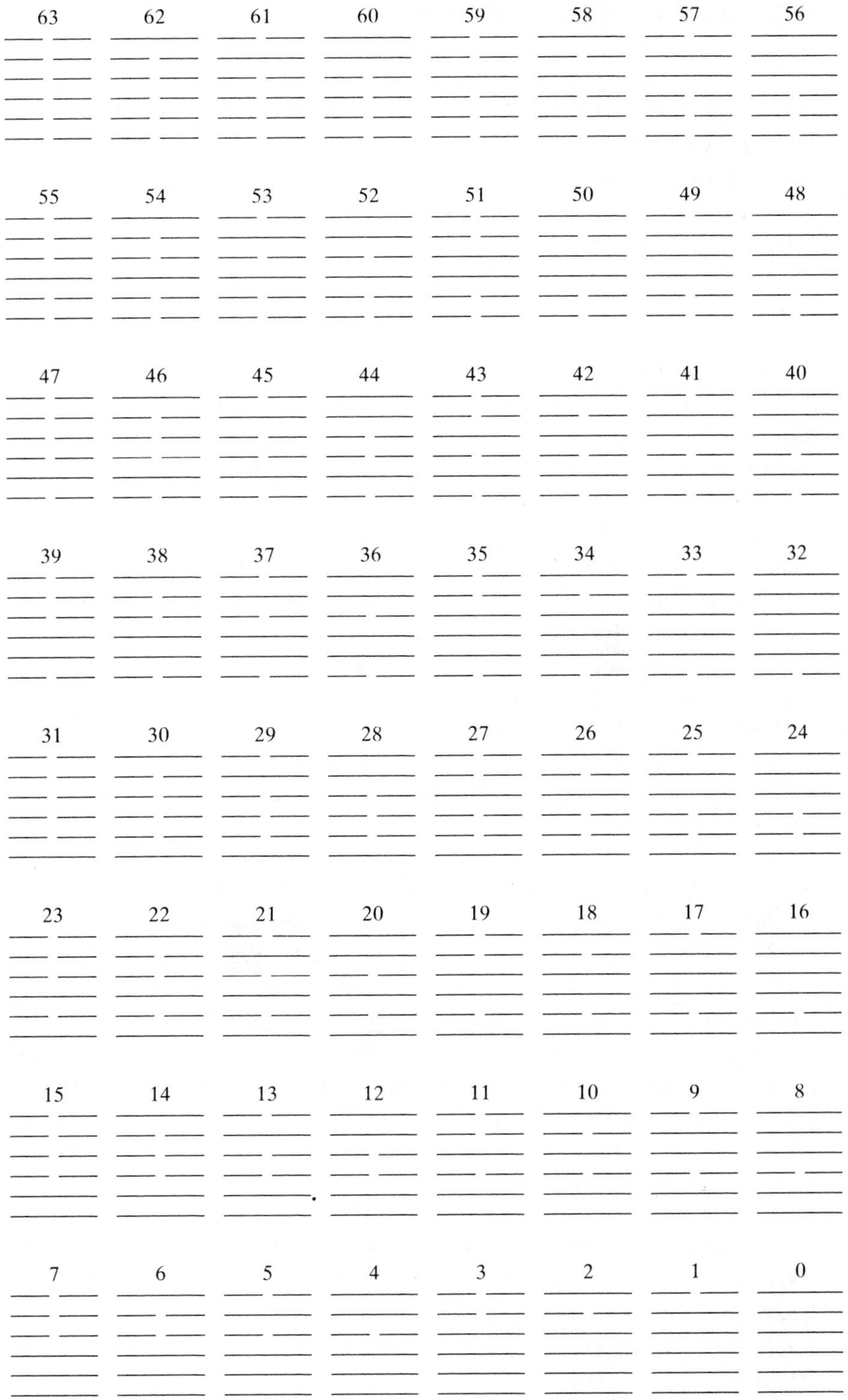

Fig. 16.11   The Fu Hsi Sequence that Corresponds to Binary Numbers 0 through 63

and are connected to it via telephone lines. As a result, the terminal-user will have direct access to the computer and will not have to wait for hours or even days to get the results.

Section 16.1 provides us with a brief history of the origin of modern computers starting with the abacus. The names Pascal, Leibniz, Babbage, Jacquard, Hollerith, Aiken, Eckert, Mauchly, Neumann, etc., will be remembered forever for their remarkable contributions that led to the development of modern high-speed computers.

In Section 16.2, we outlined the five basic components of a typical digital computer: the central processing unit, consisting of arithmetic, memory, and control units, and the I/O device consisting of input and output units.

Flow charts are a convenient vehicle to arrive at algorithms, especially for complicated problems.

We pointed out the need for artificial languages to communicate with a computer. FORTRAN, ALGOL, COBOL, and BASIC are a few of them.

In Section 16.4, we discussed algorithms to convert fractions in base ten to base two and to express a binary numeral in base eight and base sixteen. We observed that subtraction can be achieved by adding the complement of the subtrahend to the minuend and then dropping the leading digit.

One of the strongest arguments against any proliferation of the computer industry is that the government and other interested parties might use the computer as a tool to invade the privacy of man. Computers can be used to store personal and confidential data of every individual on file, so that parties that have access to these files could use the information against him directly or indirectly. There is no sense in blaming a computer for this. Although a computer can be instructed to solve highly complicated problems, it is not intelligent enough to distinguish between right and wrong; it does not have any control over itself. It executes man's instructions like a faithful slave. Consequently, it is up to man not to misuse the capabilities of this fabulous machine.

In any event, the influence of the computer on our society is so great and so rapidly growing that we cannot predict in how many different ways the computer will be used in the years to come.

SUGGESTED READING

Arsdel, J. V., and J. Lasky, "A Two-Dimensional Abacus—The Papy Minicomputer," *The Arithmetic Teacher*, vol. 19 (Oct. 1972), pp. 445–451.

Desmonde, W. H., *Computers and Their Uses*, Prentice-Hall, Inc., Englewood Cliffs, N.J. (1964).

Forsythe, A. L., and D. H. Stansbury, "Bobby and a Computer," *The Arithmetic Teacher*, vol. 18 (Feb. 1971), pp. 88–93.

Gardner, M., "The Abacus: Primitive but Effective Digital Computer," *Scientific American*, vol. 222 (Jan. 1970), pp. 124–127.

Miller, C. D., and V. E. Heeren, *Mathematical Ideas, an Introduction* (2nd ed.), Scott, Foresman and Co., Glenview, Ill. (1973), pp. 267–301.

Suppes, P., "The Uses of Computers in Education," *Scientific American*, vol. 215 (Sept. 1966), pp. 207–220.

Answers to Selected Exercises

CHAPTER 1

Exercise 1.1

1. a) yes   **b)** yes   **c)** no   **d)** no

   e) yes   **f)** no   **g)** no

3. a) $\{x \mid x$ is a nation in the world$\}$

   b) $\{x \mid x$ is a day of the week that begins with the letter $T\}$

   c) $\{x \mid x$ is a month of the year that ends in $y\}$

   d) $\{x \mid x$ is a member of the United Nations$\}$

   e) $\{x \mid x$ is a permanent member of the U.N. Security Council$\}$

5. $\{5, \text{Jim}\}$

7. a) F   **b)** F   **c)** F   **d)** F

   e) T   **f)** T   **g)** T

Exercise 1.2

1. a) T   **b)** F   **c)** T   **d)** T

   e) T   **f)** F   **g)** F   **h)** F

   i) F   **j)** T   **k)** T   **l)** T

   m) T   **n)** F   **o)** F   **p)** T

   q) T   **r)** T   **s)** F   **t)** T

   u) T   **v)** T   **w)** F   **x)** T

3. a) F   **b)** F   **c)** T   **d)** F

   e) T   **f)** T   **g)** T   **h)** T

5. $A = D, B = H, C = F,$ and $E = G$

7. a) no   **b)** no   **c)** yes   **d)** no

9. yes   **11.** yes

13. a) T   **b)** F   **c)** F   **d)** F

   e) T   **f)** T   **g)** T   **h)** F

   i) F

15. a) no   **b)** yes   **c)** no   **d)** no

17. The elements of the set $\{a,b\}$ are a and b, whereas those of $\{\{a\},\{b\}\}$ are $\{a\}$ and $\{b\}$.

19. a) $\{\{x\},\{y\}\}$

   b) $\{\{a\},\{b\},\{c\},\{a,b\},\{a,c\},\{b,c\},A\}$

   c) none

Exercise 1.4

1.

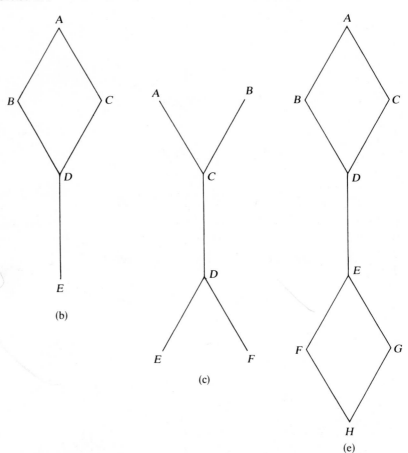

(b)

(c)

(e)

2. a) $A = \{1,2,3\}$, $B = \{1,2\}$, $C = \{2,3\}$, and $D = \{2\}$

   c) $A = \{1,2,3,4,5\}$, $B = \{2,3,4,5\}$, $C = \{1,2,3,4\}$, $D = \{2,3\}$, and $E = \{4\}$

   f) $A = \{a,b,c,d,e\}$, $B = \{a,b,c,d,f\}$, $C = \{a,b,c,d\}$, $D = \{a,b,c\}$, $E = \{a,b,d\}$, $F = \{a,b\}$ and $G = \{a\}$

3. a) $p(A) = \{\emptyset, A\}$ **b)** $p(A) = \{\emptyset, \{1\}, \{2\}, A\}$

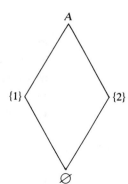

Exercise 1.5

1. a) $\{1,2,3,4,6,8\}$ **c)** A

    e) $\{1,2,3,4,6,8,9\}$ **f)** $\emptyset$

2. a) $\{a,b,n,r\}$ **c)** $\{a,b,c,d,n,r\}$

    e) B **f)** B

3. a) T   **b)** F   **c)** T   **d)** T

    e) T   **f)** T   **g)** F   **h)** T

4. a) $\{x,z\}$ **b)** $\{1\}$ **c)** $\emptyset$

Exercise 1.6

1. a) $7 \in \{1,3,5,7\}$ **b)** $A \subseteq B$ **c)** $A \subset B$

    d) $A \cap B = \emptyset$ **e)** $A \cap B \neq \emptyset$

2. a) $\{2,4\}$ **c)** $\emptyset$ **e)** $\{4\}$ **f)** $\emptyset$

3. a) $\{a,b,l,m\}$ **c)** $\{a\}$ **e)** A **f)** $\emptyset$

4. a) T   **b)** T   **c)** F   **d)** F

    e) F   **f)** F   **g)** T   **h)** F

Exercise 1.7

1. a) $\{1,2,3,4,8\}$ **c)** A

    e) $\{1,2,3,4,8\}$ **f)** $\{1,2,3,4,6,8\}$

2.

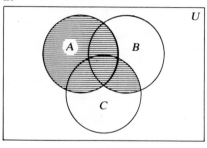

(a)

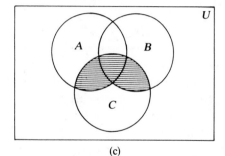

(c)

3. a) {1,3} **b)** {*a*} **c)** ∅

Exercise 1.8

1. a) {1,3} **c)** *A* **d)** ∅
  f) ∅ **g)** {1,3,4} **i)** {1,2,3}
3. a) {*c,y*} **b)** {*a,b,c*} **c)** {*b,c,x*}
  d) *U* **e)** {*a,b,c,y*} **f)** ∅
4. a) {2,6,7} **c)** {1,3} **d)** {0,2,4,5,6,7}
  f) {2} **g)** {0,2,4,5,6,7} **i)** {2,4,5,6,7}
  j) {0,1,2,3,4,5,6} **l)** {2,6,7} **m)** {0,4,5,6,7}
5. *A*′ = B and B′ = A.
6. a) *U* **c)** *A* **d)** *A*′
  f) *A* **g)** ⊘ **i)** *U*
  j) *A* **l)** *A* **m)** ∅
7. a) *A* **b)** *A* **c)** ∅
  d) ∅ **e)** *A*′ **f)** *A*′
  g) *A*′ **h)** *A*′ **i)** *A*

8.

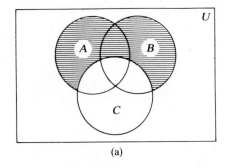

(a)

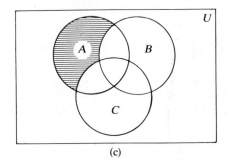

(c)

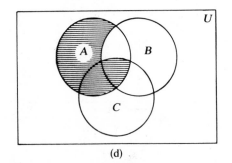

(d)

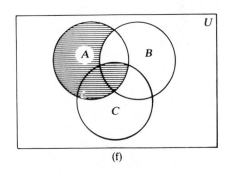

(f)

9. a) *A* ∪ *B* ∪ *C* **b)** *A* ∩ *B* ∩ *C* **c)** *A* ∪ *B* ∪ *C* − *A* ∩ *B* ∩ *C*
10. a) T **b)** F **c)** F **d)** F
  e) F **f)** F **g)** T **h)** F
  i) T **j)** T

Exercise 1.9

1. a) $A = \{a,b,c,d\}$, $B = \{c,d,e\}$

   b) $U = \{a,b,c,d,e,f,g\}$

2. 50

5. a) $A = \{1,2,3,5\}$, $B = \{2,4,5,6\}$, $C = \{3,4,5,7\}$

   b) $U = \{0,1,2,3,4,5,6,7\}$

6. a) 45	**b)** 15	**c)** 132	**d)** 147
e) 60	**f)** 35	**g)** 80	**h)** 10
i) 45	**j)** 13		

Exercise 1.10

1. a) T	**b)** F	**c)** T	**d)** F
e) T	**f)** F	**g)** T	**h)** T

3.

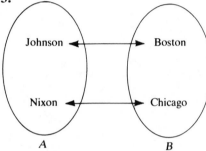

5.

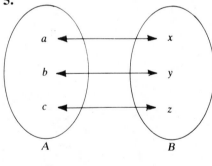

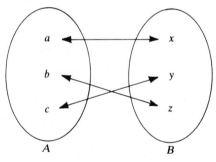

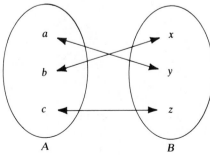

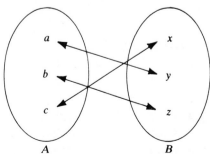

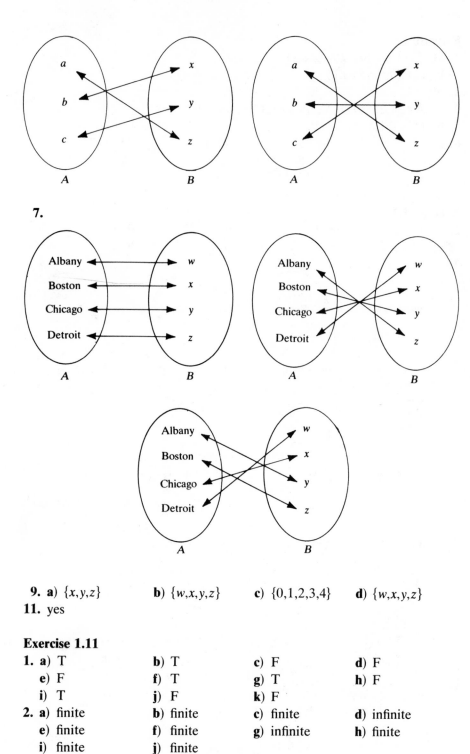

7.

9. a) $\{x,y,z\}$ **b)** $\{w,x,y,z\}$ **c)** $\{0,1,2,3,4\}$ **d)** $\{w,x,y,z\}$

11. yes

Exercise 1.11

1. a) T   **b)** T   **c)** F   **d)** F

    e) F   **f)** T   **g)** T   **h)** F

    i) T   **j)** F   **k)** F

2. a) finite   **b)** finite   **c)** finite   **d)** infinite

    e) finite   **f)** finite   **g)** infinite   **h)** finite

    i) finite   **j)** finite

Exercise 1.12

1. **a)** (1) $A \times B = \{(x,x),(x,y)\}$
   (2) $A \times B = \{(a,1),(b,1),(c,1)\}$
   (3) Let's use the first letters of the names for convenience:
   $A \times B = \{(K,M),(K,T),(K,C),(J,M),(J,T),(J,C),(N,M),(N,T),(N,C)\}.$

b)

B

×	x	y
(1) A x	(x,x)	(x,y)

B

×	1
a	$(a,1)$
(2) A b	$(b,1)$
c	$(c,1)$

B

×	M	T	C
K	(K,M)	(K,T)	(K,C)
(3) A J	(J,M)	(J,T)	(J,C)
N	(N,M)	(N,T)	(N,C)

3. **a)** $A = \{0,1\}$, $B = \{1,2\}$
   b) $A = \{a,b,c\}$, $B = \{x,y,z\}$
   c) $A = \varnothing$ or $B = \varnothing$

4. **a)** F   **b)** F   **c)** T   **d)** T
   e) F   **f)** F   **g)** F   **h)** T
   i) T   **j)** T   **k)** T   **l)** T
   m) T   **n)** T

5. Since $\{a,b\} = \{a,c\}$, we have $b = c$. Therefore, $(a,b) = (a,c)$.

9. **a)** (1) $A \times (B \cup C) = \{(a,b),(a,c),(a,d),(b,b),(b,c),(b,d)\}$
                   $= (A \times B) \cup (A \times C)$

(2) $A \times (B \cap C) = \{(a,c),(b,c)\} = (A \times B) \cap (A \times C)$

b) no
c) no

11. 6; let's, for convenience, label the boys by a and b, and the chairs by 1, 2, and 3. Then the tree diagram looks as follows.

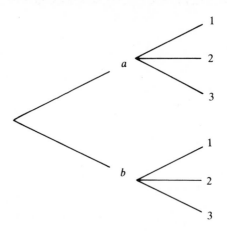

CHAPTER 2

Exercise 2.1

1. a) yes
b) no
c) yes
d) yes
e) yes
f) yes
g) no

2. a) compound
b) simple
c) simple
d) compound
e) compound
f) simple
g) compound
h) compound

3. a) Mathematics can be fun; fun can be mathematics.
b) Fred likes hunting; Fred likes swimming.
c) $2 = 3; 5 = 5$.
d) The house is blue; the car is red.
e) Chicago is in Canada; Paris is in England.

Exercise 2.2

1. a) $\sim p$
b) $p \wedge q$
c) $p \vee q$
d) $p \rightarrow q$
e) $p \leftrightarrow \sim q$
f) $p \rightarrow \sim q$
g) $\sim p \wedge \sim q$
h) $\sim p \wedge q$
i) $\sim p \rightarrow \sim q$
j) $\sim (p \vee q)$

2. a) It is not cold.
b) It is cold and it is snowing.
c) It is cold or it is snowing.
d) It is not cold and it is snowing.
e) It is cold or it is not snowing.

f) It is false that it is cold and it is snowing.

g) It is false that it is cold or it is snowing.

h) It is not cold or it is not snowing.

i) It is not cold and it is not snowing.

3. a) If $x = 2$, then $x^2 = 4$.

c) If a triangle is equiangular, then it is isosceles.

e) If $x = 4$, then $x^2 = 16$.

g) If $x = 1$, then $x^2 = 1$.

4. a) Canada is not the largest country in the world.

c) John is not tall or he is not smart.

e) Ron is handsome.

g) $1 + 1 = 0$ or $2 + 3 = 6$.

5. a) T   **b)** T   **c)** T

   d) T   **e)** T   **f)** F

6. a) T   **b)** T   **c)** F   **d)** F

   e) T   **f)** F   **g)** T   **h)** T

7. a) F   **b)** F   **c)** T   **d)** T

   e) T   **f)** F   **g)** T   **h)** T

8.

p	q	$\sim q$	$p \wedge \sim q$
T	T	F	F
T	F	T	T
F	T	F	F
F	F	T	F

(a)

p	q	$\sim p$	$\sim q$	$\sim p \vee \sim q$
T	T	F	F	F
T	F	F	T	T
F	T	T	F	T
F	F	T	T	T

(c)

p	q	$p \wedge q$	$\sim (p \wedge q)$
T	T	T	F
T	F	F	T
F	T	F	T
F	F	F	T

(e)

p	q	$\sim p$	$\sim q$	$\sim p \rightarrow \sim q$
T	T	F	F	T
T	F	F	T	T
F	T	T	F	F
F	F	T	T	T

(g)

9. a) Converse: If Paris is in France, then London is in England.

   Inverse: If London is not in England, then Paris is not in France.

   Contrapositive: If Paris is not in France, then London is not in England.

10. a) yes   **b)** yes   **c)** no   **d)** yes

   e) yes   **f)** no   **g)** yes   **h)** no

Exercise 2.3

1. F

3. q must be true, for if q is false, then $p \to q$ is false.

5. a) T   **b)** F   **c)** T   **d)** F   **e)** F

6. a) T   **b)** T   **c)** T.   **d)** F   **e)** F

    f) F   **g)** F   **h)** F   **i)** F   **j)** F

7.

p	$\sim p$	$\sim(\sim p)$
T	F	T
F	T	F

(a)

p	q	$p \to q$	$\sim(p \to q)$	$\sim q$	$p \wedge \sim q$
T	T	T	F	F	F
T	F	F	T	T	T
F	T	T	F	F	F
F	F	T	F	T	F

(e)

9. a) $\sim(\sim p \wedge \sim q) \equiv \sim(\sim p) \vee \sim(\sim q) \equiv p \vee q$

Exercise 2.4

1.

p	$\sim p$	$\sim(\sim p)$	$\sim(\sim p) \leftrightarrow p$
T	F	T	T
F	T	F	T

(a)

p	q	$p \wedge q$	$(p \wedge q) \to p$
T	T	T	T
T	F	F	T
F	T	F	T
F	F	F	T

(c)

p	q	$p \to q$	$p \wedge (p \to q)$	$[p \wedge (p \to q)] \to q$
T	T	T	T	T
T	F	F	F	T
F	T	T	F	T
F	F	T	F	T

(e)

2. a) invalid   **b)** invalid   **c)** valid   **d)** valid

    e) invalid   **f)** valid   **g)** valid   **h)** valid

Exercise 2.5

1. a) *Proof:*

$\sim q$ is true	*given*
q is false	*definition of negation*
$p \vee q$ is true	*given*
p is true	*definition of disjunction*

c) **Proof:**

r is true	*given*
$\sim r$ is false	*definition of negation*
$p \rightarrow \sim r$ is true	*given*
p is false	*definition of implication*
$p \vee q$ is true	*given*
q is true	*definition of disjunction*

h) **Proof:** Let p: Alice is rich

q: Alice is beautiful

r: Alice is married

Now the theorem can be rewritten

$$p \leftrightarrow q$$
$$r \vee \sim q$$
$$\underline{\quad \sim r \quad}$$
$$\therefore \sim p$$

$\sim r$ is true	*given*
r is false	*definition of negation*
$r \vee \sim q$ is true	*given*
$\sim q$ is true	*definition of disjunction*
q is false	*definition of negation*
$p \leftrightarrow q$ is true	*given*
p is false	*definition of biconditional*
$\sim p$ is true	*definition of negation*

Thus, that Alice is not rich is a true statement and agrees with the given conclusion. Hence the theorem.

2. b) Proof: Assume that p is false.

$p \leftrightarrow q$ is true	*given*
q is false	*definition of biconditional*
$\sim q$ is true	*definition of negation*
$\sim q \rightarrow r$ is true	*given*
r is true	*law of detachment*

But this is a contradiction since r is given to be false. Thus, our assumption is false and the theorem holds.

f) **Proof:**   Assume $\sim p$ is false; that is, p is true.

$p \rightarrow q$ is true	*given*
q is true	*law of detachment*
$q \rightarrow r$ is true	*given*
r is true	*law of detachment*

This is a contradiction since $\sim r$ is given to be true. Hence the theorem.

h) ***Proof:***   Let p: Carol is a baby

                q: Carol is illogical

                r: Carol is happy

Now the theorem can be rewritten

$$p \leftrightarrow q$$
$$q \vee {\sim}r$$
$$\underline{r}$$
$$\therefore p$$

Assume that p is false.

$p \leftrightarrow q$ is true	*given*
q is false	*definition of biconditional*
$q \vee {\sim}r$ is true	*given*
${\sim}r$ is true	*definition of disjunction*

But this gives a contradiction since we are given that r is true.

3. a) New York City   **b)** January   **c)** Canada   **d)** 3

Exercise 2.7

1. a) $(A \vee B') \vee B$ **b)** $(A \wedge B') \vee (A' \wedge B)$

    c) $[A \wedge (A' \vee B)] \vee (A \wedge B')$ **d)** $(A \vee B') \wedge (A' \vee B)$

    e) $A \wedge [(B \wedge C') \vee B'] \vee (A' \vee C)$ **f)** $(A' \wedge C) \vee (B' \vee C') \vee (A \wedge B)$

2.

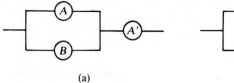

(a)   (e)

3. a) closed   **b)** closed   **c)** open

    d) open   **e)** closed   **f)** open

CHAPTER 3

Exercise 3.1

1. a) 3   **b)** 3   **c)** 7   **d)** 26

    e) 2   **f)** 5   **g)** 5   **h)** 4

    i) 3   **j)** 3

3. a) 1   **b)** 2   **c)** 2

    d) 1   **e)** 2   **f)** 3

4. a) $\{1,2,3,4\}$; 4   **b)** $\{1,2,3,4\}$; 4   **c)** $\{1,2,3\}$; 3

    d) $\{1,2,3,4,5,6\}$; 6   **e)** $\{1,2,3,4,5\}$; 5   **f)** $\{1\}$; 1

    g) $\{1,2,3,...,50\}$; 50   **h)** $\{1,2,3,4,5,6,7,8\}$; 8

5. a) 6   **b)** 4   **c)** 2   **d)** 8
   e) 0   **f)** 6   **g)** 0   **h)** 0
   i) 6
6. a) T   **b)** T   **c)** T   **d)** F
   e) F   **f)** T   **g)** T   **h)** T
   i) T   **j)** F

Exercise 3.3

1. a) Let $A = \{a\}$ and $B = \{x,y,z\}$. Then $A \cup B = \{a,x,y,z\}$, $A \cap B = \varnothing$, $c(A \cup B) = 4$, $c(A) = 1$, $c(B) = 3$. Then $c(A \cup B) = c(A) + c(B)$, and $4 = 1 + 3$.

   b) Let $A = \{a,b,c\}$ and $B = \{x,y\}$. Then $A \cup B = \{a,b,c,x,y\}$, $A \cap B = \varnothing$, $c(A \cup B) = 5$, $c(A) = 3$, $c(B) = 2$. Thus $c(A \cup B) = c(A) + c(B)$, and $5 = 3 + 2$.

2. a) $2 + (3 + 5) = 2 + 8 = 10$

     $(2 + 3) + 5 = 5 + 5 = 10$

     Thus, $2 + (3 + 5) = (2 + 3) + 5$.

   b) $1 + (2 + 3) = 1 + 5 = 6$

     $(3 + 1) + 2 = 4 + 2 = 6$

     Thus, $1 + (2 + 3) = (3 + 1) + 2$.

3. a) yes   **b)** no   **c)** yes   **d)** yes
5. a) closure property   **b)** commutative property
   c) commutative property   **d)** commutative property
   e) 0 is the additive identity   **f)** cancellation property
   g) addition property   **h)** commutative property
   i) associative property   **j)** associative property
6. a) No, since $1 + 1 = 2 \notin \{0,1\}$.
   b) Yes, since $0 + 0 = 0 \in \{0\}$.
8. a) T   **b)** T   **c)** T   **d)** F
   e) T   **f)** T   **g)** F   **h)** T
   i) T   **j)** T   **k)** T
10. a) 20   **b)** 15   **c)** 10
   d) 5   **e)** 0   **f)** 14
11. No

Exercise 3.4

1. a) Let $A = \{a,b,c\}$ and $B = \{x\}$. Then $A \times B = \{(a,x),(b,x),(c,x)\}$, $c(A \times B) = c(A) \cdot c(B)$, and so $3 = 3 \cdot 1$.

   b) Let $A = \{a,b\}$ and $B = \{a,x,y,z\}$. Then $A \times B = \{(a,a),(a,x),(a,y),(a,z),(b,a),(b,x),(b,y),(b,z)\}$, hence $c(A \times B) = c(A) \cdot c(B)$, i.e., $8 = 2 \cdot 4$.

2. a) $2 \cdot (3 \cdot 4) = 2 \cdot 12 = 24$

$(2 \cdot 4) \cdot 3 = 8 \cdot 3 = 24$

Thus, $2 \cdot (3 \cdot 4) = (2 \cdot 4) \cdot 3$.

b) $2 \cdot (3 \cdot 6) = 2 \cdot 18 = 36$

$(6 \cdot 2) \cdot 3 = 12 \cdot 3 = 36$

Thus, $2 \cdot (3 \cdot 6) = (6 \cdot 2) \cdot 3$.

4. a) $A \times A = \{(0,0),(0,1),(1,0),(1,1)\}$

$c(A \times A) = 4 = 2 \cdot 2 = c(A) \cdot c(A)$

b) $A \times A = \{(\varnothing,\varnothing)\}$

$c(A \times A) = 1 = 1 \cdot 1 = c(A) \cdot c(A)$

e) $A = \{\text{January,June,July}\}$

$A \times A = \{(\text{January,January}),(\text{January,June}),(\text{January,July}),$
$(\text{June,January}),(\text{June,June}),(\text{June,July}),(\text{July,January}),$
$(\text{July,June}),(\text{July,July})\}$

$c(A \times A) = 9 = 3 \cdot 3 = c(A) \cdot c(A)$

5. a) closure property   **b)** commutative property

c) commutative property   **d)** 1 is the multiplicative identity

e) associative property   **f)** cancellation property

g) product law   **h)** multiplication property

i) cancellation property   **j)** multiplication property

7. a) 2   **b)** 1   **c)** 0

  d) 2   **e)** any whole number   **f)** none

8. a) T   **b)** F   **c)** T   **d)** F

  e) F   **f)** T   **g)** T   **h)** T

  i) T   **j)** F

12. $x = 0$ **14.** No

Exercise 3.5

1. a) 17   **b)** 16   **c)** 11   **d)** 24

  e) 41   **f)** 16   **g)** 13   **h)** 38

2. a) $a = 2, b = 1, c = 1$

$a(b+c) = 2(1+1) = 2 \cdot 2 = 4$

$ab + ac = 2 \cdot 1 + 2 \cdot 1 = 2+2 = 4$

Thus, $a(b+c) = ab + ac$.

b) $a = 1, b = 2, c = 3$

$a(b+c) = 1(2+3) = 1 \cdot 5 = 5$

$ab + ac = 1 \cdot 2 + 1 \cdot 3 = 2+3 = 5$

Thus, $a(b+c) = ab + ac$.

4. a) commutative property of addition
   b) commutative property of multiplication
   c) distributive property of multiplication over addition
   d) distributive property of multiplication over addition
   e) commutative property of multiplication
   f) commutative property of addition

5. a) T	**b)** F	**c)** T
d) T	**e)** F	**f)** F
g) F	**h)** T	**i)** F

8. a) 1	**b)** 2	**c)** 0
d) 2	**e)** 1	**f)** 3

9. $a = 0, b = 1, c = 2; a = 0, b = 3, c = 5; a = 0, b = 6, c = 8$

10.

$a \cdot 0 + 0 = a \cdot 0$	*0 is the additive identity*
$a \cdot 0 + 0 = a(0+0)$	*0 is the additive identity*
$a \cdot 0 + 0 = a \cdot 0 + a \cdot 0$	*distributive property*
$0 = a \cdot 0$	*cancellation property of addition*

Thus, $a \cdot 0 = 0$

Exercise 3.6

1. a) $3 < 8$ since there is a natural number $x = 5$ such that $3 + 5 = 8$.
   b) $11 < 14$ since we can find a natural number $x = 3$ such that $11 + 3 = 14$.

2. $A \cup B = \{0,1,2,3,4,5\}$, $A \cap B = \{1,2,3,4\}$, $A - B = \{0\}$, $B - A = \{5\}$

4. a) 1	**b)** 1, 2	**c)** 1, 2, 3
d) 1, 2	**e)** 1, 2, 3	**f)** 1

5. a) 2	**b)** 4	**c)** 5
d) 4	**e)** 3	**f)** 3

6. a) F	**b)** T	**c)** F	**d)** F
e) F	**f)** F	**g)** F	**h)** T
i) T	**j)** T	**k)** T	**l)** T

10. $x = 0$ **11.** $x = 1$

14. a) $\{0,1\}$	**b)** $\{0\}$	**c)** $\{0,1,2,3,4\}$
d) $\{0,1,2\}$	**e)** $\{0,1,2,3\}$	**f)** $\{0,1,2,3,4,5,6,7,8\}$

15. $\{27,28,29,...,78,79\}$ **17.** 12

Exercise 3.8

1. a) $\{3\}$

   b) $\{5\}$

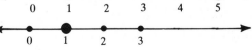

   c) $\{1\}$

d) {0}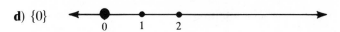

3. 26 dollars   **5.** 31 and 38
7. 5 years   **9.** 45
11. 43 miles
13. a) {0,1,2,3}

b) {0,1,2,3,4}

c) {0,1,2,3}

Exercise 3.9
1. a) 3   **b)** 2   **c)** not solvable
   d) 5   **e)** not solvable   **f)** not solvable
2. a) $2(5-2) = 2 \cdot 3 = 6$

$2 \cdot 5 - 2 \cdot 2 = 10 - 4 = 6$

Thus, $2(5-2) = 2 \cdot 5 - 2 \cdot 2$.

b) $(3+5) - 2 = 8 - 2 = 6$

$3 + (5-2) = 3 + 3 = 6$

Thus, $(3+5) - 2 = 3 + (5-2)$.

3. a) F   **b)** F   **c)** F   **d)** T
   e) F   **f)** F   **g)** T   **h)** T
   i) T   **j)** T
5. yes   **7.** 25 years
9. 42 degrees   **11.** 6

CHAPTER 4

Exercise 4.1
1. a) -5 **c)** 0   **e)** -10 **i)** -4
2. a) -3 **b)** 0   **c)** 5
   d) 8   **e)** -8 **f)** 0
3. a) T   **b)** F   **c)** T
   d) T   **e)** T

Exercise 4.2
1. a) 2   **b)** 6   **c)** 8   **d)** -15
   e) -3 **f)** 0   **g)** $a-1$ **h)** 3
2. a) $2+3+5 = 2+(3+5) = 2+8 = 10$

$5+3+2 = 5+(3+2) = 5+5 = 10$

Thus, $2+3+5 = 5+3+2$.

b) $(-2)+(-3)+4=(-2)+[(-3)+4]=-2+1=-1$

3. a) 8   **b)** -5 **c)** 8
    d) -3 **e)** 4   **f)** 15

5. a) T   **b)** T   **c)** T   **d)** F
    e) T   **f)** T   **g)** F   **h)** T
    i) T   **j)** T

7. a) 0 is the additive identity   **b)** closure property
    c) commutative property   **d)** addition property
    e) commutative property   **f)** associative property
    g) cancellation property   **h)** associative property
    i) associative property   **j)** commutative property

9. 14 degrees below zero   **11.** 50 miles

13. a) a **b)** $-b$ **c)** a
    d) $-a$ **e)** $a-c$ **f)** 0

Exercise 4.3

1. a) 0   **b)** -2 **c)** -9
    d) 4   **e)** -16 **f)** 1
    g) -10 **h)** 1   **i)** -12

2. a) 11   **b)** 3   **c)** -3 **d)** -13

4. $a=0$

6. a) T   **b)** F   **c)** F   **d)** F
    e) T   **f)** T   **g)** T   **h)** T
    i) T   **j)** F

8. $b=0$

11. a) $a+b$ **b)** $b-a$ **c)** b
    d) b **e)** a **f)** $-b$
    g) b **h)** a **i)** 0

13. 17 degrees   **15.** $185
17. 12 degrees

Exercise 4.4

1. a) -15 **b)** -14 **c)** 35   **d)** 0
    e) 1   **f)** -120 **g)** 36   **h)** 105

2. a) $2\cdot 3\cdot 5=2\cdot(3\cdot 5)=2\cdot 15=30$
      $5\cdot 3\cdot 2=5\cdot(3\cdot 2)=5\cdot 6=30$
      Thus, $2\cdot 3\cdot 5=5\cdot 3\cdot 2$.

    b) $(-2)\cdot(-3)\cdot 4=(-2)\cdot[(-3)\cdot 4]=(-2)\cdot(-12)=24$

3. a) $\{-5\}$

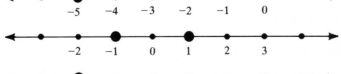

i) $\{-2\}$

5. $b = 0$

6. a) T   **b)** F   **c)** T   **d)** T

    e) T   **f)** F   **g)** T   **h)** F

    i) T   **j)** T   **k)** F   **l)** F

8. a) closure property   **b)** 1 is the multiplicative identity

    c) cancellation property   **d)** associative property

    e) commutative property   **f)** multiplication property

    g) commutative property   **h)** associative property

    i) commutative property   **j)** associative property

9. a) $2b$ **b)** $-2b$ **c)** $2a$ **d)** $2b$

12. $a = 0$ or $b = 1$ **13.** 9 degrees

15. 18 degrees   **17.** $675

Exercise 4.5

1. a) F   **b)** T   **c)** T   **d)** F

    e) F   **f)** T   **g)** T   **h)** F

    i) T   **j)** T

4. a) 8   **b)** 3   **c)** 1

    d) 5   **e)** 2   **f)** any integer

5. 12 degrees   **7.** 7, 8

9. 6 years

Exercise 4.6

1. a) $3 < 7$ since there is a positive integer $x = 4$ such that $3 + 4 = 7$.

    b) $-8 < 3$ since we can find a positive integer $x = 11$ such that $-8 + 11 = 3$.

2. $A \cup B = \{\pm 4, \pm 3, \pm 2, \pm 1, 0\}$, $A \cap B = \{\pm 2, \pm 1, 0\}$, $A - B = \{-4, -3\}$, $B - A = \{3, 4\}$

3. a) $\{-5, -4, -3, -2, -1, 0\}$ **b)** $\{-4, -3, -2, -1\}$

    c) $\{-5\}$ **d)** $\{-7, -6, 0, 1, 2, 3, 4\}$

    e) $\{-5\}$ **f)** $\{-1\}$

    g) $\{-5, -4, -3, -2\}$ **h)** $\{1, 2, 3\}$

    i) $\{-7, -6, -5, \pm 4, \pm 3, \pm 2, 1\}$

4. a) $\{x \in I \mid x < 3\}$

    e) $\{x \in I \mid x > -5\}$

    h) $\{-1, 0, 1, 2\}$

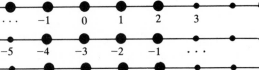

5. a) T   **b)** T   **c)** F

    d) T   **e)** T   **f)** F

    g) F   **h)** F   **i)** T

    j) F

7. a) (1) {1,2,3}   (2) {1,2}   (3) {0,1,2,3}

b) (1)

(2)

(3)

9. {−3,−2,−1,...,15,16}   **11.** 6

Exercise 4.7

1. a) 3   **b)** −5   **c)** −3   **d)** 3
2. a) 7   **b)** −7   **c)** 12   **d)** −5
3. a) F   **b)** F   **c)** F   **d)** F
   e) F   **f)** F   **g)** F   **h)** F
   i) F   **j)** T

5. 50 cents   **7.** 5 hours
9. 6 miles   **10.** 5

12. a) No, since $1 \div 0$ is undefined.
   b) No, since $1 \div 2$ does not exist in the set {1,2}.

CHAPTER 5

Exercise 5.1

1. a) 11, 22, 33   **b)** −2, −4, −6, −8, −10
   c) 3, 6, 9   **d)** −3, −6, −9, −12, −15
3. a) {0,1,2,3,4}   **b)** {0,1,2,3,4,5,6}
   c) {0,1}   **d)** {0,1,2,3,4,5,6,7,8,9,10,11}
4. $a = \pm b$
5. a) T   **b)** T   **c)** T   **d)** F
   e) F   **f)** F   **g)** F   **h)** T
   i) F   **j)** T
7. a) $q = 5, r = 3$ **b)** $q = 7, r = 1$
   c) $q = -8, r = 7$ **d)** $q = -25, r = 0$
   e) $q = -1, r = 2$ **f)** $q = 0, r = 57$

Exercise 5.2

1. a) F   **b)** F   **c)** T   **d)** F
   e) T   **f)** T   **g)** T   **h)** T
   i) T
3. a) odd   **b)** even   **c)** even   **d)** odd
   e) even   **f)** even   **g)** even   **h)** odd
   i) odd   **j)** odd   **k)** even   **l)** even

4. $\{0,1,2\}$
5. ***Proof:***   Let $a = 2x$ and $b = 2y$ be two even integers. Then $a + b = 2x + 2y = 2(x + y)$ is also an even integer.
7. ***Proof:***   Let $a = 2x$ and $b = 2y$ be two even integers. Then $ab = 2x \cdot 2y = 2(2xy)$, an even integer.

Exercise 5.3
1. a) F   **b)** F   **c)** F   **d)** F
   e) F   **f)** F   **g)** F   **h)** T
   i) T   **j)** T   **k)** T
2. a) no   **b)** yes   **c)** yes   **d)** no
   e) no   **f)** no   **g)** no   **h)** no
3. a) $2 \cdot 3^2 \cdot 5$ **b)** $2 \cdot 3 \cdot 5 \cdot 7$ **c)** 2^{10}
   d) $2^3 \cdot 3^3 \cdot 5$ **e)** $2 \cdot 3^2 \cdot 5^2$ **f)** $2^4 \cdot 3 \cdot 37$
4. a) $\{1,p\}$ **b)** $\{1,p,p^2\}$ **c)** $\{1,p,q,pq\}$
5. 5, 13, 17, 29, 37
7. 2, 3, 5, 7, 11, 13, 17, 19, 23, 29, 31, 37, 41, 43, 53, 59, 61, 67, 71, 73, 79, 83, 89, 97
8. 90, 91, 92, 93, 94, 95, 96

Exercise 5.4
1. a) 30   **b)** 35   **c)** 77   **d)** 125
2. a) 4   **b)** 6   **c)** 36
   d) 6   **e)** 1   **f)** 5
3. a) 39   **b)** 11   **c)** 3
   d) 1   **e)** 2   **f)** 16
5. a) a **b)** b **c)** 1
   d) a **e)** 1   **f)** 1
6. a) T   **b)** T   **c)** F   **d)** T
   e) F   **f)** T   **g)** F   **h)** F
   i) F   **j)** F
8. 6 inches

Exercise 5.5
1. a) $2^3 \cdot 3^2 \cdot 5$ **b)** $2 \cdot 3^2 \cdot 7^2$
   c) $3^5 \cdot 5^5 \cdot 7$ **d)** $2^3 \cdot 3^2 \cdot 5^2 \cdot 7$
2. a) 120   **b)** 252   **c)** 864
   d) 96   **e)** 2310   **f)** 4290
3. a) 144   **b)** 600   **c)** 864
   d) 360   **e)** 1092   **f)** 140,712
4. a) b **b)** a **c)** b
   d) a **e)** ab **f)** ab
5. 231   **7.** 1

9. a) F   **b)** F   **c)** T
   d) T   **e)** T   **f)** T
   g) T   **h)** F   **i)** T
12. 11
13. a) 120, 7200   **b)** 300, 9000
14. Fred 7; Frank 6; 252 yards

Exercise 5.6
1. a) T   **b)** F   **c)** F   **d)** F
   e) F   **f)** F   **g)** F   **h)** F
   i) T   **j)** T
4. 1, 3, 7, 15, 63; 3 and 7 are primes
5. 3, 5, 17, 257, 65,537; all are primes
7. 1, 3, 6, 10, 15, 21, 28, 36, 45, 55
9. $64 = 28 + 36$, $81 = 36 + 45$, $100 = 45 + 55$
10. a) 11, 22, 33, 44, 55, 66, 77, 88, 99
    b) 101, 111

Exercise 5.7
2. 16, 2048

Exercise 5.8
1.

(a)

5	19	15
23	13	3
11	7	21

(b)

7	23	21
31	17	3
13	11	27

4. a) no   **b)** yes   **c)** no   **d)** no

6.

	1	7	
5	11	4	12
2	8	6	9
	10	3	

CHAPTER 6

Exercise 6.1

1. a) numerator $= 2$, denominator $= 5$
  b) numerator $= -3$, denominator $= 8$
  c) numerator $= -4$, denominator $= -3$

2. a) yes   **b)** yes   **c)** no   **d)** yes

4. a) 1   **b)** $\frac{4}{3}$ **c)** $\frac{-15}{4}$
  d) -2 **e)** $\frac{15}{7}$ **f)** $\frac{8}{3}$

5. a) $\frac{2}{3}$ **b)** $\frac{6}{5}$ **c)** $\frac{-9}{22}$ **d)** $\frac{8}{-11}$
  e) $\frac{13}{24}$ **f)** $\frac{11}{111}$ **g)** $\frac{-41}{77}$ **h)** $\frac{19}{444}$

6. a) T   **b)** T   **c)** T   **d)** T
  e) T   **f)** F   **g)** F   **h)** T
  i) T   **j)** T

9. a) 2   **b)** 8   **c)** -3

Exercise 6.2

1. a) $\frac{5}{6}$ **b)** $\frac{74}{35}$ **e)** $\frac{133}{60}$ **g)** $\dfrac{3a+2b}{ab}$

3. a) $-\frac{1}{3}$ **b)** $\frac{55}{24}$ **c)** $\frac{4}{7}$
  d) 1   **e)** $-\frac{1}{8}$ **f)** $-\frac{1}{70}$

4. a) 3   **b)** 1   **c)** 7   **d)** -4

5. $x = 0$ **6.** $x = 0$

7. a) T   **b)** F   **c)** T   **d)** F
  e) T   **f)** F   **g)** T   **h)** T
  i) T   **j)** T   **k)** F

9. a) closure property   **b)** 0 is the additive identity
  c) addition property   **d)** cancellation property
  e) commutative property   **f)** commutative property
  g) associative property   **h)** commutative property
  i) cancellation property   **j)** 0 is the additive identity

11. a) $\frac{19}{5}$ **b)** $\frac{37}{7}$ **c)** $-\frac{47}{11}$ **d)** $-\frac{29}{6}$

13. $a = 1, b = 2, c = 3$ **15.** $\frac{8}{15}, \frac{7}{15}$

17. $\frac{281}{390}$

19. a) yes   **b)** yes   **c)** yes
  d) yes, $-\frac{1}{2}$ **e)** yes, $-(x+1)$

Exercise 6.3

1. a) $\frac{1}{6}$ **c)** $-\frac{59}{56}$ **e)** $-\frac{1}{12}$ **h)** $-\frac{16}{15}$

3. a) 4   **b)** -3 **c)** 42   **d)** 7

5. a) T   **b)** F   **c)** F   **d)** T
  e) F   **f)** T   **g)** T   **h)** T
  i) T   **j)** T

7. a) $\dfrac{1}{a(a+1)}$ **b)** $\dfrac{a-b}{(a+1)(b+1)}$ **c)** $\dfrac{b-a}{ba}$

9. $\frac{7}{44}$ **11.** $\frac{25}{84}$

12. $\frac{2}{3}, \frac{3}{2}$

Exercise 6.4

1. a) $-\frac{9}{14}$ **c)** $-\frac{11}{2}$ **d)** $\frac{31}{30}$ **h)** $\frac{13}{28}$

3. a) $\frac{7}{5}$ **b)** $-\frac{11}{8}$ **c)** $-\frac{10}{7}$ **d)** $\frac{3}{17}$

4. $3\frac{1}{5} = 3 + \frac{1}{5} = \frac{16}{5}, \; 3 \cdot \dfrac{1}{5} = \dfrac{3 \cdot 1}{5} = \dfrac{3}{5}$

6. a) $\frac{5}{7}$ **b)** $\frac{8}{5}$ **c)** $\frac{3}{2}$ **d)** $\frac{3}{4}$

7. a) $\dfrac{a}{b}$ **b)** $\dfrac{1-b}{1+b}$ **c)** $a-1$

8. $a = 0$ or $c = 0$

12. a) T   **b)** F   **c)** T   **d)** T

   e) F   **f)** T   **g)** T   **h)** T

   i) F   **j)** T   **k)** T   **l)** T

15. $x = \pm 1$ **19.** 1 foot

21. $\frac{1}{12}$ **23.** 40 cents

25. 200   **27.** $8\frac{2}{5}$ dollars

Exercise 6.5

1. a) $\frac{7}{5}$ **c)** $\frac{-9}{11}$ **e)** $\frac{1}{5}$ **h)** $\frac{23}{5}$

2. a) $\frac{20}{21}$ **c)** $\frac{75}{64}$ **e)** 0

5. yes

6. a) $\dfrac{a-b}{a+b}$ **b)** $b-a$ **c)** $\dfrac{a^2-b^2}{a^2+b^2}$

7. a) F   **b)** T   **c)** T

   d) T   **e)** T

9. $\frac{1}{10}$ **11.** 5

13. 21

Exercise 6.6

1. a) yes   **b)** no   **c)** yes

3. $A \cup B = \{x \in Q \mid -\frac{3}{2} \le x \le \frac{5}{6}\}, \quad A \cap B = \{x \in Q \mid \frac{3}{8} < x < \frac{3}{5}\}$

   $A - B = \{x \in Q \mid -\frac{3}{2} \le x \le \frac{3}{8}\}, \quad B - A = \{x \in Q \mid \frac{3}{5} \le x \le \frac{5}{6}\}$

4. a) $\{x \in Q \mid x < \frac{3}{10}\}$ **c)** $\{x \in Q \mid -\frac{29}{42} < x < \frac{1}{42}\}$

5. $\frac{30}{19}$

6. a) $\frac{29}{40}$ **c)** $\frac{1}{3}$ **f)** $-\frac{17}{24}$

9. No, since there are infinitely many rational numbers still closer to $\frac{1}{3}$ than $\frac{1}{4}$ and $\frac{1}{2}$.

11. a) F   **b)** F   **c)** T   **d)** F
   e) F   **f)** T   **g)** T   **h)** F
   i) T   **j)** T

13. $\{x \in Q \mid x \geq \frac{3}{5}\}$

15. $\{x \in Q \mid 9 < x \leq \frac{54}{5}\}$

16. $\{x \in Q \mid \frac{-15}{4} \leq x \leq \frac{187}{5}\}$

Exercise 6.7

1. a) T   **b)** T   **c)** T
   d) F   **e)** F   **f)** F

2. d) $a = 4$

   e) $\sqrt{2}$ is not a rational number.

   f) π is not a rational number and hence cannot be equal to $\frac{22}{7}$.

CHAPTER 7

Exercise 7.1

1. a) T   **b)** T   **c)** F

3. a) $\frac{75}{100}$ **c)** $\frac{115}{100}$ **g)** $\frac{45}{100}$ **i)** $\frac{38}{10}$

4. a) 0.25   **b)** 0.35   **c)** 0.325   **d)** 0.415

5. a) $\frac{1}{2}$ **c)** $\frac{11}{4}$ **e)** $\frac{3}{250}$ **g)** $\frac{111}{100}$

6. a) 111.291   **c)** 36.36   **e)** 136.538787   **g)** 26.801

Exercise 7.2

1. a) 0.06   **c)** 0.123   **e)** 1.2

2. a) 270%   **c)** 304%   **e)** 3%

3. a) 35%   **c)** 44%   **e)** 26.7%

4. b) 9   **d)** 2.52   **f)** 23%   **h)** 740

5. $7.50, $132.50, $140.45

7. 34, 51   **9.** $860

11. $2550   **13.** $200

Exercise 7.3

1. a) 0.915   **b)** 1609   **c)** 4941.4

3. a) 0.473   **b)** 2.838   **c)** 3.785

5. 2340   **7.** 3050 centimetres

9. 1368   **11.** 3.9×10^5

13. 9.46×10^{12} kilometres   **15.** 24 litres

17. a) 100 metres   **b)** 100 miles
   c) 100 gallons   **d)** 100 kilometres

19. a) 12.87 kilometres per second   **b)** 9.65 kilometres per second
   c) 16.09 kilometres per second

21. 0.012345 square metres   **23.** 144

Exercise 7.4

1. a) T   **b)** T   **c)** F
   d) T   **e)** T

3. a) $0.\overline{012}$ **d)** $0.\overline{1176470588235294}$ **g)** $1.1\overline{25}$
4. a) 1   **b)** 22   **c)** 6
   d) 18   **e)** 22
5. a) rational   **c)** nonrational   **e)** nonrational
6. a) $\frac{1}{4}$ **d)** $\frac{34}{99}$ **f)** $\frac{1106}{495}$ **h)** $\frac{901}{900}$

Exercise 7.5

 1. a) irrational   **e)** rational   **h)** irrational
 3. a) $3+a<3+b$ **b)** $2a<2b$ **f)** $2-3a>2-3b$

 4. a) ± 2 **b)** $\pm\sqrt{5}$ **c)** ± 4 **d)** $\pm\sqrt{6}$
 5. a) F   **b)** F   **c)** T   **d)** F
    e) T   **f)** F   **g)** T   **h)** T
    i) T   **j)** T   **k)** T
 7. a) 2.5   **b)** 1.235   **c)** 3
13. a) 2.124   **b)** 5.12335

Exercise 7.7

1. a) T   **b)** F   **c)** T
   d) T   **e)** F   **f)** T
2. a) 0   **b)** 1   **c)** 1   **d)** 1

4. a) $[-2,5)$

c) $[-2,1]$

 i) $\{x \in R \mid x < -2 \text{ or } x > 1\}$

j) $[3,5)$

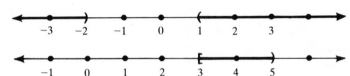

Exercise 7.8

1. no

2. No, the decimal expansion of $\sqrt{2}$ never terminates.

3. d)

$$2 < \sqrt{7} < 3$$

$$2.6 < \sqrt{7} < 2.7$$

$$2.64 < \sqrt{7} < 2.65$$

$$2.645 < \sqrt{7} < 2.646$$

$$2.6457 < \sqrt{7} < 2.6458$$

Exercise 7.9

1. a) F   **b)** F   **c)** T   **d)** T

   e) T   **f)** T   **g)** T   **h)** F

   i) T   **j)** T   **k)** T

3. a) 1.732   **d)** 9.045   **g)** 44.452

   i) 42.145   **j)** 0.224   **l)** 0.084

Exercise 7.10

  1. b) 2   **c)** 6   **f)** 3   **h)** 8

  3. a) 3   **c)** 4   **e)** 2   **g)** 7

  4. a) 1   **b)** -1 **c)** 1   **d)** -1

  6. $x = 0$

  8. a) F   **b)** F   **c)** T   **d)** F

    e) F   **f)** F   **g)** F   **h)** T

    i) F   **j)** F

11. a) $\{5,9\}$ **c)** $\{-5,-9\}$ **d)** $\emptyset$

12. a) $\{-2,9\}$ **c)** $[-3,3]$ **e)** $[-1,6]$

13. a) $\{2\}$ **c)** $\{1,5\}$ **e)** $\{2,\tfrac{2}{3}\}$ **g)** $\{\tfrac{3}{2}\}$

Exercise 7.11

1. a) T   **b)** F   **c)** F   **d)** F

   e) F   **f)** T   **g)** T   **h)** T

   i) F   **j)** F

3. a) 2^3 **c)** 3^4 **e)** 3^5

5. a) 1024   **b)** 1296   **c)** 4   **d)** $\tfrac{1}{8}$

   e) 2187   **f)** 4096   **g)** 625   **h)** 1

6. a) $\tfrac{7}{5}$ **c)** -1 **e)** 1350

7. a) a **c)** a^{18} **e)** a^{20} **g)** a^4

8. a) 5   **c)** 2   **e)** 1   **g)** 3

CHAPTER 8

Exercise 8.1

1. a) 121   **c)** 10,121   **e)** 1,000,020

2. a) ꟼ∩ ∩|

  c) 𒐊𒐊ꟼꟼꟼꟼꟼ∩ ∩ ∩ ∩ ∩ ∩ ∩|||||

Exercise 8.2

1. a) 10   **c)** 140   **e)** 36,070

2. a) Y ⪡⪡⪡ YYYY YYYY   **c)** YY YY   ⪡ YYY YYY

Exercise 8.3

1. a) 400   **c)** 16   **e)** 466
2. a) LIX   **c)** DCCCXCI
3. a) DCLXXX   **c)** MDLIV   **e)** LXII

Exercise 8.4

1. a) 1   **c)** 23   **e)** 14,501

2. a) • •••   **c)** ═ •••• ═

Exercise 8.5

1. a) $1(10)+3(10^0)$
  c) $2(10)+3(10^0)+5(10^{-1})+8(10^{-2})$
2. a) 25   **c)** 4708   **e)** 200.43
3. a) 10^0 **c)** 10   **e)** 10
5. a) 2   **b)** 0   **c)** 0

Exercise 8.6

1. a) $7(10)+4(10^0)$

           $2(10)+5(10^0)$
    sum = $\overline{(7+2)(10)+(4+5)(10^0)}$

        = $9(10)+9(10^0)$

  d) $0(10)+2(10^0)+3(10^{-1})+4(10^{-2})$

           $7(10)+0(10^0)+2(10^{-1})+0(10^{-2})$
    sum = $\overline{(0+7)(10)+(2+0)(10^0)+(3+2)(10^{-1})+(4+0)(10^{-2})}$

        = $7(10)+2(10^0)+5(10^{-1})+4(10^{-2})$

2. b) $46=4(10)+6(10^0)$

     $64=6(10)+4(10^0)$
    sum = $\overline{(4+6)(10)+(6+4)(10^0)}$

        = $10(10)+10(10^0)$

        = $1(10^2)+1(10)+0(10^0)$

        = 110

d) $21.03 = 2(10) + 1(10^0) + 0(10^{-1}) + 3(10^{-2})$

$\underline{0.54 = 0(10) + 0(10^0) + 5(10^{-1}) + 4(10^{-2})}$

$\text{sum} = (2+0)(10) + (1+0)(10^0) + (0+5)(10^{-1}) + (3+4)(10^{-2})$

$\phantom{d)\text{sum}} = 2(10) + 1(10^0) + 5(10^{-1}) + 7(10^{-2})$

$\phantom{d)\text{sum}} = 21.57$

Exercise 8.7

1. a) $x = 2(10) + 5(10^0)$

$\underline{y = \times 4(10^0)}$

$xy = 8(10) + 20(10^0)$

$ = 8(10) + 2(10) + 0(10^0)$

$ = (8+2)(10) + 0(10^0)$

$ = 10(10) + 0(10^0)$

$ = 1(10^2) + 0(10) + 0(10^0)$

2. d) $1.23 = 1(10^0) + 2(10^{-1}) + 3(10^{-2})$

$\times 13 = 1(10) + 3(10^0)$

$\overline{}$

$3(10^0) + 6(10^{-1}) + 9(10^{-2})$

$1(10) + 2(10^0) + 3(10^{-1})$

$\overline{}$

$\text{product} = 1(10) + 5(10^0) + 9(10^{-1}) + 9(10^{-2})$

$\phantom{2. d)\text{product}} = 15.99$

3. a)

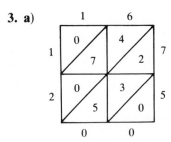

c)

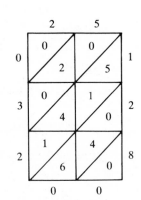

Exercise 8.8

1. a) $x = 5(10^2) + 7(10) + 4(10^0)$

$\underline{y = 3(10^2) + 6(10) + 2(10^0)}$

$x - y = 2(10^2) + 1(10) + 2(10^0)$

2. a)
$$x = 6(10) + 7(10^0)$$
$$y = 3(10) + 4(10^0)$$

$$x - y = 3(10) + 3(10^0)$$
$$= 33$$

3. a)

$$5(10^0)$$

$$1(10) + 3(10^0) \,\Big|\, \begin{array}{l} 6(10) + 5(10^0) \\ 6(10) + 5(10^0) \end{array}$$

$$q = 5, r = 0$$

c)

$$1(10) + 2(10^0)$$

$$2(10) + 3(10^0) \,\Big|\, \begin{array}{l} 2(10^2) + 8(10) + 4(10^0) \\ 2(10^2) + 3(10) \end{array}$$

$$5(10) + 4(10^0)$$
$$4(10) + 6(10^0)$$

$$8(10^0)$$

$$q = 12, r = 8$$

4. a) $8 + 2 + 1$ **c)** $16 + 8 + 4 + 2 + 1$

5. a)   **b)**

1	56
2	112
4	224
8	448*

$$8 \times 56 = 448$$

1	91*
2	182*
4	364
8	728
16	1456*

$$19 = 16 + 2 + 1$$
$$19 \times 91 = 91 + 182 + 1456 = 1729$$

6. a)   **c)**

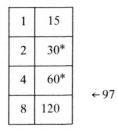

1	15
2	30*
4	60*
8	120

← 97

1	41
2	82
4	164
8	328*

← 328

$97 = 60 + 30 + \boxed{7}$ $328 = 328 + \boxed{0}$

$q = 4 + 2 = 6$ $q = 8$

$r = 7$ $r = 0$

Exercise 8.9

1. a) $2(7^2) + 5(7^1) + 6(7^0)$
  b) $1(2^3) + 0(2^2) + 1(2^1) + 1(2^0)$
  e) $2(5^3) + 0(5^2) + 3(5^1) + 0(5^0)$
  f) $1(8^2) + 2(8^1) + 1(8^0)$

2. a) $(10304)_5$ **c)** $(30540)_7$ **e)** $(11001)_2$
3. a) 38   **c)** 1976   **e)** 2000
4. a) $(11110111000)_2$ **c)** $(5522)_7$
  e) $(85)_{12}$ **g)** $(100111)_3$
6) a) $(11)_5$ **c)** $(110)_2$ **e)** $(1101011)_2$
7. $(1234)_5 < (1432)_5 < (2431)_5 < (3214)_5$
9. a) $(1121)_3$ **c)** $(10110101011)_2$
10. a) 12   **c)** 5   **e)** 4
12. a) 0   **b)** 1

Exercise 8.10

1. a) $(10)_3$ **b)** $(10)_5$ **c)** $(13)_7$
  d) $(11)_5$ **e)** $(10)_{12}$ **f)** $(19)_{12}$

2.

+	0	1	2
0	0	1	2
1	1	2	10
2	2	10	11

5. a) $(124)_5$ **c)** $(122)_3$ **e)** $(280)_{12}$
6. a) $(1020)_3$ **c)** $(1000)_2$ **f)** $(1285)_{12}$

8. a) $(122)_5$ **c)** $(11)_2$ **e)** $(100)_3$
9. a) $(11)_2$ **c)** $(32)_5$ **e)** $(331)_7$
10. a) $(11000)_2$ **c)** $(301)_5$

Exercise 8.11

1. a) $(11)_5$ **b)** $(6)_7$ **c)** $(11)_3$
   d) $(42)_7$ **e)** $(34)_{12}$ **f)** $(92)_{12}$

3.

·	0	1	2	3	4	5
0	0	0	0	0	0	0
1	0	1	2	3	4	5
2	0	2	4	10	12	15
3	0	3	10	13	20	23
4	0	4	12	20	24	32
5	0	5	14	23	32	41

5. a) $(11110)_2$ **c)** $(22222)_3$ **f)** $(tt20)_{12}$
6. a)   **c)**

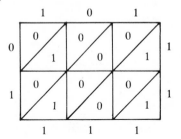 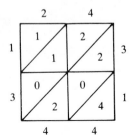

7. a) $q = (2)_5$, $r = (3)_5$ **b)** $q = (100)_2$, $r = (11)_2$
   c) $q = (100)_3$, $r = (11)_3$ **e)** $q = (17)_{12}$, $r = (14)_{12}$
9. a) $(10100)_2$ **b)** $(22220)_3$ **c)** $(142566)_7$

CHAPTER 9

Exercise 9.1

1. a) $5 + 3i$ **b)** $2 + 2i$ **c)** $-1 + 0i$
   d) $-2 + 2i$ **e)** $-1 + 0i$ **f)** $-\sqrt{3} + \sqrt{3}\, i$
3. a) $\pm 3i$ **b)** $\pm 5i$ **c)** $\pm 3i$
   d) $\pm\sqrt{3}i$ **e)** $\pm\frac{3}{2}i$ **f)** $\pm\sqrt{7}\, i$

5. a) T $\qquad$ **b)** F $\qquad$ **c)** T
$\quad$ **d)** T $\qquad$ **e)** F $\qquad$ **f)** T
$\quad$ **g)** F $\qquad$ **h)** F $\qquad$ **i)** T

Exercise 9.3

1. a) $4+3i$ $\qquad$ **c)** $3+\sqrt{3}\,i$ $\qquad$ **e)** 2
$\quad$ **g)** $1+3i$ $\qquad$ **i)** $10i$ $\qquad$ **j)** $3+\sqrt{5}\,i$
2. a) $26+7i$ $\qquad$ **c)** 10 $\qquad$ **e)** $2i$
3. a) $\frac{7}{29}-\frac{26}{29}i$ $\qquad$ **c)** $\frac{2}{3}+\frac{\sqrt{2}}{3}i$ $\qquad$ **e)** $-i$
$\quad$ **g)** $-i$ $\qquad$ **h)** 1 $\qquad$ **i)** i
4. a) $\frac{1}{5}-\frac{2}{5}i$ $\qquad$ **c)** $\frac{1}{2}-\frac{1}{2}i$ $\qquad$ **e)** $\frac{2}{7}-\frac{\sqrt{3}}{7}i$
7. a) $0,\pm 1$ $\qquad$ **b)** $0,\pm i$ $\qquad$ **c)** $\pm 1,\pm i$
$\quad$ **d)** $1,\pm i$ $\qquad$ **e)** $i,\pm i$ $\qquad$ **f)** $-1,\pm i$
10. a) T $\qquad$ **b)** F $\qquad$ **c)** F
$\quad$ **d)** T $\qquad$ **e)** F $\qquad$ **f)** F
12. a) Not closed under addition since $i+(-i)=0$ does not belong to $\{\pm i\}$; not closed under multiplication since $i \cdot i = -1$ is not in $\{\pm i\}$.
$\quad$ **b)** Not closed under addition since $1+(-1)=0$ does not belong to $\{\pm 1,\pm i\}$; closed under multiplication as seen by the following table.

$\cdot$	1	-1	i	$-i$
1	1	-1	i	$-i$
-1	-1	1	$-i$	i
i	i	$-i$	-1	1
$-i$	$-i$	i	1	-1

13. a) 1 $\qquad$ **b)** 0 $\qquad$ **c)** $\sqrt{3}\,i$ $\qquad$ **d)** $4i$

14. a) $-2,-3$ $\qquad$ **c)** $\dfrac{-1\pm\sqrt{3}\,i}{2}$ $\qquad$ **e)** $1\pm i$

CHAPTER 10

Exercise 10.1

1. a) yes $\qquad$ **b)** yes $\qquad$ **c)** no
$\quad$ **d)** no $\qquad$ **e)** no $\qquad$ **f)** no
3. a) domain $=\{a,b,c\}$ $\qquad$ **c)** domain $= A$
$\qquad$ range $=\{1,2,4\}$ $\qquad$ range $=\{1,2,3,5\}$
$\quad$ **e)** domain $= \varnothing$ $\qquad$ **f)** domain $= A$
$\qquad$ range $= \varnothing$ $\qquad$ range $= B$

4. b) {(2,2),(4,4)}   **d)** {(1,2),(3,4),(4,5)}

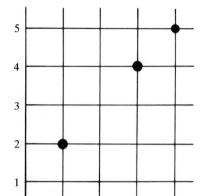

5. a) T   **b)** T   **c)** F
   d) F   **e)** T
7. a) T   **b)** F   **c)** T   **d)** T
9. t = {(Augusta,Maine),(Concord,New Hampshire),(Montpelier,Vermont),
      (Boston,Massachusetts),(Hartford,Connecticut),(Providence, Rhode
      Island)}

Exercise 10.2
1. a) {(3,0),(3,1),(6,2),(6,3)}
       domain = {3,6}
       range   = {0,1,2,3}

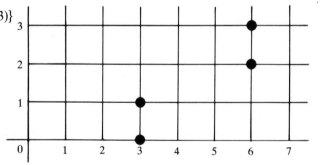

   e) {(7,1),(7,2),(6,2)}
       domain = {6,7}
       range   = {1,2}

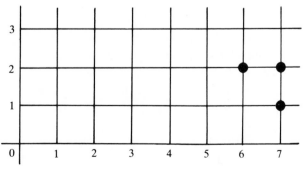

2. a) R **c)** R **e)** S
   f) S **g)** $\{(1,1),(2,2)\}$ **h)** $\{(1,1)(2,2)\}$

3. a) domain $= \{2,4,6\}$ **c)** domain $= \{2,4,6\}$
   range $= \{1,3,5\}$ range $= \{1,3\}$

   e) domain $= \varnothing$ **f)** domain $= \{2,4,6\}$
   range $= \varnothing$ range $= \{1,3,5\}$

4. a) F   **b)** T   **c)** T
   d) T   **e)** F   **f)** T

Exercise 10.3

1. a) irreflexive, symmetric, transitive, not an equivalence relation
   c) reflexive, asymmetric, intransitive, not an equivalence relation
   e) irreflexive, asymmetric, intransitive, not an equivalence relation
   g) reflexive, symmetric, transitive, equivalence relation

2. a) reflexive, asymmetric, transitive, not an equivalence relation
   c) irreflexive, asymmetric, intransitive, not an equivalence relation

3. reflexive, asymmetric, transitive, not an equivalence relation

5. irreflexive, symmetric, intransitive, not an equivalence relation

6. a) F   **b)** T   **c)** T   **d)** T
   e) T   **f)** T   **g)** T   **h)** T

8. r is irreflexive if $(x,x) \notin r$ for some $x \in r$;
   r is asymmetric if $(y,x) \notin r$ for some $(x,y) \in r$;
   r is intransitive if $(x,y) \in r$ and $(y,z) \in r$, but $(x,z) \notin r$.

9. a) $\{(1,1),(2,2),(3,3),(1,2),(2,1)\}$
   c) $\{(1,1),(2,2),(3,3),(1,2)\}$
   e) $\{(1,1),(2,2),(3,3),(1,2),(2,3)\}$
   g) $\{(1,1),(1,2)\}$

Exercise 10.4

1. a) yes, $\{1,2,3\}$ **c)** no, $\{1,3\}$
   e) no, $\{1,2,3\}$ **f)** yes, $\{1,4\}$

3. a) $\{(2,a),(3,b),(2,c),(1,d)\}$, no
   c) $\{(3,a),(3,b),(1,c)\}$, no
   e) $\{(2,a),(1,b),(2,c),(3,a)\}$, no
   f) $\{(4,a),(1,b),(4,c),(1,d)\}$, no

4. a) T   **b)** F   **c)** F   **d)** F
   e) F   **f)** F   **g)** T   **h)** F

6. a) $g(0) = -1$, $g(\frac{1}{2}) = -\frac{3}{4} = g(-\frac{1}{2})$, $g(\frac{2}{3}) = -\frac{5}{9}$
   b) $\{\pm 1\}$, $\{\pm 2\}$

7. $h(0) = 1$, $h(-1) = 3$, $h(-\frac{2}{3}) = \frac{19}{9}$, $h(\frac{3}{5}) = \frac{19}{25}$

9. a) -1 **b)** -1 **c)** 1   **d)** 1

11. a) $\frac{2}{3}$ **b)** $\frac{4}{5}$ **c)** 1   **d)** 2

13. $f(0) = f(1) = f(-\frac{1}{2}) = f(\sqrt{4}) = 1,\ f(\sqrt{2}) = 0$; range $(f) = \{0, 1\}$
15. $t(0) = t(1) = t(-\frac{1}{2}) = t(\frac{2}{3}) = 5$; range $(t) = \{5\}$

Exercise 10.5

1. a) T	**b)** F	**c)** F	**d)** T
e) F	**f)** T	**g)** F	**h)** F
i) F	**j)** F		

3. two; $\{(a,1),(b,1),(c,1)\},\ \{(a,2),(b,2),(c,2)\}$

5. a) yes	**b)** no	**c)** no	**d)** yes
6. a) no	**b)** yes	**c)** yes	**d)** no
7. a) yes	**b)** no	**c)** yes	**d)** no

9. If $c(A) = c(B) = 1$.
11. No, since, for example, the element 4 in B has two preimages, ± 2.

Exercise 10.6

1.

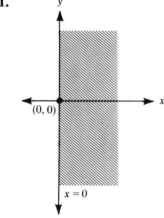

(a)

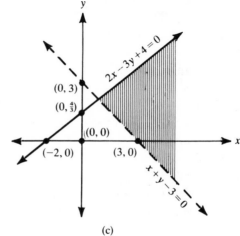

(e)

2.

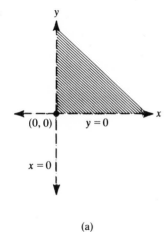

(a)

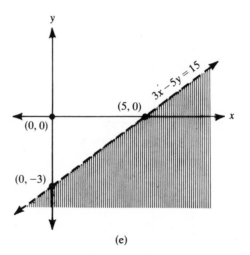

(c)

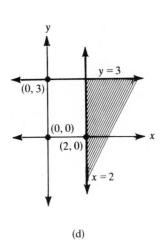

(d)

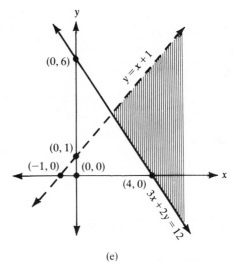

(e)

3.

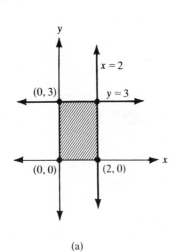

(a)

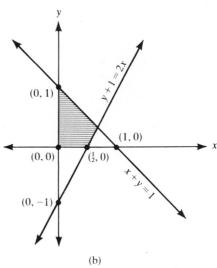

(b)

4. a) $(1,2)$ **b)** $(2,3)$ **c)** $(2,3)$ **d)** $(-1,0)$

Exercise 10.7

1. a) -15 **b)** -4 **c)** $\frac{14}{3}$

2. a) $(0,0),(2,0),(2,2),(0,6)$

   d) $(0,0),(6,0),(6,4),(3,6),(0,6)$

3. a) 42 for $x=0$, $y=6$ **d)** 94 for $x=6$, $y=4$

4. b) -25 for $x=0$, $y=5$ **c)** -10 for $x=0$, $y=2$

5. $x=$ number of sweaters of type A

   $y=$ number of sweaters of type B

   a) $x\geq0$, $y\geq0$, $x+y\leq300$, $9x+3y\leq1800$, $y\leq x+100$

b) $(0,0),(200,0),(150,150),(100,200),(0,100)$

c) $p(x,y) = 5x + 3y$

d) 150 sweaters of type A and 150 sweaters of type B

e) 1200 dollars

7. 100 batteries of type A and 500 batteries of type B

8. 0 gallons of gasoline from place A to Boston

20,000 gallons of gasoline from place B to Boston

30,000 gallons of gasoline from place A to New York

20,000 gallons of gasoline from place B to New York

Minimum cost $= \$3,300$

CHAPTER 11

Exercise 11.1

1. a) T   **b)** T   **c)** T   **d)** T

    e) T   **f)** T   **g)** T   **h)** F

    i) T   **j)** T

3. a) 1, 5, 9   **c)** 7, 15, 23   **d)** 2, 9, 16

    f) 0, 5, 10   **g)** 2, 8, 14   **i)** 21, 46, 71

4. a) $\{2 + 3k \mid k \in I\}$ **c)** $\{6k \mid k \in I\}$

    e) $\{3 + 10k \mid k \in I\}$ **f)** $\{9 + 11k \mid k \in I\}$

5. True   **6.** False

Exercise 11.2

1. 7 o'clock, 6 o'clock, 5 o'clock

3. 6 o'clock, 5 o'clock, 2 o'clock

5. a) $2 + (3 + 8) = 2 + 11 = 1$

     $(2 + 3) + 8 = 5 + 8 = 1$

     Thus, $2 + (3 + 8) = (2 + 3) + 8$.

   c) $5 + (7 + 9) = 5 + 4 = 9$

     $(5 + 7) + 9 = 0 + 9 = 9$

     Thus, $(5 + 7) + 9 = 5 + (7 + 9)$.

6. a) $2 \cdot 3 = 6 = 3 \cdot 2$ **c)** $8 \cdot 10 = 8 = 10 \cdot 8$

7. a) $2 \cdot (3 \cdot 8) = 2 \cdot 0 = 0$

     $(2 \cdot 3) \cdot 8 = 6 \cdot 8 = 0$

     Thus, $2 \cdot (3 \cdot 8) = (2 \cdot 3) \cdot 8$.

   c) $5 \cdot (7 \cdot 9) = 5 \cdot 3 = 3$

     $(5 \cdot 7) \cdot 9 = 11 \cdot 9 = 3$

     Thus, $5 \cdot (7 \cdot 9) = (5 \cdot 7) \cdot 9$.

9. $-1 = 11,\ -4 = 8,\ -5 = 7,\ -6 = 6,\ -8 = 4,\ -9 = 3,\ -11 = 1$

11. $1^{-1} = 1$, $5^{-1} = 5$, $7^{-1} = 7$; 3^{-1}, 6^{-1}, and 8^{-1} do not exist.

13. a) 5   **c)** 3   **d)** 1, 4, 7, 10
  f) 8   **g)** 0, 6   **i)** 0, 1, 4, 9

14. a) T   **b)** T   **c)** T   **d)** T
  e) T   **f)** T   **g)** T   **h)** F
  i) F   **j)** T   **k)** T

16. b) yes   **d)** yes   **f)** 1, 5   **h)** no

Exercise 11.3

1. Tuesday

2. Wednesday, Tuesday, Tuesday

3. a) $5 + 4 = 2 = 4 + 5$ **e)** $4 + (2 + 3) = 4 + 5 = 2$
                                                    $(4 + 2) + 3 = 6 + 3 = 2$

4. a) $5 \cdot 4 = 6 = 4 \cdot 5$ **e)** $4 \cdot (2 \cdot 3) = 4 \cdot 6 = 3$
                                                    $(4 \cdot 2) \cdot 3 = 1 \cdot 3 = 3$

6. 4, 3, 1   **7.** 4, 3, 6

8. a) 5   **b)** 2   **e)** 5   **f)** 3

9. a) 4   **c)** 3   **d)** 1
  f) 1   **g)** 0   **i)** 0, 1

10. a) T   **b)** F   **c)** T   **d)** T
  e) T   **f)** T   **g)** F   **h)** T
  i) F   **j)** T   **k)** T   **l)** T
  m) T

12. b) yes   **d)** yes   **f)** 1, 2, 3, 4   **h)** yes

Exercise 11.4

1. a) 1   **b)** 4   **c)** 4   **d)** 5
  e) 3   **f)** 4   **g)** 7   **h)** 5

3. a) probably correct
  c) definitely wrong
  e) probably correct
  g) probably correct

4. a) 8   **b)** 9   **c)** 1   **d)** 5

Exercise 11.5

1. a) T   **b)** F   **c)** T
  d) F   **e)** F   **f)** T
  g) F   **h)** T   **i)** T

3.

Number	Divisible by 2	Divisible by 4	Divisible by 8
248	yes	yes	yes
425	no	no	no
1948	yes	yes	no
1776	yes	yes	yes
1976	yes	yes	yes
327	no	no	ño
2000	yes	yes	yes
31264	yes	yes	yes
232426	yes	no	no
427364	yes	yes	no
2380	yes	yes	no
3240	yes	yes	yes

5. a) yes **b)** yes **c)** no
  d) yes **e)** no
7. a) no **b)** yes **c)** no
  d) yes **e)** yes
9. 100,020 **11.** 300,240
13. 100,620

CHAPTER 12

Exercise 12.1

1. a) yes **b)** yes **c)** yes **d)** yes, 1
  e) yes, the inverse of any element a is $2 - a$.
3. a) yes **b)** yes **c)** yes **d)** yes, 0
  e) No, every integer $a(\neq 1)$ has an inverse $\dfrac{a}{a-1}$.
5. a) yes **b)** no **c)** yes **d)** no
  e) no inverses, since no identity exists
7. a) yes **b)** yes **c)** yes **d)** yes, 1
  e) Yes, the inverse of every integer is 1.
8. a) F **b)** T **c)** F
  d) F **e)** T **f)** F
11. a) yes **b)** yes **c)** no **d)** no
13. a) yes **b)** yes **c)** yes **d)** yes
15. a) yes **b)** yes **c)** yes **d)** no
  e) no inverses, since no identity exists

Exercise 12.2

1. yes   **3.** yes

5. no, since $1+1=2\notin\{-1,0,1\}$

7. That $\{1,2,3,4\}$ is an abelian group under multiplication mod 5 follows from the following table.

·	1	2	3	4
1	1	2	3	4
2	2	4	1	3
3	3	1	4	2
4	4	3	2	1

9. The following table shows that $\{1,5\}$ is an abelian group under multiplication mod 6.

·	1	5
1	1	5
5	5	1

11. No, since not every rational number has an inverse with respect to ∗.

12. a) F   **b)** F   **c)** T   **d)** T

  e) T   **f)** T   **g)** F   **h)** F

  i) T   **j)** F   **k)** F   **l)** F

  m) F

15.

∗	e	a	b
e	e	a	b
a	a	b	e
b	b	e	a

17. $I_3=\{0,1,2\}$, under addition mod 3, is a group with three elements.

  $I_4=\{0,1,2,3\}$, under addition mod 4, is a group with four elements.

  $I_5=\{0,1,2,3,4\}$, under addition mod 5, is a group with five elements.

19. yes, $x=a^{-1}*b$

Exercise 12.3

1. yes, abelian

3. a) α_1 **b)** e **c)** α_2 **d)** e

  e) β_1 **f)** e **g)** α_2 **h)** e

  i) $\beta_1^0=e,\ \beta_1^1=\beta_1,\ \beta_2^2=e$

5. a) T   **b)** F   **c)** T
   d) F   **e)** F

7. a) $\begin{pmatrix} 1 & 2 & 3 & 4 & 5 \\ 2 & 5 & 1 & 4 & 3 \end{pmatrix}$ **c)** $\begin{pmatrix} 1 & 2 & 3 & 4 & 5 \\ 1 & 2 & 5 & 4 & 3 \end{pmatrix}$

   e) $\begin{pmatrix} 1 & 2 & 3 & 4 & 5 \\ 4 & 2 & 1 & 5 & 3 \end{pmatrix}$ **g)** $\begin{pmatrix} 1 & 2 & 3 & 4 & 5 \\ 3 & 1 & 5 & 2 & 4 \end{pmatrix}$

Exercise 12.4

1. yes   **3.** no

5. a) $\alpha_1 \circ (\alpha_2 \circ \alpha_3) = \alpha_1 \circ \alpha_1 = \alpha_2$
   $(\alpha_1 \circ \alpha_2) \circ \alpha_3 = \alpha_3 \circ \alpha_3 = \alpha_2$
   c) $\alpha_2 \circ (\beta_2 \circ \gamma_2) = \alpha_2 \circ \alpha_3 = \alpha_1$
   $(\alpha_2 \circ \beta_2) \circ \gamma_2 = \beta_1 \circ \gamma_2 = \alpha_1$

6. a) α_3 **c)** γ_1 **e)** α_2 **g)** α_3
7. a) T   **b)** F   **c)** F
9. No; D_4 contains only eight elements, whereas S_4 contains 24 elements.

Exercise 12.5

1. No, since K is not closed under the operation
2. These tables show that both K and L are closed under addition mod 8. Also, each element has an inverse. Consequently, each is a subgroup of I_8.

+	0	4
0	0	4
4	0	0

+	0	2	4	6
0	0	2	4	6
2	2	4	6	0
4	4	6	0	2
6	6	0	2	4

5. This table shows that $\{E, Q_1, Q_2\}$ is closed under composition and every element has an inverse. Thus it is a subgroup of the group of rotations of an equilateral triangle.

$\circ$	E	Q_1	Q_2
E	E	Q_1	Q_2
Q_1	Q_1	Q_2	E
Q_2	Q_2	E	Q_1

7. $\{1\}$ and $\{\pm 1\}$ are finite subgroups of the multiplicative group of nonzero rational numbers.

8. a) F   **b)** F   **c)** T
    d) T   **e)** T   **f)** F
    g) F   **h)** F   **i)** T

11. two   **13.** 1, 2, 3, 4, 6, or 12

15. $\{1\}, \{\pm 1\}$

Exercise 12.6

1. No, since multiplication is not distributive over addition

3. yes, yes

5. yes, no unit element, not an integral domain

6. a) T   **b)** F   **c)** F   **d)** T
    e) F   **f)** T   **g)** T   **h)** F
    i) T   **j)** F   **k)** T   **l)** T
    m) F   **n)** T

9. yes

11. no, since not every nonzero element in L has a multiplicative inverse

12. yes

CHAPTER 13

Exercise 13.1

1. a) $\overline{PR}$ **b)** $\overline{QR}$ **c)** $\{Q\}$
    d) $\varnothing$ **e)** $\overline{PQ}$ **f)** $\{Q\}$
    g) $\overline{PR}$ **h)** $\overline{QR}$ **i)** $\{Q\}$
    j) $\overline{QR}$ **k)** $\overline{PQ}$ **l)** $\overleftrightarrow{RS}$

3. a) $\{D\}$ **b)** $\{B\}$ **c)** $\varnothing$ **d)** $\{A,C\}$

5. yes

7. a) T   **b)** T   **c)** T   **d)** F
    e) F   **f)** F   **g)** T   **h)** T
    i) T   **j)** F   **k)** F   **l)** T
    m) F   **n)** T   **o)** T   **p)** T
    q) F   **r)** F   **s)** T   **t)** T
    u) T   **v)** F   **w)** F   **x)** T
    y) T

9. no   **11.** yes, yes

13. a) 1   **b)** 1   **c)** 3   **d)** 6

Exercise 13.2

1. a) F   **b)** T   **c)** F   **d)** F
    e) T   **f)** F   **g)** T   **h)** T
    i) F   **j)** F

3. a) 1, 4; 1, 2; 2, 3; 3, 4;   **b)** 1, 2; 2, 3; 3, 4; 4, 1;
    5, 6; 6, 7; 7, 8; 8, 5;   5, 6; 6, 7; 7, 8; 8, 5;
    9, 10; 10, 11; 11, 12; 12, 9.   9, 10; 10, 11; 11, 12; 12, 9.
  c) 1, 3; 2, 4; 5, 7; 6, 8; 9, 11; 10, 12.

5. 12

7. a) T   **b)** F   **c)** T   **d)** F
  e) T   **f)** T   **g)** T   **h)** T
  i) F   **j)** F

9. a) yes   **b)** no   **c)** yes   **d)** no
  e) yes   **f)** yes   **g)** no

Exercise 13.3

1. a) yes   **b)** no   **c)** no   **d)** no
3. a) yes   **b)** yes   **c)** yes   **d)** yes
5. a) yes   **b)** no   **c)** no   **d)** yes

7.

9.

11. a) {*A,B*}   **b)** {*A*}   **c)** {*A,B*}
13. a) {*O*}   **b)** {*A,B*}   **c)** {*P,Q*}
15. {*P,Q*}
17. a) ∅   **b)** $\overline{PQ}$ **c)** ∅
   d) ∅
18. a) F   **b)** F   **c)** F
   d) F   **e)** T   **f)** F
   g) T   **h)** F   **i)** T
   j) T   **k)** T   **l)** T
   m) T   **n)** F   **o)** F
19. $C \cup$ Int C

Exercise 13.4

1. a) T   **b)** T   **c)** F
  d) T   **e)** F   **f)** F
3. a) yes   **b)** no   **c)** yes
  d) no   **e)** yes

Exercise 13.5

1. a) T   **b)** T   **c)** T   **d)** T
   e) T   **f)** T   **g)** F   **h)** T
   i) F   **j)** T   **k)** T   **l)** F

3.

Pyramid	V	F	E	V + F − E
Triangular	4	4	6	2
Square	5	5	8	2
Pentagonal	6	6	10	2
Hexagonal	7	7	12	2

5. 20   **7.** 10
9. 18   **11.** circle
13. circle

Exercise 13.6

1. a) 23   **b)** 19   **c)** 24.5   **d)** 17
3. a) T   **b)** T   **c)** F   **d)** T
   e) F   **f)** F   **g)** T
5. 12 feet   **7.** 10.5 centimetres
9. $\frac{5}{7}$

Exercise 13.7

1. a) F   **b)** T   **c)** T   **d)** F
   e) T   **f)** T   **g)** T
3. 2.5 feet   **5.** 30
7. 40 square inches   **9.** 8 centimetres

Exercise 13.8

1. a) 162 cubic centimetres   **b)** 540 cubic inches
3. 452.6   **5.** 113 cubic centimetres

CHAPTER 14

Exercise 14.1

1. 120   **3.** 16
5. 8
7. a) 48   **b)** 36   **c)** 12
   d) 12   **e)** 12

9. 125

13. 54,000

15. 9,989,001

17. a) 32

11. 120

16. 660

b) 243

Exercise 14.2

1. a) 210   **b)** 17,160   **c)** 20,160   **d)** 11

2. a) F   **b)** F   **c)** T   **d)** F

   e) F   **f)** F   **g)** F   **h)** F

   i) F   **j)** F   **k)** F   **l)** F

   m) T

3. a) 2   **b)** 24   **c)** 120   **d)** 24

   e) 720   **f)** 40,320   **g)** 40,320   **h)** 720

5. 120   **7.** 17,160

9. 24

11. a) 2   **b)** 6   **c)** 12   **d)** 30

13. a) 40,320   **b)** 1440

15. 34,560

Exercise 14.3

1. a) 165   **b)** 1   **c)** 1   **d)** 165

4. 126   **5.** 270,725

7. 462   **9.** 845,000

11. 150   **13.** 21

15. 560   **16.** 243

17. 260

Exercise 14.4

1. $\{HH,HT,TH,TT\}$, 4   **3.** $\{1,2,3,4,5,6\}$, 6

5. 216   **7.** $\frac{1}{2}$

9. $\frac{9}{11}$

11. a) $\frac{1}{13}$ **b)** $\frac{1}{4}$ **c)** $\frac{2}{13}$ **d)** $\frac{1}{2}$

14. a) $\frac{55}{136}$ **b)** $\frac{275}{1224}$

15. a) $\frac{1}{22}$ **b)** $\frac{7}{44}$ **c)** $\frac{7}{22}$

17. a) $\frac{7}{429}$ **b)** $\frac{2}{429}$ **c)** $\frac{175}{429}$ **d)** $\frac{140}{429}$

20. a) $\frac{1}{36}$ **b)** $\frac{1}{18}$ **c)** $\frac{1}{6}$ **d)** $\frac{1}{12}$

Exercise 14.5

1. $\frac{5}{6}$ **3.** 1

5. a) $\frac{4}{15}$ **b)** $\frac{1}{3}$ **c)** $\frac{3}{5}$

   d) $\frac{11}{15}$ **e)** $\frac{2}{3}$ **f)** 1

6. a) $\frac{1}{13}$ **b)** $\frac{1}{13}$ **c)** $\frac{2}{13}$

    d) $\frac{2}{13}$ **e)** $\frac{1}{2}$ **f)** $\frac{3}{4}$

7. a) $\frac{1}{6}$ **b)** $\frac{1}{3}$ **c)** $\frac{1}{3}$ **d)** $\frac{1}{2}$

Exercise 14.6

1. $\frac{6}{35}$ **2.** $\frac{1}{5}$

3. a) $\frac{1}{17}$ **b)** $\frac{4}{51}$ **c)** $\frac{13}{51}$

5. a) $\frac{1}{36}$ **b)** $\frac{1}{36}$

7. $\frac{1}{156}$ **8.** $\frac{12}{35}$

9. a) $\frac{5}{33}$ **b)** $\frac{7}{22}$ **c)** $\frac{35}{132}$

11. a) $\frac{1}{36}$ **b)** $\frac{1}{6}$ **c)** $\frac{1}{12}$

     d) $\frac{1}{12}$ **e)** $\frac{1}{8}$

13. a) $\frac{1}{220}$ **b)** $\frac{1}{22}$ **c)** $\frac{1}{55}$

     d) $\frac{3}{110}$ **e)** $\frac{1}{22}$

15. a) $\frac{1}{221}$ **b)** $\frac{1}{17}$ **c)** $\frac{13}{204}$

     d) $\frac{1}{1326}$ **e)** 0

CHAPTER 15

Exercise 15.1

1.

Height	Tally	Frequency
60	||	2
63	||||	4
64	||	2
65	||	2
66	|	1
68	|	1
69	ⅢЛ	5
70	|||	3
71	||	2
72	|||	3
78	||	2
80	||	2
82	|	1

a) 22   **b)** 69   **c)** 18

d) 26.8%   **e)** 3

2.

Class	Tally	Frequency
58–63	𝍷𝍷	6
63–68	𝍷𝍷	6
68–73	𝍷𝍷𝍷	13
73–78	‖	2
78–83	‖‖	3

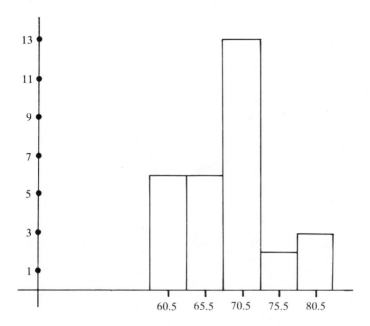

5. a)

Class	Tally	Frequency
46–52			1
52–58	‖	2
58–64	‖‖	3
64–70	𝍷𝍷	6
70–76	𝍷𝍷𝍷	9
76–82	‖‖‖	4
82–88	𝍷𝍷	6
88–94	𝍷	5
94–100	‖‖‖	4

b) 48   **c)** 73   **d)** 2   **e)** 1

f) 7   **g)** 37   **h)** 40%

6. b)

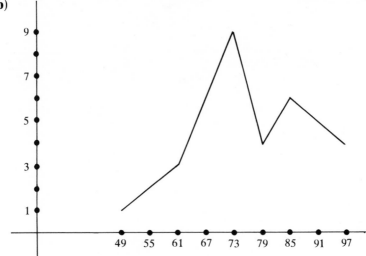

7. a) 44   **b)** 7   **c)** 46.67%   **e)** 35–40

9. a)

b) 12

c) 1

d) 54.3%

Class	Tally	Frequency
40–50	||	2
50–60	|||||	5
60–70	||||| ||||| ||	12
70–80	||||| |||||	10
80–90	|||||	5
90–100			1

Exercise 15.2

1. a) 4, 4, 5   **b)** 8.67, 7, 3

   c) 2.7, 3, 4   **d)** 2263, 1787.5, 1234

   e) *a, a,* no mode   **f)** *a, a,* no mode

3. a) 90.5   **b)** 90

5. 950,000   **7.** 14,000–17,000; 15,500

9. mean = 14,971; median = 8700; mode = 8600

   a) 14,971   **b)** 8600

Exercise 15.3

1. a) 6, 1.7, 2   **b)** 20, 5.3, 6.3

   c) 4, 1.1, 1.3   **d)** $6d, 2d, \sqrt{5}d$

   e) $4d, \frac{6}{5}d, \sqrt{2}d$

3. yes, when all values are equal   **5.** yes

7. a) $\bar{x} = 81$, 8.2, 9.9   **b)** $\bar{x} = 75.7$, 11.4, 12.5   **c)** second exam

8. $\bar{x} = 27$, 10.4, 12.1

Exercise 15.4

1. a) 102   **b)** 41   **c)** 47.7%
  d) 7   **e)** 48   **f)** 0.682
2. a) 48   **b)** 2.2%   **c)** 68   **d)** 0.136
3. a) 45   **b)** 8   **c)** 45
  d) 45   **e)** 13.6%   **f)** 13.6%
4. a) bell-shaped   **b)** normal curve   **c)** 65
  d) 168   **e)** 2.2%   **f)** 32

CHAPTER 16

Exercise 16.3

1. a) *Step 1*   If p is false, then $p \wedge q$ is false. If p is not false, proceed to step 2.
      Step 2   If q is false, then also $p \wedge q$ is false; otherwise $p \wedge q$ is true.

  b)

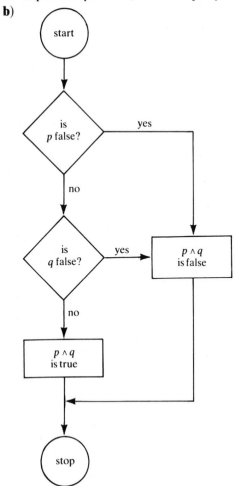

3. a) *Step 1*   If p is false, then $p \to q$ is true; otherwise, go to step 2.
   Step 2   If q is true, then $p \to q$ is true; if not, $p \to q$ is false.

b)

```
                    ┌─────────┐
                    │  start  │
                    └─────────┘
                         │
                         ▼
                      ╱     ╲
                    ╱   is    ╲        no
                    ╲ p true? ╱ ──────────────┐
                      ╲     ╱                  │
                        │                      │
                       yes                     │
                        │                      │
                        ▼                      │
                     ╱     ╲      yes    ┌───────────┐
                   ╱   is    ╲ ────────▶ │   p → q   │
                   ╲ q true? ╱           │  is true  │
                     ╲     ╱             └───────────┘
                       │                      │
                       no                     │
                       │                      │
                       ▼                      │
                 ┌───────────┐                │
                 │   p → q   │                │
                 │  is false │                │
                 └───────────┘                │
                       │                      │
                       │◀─────────────────────┘
                       ▼
                  ┌─────────┐
                  │  stop   │
                  └─────────┘
```

5. a) *Step 1*   Is $x \in A$? If yes, then $x \in A \cup B$; otherwise, go to step 2.
  Step 2   Is $x \in B$? If yes, then also $x \in A \cup B$; if not, $x \notin A \cup B$.

b)

```
                    ┌─────────┐
                    │  start  │
                    └─────────┘
                         │
                         ▼
                   ╱─────────╲
                  ╱    is      ╲          yes
                 ╱   x ∈ A ?    ╲─────────────────┐
                 ╲              ╱                  │
                  ╲            ╱                   │
                   ╲──────────╱                    │
                        │                          │
                        │ no                       │
                        ▼                           │
                   ╱─────────╲                      │
                  ╱    is      ╲   yes       ┌──────▼──────┐
                 ╱   x ∈ B ?    ╲──────────▶ │  x ∈ A ∪ B  │
                 ╲              ╱            └─────────────┘
                  ╲            ╱                    │
                   ╲──────────╱                     │
                        │                           │
                        │ no                        │
                        ▼                            │
                ┌──────────────┐                     │
                │  x ∉ A ∪ B   │                     │
                └──────────────┘                     │
                        │◀───────────────────────────┘
                        ▼
                    ┌─────────┐
                    │  stop   │
                    └─────────┘
```

8.

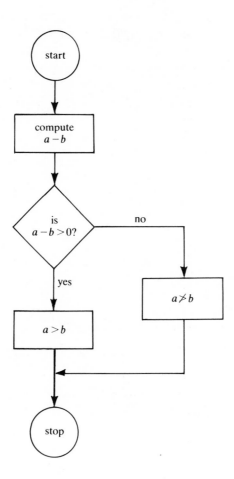

10. a) *Step 1*   First, we let $a = 0$, $b = 1$, and $n = 0$.

Step 2   Now, write the values of b and n.

Step 3   Increase the value of n by 1.

Step 4   Compute $c = b + a$.

Step 5   Write the values of c and n.

Step 6   Is $n \leq 9$? If yes, replace the value of a by b and that of b by c, and go to step 3; if not, we are done.

Step 7   The first value of b and all the values of c give the first ten Fibonacci numbers.

b)

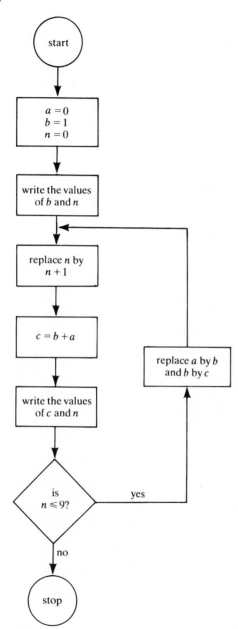

Exercise 16.4

1. a) 19   **b)** 35   **c)** 88.625

   d) 0.25   **e)** 2.625   **f)** 2989.875

3. a) $(13)_8$ **b)** $(26)_8$ **c)** $(3360)_8$
5. a) $(27)_8$ **b)** $(145)_8$ **c)** $(527)_8$
 d) $(15.2)_8$ **e)** $(27.1)_8$ **f)** $(51.5)_8$
7. a) 70 **b)** 743 **c)** 8750 **d)** 8024
9. a) 16 **b)** 38 **c)** 726
11. a) $(221)_8$ **b)** $(35)_8$ **c)** $(8EF)_{16}$
 d) $(1FE)_{16}$ **e)** $(1277)_8$ **f)** $(1F0E)_{16}$

SUGGESTED COURSE OUTLINES

One Semester Course

Chapter	Liberal Arts A	Liberal Arts B	Elementary Education A	Elementary Education B	Liberal Arts and Elementary Education A	Liberal Arts and Elementary Education B
1	all	1.0–1.3, 1.5–1.13	all	1.9–1.3, 1.5–1.13	all	1.0–1.3, 1.5–1.13
2	2.0–2.4, 2.7, 2.8	2.0–2.4, 2.8	all	2.0–2.4, 2.8	2.0–2.4, 2.7, 2.8	2.0–2.4
3	all	3.0–3.5, 3.8–3.10	all	all	all	all
4	all	4.0–4.5, 4.7, 4.8	all	all	all	all
5	all	5.0–5.2, 5.4, 5.5	all	all	all	all
6	6.0–6.5, 6.8	6.0–6.5, 6.8	all	all	all	all
7	7.0–7.5, 7.11	7.0–7.5, 7.11	all	all	7.0–7.5	all
8	8.0–8.5, 8.9–8.12	8.0–8.5	all	all	8.0–8.5	all
9	omit	omit	omit	omit	omit	omit
10	10.0–10.1, 10.4–10.8	10.0–10.1, 10.4	10.0–10.1, 10.4–10.5	10.0–10.1, 10.4–10.5	omit	omit
11	omit	omit	omit	omit	omit	omit
12	omit	omit	omit	omit	omit	omit
13	omit	omit	all	all	omit	omit
14	all	all	omit	omit	all	all
15	all	all	omit	omit	all	all
16	omit	omit	omit	omit	omit	omit

Note: Column A suggests a course more rigorous than the standard suggested by Column B.

SUGGESTED COURSE OUTLINES

Two Semester Courses

Chapter	Liberal Arts A	Liberal Arts B	Elementary Education A	Elementary Education B	Liberal Arts and Elementary Education A	Liberal Arts and Elementary Education B
1	all	1.0–1.3, 1.5–1.13	all	1.0–1.3, 1.5–1.13	all	1.0–1.3, 1.5–1.13
2	2.0–2.4, 2.7–2.8	2.0–2.4	all	2.0–2.4, 2.8	all	2.0–2.4, 2.8
3	all	all	all	all	all	all
4	all	all	all	all	all	all
5	all	5.0–5.5, 5.7–5.9	all	all	all	5.0–5.5, 5.7–5.9
6	all	all	all	all	all	all
7	all	all	all	all	all	all
8	all	all	all	all	all	all
9	omit	omit	all	omit	all	omit
10 ′	all	all	all	all	all	all
11	all	all	all	all	all	all
12	omit	12.0–12.4	all	12.0–12.4	all	12.0–12.4
13	all	all	all	all	all	all
14	all	all	all	all	all	all
15	all	all	all	all	all	all
16	all	all	all	all	all	all

Note: Column A suggests a course more rigorous than the standard suggested by Column B.

SUGGESTED COURSE OUTLINES

One Quarter Courses

Chapter	Liberal Arts		Elementary Education		Liberal Arts and Elementary Education	
	A	B	A	B	A	B
1	all	1.0–1.3, 1.5–1.13	all	1.0–1.3, 1.5–1.13	all	1.0–1.3, 1.5–1.13
2	2.0–2.4, 2.7–2.8	2.0–2.4, 2.8	2.0–2.4, 2.7–2.8	2.0–2.4, 2.8	2.0–2.4, 2.7–2.8	2.0–2.4, 2.8
3	3.0–3.5, 3.8–3.10	3.0–3.5, 3.8–3.10	3.0–3.5, 3.8–3.10	3.0–3.5, 3.8–3.10	3.0–3.5, 3.8–3.10	3.0–3.5, 3.8–3.10
4	4.0–4.5, 4.7–4.8	4.0–4.5, 4.7–4.8	4.0–4.5, 4.7–4.8	4.0–4.5, 4.7–4.8	4.0–4.5, 4.7–4.8	4.0–4.5, 4.7–4.8
5	5.0–5.5, 5.9	5.0–5.2, 5.4–5.5	5.0–5.5, 5.9	5.0–5.2, 5.4–5.5	5.0–5.5, 5.9	5.0–5.2, 5.4–5.5
6	6.0–6.5, 6.8	6.0–6.5	6.0–6.5, 6.8	6.0–6.5	6.0–6.5, 6.8	6.0–6.5
7	7.0–7.5	7.0–7.3, 7.5	7.0–7.3, 7.5, 7.11	7.0–7.3, 7.5, 7.11	7.0–7.3, 7.5, 7.11	7.0–7.3, 7.5
8	8.0–8.5, 8.9	8.0–8.5	8.0–8.9	8.0–8.8	8.0–8.9	8.0–8.8
9	omit	omit	omit	omit	omit	omit
10	10.0–10.1, 10.4–10.8	10.0–10.1, 10.4	10.0–10.1, 10.4	10.0–10.1, 10.4	10.0–10.1, 10.4	10.10–10.1, 10.4
11	omit	omit	omit	omit	omit	omit
12	omit	omit	omit	omit	omit	omit
13	omit	omit	all	all	all	all
14	14.0–14.4	14.0–14.4	omit	omit	14.0–14.4	14.0–14.4,
15	all	all	omit	omit	all	all
16	omit	omit	omit	omit	omit	omit

Note: Column A suggests a course more rigorous than the standard suggested by Column B.

SUGGESTED COURSE OUTLINES

Two Quarter Courses

Chapter	Liberal Arts A	Liberal Arts B	Elementary Education A	Elementary Education B	Liberal Arts and Elementary Education A	Liberal Arts and Elementary Education B
1	all	1.0–1.3, 1.5–1.13	all	1.0–1.3, 1.5–1.13	all	1.0–1.3, 1.5–1.13
2	2.0–2.4, 2.7–2.8	2.0–2.4, 2.8	2.0–2.4, 2.7–2.8	2.0–2.4, 2.8	2.0–2.4, 2.7–2.8	2.0–2.4, 2.8
3	all	all	all	all	all	all
4	all	all	all	all	all	all
5	all	5.0–5.5, 5.7–5.9	all	5.0–5.5, 5.7–5.9	all	5.0–5.5, 5.7–5.9
6	all	all	all	all	all	all
7	all	all	all	all	all	all
8	all	all	all	all	all	all
9	omit	omit	omit	omit	omit	omit
10	all	all	all	10.0–10.5	all	all
11	all	11.0–11.5	all	11.0–11.5	all	11.0–11.5
12	omit	omit	12.0–12.4	12.0–12.4	12.0–12.4	12.0–12.4
13	omit	omit	all	all	all	all
14	all	all	all	all	all	all
15	all	all	all	all	all	all
16	omit	omit	omit	omit	omit	omit

Note: Column A suggests a course more rigorous than the standard suggested by Column B.

SUGGESTED COURSE OUTLINES

Three Quarter Courses

Chapter	Liberal Arts		Elementary Education		Liberal Arts and Elementary Education	
	A	B	A	B	A	B
1	all	1.0–1.3, 1.5–1.13	all	1.0–1.3, 1.5–1.13	all	1.0–1.3, 1.5–1.13
2	2.0–2.4, 2.7–2.8	2.0–2.4, 2.8	all	2.0–2.4, 2.8	all	2.0–2.4, 2.8
3	all	all	all	all	all	all
4	all	all	all	all	all	all
5	all	5.0–5.5, 5.7–5.9	all	all	all	5.0–5.5, 5.7–5.9
6	all	all	all	all	all	all
7	all	all	all	all	all	all
8	all	all	all	all	all	all
9	omit	omit	all	omit	all	omit
10	all	all	all	all	all	all
11	all	all	all	all	all	all
12	omit	omit	all	12.0–12.4	all	12.0–12.4
13	all	all	all	all	all	all
14	all	all	all	all	all	all
15	all	all	all	all	all	all
16	all	all	all	all	all	all

Note: Column A suggests a course more rigorous than the standard suggested by column B.

Index